The SAGE Handbook of

Regression Analysis and Causal Inference

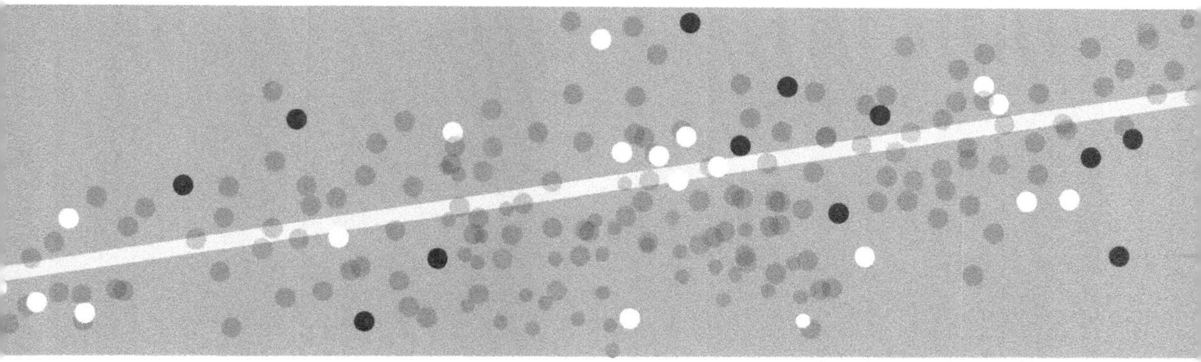

The SAGE Handbook of

Regression Analysis and Causal Inference

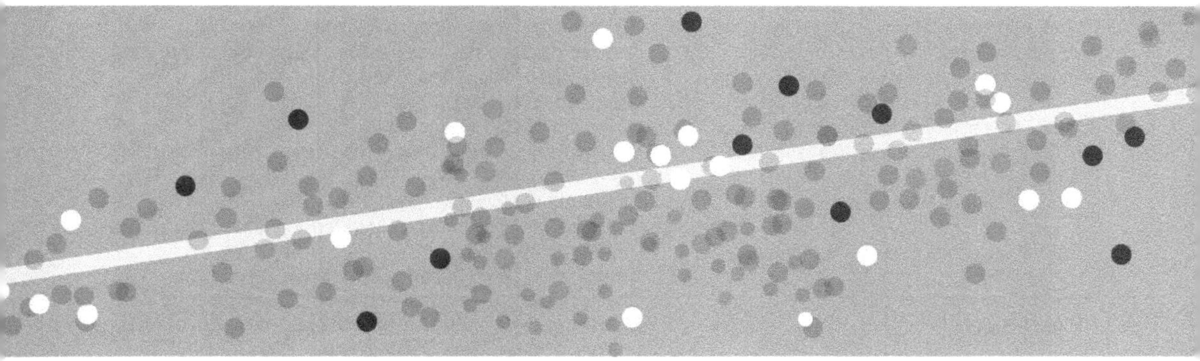

Edited by

Henning Best and Christof Wolf

SAGE reference

Los Angeles | London | New Delhi
Singapore | Washington DC

SAGE

Los Angeles | London | New Delhi
Singapore | Washington DC

SAGE Publications Ltd
1 Oliver's Yard
55 City Road
London EC1Y 1SP

SAGE Publications Inc.
2455 Teller Road
Thousand Oaks, California 91320

SAGE Publications India Pvt Ltd
B 1/I 1 Mohan Cooperative Industrial Area
Mathura Road
New Delhi 110 044

SAGE Publications Asia-Pacific Pte Ltd
3 Church Street
#10-04 Samsung Hub
Singapore 049483

Editor: Katie Metzler
Assistant editor: Lily Mehrbod
Production editor: Ian Antcliff
Copyeditor: Richard Leigh
Proofreader: Clare Weaver
Indexer: Silvia Benvenuto
Marketing manager: Sally Ransom
Cover design: Wendy Scott
Typeset by: C&M Digitals (P) Ltd, Chennai, India
Printed and bound by CPI Group (UK) Ltd, Croydon, CR0 4YY

MIX
Paper from responsible sources
FSC www.fsc.org FSC® C013604

First published 2015

Library of Congress Control Number: 2014931157

British Library Cataloguing in Publication data

A catalogue record for this book is available from the British Library

ISBN 978-1-4462-5244-4

Contents

Contributors

Gerrit Bauer is a postdoctoral researcher at the Deparment of Sociology, Ludwig Maximilian University of Munich. His research focuses on family sociology, the life course and social stratification.

Patricia A. Berglund is a Senior Research Associate in the Survey Methodology Program at the Institute for Social Research. Her research interests include survey data analysis and mental/physical health research.

Henning Best is Professor of Quantitative Methods in the Social Sciences at the University of Würzburg. His research interests include survey methodology, rational choice, environmental sociology.

Gwendolin J. Blossfeld is a DPhil student at Nuffield College at the University of Oxford. Her research interests include longitudinal data analysis, labor market dynamics, family sociology, social inequality and demography.

Hans-Peter Blossfeld is is Professor of Sociology at the European University Institute (EUI) in Florence, Italy, and Professor of Sociology at Bamberg University since 2002, where he is on leave. He has published 35 books and over 240 articles on life course research, social inequality, family and educational sociology, labor market research, and statistical methods for longitudinal data analysis – which have been cited more than 17,000 times (Google Scholar, 2014).

Josef Brüderl is Professor of Sociology at the Ludwig Maximilian University of Munich. His research interests include methods of social research, especially longitudinal methods, family research, organizational research.

Martin Elff is a Senior Lecturer at the Department of Political and Administrative Sciences at the University of Konstanz. His research interests are in comparative politics, political behavior, and political methodology.

Viktor Emonds is a doctoral student at the Centre for Sociological Research, KU Leuven. His main research interests include ethnic inequalities and educational sociology.

Jessica Fortin-Rittberger is Professor of Comparative Politics at the University of Salzburg. Her main areas of research interest include political developments in former communist countries, political institutions and their measurement, women's political representation, as well as the impact of state capacity on democratization.

Markus Gangl is Professor of Sociology and Chair for Social Stratification and Social Policy at the Goethe University of Frankfurt, and Permanent Honorary Fellow of the Department of

Sociology at the University of Wisconsin-Madison. Besides his interest in the methodology of quantitative social science, his main area of research is the interplay of public policy, economic inequality and social stratification in affluent countries.

Steve G. Heeringa is a Research Scientist at the Institute for Social Research, University of Michigan. His research interests are focused in methods of sample design and inference for large-scale population studies.

Joop Hox is Professor of Social Science Methodology at the Faculty of Social Sciences of Utrecht University. His research interests are data quality in surveys and analysis models for complex data.

David S. Lee is Professor of Economics and Public Affairs at Princeton University. His main research interests are labor economics and the econometrics of program evaluation.

Thomas Lemieux is Professor of Economics at the University of British Columbia. His research focuses on the determinants of income inequality and applied econometrics.

Henning Lohmann is Professor of Sociology, in particular Social Research Methods at Hamburg University. His research focuses on poverty and social inequality in a comparative perspective.

J. Scott Long is Distinguished Professor and Chancellor's Professor of Sociology and Statistics at Indiana University, Bloomington.

Geert Loosveldt is Professor at the Department of Sociology of the Catholic University of Leuven. His research focuses on research methodology in general and evaluation of survey data quality in particular.

Volker Ludwig is a researcher at the Institute of Sociolology of the Ludwig Maximilian University of Munich. He is interested in social research methods, family sociology and labor market research.

Bart Meuleman is Assistant Professor at the Centre for Sociological Research, KU Leuven, where he teaches research methodology and statistics. His current research focuses on cross-national comparisons of welfare support and anti-immigration attitudes.

Stephen L. Morgan is the Jan Rock Zubrow '77 Professor in the Social Sciences at Cornell University. His main research interests are sociology of education, social stratification, and the methodology of social inquiry.

Christopher Muller is a PhD candidate in sociology and a doctoral fellow in the Multidisciplinary Program in Inequality and Social Policy at Harvard University. His research interests include historical sociology, inequality, incarceration, and slavery.

Susumu Shikano is a Professor of Political Methodology at the Department of Politics and Public Administration of the University of Konstanz, Germany. His research interests include electoral politics, coalition formation and bureaucratic behavior.

Brady T. West is a Research Assistant Professor in the Survey Methodology Program, located within the Survey Research Center at the Institute for Social Research on the University of Michigan, Ann Arbor campus. His research interests include the analysis of complex sample survey data and regression models for longitudinal and clustered data.

Leoniek Wijngaards-de Meij is a Lecturer at the Department of Methodology and Statistics, University of Utrecht.

Christopher Winship is the Diker-Tishman Professor of Sociology and a member of the senior faculty in the Harvard Kennedy School of Government. Since 1995, he has been editor of *Sociological Methods & Research*. His research interests include quantitative methodology, pragmatism, and applications of cognitive psychology to sociology.

Christof Wolf is Scientific Director at GESIS – Leibniz Institute for the Social Sciences and Professor of Sociology at Mannheim University. His research focuses on sociology of religion, social stratification, methodology, and data analysis.

Preface

A book like this one cannot be written without the help, support, and collaboration of many persons. We are most grateful to the authors for their enthusiasm in contributing to this volume, and the reviewers for devoting their valuable time to helping further improve the quality of the contributions. We also thank Julia Khorshed for helping with the preparation of the LATEX typescript and Heike Antoni and Désirée Nießen for proofreading. Finally, this book would not have seen the light of a well-illuminated desk without the excellent work of Katie Metzler and the Sage publishing team who supported the idea for this book from the beginning. We thank them very much for their enthusiasm and support.

Henning Best and Christof Wolf

1

Introduction

Christof Wolf and Henning Best

In recent years, the social sciences have made tremendous progress in quantitative methodology and data analysis. The classical linear model, while still remaining an important foundation for more advanced methods, has been increasingly complemented by specialized techniques. Major improvements include the widespread use of non-linear models, advances in multilevel modeling and Bayesian estimation, the diffusion of longitudinal analyses and, more recently, the focus on novel methods for causal inference.

The interested reader can chose from a number of excellent textbooks on a wide range of topics: starting from general econometrics books such as Wooldridge (2009, 2010) or Greene (2012), ranging over volumes on regression and Bayesian methods (Gelman et al., 2003; Fox, 2008; Gelman and Hill, 2007), multilevel modeling (Hox, 2010), non-linear models for limited dependent variables (Long, 1997; Train, 2009), event history techniques (Blossfeld et al., 2007), right up to trend-setting textbooks on causal inference (Pearl, 2009; Angrist and Pischke, 2009; Morgan and Winship, 2007) or specialized handbooks like the one edited by Morgan (2013).

Having so many excellent monographs on matters of regression analysis and causal inference makes it difficult for scholars and researchers to obtain an overview of these different approaches. Our aim with this *Sage Handbook of Regression and Causal Inference* is to give readers an accessible outline of a broad set of regression techniques and methods for causal inference written by international experts in the field. Many students and researchers in the social sciences will find this handbook useful as it provides an overview of a range of different methods: ordinary least squares and logistic regression, multilevel and panel regression, time-series cross-section models as well as methods for causal inference – for example, instrumental variables regression, regression discontinuities or propensity score matching. Hence, this volume covers the most commonly used techniques for the statistical analysis of cross-sectional and longitudinal data as well as a number of newer and advanced regression models. Each chapter provides an accessible yet at the same time rigorous presentation of a statistical method. With few exceptions, the contributions follow a common structure, making it easy for readers to navigate through the text. Each chapter begins with an easily accessible, non-technical introduction to the respective method, providing a basic understanding of the method's logic, scope and unique features. The introduction is followed by a presentation of the statistical foundations of the method. To give readers a better understanding of how a particular method can be applied, the next step consists of a comprehensive discussion of the method's application in an example analysis based on publicly available real-world data. Whenever possible, authors used the European Social Survey (see `http://www.europeansocialsurvey.org/`). Readers can download Stata or R code from the companion website to this book and reproduce the analyses (see `https://study.sagepub.com/bestwolf`). The example is followed by discussion of frequently made errors and caveats of the methods and their applications. Each chapter ends with a brief annotated list of references for further reading.

The book is divided into three major blocks: two chapters on estimation techniques, eight chapters on regression models for cross-sectional data, and six chapters focusing on causal inference and the analysis of longitudinal data.

The volume opens with two chapters on different estimation techniques used in regression analysis. In the first of these Martin Elff discusses ordinary least squares and maximum likelihood methods for the estimation of parameters of linear regression and other statistical models. One of the caveats discussed by Elff is that maximum likelihood estimation can become very difficult if sample sizes are small. A technique particularly suited to this situation is Bayesian estimation, which Susumu Shikano presents in the following chapter. After an introduction to the general idea of Bayesian analysis, Shikano shows how the coefficients of a regression model are estimated in the Bayesian framework.

The second block of chapters in this volume deals with regression analysis for cross-sectional data. Linear regression, a powerful tool often termed the workhorse of the social sciences, is introduced by Christof Wolf and Henning Best. Sound applications can only be expected if the assumptions underlying this model are understood. These are elaborately discussed in the next chapter by Bart Meuleman, Geert Loosveldt and Viktor Emonds. They also present the tools used to diagnose deviations from the assumptions. In the following chapter Henning Lohmann shows how we can incorporate non-linear and non-additive effects into linear regression models. In great detail he discusses interaction effects, polynomials and splines and demonstrates how flexible multiple linear regression is. Joop Hox and Leoniek Wijngaards-de Meij's contribution focuses on regression models for hierarchical, multilevel data. These models are suitable if the units of observations are 'nested' within higher-level units (e.g. students in schools, residents in neighborhoods or employees in firms). The authors discuss these models for both metric and binary dependent variables. An in-depth coverage of regression models for binary outcomes can be found in the next chapter by Henning Best and Christof Wolf on logistic regression. This is directly followed by a presentation of regression models for multinomial and ordinal variables authored by Scott Long. In both chapters dealing with non-metric outcome variables the authors emphasize that interpreting the results of these kinds of models is anything but straightforward. An indispensable tool to successfully meet the challenge to correctly interpret regression results are graphical displays. These are presented and discussed in the subsequent chapter by Gerrit Bauer. The block on regression analysis for cross-sectional data closes with a contribution by Steven Heeringa, Brady West, and Patricia Berglund who address regression modeling for complex sample survey data.

The third block of chapters is devoted to methods for longitudinal data analysis and causal inference that are based on a counterfactual model of causality. Markus Gangl opens this part with a contribution on matching estimators for treatment effects. The chapter discusses analytical goals and mathematical foundations that underlie the use of matching estimators for causal inference. As the name of the method suggests, two types of units – the 'treated' and 'non-treated' – are matched based on some common characteristic. An alternative method for causal inference is introduced in the chapter by Christopher Muller, Christopher Winship, and Stephen Morgan. They provide a non-technical introduction to instrumental variables regression. This kind of regression helps in dealing with endogeneity by using an additional instrument variable that is correlated with the causal factor of interest, but otherwise exogenous. Another important method, regression discontinuity designs, is presented by Thomas Lemieux and David Lee. They present the conceptual framework behind this research design and draw a parallel between regression discontinuity and randomized experiments. The next chapter, by Josef Brüderl and Volker Ludwig, offers a description of fixed-effects panel regression which they compare to random-effects models and models including a lagged dependent variable. In addition to the basic model of fixed-effects panel regression, the authors discuss a more advanced variant of this approach allowing for heterogeneous change, that is, a model with individual slopes. Another form of longitudinal data is event history data that provides information on a sequence of different states occupied by each unit of analysis and the timing of changes among these states. Hans-Peter Blossfeld and Gwendolin Blossfeld present regression models to analyze such data structures. For them event history models are closely linked to an understanding of causation as a generative process. The book closes with a contribution by Jessica Fortin-Rittberger on models for time-series cross-section. These models are particularly useful if we have data on a comparatively small number of units for a comparatively large number of time points. This type of data structure arises often in comparative political science applications.

We hope that readers will find this *Sage Handbook* useful for their daily practice in social science teaching and research. We are confident that the book will help students and researchers in conducting quantitative social research and contribute to the further diffusion of important methods for causal inference. If the book helps advance the methodologically sound analysis of society, the time invested will have been well spent.

REFERENCES

Angrist, J. D. and Pischke, J.-S. (2009). *Mostly Harmless Econometrics*. Princeton: Princeton University Press.

Blossfeld, H.-P., Golsch, K., and Rohwer, G. (2007). *Event History Analysis with Stata*. Mahwah: Erlbaum.

Fox, J. (2008). *Applied Regression Analysis and Generalized Linear Models*. Thousand Oaks: Sage.

Gelman, A., Carlin, J. B., Stern, H. S., and Rubin, D. B. (2003). *Bayesian Data Analysis*, Second Edition. Chapman and Hall/CRC Texts in Statistical Science. Taylor & Francis.

Gelman, A. and Hill, J. (2007). *Data Analysis Using Regression and Multilevel/hierarchical Models*. Cambridge: Cambridge University Press.

Greene, W. H. (2012). *Econometric Analysis*. New York: Prentice Hall.

Hox, J. J. (2010). *Multilevel Analysis. Techniques and Applications*. New York: Routledge.

Long, J. S. (1997). *Regression Models for Categorical and Limited Dependent Variables*. Thousand Oaks: Sage.

Morgan, S. L., (Ed.) (2013). *Handbook of Causal Analysis for Social Research*. New York: Springer.

Morgan, S. L. and Winship, C. (2007). *Counterfactuals and Causal Inference*. New York: Cambridge University Press.

Pearl, J. (2009). *Causality: Models, reasoning, and inference*. Cambridge: Cambridge University Press.

Train, K. (2009). *Discrete Choice Methods with Simulation*. Cambridge/New York: Cambridge University Press.

Wooldridge, J. M. (2009). *Introductory Econometrics: A modern approach*. Mason: Thomson/South-Western.

Wooldridge, J. M. (2010). *Econometric Analysis of Cross Section and Panel Data*. Cambridge: MIT Press.

PART I

Estimation and Inference

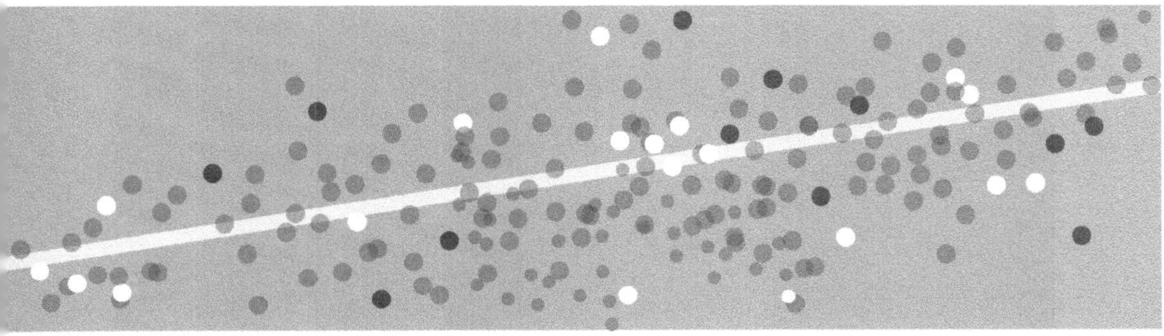

2

Estimation techniques: Ordinary least squares and maximum likelihood

Martin Elff

INTRODUCTION

A major task in regression analysis and in much of data analysis in the social sciences in general is the construction of a model that best represents (1) substantial assumptions and hypotheses a researcher may entertain and (2) auxiliary information or assumptions about the way the data under analysis are generated. To complete this task of *model specification* successfully, a researcher will need a fair knowledge of a variety of statistical models and their assumptions. Introducing these is one of the main purposes of this volume. In contrast to most other chapters, the present one presumes all questions with regard to model specification as already addressed and focuses on the theoretical foundations of a step that comes thereafter, the step of *estimating model parameters*.

While model specification sometimes appears to be something of an art, estimation clearly is a technique, the application of which researchers often gladly delegate to their computers. But for scholars intent on gaining a full understanding of the research process it is important to know the foundations of estimation. Therefore it is the purpose of this chapter to introduce these foundations, to provide an understanding of what it means to estimate parameters and to give some idea of what a 'good' estimator is.

The task of model specification usually leads us to a *probability model* of the process by which the data under analysis are generated. That is, we assume that each piece of data that we have observed, could have observed or may observe in the future has done or will do so with a particular probability. In other words, our data are *observations* of *random variables*. Roughly speaking, a random variable is a set of numbers, called the *sample space*, together with probabilities assigned to them or to subsets of the sample space. The set of rules by which probabilities are assigned to numbers or sets of numbers is the *probability distribution* of the random variable. For example, if we roll a die, then the number it shows is an observation of a random variable that can take the values 1, 2, 3, 4, 5, 6, each with a particular probability. If the die is fair, the random variable has a discrete uniform distribution in so far as the same probability $\frac{1}{6}$ is assigned to each of the possible six numbers.

We usually do not know the exact probability with which values of the random variables occur or of which our data are observations. Model specification does not lead in general to a specific probability distribution, but to the assumption that our data come from a member of a *parametric family* of probability distributions, where each member of the family can be identified by one or more numeric values. The set of numbers that can be used to distinguish probability distributions from one another is the *parameter space* of the family of distributions, and the values that identify each particular distribution within the family are its *parameter values*. For example, the members of the family of normal distributions are distinguished by the values of the mean parameter μ and the variance parameter σ^2, so that the standard normal distribution is the particular member of this family identified by the parameter values $\mu = 0$ and $\sigma^2 = 1$. If we make a (hopefully educated) guess about the values of the parameters of the distribution that describes the probability with which our data have occurred, we are engaging in the *estimation* (or, more precisely, point estimation) of these parameters. A (point) *estimator* is, roughly speaking, a rule for obtaining estimates from observed data. The next section gives a more formal treatment of these concepts and introduces essential notation.

If we run a linear regression analysis on our computer with a social science data set and if statistical software gives us coefficient values, then these values are estimates in the sense of the previous paragraph. A common rule by which estimates for linear regression coefficients are computed is the *ordinary least squares* (OLS) estimator. This estimator is a relatively simple one, which nevertheless has a couple of particular desirable properties, such as that it is, under appropriate conditions, the *best linear unbiased estimator* (BLUE). Details of the OLS estimator and its properties are given in the third section of this chapter.

A more general class of estimators is based on the principle of maximum likelihood (ML). These estimators are used to obtain coefficient estimates for various generalizations of the linear regression model, such as logistic regression or regression for multinomial and ordinal dependent variables (see Chapter 9 in this volume). One advantage of ML estimators is that they are relatively easy to derive, once the model specification has led to the selection of a family of probability distributions. Another advantage is that they can be shown to have, at least with large samples and under appropriate conditions, some desirable properties, such as consistency, asymptotic efficiency, and asymptotic normality. The fourth section of the present chapter discusses the ML criterion from which these estimators are derived as well as their properties.

As already stated, parameter estimation means selecting a member from a family of probability distributions. Yet the assumption about the parametric family may be wrong and the data may in fact be generated from a distribution outside this family. For example, we might erroneously assume that our data come from a normal distribution when in fact they come from an exponential distribution (the latter often arises in duration data; see Chapter 16 in this volume) or some other, unknown family of distributions. More generally, the conditions under which OLS and ML estimators exhibit their desirable characteristics may be violated. The consequences of such model violations are discussed in the fifth section of the present chapter, as are the consequences of small sample sizes for the performance of ML estimators.

MATHEMATICAL FOUNDATIONS

Random variables and their distributions

A concept that is fundamental to the discussion of estimators is that of a random variable and its distribution. A *random variable* is a mathematical construct used to describe, in quantitative terms, events that can occur with a specific probability. In the following, a random variable will be represented by an upper-case letter (e.g. X), whereas the values that it can take 'at random' will be represented by lower-case letters, (e.g. $x_1, x_2, \ldots$). Any set of possible values of a random variable will be represented by a 'calligraphic' letter (e.g. $\mathcal{A}$). The set of all possible values of the random variable, its *range*, will be denoted by the calligraphic version of the letter that denotes the random variable (e.g. the range of X will be written as $\mathcal{X}$). There is no place for an explicit introduction to the concept of probability in this chapter. Suffice it to state that a probability is a number between zero and one, such that $\Pr(X \in \mathcal{A}) = 0$ if the event $X \in \mathcal{A}$ cannot happen and that $\Pr(X \in \mathcal{A}) = 1$ if the event $X \in \mathcal{A}$ inevitably happens, that for two events $X \in \mathcal{A}$ and $X \in \mathcal{B}$,

$$\Pr(X \in \mathcal{A} \text{ or } X \in \mathcal{B}) = \Pr(X \in \mathcal{A}) + \Pr(X \in \mathcal{B}) - \Pr(X \in \mathcal{A} \text{ and } X \in \mathcal{B}), \tag{2.1}$$

and that for any event $X \in \mathcal{A}$,

$$\Pr(\text{ not } X \in \mathcal{A}) = 1 - \Pr(X \in \mathcal{A}). \tag{2.2}$$

The rules by which such probabilities are assigned to sets of values of a random variable are its *distribution.*

Among random variables that have real numbers as values, one can distinguish between two types, discrete and continuous random variables. *Discrete* random variables take only finitely many different values or countably infinitely many different values.[1] When a die is rolled, it can show only one of the numbers 1, 2, 3, 4, 5, or 6. Hence this 'random experiment' can be represented by a random variable with a finite number of potential values. If the die is fair, each of these numbers has identical probability of being shown:

$$\Pr(X = x) = \frac{1}{6}, \quad \text{for } x = 1, 2, 3, 4, 5, 6.$$

Another example of a discrete random variable is the number of traffic accidents on a particular day on a busy street. The number will be integer, but in principle there is no upper limit to the possible number of accidents.

The distribution of a discrete random variable, a *discrete distribution*, is characterized by its *probability mass function* (PMF), which assigns to each possible value of the random variable the probability of its occurrence:

$$f(X) = \Pr(X = x). \tag{2.3}$$

A particularly simple example of a PMF is that of a Bernoulli distribution, which occurs, for example, if we model the toss of a coin. If the probability of getting 'heads' is $p = \Pr(X = 1)$ and the probability of getting 'tails' is $1 - p = \Pr(X = 0)$, then the PMF, indexed by the parameter p, is

$$f_{\mathrm{B}}(X; p) = p^x (1 - p)^{1-x}. \tag{2.4}$$

A real-valued random variable is *continuous* if it has a *continuous distribution* that is characterized by a *probability density function* (PDF) $f(X)$ such that

$$\Pr(a \leq X \leq b) = \int_a^b f(X)\,\mathrm{d}x. \tag{2.5}$$

For example, the normal distribution with mean parameter μ and variance parameter σ^2 has the PDF

$$f_{\mathrm{N}}(X; \mu, \sigma^2) = \frac{1}{\sqrt{2\pi\sigma^2}} \exp\left(-\frac{1}{2}\frac{(X - \mu)^2}{\sigma^2}\right). \tag{2.6}$$

Other continuous distributions are the t, F and χ^2 distributions.

Distributions of random variables can be grouped into families of distributions with PDFs or PMFs of the same functional form but with different parameter values. Thus the distribution that describes the tossing of a coin is just one of the members of the family of Bernoulli distributions (that differ from one another in the value of the 'success probability' parameter p). In a similar vein, one can speak of the family of normal distributions (each of which is characterized by its respective values of the parameters μ and σ^2).

Based on the PDF or PMF of a random variable one can define its *expectation* or *expected value*:

$$\begin{aligned} E(X) &:= \sum_{x \in \mathcal{X}} xf(X;\theta) = \sum_{x \in \mathcal{X}} x \Pr(X = x) \quad \text{if } X \text{ is discrete,} \\ E(X) &:= \int_{-\infty}^{\infty} xf(X;\theta)\,\mathrm{d}x \quad \text{if } X \text{ is continuous.} \end{aligned} \tag{2.7}$$

One of the most important properties of the expectation is its additivity: the expectation of the sum of two random variables X and Y equals the sum of their expectations, and, more generally, for constants a and b,

$$E(aX + bY) = a\,E(X) + b\,E(Y). \tag{2.8}$$

Based on the concept of expectation, one can define the *variance* of a random variable as a measure of its 'spread' or dispersion:

$$\operatorname{Var}(X) := E[(X - E(X))^2] = E(X^2) - (E(X))^2. \tag{2.9}$$

The expected value of a random variable representing the rolling of a die is $E(Y) = (1 + 2 + 3 + 4 + 5 + 6)/6 = 3.5$. For a random variable X with Bernoulli distribution the expectation is $E(X) = p$ and the variance is $\operatorname{Var}(X) = p(1 - p)$; for a random variable with normal distribution with parameters μ and σ^2 the expectation is $E(X) = \mu$ and the variance is $\operatorname{Var}(X) = \sigma^2$. A *standard normal* random variable has of course $\mu = 0$ and $\sigma^2 = 1$.

Joint distributions, stochastic independence and conditional distributions

It is often of interest to look at several random variables simultaneously. Thus one can describe the degree of interdependence of two random variables X and Y by their *joint distribution* with PDF or PMF $f(X, y)$. A simple summary of this interdependence is the *covariance* between the two random variables:

$$\operatorname{Cov}(X, Y) = E[(X - E(X))(Y - E(Y))] = E(XY) - E(X)E(Y). \tag{2.10}$$

(Note that $\operatorname{Cov}(X, X) = \operatorname{Var}(X)$.) An important property of variances and covariances of random variables is the following equality (for random variables X and Y and constants a and b), which is reminiscent of the first binomial formula:

$$\operatorname{Var}(aX + bY) = a^2 \operatorname{Var}(X) + b^2 \operatorname{Var}(Y) + 2ab \operatorname{Cov}(X, Y). \tag{2.11}$$

Two discrete random variables or two continuous random variables X and Y are *(stochastically) independent* if the PDF or PMF of their joint distribution can be factored into those of the *marginal* distributions of the two random variables:

$$f(X, y) = f(X)f(y). \tag{2.12}$$

If two random variables X and Y are independent they are also uncorrelated, that is, $E(XY) = E(X)E(Y)$ and thus $\operatorname{Cov}(X, Y) = 0$. However, uncorrelated random variables are not always independent of each other.

The pattern of interdependence of several (more than two) random variables, $X_1, \dots, X_n$ is represented by their *multivariate* joint distribution, which can be described by their multivariate joint PMF or PDF $f(X_1, \dots, x_n)$. It is often convenient to view such random variables as components of an n-dimensional *random vector* $\mathbf{X}$ the distribution of which has a PDF or PMF $f(\mathbf{x})$ that takes a vector argument $\mathbf{x}$. The expectation of such a random vector will be an n-dimensional vector. The variances and covariances of all pairs of components of such a random vector are usually arranged into a matrix, its *variance–covariance matrix* or, more simply, its *variance matrix* $\text{Var}(\mathbf{X})$.

A multivariate distribution of particular importance is the *multivariate normal distribution*. The PDF of a random vector having a n-dimensional multivariate distribution with mean vector $\boldsymbol{\mu}$ and variance matrix $\boldsymbol{\Sigma}$ is

$$f_{\text{MVN}}(\mathbf{x}; \boldsymbol{\mu}, \boldsymbol{\Sigma}) = (2\pi)^{-n/2} (\det \boldsymbol{\Sigma})^{-1/2} \exp\left(-\frac{1}{2}(\mathbf{x} - \boldsymbol{\mu})' \boldsymbol{\Sigma}^{-1} (\mathbf{x} - \boldsymbol{\mu}) \right). \tag{2.13}$$

A random vector $\mathbf{X}$ with such a distribution has expectation $E(\mathbf{X}) = \boldsymbol{\mu}$ and variance $\text{Var}(\mathbf{X}) = \boldsymbol{\Sigma}$.

Another concept that is important in the context of regression models is that of a *conditional distribution*. In case of two discrete random variables X and Y with joint PMF $f(X, y) = \Pr(X = x \text{ and } Y = y)$, the probability that $Y = y$ *conditional on* (or given that) $X = x$ is

$$\Pr(Y = y | X = x) = \frac{\Pr(X = x \text{ and } Y = y)}{\Pr(X = x)} = \frac{f(X, y)}{\sum_{y \in \mathcal{Y}} f(X, y)} \tag{2.14}$$

(where $\mathcal{Y}$ is the range of the values of Y) which gives the PMF $f(y|x)$ of the conditional distribution. Similarly, in the case of two continuous random variables X and Y, the PDF of the conditional distribution of Y given the values of X is defined as

$$f(y|x) = \frac{f(X, y)}{f(X)} = \frac{f(X, y)}{\int_{-\infty}^{\infty} f(X, y)\, \mathrm{d}y}. \tag{2.15}$$

The concept of conditional distributions generalizes to the multivariate case: if $X_1, \dots, X_m, Y$ are random variables with joint PDF or PMF $f(X_1, \dots, x_m, y)$ then the PDF or PMF of the conditional distribution (or the *conditional* PDF or PMF) is defined as

$$f(y|x_1, \dots, x_m) = \frac{f(X_1, \dots, x_m, y)}{f(X_1, \dots, x_m)} = \begin{cases} \dfrac{f(X_1, \dots, x_m, y)}{\sum_{y \in \mathcal{Y}} f(X_1, \dots, x_m, y)} & \text{if } Y \text{ is discrete,} \\[2ex] \dfrac{f(X_1, \dots, x_m, y)}{\int_{-\infty}^{\infty} f(X_1, \dots, x_m, y)\, \mathrm{d}y} & \text{if } Y \text{ is continuous.} \end{cases} \tag{2.16}$$

Based on the conditional distribution of a random variable Y given the values of the random variables $X_1, \dots, X_m$, the concept of the *conditional expectation* given the values of $X_1, \dots, X_m$ is defined as the expectation of Y based on the conditional distribution given the values of $X_1, \dots, X_m$. Such conditional expectations form the core of linear regression models as well as of generalized linear models, which include logistic regression and other logit models.

Samples, the law of large numbers and the central limit theorem

From the stochastic point of view, a *random sample* is a sequence of (observations from) *stochastically independent, identically distributed* (i.i.d.) random variables (Casella and Berger, 2002,

p. 208). That is, if a die is rolled a number of times and in each instance the number the die shows is recorded as an observation of a random variable, these numbers will also be a sample.

In the social sciences one often thinks of random samples in terms of survey samples from a population, where the members of the population have some fixed values, and one attempts to estimate some population averages or population totals. For example, one may want to estimate the average income of the economically active population of the UK by asking the respondents of a survey sample about their income. If every member of this population has the same chance of getting into the the sample, the incomes of the respondents in the sample are i.i.d. random variables with a distribution that reflects the distribution of incomes in the population. Thus a survey sample from a (finite) population is a special case of a random sample in the stochastic sense.

From the stochastic perspective a *statistic* is any function of a random sample and is therefore a random variable itself. Based on this idea, the *arithmetic mean* of the sample (or *sample average*) and the *sample variance* can be introduced as random variables and their relation to the mean and variance as characteristics or parameters of a distribution can be elucidated. Suppose that the random variables $X_1, \ldots, X_n$ are i.i.d. and therefore a sample. The *sample mean* is then defined as

$$\bar{X} := \frac{1}{n}\sum_{i=1}^{n} X_i = \frac{1}{n}(X_1 + \cdots + X_n). \tag{2.17}$$

Now if μ is the common expectation of the random variables X_i ($i = 1, \ldots, n$) then the expected value of the sample mean equals μ:

$$E(\bar{X}) = \frac{1}{n}\sum_{i=1}^{n} E(X_i) = \frac{1}{n}\cdot n \cdot \mu = \mu. \tag{2.18}$$

For the *sample variance* there are actually two common definitions:

$$V := \frac{1}{n-1}\sum_{i=1}^{n}(X_i - \bar{X})^2 \quad \text{and} \quad V^* := \frac{1}{n}\sum_{i=1}^{n}(X_i - \bar{X})^2 . \tag{2.19}$$

While the definition of V^* looks more 'natural', V is the preferred one for small samples because, if σ^2 is the common variance of the random variables X_i in the sample,

$$\begin{aligned} E(V) &= \frac{1}{n-1}\sum_{i=1}^{n} E[(X_i - \bar{X})^2] = \frac{1}{n-1}\sum_{i=1}^{n}\left[E(X_i^2) - 2E(X_i\bar{X}) + E(\bar{X}^2)\right] \\ &= \frac{n}{n-1}\left[\mu^2 + \sigma^2 - 2\left(\mu^2 + \frac{1}{n}\sigma^2\right) + \mu^2 + \frac{1}{n}\sigma^2\right] \\ &= \frac{n}{n-1}\left[\sigma^2 - \frac{1}{n}\sigma^2\right] = \sigma^2. \end{aligned} \tag{2.20}$$

A fundamental result of probability theory is the *law of large numbers*. In its weak version it states that the larger a sample (of i.i.d. random variables), the less likely it is that the arithmetic mean shows a (however small) difference from the expectation of the random variables, or, more precisely, that the arithmetic mean *converges in probability* to the expected value:

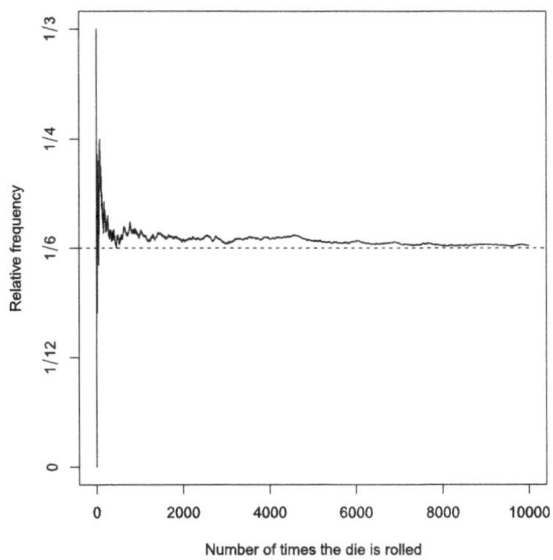

Figure 2.1 An illustration of the law of large numbers: the relative frequency (average occurrence) of obtaining a six when rolling a die

Theorem 1 (Weak law of large numbers). *Suppose there is an infinite sequence of i.i.d. random variables* $X_1, X_2, \ldots$ *with common expectation* μ *and finite variance. Then the arithmetic mean* $\bar{X}_n = \frac{1}{n}\sum_{i=1}^{n} X_i$ *of n random variables converges in probability to* μ*, that is, for any* $\epsilon > 0$*,*

$$\lim_{n\to\infty}\left(\Pr(|\bar{X}_n - \mu| > \epsilon)\right) = 0, \tag{2.21}$$

or, more briefly,

$$\operatorname*{plim}_{n\to\infty} \bar{X}_n = \mu. \tag{2.22}$$

There is also a strong version of the law of large numbers that states that the arithmetic mean *converges almost surely* to the expected value; however, the interpretation of the strong law is less intuitive.[2] Convergence in probability, of which the law of large numbers is a particular case, is an important notion for the discussion of properties of estimators. But it also plays a role in the common interpretation of expected values as 'long-term averages'. For example, the more often one rolls a die, the more the relative frequency of winning by getting a six tends to be close to $\frac{1}{6}$.

The law of large numbers is illustrated by Figure 2.1. It shows the relative frequency of obtaining a six in 10,000 simulated rolls of a die. Since the event of obtaining a six is represented by a binary random variable each time the die is rolled ($X_i = 1$ for a six and $X_i = 0$ otherwise), the relative frequency of obtaining a six is identical to the average or arithmetic mean $\bar{X}_n$ of the binary random variables. The figure shows that after the first couple of times the relative frequency can depart quite substantially from the expected value, but that it gets closer and closer to the expected value of $\frac{1}{6}$ the more often the die is rolled.

Another important result of probability theory that is pertinent to the discussion of estimators is the *central limit theorem*. In informal terms, it states that (under suitable conditions) the larger the sample, the more closely the sample average follows a normal distribution. More precisely, it states that the arithmetic mean *converges in distribution* to a normal distribution:

Theorem 2 (Central limit theorem). *Suppose there is an infinite sequence of i.i.d. random variables* $X_1, X_2, \ldots$ *with common expectation* μ *and common variance* $\sigma^2 < \infty$*. Then the* *arithmetic mean* $\bar{X}_n = \frac{1}{n}\sum_{i=1}^{n} X_i$ *converges in distribution to a normal distribution with mean* μ *and variance* $n^{-1}\sigma^2$*, that is,*

$$\lim_{n\to\infty} \Pr\left(\sqrt{n}(\bar{X}_n - \mu) \le z\right) = F_{\mathrm{N}}(z; 0, \sigma^2) \tag{2.23}$$

where $F_{\mathrm{N}}(X; 0, \sigma^2)$ *is the cumulative distribution function of the normal distribution with mean* 0 *and variance* σ^2.

The central limit theorem is one of the (many) reasons why normal distributions play such an important role in probability theory and statistics. It is also instrumental in the construction of large-sample confidence intervals and of some large-sample test statistics.

Estimators and their properties

In the introduction it was stated that an estimator is a rule for making a guess about a parameter of a probability distribution. More formally, an estimator is a statistic (as defined above) of the same dimension as the parameter of interest. For example, the arithmetic mean $\bar{X}$ of a sample can be seen as an estimator of the expected value μ of the members of the sample, so that $\hat{\mu} = \bar{X}$.

An estimator is a function of a sample and therefore itself a random variable. From its distribution, the *sampling distribution* of the estimator, one can derive certain properties that it may or may not have. Such stochastic properties are usually the guideline for choosing among several estimators of the same parameter. The first desirable property is the absence of bias. The *bias* of an estimator is the difference between the expected value $E(\hat{\theta})$ of an estimator and the true value θ_0 of a parameter:

$$\mathrm{bias}(\hat{\theta}) = E(\hat{\theta}) - \theta_0. \tag{2.24}$$

If this bias is zero, that is, $E(\hat{\theta}) = \theta_0$, then an estimator is said to be *unbiased*. Clearly, the arithmetic mean $\bar{X}$ is an unbiased estimator of the expectation of the elements of a sample, as is the variance as defined as V but not as defined by V^* in equation (2.19).

If an estimator is unbiased this means that it will be correct on average. But that does not mean that it is necessarily close to the true value of the parameter it is supposed to estimate. So even an unbiased estimator can be almost useless if it has a very large variance. Thus one compares estimators not only in terms of their bias, but also in terms of their variance. An unbiased estimator $\hat{\theta}$ is said to be *more efficient* than another unbiased estimator $\tilde{\theta}$ if $\mathrm{Var}(\hat{\theta}) < \mathrm{Var}(\tilde{\theta})$. If an estimator is more efficient than any other unbiased estimator it is called the *minimum variance unbiased* or *best unbiased* estimator or an *efficient unbiased* estimator. OLS estimators have this property under suitable conditions, as do, for specific parameters of distributions of a particular type, ML estimators.

When practitioners compute estimates or have them computed by software, they expect a measure of uncertainty about them. Such a measure is the standard error, which is nothing more than the square root of the variance of the estimator:

$$\mathrm{SE}(\hat{\theta}) := \sqrt{\mathrm{Var}(\hat{\theta})}. \tag{2.25}$$

Since the variance of an estimate usually depends on the true value of the parameter to be estimated, which in practice is usually unknown, the standard error itself has to be estimated.

Another measure of uncertainty about an estimate is the confidence interval associated with it. A *confidence interval* is the set of numbers between two values $\hat{\theta}_{\mathrm{lower}}$ and $\hat{\theta}_{\mathrm{upper}}$ such that

$$\Pr(\hat{\theta}_{\mathrm{lower}} \le \theta_0 \le \hat{\theta}_{\mathrm{upper}}) = 1 - \alpha, \tag{2.26}$$

where a 95% confidence interval has $\alpha = 0.05$.[3] The construction of an exact confidence interval usually requires the exact knowledge of the sampling distribution of an estimator, which is not always available.

Another desirable property that an estimator may or may not have is consistency – or, more precisely, the consistency of a sequence of estimators of the same type. An estimator – or, better, a sequence of estimators $\hat{\theta}_n$ $(n = 1, 2, \ldots)$ – is called *consistent* if it converges in probability to the true value of the parameter it is supposed to estimate, that is,

$$\operatorname*{plim}_{n\to\infty} \hat{\theta}_n = \theta_0. \tag{2.27}$$

More informally speaking, if estimators based on ever larger samples are likely to be closer and closer to the true value of the parameter they are supposed to estimate then they constitute a consistent sequence of estimators. Also one usually says in this case that an estimator from such a sequence is consistent. While one may intuitively have the idea that 'larger samples give better results', this is not a matter of course, because there are in many situations (theoretically conceivable) estimators that are not consistent. For example, the arithmetic mean of a sample of size n is a consistent estimator of the expected value, as the law of large numbers states. Also, both V and V^* in equation (2.19) are consistent estimators of the variance.

In the previous section we saw, with the central limit theorem, an example of a statement of convergence in distribution. The distribution to which a sequence of estimators converges is called its *asymptotic* distribution, and if the asymptotic distribution is a normal distribution one calls this sequence of estimators *asymptotically normal*, which is usually expressed in a formula as

$$\sqrt{n}(\hat{\theta}_n - \theta_0) \xrightarrow{d} \mathrm{N}(0, v), \tag{2.28}$$

with v being a constant that may depend on θ_0 and $N(0, v)$ symbolizing the normal distribution with mean 0 and variance v. The arithmetic mean clearly is asymptotically normal under suitable conditions, and as we will see later, if the appropriate conditions hold, OLS and ML estimators are also asymptotically normal.

LINEAR REGRESSION AND ORDINARY LEAST SQUARES

From a probability point of view a linear regression model for a data set with n observations poses a *linear relation* between an n-dimensional random vector $\mathbf{Y}$ (with components $Y_1, \ldots, Y_n$) that forms the *regressand* or *response vector* and n-dimensional random vectors $\mathbf{X}_1, \ldots, \mathbf{X}_m$ (where each random vector $\mathbf{X}_j$ has the components $X_{1j}, \ldots, X_{nj}$) that form the *regressors* or *predictors* and an *error term*, the random vector $\boldsymbol{\epsilon}$ (with components $\epsilon_1, \ldots, \epsilon_n$ each having a zero expectation, i.e. $E(\epsilon_i) = 0$):

$$Y_i = \beta_0 + \beta_1 X_{1i} + \cdots + \beta_m X_{mi} + \epsilon_i, \quad \text{for } i = 1, \ldots, n, \tag{2.29}$$

where each i $(i = 1, \ldots n)$ corresponds to a 'case' in the data set.

In regression models, one is primarily interested the regression coefficients. The distribution of the predictor variables and of the error term is usually of minor interest. Therefore, one looks at the *conditional expectation* of the random vector $\mathbf{Y}$ given the values of the random vectors $\mathbf{X}_1 = \mathbf{x}_1, \ldots, \mathbf{X}_m = \mathbf{x}_m$. This conditional expectation takes the form

$$E(\mathbf{Y}|\mathbf{X}_1 = \mathbf{x}_1, \ldots, \mathbf{X}_m = \mathbf{x}_m) = \beta_0 \mathbf{1} + \beta_1 \mathbf{x}_1 + \cdots + \beta_m \mathbf{x}_m = \mathbf{X}\boldsymbol{\beta}, \tag{2.30}$$

where $\mathbf{1}$ is a n-dimensional vector with all elements equal to one, $\mathbf{X}$ is a matrix, the *regressor matrix*, with columns equal to $\mathbf{1}, \mathbf{x}_1, \ldots, \mathbf{x}_m$, and $\boldsymbol{\beta}$ is the *coefficient vector* with elements $\beta_0, \beta_1, \ldots, \beta_m$.

Definition of ordinary least squares

An OLS estimator is a straightforward way to obtain estimates of the coefficient vector based on the matrix $\mathbf{X}$ and an observation $\mathbf{y}$ of the random vector $\mathbf{Y}$. For given observed response vector $\mathbf{y}$ and regressor matrix $\mathbf{X}$ the OLS estimate of the coefficient vector $\boldsymbol{\beta}$ is the one that minimizes the residual sum of squares

$$SSQ(\boldsymbol{\beta}) = (\mathbf{y} - \mathbf{X}\boldsymbol{\beta})'(\mathbf{y} - \mathbf{X}\boldsymbol{\beta}). \tag{2.31}$$

While finding a coefficient vector $\boldsymbol{\beta}$ that minimizes the residual sum of squares SSQ may seem to be a formidable task, a little linear algebra and vector calculus can be used to derive a simple and elegant formula for the computation of OLS estimates.

First, a necessary condition for a coefficient vector $\boldsymbol{\beta}_{\text{OLS}}$ to minimize the sum of squares is that the gradient (the vector of partial derivatives) of the sum of squares with respect to $\boldsymbol{\beta}$ is zero at $\boldsymbol{\beta} = \boldsymbol{\beta}_{\text{OLS}}$, which implies

$$\frac{\partial SSQ(\boldsymbol{\beta}_{\text{OLS}})}{\partial \boldsymbol{\beta}} = 2\mathbf{X}'(\mathbf{X}\boldsymbol{\beta}_{\text{OLS}} - \mathbf{y}) = 0 \Leftrightarrow \mathbf{X}'\mathbf{X}\boldsymbol{\beta}_{\text{OLS}} = \mathbf{X}'\mathbf{y}. \tag{2.32}$$

Another necessary condition (which together with the first condition is necessary and sufficient) is that the Hessian, the matrix of second partial derivatives with respect to $\boldsymbol{\beta}$, is positive definite (which means that it is 'larger than zero' in a matrix sense) at $\boldsymbol{\beta} = \boldsymbol{\beta}_{\text{OLS}}$.[4] This matrix equals $\mathbf{X}'\mathbf{X}$ for any value of $\boldsymbol{\beta}$ and as a cross-product matrix (which is the matrix analogue of a squared number) is always positive definite if $\mathbf{X}$ has full column rank, that is, if the columns $\mathbf{x}_0, \mathbf{x}_1, \ldots \mathbf{x}_m$ of the matrix are linearly independent.[5] For this reason one can simply solve equation (2.32) for $\boldsymbol{\beta}_{\text{OLS}}$ to obtain the well-known formula

$$\boldsymbol{\beta}_{\text{OLS}} = (\mathbf{X}'\mathbf{X})^{-1}\mathbf{X}'\mathbf{y}. \tag{2.33}$$

As has become obvious, OLS estimators can be computed without knowledge of the variance of ϵ_i. However, an estimate of this error variance is required for the computation of standard errors for the regression intercept and coefficients. Based on arguments similar to equation (2.20), it can be shown that, if $\text{Var}(\epsilon_i) = \sigma^2$ (the error variance is constant for all observations),

$$\hat{\sigma}^2_{\text{OLS}} = \frac{(\mathbf{y} - \mathbf{X}\boldsymbol{\beta}_{\text{OLS}})'(\mathbf{y} - \mathbf{X}\boldsymbol{\beta}_{\text{OLS}})}{n - m - 1} \tag{2.34}$$

is an unbiased estimator of σ^2.

Statistical properties of OLS estimators

OLS estimators have several of the desirable properties described earlier. Most of these properties derive from the fact that OLS estimators are linear. An estimator based on a random vector $\mathbf{Y}$ is *linear* if it has for some matrix $\mathbf{A}$ that does not depend on the values of the response the form

$$\hat{\boldsymbol{\beta}}(\mathbf{Y}) = \mathbf{A}\mathbf{Y}. \tag{2.35}$$

An OLS estimator is a linear estimator with $\mathbf{A} = (\mathbf{X}'\mathbf{X})^{-1}\mathbf{X}'$, for the fixed regressor matrix $\mathbf{X}$. From the linearity of OLS estimators it can be easily established that they are unbiased. If a linear regression model is correctly specified, then $E(\mathbf{Y}) = \mathbf{X}'\boldsymbol{\beta}_0$. Furthermore, from the linearity property of expectations (see equation (2.8)), it follows that for any constant matrix $\mathbf{A}$,

$$E(\mathbf{A}\mathbf{Y}) = \mathbf{A}\,E(\mathbf{Y}). \tag{2.36}$$

Thus for OLS estimators we obtain

$$E(\boldsymbol{\beta}_{\text{OLS}}) = E((\mathbf{X}'\mathbf{X})^{-1}\mathbf{X}'\mathbf{y}) = (\mathbf{X}'\mathbf{X})^{-1}\mathbf{X}'E(\mathbf{y}) = (\mathbf{X}'\mathbf{X})^{-1}\mathbf{X}'\mathbf{X}'\boldsymbol{\beta}_0 = \boldsymbol{\beta}_0. \tag{2.37}$$

A similar argument based on equation (2.11) allows us to derive the variance of OLS estimators, if $\text{Var}(\epsilon_i) = \sigma^2$:

$$\begin{aligned}\text{Var}(\boldsymbol{\beta}_{\text{OLS}}) &= \text{Var}((\mathbf{X}'\mathbf{X})^{-1}\mathbf{X}'\mathbf{y}) = (\mathbf{X}'\mathbf{X})^{-1}\mathbf{X}'\,\text{Var}(\mathbf{y})\,\mathbf{X}(\mathbf{X}'\mathbf{X})^{-1} \\ &= \sigma^2(\mathbf{X}'\mathbf{X})^{-1}\mathbf{X}'\mathbf{X}(\mathbf{X}'\mathbf{X})^{-1} = \sigma^2(\mathbf{X}'\mathbf{X})^{-1}\end{aligned} \tag{2.38}$$

OLS estimators are not only unbiased, but they are *best linear unbiased estimators* (BLUE) under the conditions of the *Gauss–Markov theorem* (Greene, 2011, p. 100):

Theorem 3 (Gauss–Markov theorem). *If the linear regression model of equation* (2.29) *holds (which in particular means that the errors have a zero expectation* $E(\epsilon_i) = 0$*) and if* $\text{Cov}(\epsilon_i, \epsilon_j) = 0$ *for* $i \neq j$ *(the errors are uncorrelated) and* $\text{Var}(\epsilon_i) = \sigma^2$ *(they have a constant variance) and if the matrix* $\mathbf{X}$ *containing the constant and the independent variables is of full column rank, then:*

1. *the OLS estimator is unbiased;*
2. *its variance is* $\text{Var}(\boldsymbol{\beta}_{OLS}) = \sigma^2(\mathbf{X}'\mathbf{X})^{-1}$*;*
3. *there is no linear unbiased estimator with a smaller variance than the OLS estimator.*

It should be noted that the Gauss–Markov theorem holds under fairly general conditions. No assumptions are made with respect to the particular distribution which the observations $y_1, \ldots, y_n$ of the dependent variable come from. All that is required is that the errors ϵ_i have zero expectation and constant variance and are uncorrelated with one another.

The Gauss–Markov theorem tells us that under quite general conditions, an OLS estimator has the lowest variance of all linear unbiased estimators. Less formally speaking, among this class of estimators, OLS provides estimates that are on average as close as possible to the true regression coefficients based on a sample of fixed size n. However, this alone would not satisfy practitioners if they did not get the better estimates the larger the sample size is. That is, they would usually expect an estimator to be *consistent* in order to be really useful. Fortunately, OLS estimators are consistent (Greene, 2011, p. 105).

Similar, but somewhat more restrictive, conditions lead to the *asymptotical normality* of OLS estimates. Stated formally, the OLS estimator $\hat{\boldsymbol{\beta}}_n$ based on n observations is asymptotically normal in so far as $\sqrt{n}(\hat{\boldsymbol{\beta}}_n - \boldsymbol{\beta}_0)$ *converges in distribution* to a normal distribution with mean zero and covariance matrix $\sigma^2\mathbf{S}_X^{-1}$:

$$\sqrt{n}(\hat{\boldsymbol{\beta}}_n - \boldsymbol{\beta}_0) \xrightarrow{d} N(0; \sigma^2\mathbf{S}_X^{-1}) \tag{2.39}$$

where $\mathbf{S}_X$ is the limit of $\frac{1}{n}\mathbf{X}'\mathbf{X}$ as $n \to \infty$ if the elements of the regressor matrix $\mathbf{X}$ are fixed values or the probability limit of $\frac{1}{n}\mathbf{X}'\mathbf{X}$ as $n \to \infty$ if the elements of the regressor matrix are random (Greene, 2011, p. 107). Note that if the regression errors ϵ_i are i.i.d. normally distributed, the OLS estimator has an exact normal distribution even in finite samples, with mean $\boldsymbol{\beta}_0$ and variance matrix $\sigma^2(\mathbf{X}'\mathbf{X})^{-1}$.

MAXIMUM LIKELIHOOD

OLS estimators as discussed in the previous section are especially designed for linear regression models. A more general class of estimators is that of maximum likelihood estimators. As soon as one has constructed a statistical model in terms of a particular family of distributions, whether it be the normal, binomial, or Poisson distributions, the construction of an ML estimator is straightforward, because all that is required is the PDF or PMF of the observations.

Likelihood function and maximum likelihood

Suppose one has n observations $x_1, \ldots, x_n$ that one considers as values of n i.i.d. random variables $X_1, \ldots, X_n$, with a probability distribution that has PDF or PMF $f(X; \boldsymbol{\theta})$, where $\boldsymbol{\theta}$ is a vector of parameter values. The *likelihood function* of $\boldsymbol{\theta}$ with respect to the observations $x_1, \ldots, x_n$ is simply the product of the values of the PDF or PMF for these observations:

$$L(\boldsymbol{\theta}; \mathbf{x}) = f(X_1; \boldsymbol{\theta}) \times \cdots \times f(X_n; \boldsymbol{\theta}) \tag{2.40}$$

where $\mathbf{x}$ is the vector with the elements $x_1, \ldots, x_n$. The *maximum likelihood estimate* (MLE) $\hat{\boldsymbol{\theta}}$ then is simply the value of $\boldsymbol{\theta}$ for which the likelihood function is maximal. Care should be taken not to misinterpret the likelihood function: the MLE is *not* the 'most likely' parameter value – it is the parameter value under which the *observations* $x_1, \ldots, x_n$ of the random variables $X_1, \ldots, X_n$ are more likely than with any other parameter value.

Since the likelihood function is a product of several terms in the observations and taking derivatives (needed for the computation of MLEs) of products is tedious and difficult, one generally looks at the *log-likelihood function*,

$$\ell(\boldsymbol{\theta}; \mathbf{x}) = \ln L(\boldsymbol{\theta}; \mathbf{x}) = \ln f(X_1; \boldsymbol{\theta}) + \cdots + \ln f(X_n; \boldsymbol{\theta}), \tag{2.41}$$

which takes the form of a sum of several terms rather than a product, so that derivatives are much easier to handle. The use of the log-likelihood function is valid because the logarithm is a strictly increasing function, with the consequence that the value of $\boldsymbol{\theta}$ that maximizes the log-likelihood also maximizes the likelihood function.

If the parameter is a scalar $\theta = \boldsymbol{\theta}$, then the two conditions that an MLE $\hat{\theta}$ must satisfy are that the first derivative of the log-likelihood is zero and that the second derivative is negative:

$$\frac{\mathrm{d}\,\ell(\hat{\theta}; \mathbf{x})}{\mathrm{d}\,\theta} = 0 \quad \text{and} \quad \frac{\mathrm{d}^2\,\ell(\hat{\theta}; \mathbf{x})}{\mathrm{d}\,\theta^2} < 0. \tag{2.42}$$

An MLE $\hat{\boldsymbol{\theta}}$ for a parameter vector has to satisfy that all elements of the vector of the first derivative of log-likelihood function, the *gradient*, are zero and that the matrix of the second derivatives, the *Hessian*, is negative definite:

$$\frac{\partial \ell(\hat{\boldsymbol{\theta}}; \mathbf{x})}{\partial \boldsymbol{\theta}} = \mathbf{0} \quad \text{and} \quad \mathbf{a}' \left(\frac{\partial^2 \ell(\hat{\boldsymbol{\theta}}; \mathbf{x})}{\partial \boldsymbol{\theta} \partial \boldsymbol{\theta}'} \right) \mathbf{a} < 0 \text{ for all vectors with } \mathbf{a}'\mathbf{a} > 0. \tag{2.43}$$

Obtaining MLEs is best demonstrated by an example. Suppose one tosses a coin four times, where the 'heads' probability p is a parameter to be estimated. This gives four i.i.d. random variables X_1, X_2, X_3, X_4, where $X_i = 1$ means 'heads' and $X_i = 0$ means 'tails'. If one makes observations x_1, x_2, x_3, x_4 of these random variables, the log-likelihood function becomes

$$\ell(p; \mathbf{x}) = \sum_{i=1}^{4} \ln\left(p^{x_i}(1-p)^{1-x_i}\right) = \sum_{i=1}^{4} x_i \ln p + \sum_{i=1}^{4} (1 - x_i) \ln(1 - p). \tag{2.44}$$

If two of the observations are 'heads' ($X_i = 1$) and two are 'tails' ($X_i = 1$), the log-likelihood is

$$\ell(p;\mathbf{x}) = 2\ln p + 2\ln(1-p).$$

The first condition for $\hat{p}$ to be a maximum is that the first derivative equals zero:

$$\frac{\mathrm{d}\,\ell(\hat{p};\mathbf{x})}{\mathrm{d}p} = \frac{2}{\hat{p}} - \frac{2}{1-\hat{p}} = 0 \Leftrightarrow 2(1-\hat{p}) = 2\hat{p} \Leftrightarrow 2 = 4\hat{p} \Leftrightarrow \hat{p} = \frac{1}{2}.$$

The second condition is that the second derivative is negative, which is the case here:

$$\frac{\mathrm{d}^2\,\ell(\hat{p};\mathbf{x})}{\mathrm{d}p^2} = -\frac{2}{\hat{p}^2} - \frac{2}{(1-\hat{p})^2} = -16 < 0.$$

So we find that the MLE for p is indeed $\frac{1}{2}$.

That OLS estimators can be considered as a special case of MLEs becomes clear if one considers the log-likelihood function of a linear regression model defined by equation (2.29) with errors being i.i.d. normally distributed with common error variance σ^2, which is

$$\begin{aligned}\ell(\boldsymbol{\beta},\sigma^2;\mathbf{y}|\mathbf{X}) &= -\frac{n}{2}\ln(2\pi\sigma^2) - \frac{1}{2\sigma^2}(\mathbf{y}-\mathbf{X}\boldsymbol{\beta})'(\mathbf{y}-\mathbf{X}\boldsymbol{\beta})\\ &= -\frac{n}{2}\ln(2\pi\sigma^2) - \frac{1}{2\sigma^2}SSQ(\boldsymbol{\beta}).\end{aligned} \tag{2.45}$$

It is obvious that, whatever the value of σ^2, the value of $\boldsymbol{\beta}$ that maximizes this likelihood – the MLE – is the one that minimizes the residual sum of squares $SSQ(\boldsymbol{\beta})$ and therefore is the OLS estimate. It is also straightforward to show that the MLE of the error variance σ^2 is $\hat{\sigma}^2_{\mathrm{ML}} = SSQ(\hat{\boldsymbol{\beta}})/n$ – thus different from the usually recommended, unbiased estimator $\hat{\sigma}^2_{\mathrm{OLS}} = SSQ(\hat{\boldsymbol{\beta}})/(n-m-1)$ – if one takes the derivative of the log-likelihood function for σ^2,

$$\frac{\partial\ell(\boldsymbol{\beta},\sigma^2;\mathbf{y},\mathbf{X})}{\partial\sigma^2} = -\frac{n}{2}\frac{1}{\sigma^2} + \frac{1}{2\sigma^4}SSQ(\hat{\boldsymbol{\beta}}),$$

and sets it to zero. While the MLE of the error variance is biased, the difference between the MLE and the unbiased estimator is negligible if the number of observations n is large relative to the number of independent variables m.

Computing maximum likelihood estimates

In most cases there is no direct way to compute MLEs for given observations. Regression coefficients in normal linear models and success probabilities in binomial count models are rare exceptions. When there is no solution formula available for an MLE, one has to resort to an *iterative algorithm*. One of the most widely used algorithms for the computation of MLEs in statistical software is the *Newton–Raphson* algorithm. The Newton–Raphson algorithm generally needs a set of starting values as an initial approximation and refines the approximation of the MLE step by step. For example, if $\boldsymbol{\theta}^{(s)}$ is the current approximation of the MLE after s iterations of the algorithm, the approximation for the next step is

$$\boldsymbol{\theta}^{(s+1)} = \boldsymbol{\theta}^{(s)} - \left(\frac{\partial^2\ell(\boldsymbol{\theta}^{(s)};\mathbf{y},\mathbf{X})}{\partial\boldsymbol{\theta}\,\partial\boldsymbol{\theta}'}\right)^{-1}\frac{\partial\ell(\boldsymbol{\theta}^{(s)};\mathbf{y},\mathbf{X})}{\partial\boldsymbol{\theta}} \tag{2.46}$$

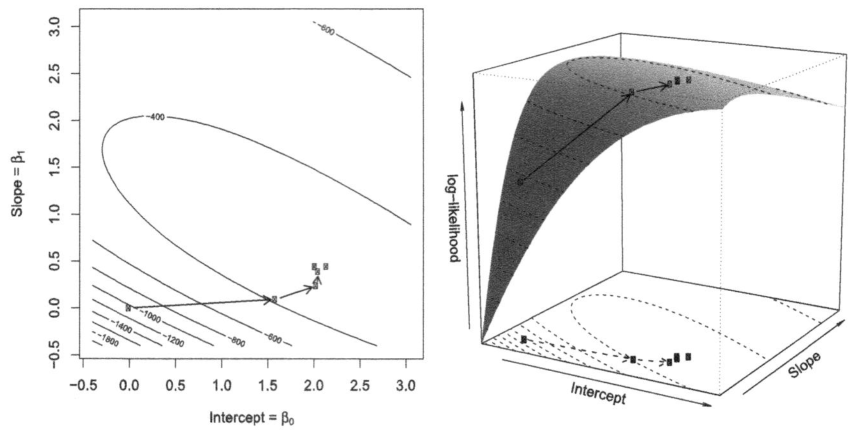

Figure 2.2 An illustration of the Newton–Raphson algorithm to estimate the intercept (α) and slope (β) of a logistic regression model with one independent variable. The algorithm is started with $\beta_0 = 0$ and $\beta_1 = 0$, each of the dots in the diagrams represents the estimate values at a particular step of the algorithm, and the arrows connect estimate values of successive iterations

and these iterations are repeated until the improvements are negligible, that is, until

$$||\boldsymbol{\theta}^{(s+1)} - \boldsymbol{\theta}^{(s)}|| < a \tag{2.47}$$

or until

$$\ell(\boldsymbol{\theta}^{(s+1)}; \mathbf{y}, \mathbf{X}) - \ell(\boldsymbol{\theta}^{(s)}; \mathbf{y}, \mathbf{X}) < a \tag{2.48}$$

for a very small number $a > 0$. (A typical choice for such a number is $a = 10^{-7}$). If the algorithm has been stopped because this criterion is satisfied, the algorithm is said to have *converged*. When convergence of the algorithm has been achieved, the current value $\boldsymbol{\theta}^{(s)}$ is declared the MLE for the model, or rather a 'good enough' approximation of the estimate. The iterative procedure to compute MLEs using the Newton–Raphson algorithm is illustrated by Figure 2.2.

Sometimes the Hessian matrix is difficult to evaluate or is positive semidefinite for some values $\boldsymbol{\theta}^{(s)}$ of the parameter. In this case one will need alternatives to the Newton–Raphson algorithm. Sometimes it is sufficient to use the negative of the Fisher information matrix (the expected value of the Hessian), which leads to a Fisher scoring algorithm. In other cases one will need recourse to quasi-Newton methods, which only use the first derivative of the log-likelihood function (see Greene, 2011, Chap. 4). Nevertheless, in many standard applications of ML estimation, such as for estimating parameters of logit and probit models, the Newton–Raphson algorithm works well, because the Hessian is negative definite for all possible values of the model parameters.

Invariance of maximum likelihood estimators

An important property of ML estimators is that they are *invariant*. To understand this, recall the example of flipping a coin, where the 'heads' probability p was the parameter of interest. The Bernoulli distribution can alternatively be written in terms of the odds $\psi = p/(1-p)$:

$$f(X;p) = p^x(1-p)^{1-x} = \left(\frac{p}{1-p}\right)^x (1-p) = \frac{\psi^x}{1+\psi} = f(X;\psi). \tag{2.49}$$

If the parameter of interest is the odds ψ instead of p, but an MLE $\hat{p}$ is available, then the MLE for the former is simply $\hat{\psi} = \hat{p}/(1-\hat{p})$. Further, if the parameter of interest is the log-odds $\eta = \ln(\psi)$, then its MLE is $\hat{\eta} = \ln(\hat{\psi}) = \ln(\hat{p}/[1-\hat{p}])$. In the case of the numerical example, where it was found that $\hat{p} = \frac{1}{2}$, the MLE of the odds is $\hat{\psi} = 1$ and the MLE of the log-odds is $\hat{\eta} = \ln(1) = 0$.

More generally, if a statistical model can be expressed both in terms of one set of parameters $\boldsymbol{\theta}$ and another set of parameters $\boldsymbol{\psi}$ and there is a smooth one-to-one mapping such that a model expressed in terms of $\boldsymbol{\theta}$ leads to the same density values or probabilities as a model in terms of $\boldsymbol{\psi} = g(\boldsymbol{\theta})$ then for their MLEs $\hat{\boldsymbol{\psi}}$ and $\hat{\boldsymbol{\theta}}$,

$$\hat{\boldsymbol{\psi}} = g(\hat{\boldsymbol{\theta}}). \tag{2.50}$$

The invariance property has obvious practical benefits. Suppose one can express a model both in terms of parameters $\boldsymbol{\theta}$ and $\boldsymbol{\psi}$, and $\boldsymbol{\psi}$ is the parameter of interest, while the MLE of $\boldsymbol{\theta}$ is easier to compute. In that case, the invariance property of MLEs opens up the convenient route to compute $\hat{\boldsymbol{\psi}}$ based on the simpler MLE $\hat{\boldsymbol{\theta}}$. However, invariance is not really a statistical property of MLEs, but a mathematical one that is implied by the rules of calculus. The statistical properties will be discussed in what follows.

Finite-sample properties of maximum likelihood estimators

The Gauss–Markov theorem introduced in the third section of this chapter states that the OLS estimator is, under fairly general conditions, the best linear unbiased estimator of the coefficients of a regression model. As shown previously, the OLS estimator is an ML estimator if the regression errors ϵ_i in equation (2.29) are i.i.d. normally distributed. Under this additional condition, the OLS estimator as ML estimator is the estimator with minimal variance not only among all linear unbiased but *among all unbiased estimators*, that is, OLS is *efficient* in this case. Such a statement is possible because there is a lower bound to the variance of the function of any random variable, including unbiased estimators – the Cramér–Rao lower bound (e.g. Lehmann and Casella, 1998, p. 127):[6]

Theorem 4 (Cramér–Rao lower bound)**.** *Let X be a random variable with PDF or PMF $f(X,\boldsymbol{\theta}_0)$, of which the first and second derivatives exist, and let $T(X)$ be a function of the random variable with expectation $E(T(X)) = \boldsymbol{\psi}(\boldsymbol{\theta}_0)$. Then the variance of $T(X)$ is bounded from below by*

$$\text{Var}(T(X)) \geq \frac{\partial \boldsymbol{\psi}}{\partial \boldsymbol{\theta}'} \mathcal{I}(\boldsymbol{\theta}_0)^{-1} \frac{\partial \boldsymbol{\psi}'}{\partial \boldsymbol{\theta}}, \tag{2.51}$$

where $\mathcal{I}(\boldsymbol{\theta}_0)$ is the Fisher information matrix

$$\mathcal{I}(\boldsymbol{\theta}_0) = \text{Var}\left(\frac{\partial \ell(\boldsymbol{\theta}_0;X)}{\partial \boldsymbol{\theta}}\right) = -E\left(\frac{\partial^2 \ell(\boldsymbol{\theta}_0;X)}{\partial \boldsymbol{\theta} \partial \boldsymbol{\theta}'}\right) \tag{2.52}$$

with $\ell(\boldsymbol{\theta};X) = \ln f(X;\boldsymbol{\theta})$. In particular, if $\hat{\boldsymbol{\theta}} = T(X)$ is an unbiased estimator of $\boldsymbol{\theta}$ then

$$\text{Var}(\hat{\boldsymbol{\theta}}) \geq \mathcal{I}(\boldsymbol{\theta}_0)^{-1}. \tag{2.53}$$

It is easy to verify, by taking the derivatives of the log-likelihood in equation (2.45), that in the case of a linear regression with i.i.d. normal errors the Fisher information matrix is

$$\mathcal{I}(\boldsymbol{\beta}_0) = \frac{1}{\sigma^2}\mathbf{X}'\mathbf{X}, \tag{2.54}$$

which is the inverse of the covariance matrix of $\boldsymbol{\beta}_{\text{OLS}}$ according to the Gauss–Markov theorem. That is, as the ML estimator in the normal linear regression model, the OLS estimator reaches the Cramér–Rao lower bound and therefore is efficient.

OLS estimators are not the only efficient unbiased ML estimators. Another example is the ML estimator of the success probability in Bernoulli experiments and binomial trials. Recall the coin-tossing experiment discussed previously. Suppose now that the coin is tossed n times. Then the number of times K that the result is 'heads' has a binomial distribution with the same success probability parameter p and with denominator n. (The Bernoulli distribution discussed earlier is a special case with $n = 1$.) The PMF of such a binomial distribution is

$$f_{\text{Bn}}(k;p,n) = \Pr(K = k) = \binom{n}{k} p^k (1-p)^{n-k}, \tag{2.55}$$

and it is well known that the expectation and the variance of a random variable with this distribution are

$$E(K) = np \quad \text{and} \quad \text{Var}(K) = np(1-p). \tag{2.56}$$

The log-likelihood function for p under this distribution is

$$\ell(p;k) = \ln\binom{n}{k} + k\ln p + (n-k)\ln(1-p) \tag{2.57}$$

which is, up to a constant not dependent on the parameters, identical to the log-likelihood function of n Bernoulli trials where a positive outcome ($X_i = 1$) has been observed k times. Generalizing the argument following equation (2.44) it is easy to see that the MLE of the success probability is equal to $\hat{p} = K/n$. The ML estimator is therefore unbiased,

$$E(\hat{p}) = E\left(\frac{K}{n}\right) = \frac{E(K)}{n} = p, \tag{2.58}$$

and has the variance

$$\text{Var}(\hat{p}) = \text{Var}\left(\frac{K}{n}\right) = \frac{\text{Var}(K)}{n^2} = \frac{p(1-p)}{n}. \tag{2.59}$$

The Fisher information is

$$\mathcal{I}(p) = -E\left(\frac{\mathrm{d}^2 \ln f_{\text{Bn}}(K;p,n)}{(\mathrm{d}p)^2}\right) = E\left(\frac{K}{p^2} + \frac{n-K}{(1-p)^2}\right) = \frac{np}{p^2} + \frac{n-np}{(1-p)^2} = \frac{n}{p(1-p)}, \tag{2.60}$$

so that $\text{Var}(\hat{p}) = \mathcal{I}(p)^{-1}$ and the Cramér–Rao lower bound is attained, that is, the ML estimator $\hat{p}$ is efficient.

Unfortunately not all ML estimators are unbiased. A straightforward example, already encountered earlier, is the ML estimator $\hat{\sigma}^2_{\text{ML}} = SSQ(\hat{\boldsymbol{\beta}})/n$ of the regression error variance, which differs from the unbiased estimator $\hat{\sigma}^2_{\text{OLS}} = SSQ(\hat{\boldsymbol{\beta}})/(n-m-1)$. Even ML estimators that are functions of unbiased ML estimators are not always unbiased. Let τ be a parameter of a family of probability distributions for which the ML estimator is unbiased, that is, $E(\hat{\tau}) = \tau_0$, where τ_0 is the true value of the parameter. Further, let θ be a parameter of the same family of distributions defined as $\theta = g(\tau)$. Then the MLE of θ, $\hat{\theta} = g(\hat{\tau})$, is unbiased if and only if g is a linear function. If the true value of θ is θ_0 then, due to Jensen's inequality,

$$E(\hat{\theta}) = E(g(\hat{\tau}))\begin{cases} \geq g(E(\hat{\tau})) = g(\tau_0) = \theta_0 & \text{if } g \text{ is convex,} \\ \leq g(E(\hat{\tau})) = g(\tau_0) = \theta_0 & \text{if } g \text{ is concave,} \end{cases}$$

with equality only if g is both convex *and* concave, that is, linear.[7] For example, while the MLE of the success probability parameter p of a binomial distribution is unbiased and efficient, the MLEs of the odds $\psi = p/(1-p)$ and of the log-odds $\eta = \ln(p/[1-p])$ are biased.

Such biases are not a particular shortcoming of MLEs. In fact the existence of efficient unbiased estimators hinges on quite restrictive conditions. They exist if and only if (Lehmann and Casella, 1998, p. 121):

1. the probability distribution of the random variable in question belongs to an exponential family;
2. the estimator is a linear function of a sufficient statistic with respect to the parameters of the exponential family;
3. the parameter to be estimated is the expected value of a linear function of the sufficient statistic.

Notably, in those cases where efficient unbiased estimates exist, they are also MLEs (Lehmann and Casella, 1998, p. 121). The concepts involved in these conditions are explained in the following.

A statistic $T = t(Y_1, \ldots, Y_n)$ of a sample $Y_1, \ldots, Y_n$ of i.i.d. random variables with common PMF or PDF $f(y; \boldsymbol{\theta})$ is a *sufficient statistic* if the distribution of Y_i *conditional on* T does not depend on the parameter values $\boldsymbol{\theta}$ (Casella and Berger, 2002, p. 272). A family of probability distributions is an *exponential family* with natural parameters $\eta_1, \ldots, \eta_d$, if the PDF or PMF of each of the i.i.d. random variables $Y_1, \ldots, Y_n$ can be written in the form (Casella and Berger, 2002, p. 111)

$$f(y_i; \boldsymbol{\eta}) = \exp\left(\sum_{k=1}^{d} t_k(y_i)\,\eta_k - a(\boldsymbol{\eta}) + b(y_i)\right), \tag{2.61}$$

and the PDF or PMF of the sufficient statistics $T_k = t_k(Y_1, \ldots, Y_n)$ takes the form (Casella and Berger, 2002, p. 279)

$$f(t(y_1 \ldots, y_n); \boldsymbol{\eta}) = \exp\left(\sum_{k=1}^{d} t_k(y_1, \ldots, y_n)\,\eta_k - na(\boldsymbol{\eta}) + b(n, t(y_1 \ldots, y_n))\right). \tag{2.62}$$

The first two (and also the higher) central moments of these sufficient statistics can be obtained by taking derivatives (Casella and Berger, 2002, p. 112):

$$\tau_k = E(T_k) = \frac{\partial a}{\partial \eta_k}, \quad \mathrm{Var}(T_k) = \frac{\partial^2 a}{(\partial \eta_k)^2}, \quad \mathrm{Cov}(T_k, T_l) = \frac{\partial^2 a}{\partial \eta_k \partial \eta_l}. \tag{2.63}$$

From these properties it can be shown easily that T_k is an unbiased estimator of τ_k that attains the Cramér–Rao lower bound of equation (2.53) and is therefore efficient.

The most common example of an exponential family is the family of normal distributions with mean parameter μ and variance parameter σ^2, with natural parameters $\eta_1 = \mu/\sigma^2$ and $\eta_2 = -1/(2\sigma^2)$ and sufficient statistics $T_1 = Y_1 + \cdots + Y_n$ and $T_2 = Y_1^2 + \cdots + Y_n^2$ with expectations $\tau_1 = E(T_1) = n\mu$ and $\tau_2 = E(T_2) = n(\mu^2 + \sigma^2)$. The family of Bernoulli distributions is an exponential family with natural parameter $\eta = \ln(p/(1-p))$, the log-odds, and the sufficient statistic for a sample of i.i.d. Bernoulli distributed random variables $Y_1, \ldots, Y_n$ is $T_1 = Y_1 + \cdots + Y_n$. This sufficient statistic has a binomial distribution with the same success parameter p and natural parameter $\eta = \ln(p/(1-p))$, but with denominator n, and has the expectation $\tau_1 = E(T_1) = np$.

Asymptotic properties of maximum likelihood estimators

While the finite-sample properties of OLS estimators do not generalize to ML estimators, their large-sample properties do so. That is, ML estimators are consistent and asymptotically normal, provided that the distribution of the observations on which they are based satisfies certain *regularity conditions*. There is not enough room to discuss these regularity conditions here, but they are discussed in Lehmann and Casella (1998) and Gourieroux and Monfort (1995a,b). Suffice it to state here that these regularity conditions are satisfied in the case of normal linear regression, logistic regression and most other regression models discussed in this volume.

In the previous subsection it was stated that ML estimators may be unbiased and efficient, yet only under particular circumstances. However, under the same regularity conditions that ensure the consistency and asymptotic normality of MLEs, the 'large-sample analogues' of unbiasedness and efficiency apply to them in general. That is, ML estimators are *asymptotically* unbiased and *asymptotically* efficient. An estimator $\hat{\boldsymbol{\theta}}_n$ based on n observations is *asymptotically unbiased* if

$$\lim_{n\to\infty} E(\hat{\boldsymbol{\theta}}_n) = \boldsymbol{\theta}_0. \tag{2.64}$$

An asymptotically normal and unbiased estimator is *asymptotically efficient* if its asymptotic distribution attains the Cramér–Rao lower bound, that is, if

$$\lim_{n\to\infty} \operatorname{Var}\left(\hat{\boldsymbol{\theta}}_n - \boldsymbol{\theta}_0\right) - \mathcal{I}_n(\boldsymbol{\theta}_0)^{-1} = 0, \tag{2.65}$$

where $\mathcal{I}_n(\boldsymbol{\theta}_0)$ denotes the Fisher information matrix as in the previous subsection (and the n subscript indicates that the information matrix is based on the log-likelihood of n observations).

These asymptotic properties of ML estimators mean that, while they may be biased in small samples, they are approximately unbiased and efficient in large samples. Their consistency means that estimates tend to be closer to the true parameter values the larger a sample is, while their asymptotic normality means that it is possible to obtain confidence intervals and significance tests that are at least approximately correct.

CAVEATS

The present chapter is concerned with the main estimation techniques in use for regression analysis and for related methods of causal inference and their statistical aspects. It has been taken for granted throughout the discussion of these techniques that the statistical models, the parameters of which are to be estimated, are correctly specified in their systematic aspects. That is, it is assumed that they correctly reflect the relation between independent variables and the expected values of the dependent variable and that the models are identified in so far as there is only a single 'true' parameter value or a single 'true' coefficient vector. These assumptions may or may not be satisfied in particular applications, but they are pertinent to *model construction* and therefore beyond the scope of this chapter, though they are discussed in the other chapters of this volume. Instead, this section discusses the statistical consequences of violations of distributional assumptions and of small sample sizes.

Departures from distributional assumptions

It is often difficult to check whether the observed data really follow the assumed distribution and sometimes the reconstruction of a correct distribution of the observed data is not feasible.

For this reason it is important to know what will happen if the actual distribution of the data differs from the distribution assumed by a particular model. Fortunately, under certain conditions, the 'damage' done by departures from the distributional assumption of a model is limited. To be specific, under suitable conditions ML estimators retain their consistency and asymptotic normality even under the misspecification of the conditional distribution of the response variable. But under almost any circumstances the consequence of misspecification is the loss of (asymptotic) efficiency.

Statistical misspecification in regression-type models (linear regression, logistic regression and similar models) basically means that the random variables that form the response have a distribution characterized by a PMF or PDF $f(y_i|\mathbf{x}_i;\boldsymbol{\theta}_0)$ but the distribution that is used for the construction of the log-likelihood function has a different and therefore incorrect PMF or PDF $f^*(y_i|\mathbf{x}_i;\boldsymbol{\phi})$. In the discussion of the consequences of misspecification, the (incorrect) log-likelihood function based on $f^*(y_i|\mathbf{x}_i;\boldsymbol{\phi})$,

$$\ell^*(\mathbf{y}|\mathbf{X};\boldsymbol{\phi}) = \sum_i \ln f^*(y_i|\mathbf{X}_i;\boldsymbol{\phi}), \tag{2.66}$$

is called the log-*pseudo-likelihood* and the estimator $\tilde{\boldsymbol{\phi}}$ maximizing it is called a pseudo-maximum likelihood (PML) estimator. Under suitable regularity conditions the PML estimator has a probability limit, the so-called *pseudo-true value* $\boldsymbol{\phi}_0^*$ (Gourieroux and Monfort, 1995a, p. 234). Furthermore, the PML estimator is asymptotically normal with mean $E_{\theta_0}(\tilde{\boldsymbol{\phi}}) = \boldsymbol{\phi}_0^*$ and with variance matrix

$$\mathrm{Var}_{\theta_0}(\tilde{\boldsymbol{\phi}}) = \left(\frac{\partial^2\ell^*(\mathbf{y}|\mathbf{X};\boldsymbol{\phi}_0^*)}{\partial\boldsymbol{\phi}\partial\boldsymbol{\phi}'}\right)^{-1} \mathrm{Var}_{\theta_0}\left(\frac{\partial\ell^*(\mathbf{y}|\mathbf{X};\boldsymbol{\phi}_0^*)}{\partial\boldsymbol{\phi}}\right)\left(\frac{\partial^2\ell^*(\mathbf{y}|\mathbf{X};\boldsymbol{\phi}_0^*)}{\partial\boldsymbol{\phi}\partial\boldsymbol{\phi}'}\right)^{-1} \tag{2.67}$$

where the subscript in E_{θ_0} and Var_{θ_0} means that these are the expectation and the variance based on the correct PMF or PDF $f(\mathbf{y}|\mathbf{x};\boldsymbol{\theta}_0)$.

Of course, a PML estimator $\tilde{\boldsymbol{\phi}}$ will be useless if it is completely unrelated to the true parameter value $\boldsymbol{\theta}_0$. However, a PML estimator is consistent in situations that are common in regression-type models (linear regression, logistic regression, etc.). Suppose that the parameter vector $\boldsymbol{\theta}$ of the correct distribution with PMF or PDF $f(y|\mathbf{x};\boldsymbol{\theta})$ can be split into sub-vectors $\boldsymbol{\beta}$ and $\boldsymbol{\omega}$ such that the conditional expectation of the dependent variable takes the form

$$E_{\theta}(Y|\mathbf{X} = \mathbf{x}) = h(\mathbf{x};\boldsymbol{\beta}), \tag{2.68}$$

where the true value of $\boldsymbol{\beta}$ is $\boldsymbol{\beta}_0$. Suppose, furthermore, that the misspecified family of distributions is an exponential family so that $f^*(y_i|\mathbf{x},\boldsymbol{\phi})$ has the form of equation (2.61) with $t_1(y_i) = y_i$ and that $\boldsymbol{\phi}$ can be split into sub-vectors $\boldsymbol{\beta}$ and $\boldsymbol{\psi}$ so that

$$E_{\phi}(Y|X = \mathbf{x}) = h(\mathbf{x};\boldsymbol{\beta}). \tag{2.69}$$

The PML estimator $\tilde{\boldsymbol{\beta}}$ is consistent, that is, $\mathrm{plim}_{n\to\infty}\tilde{\boldsymbol{\beta}} = \boldsymbol{\beta}_0$, if and only if these conditions are satisfied, that is, if the functional form of the conditional expectation of the response variable is correctly specified (Gourieroux and Monfort, 1995a, p. 239).

It is especially noteworthy about these conditions for consistency that they do not depend on aspects of the correctly specified distribution but on the distribution that is used to construct the PML estimator. And one could call it reassuring that this consistency result applies to common regression-type models, such as linear regression, logistic regression and Poisson regression. It essentially means that for the consistent estimation of the coefficients of these models, it is paramount to correctly specify the link between the independent variables and the (expected values) of the dependent variable. Whether the distribution of the dependent variable, apart from its expected values, is correctly specified is often inconsequential for the consistency of coefficient estimates.

While PML estimators may under certain circumstances remain consistent and asymptotically normal if the distribution used to construct them is misspecified, they will lose the property of asymptotic efficiency, because their variance given by equation (2.67) will be larger than the Cramér–Rao lower bound in equation (2.53). Standard errors based on the second derivatives of the pseudo-log-likelihood function will usually be too small. However, asymptotically correct estimates of standard errors are provided in the form of so-called 'robust' standard errors, also known as 'Huber–White' or 'sandwich' estimators (Huber, 1967; White, 1982), which are available as an option in some statistical software packages. One should not, however, be tempted to use robust standard errors by default. Such standard errors are themselves estimates and, if the probability distribution on which the model is based is correctly specified, are much less precise than those obtained from the Fisher information. Fortunately, White (1982) devised a statistical test that helps to decide whether robust standard errors are needed.

Small samples

In the previous section not only the finite-sample properties of OLS and ML estimators were discussed but also their asymptotic properties, such as consistency, asymptotic normality, asymptotic unbiasedness and asymptotic efficiency. These are properties that estimators do not exhibit in any finite sample, but only in samples of a size that approaches infinity. In practice this means that these properties approximately apply to OLS and ML estimators in large samples, say when n is of the order of a few thousand. The asymptotic properties are especially relevant for ML estimation of parameters of models other than linear regression, such as logistic regression, probit regression and the like. For example, MLEs of logistic regression coefficients can be heavily biased, and in some unfortunate instances they do not even exist.

The behaviour of MLEs in samples of different sizes is best illustrated by a simulation study, the results of which are presented in Figure 2.3. The left-hand diagram shows the distribution of estimates of a logistic regression coefficient based on 5000 simulated samples of size $n = 2000$. Each of these estimates is based on a simulated data set with an independent variable x, which has a standard normal distribution, and a binary dependent variable y, which is constructed such that it follows a logistic regression model with an intercept equal to zero and a regression coefficient of x that is equal to one. The true value of the logistic regression coefficient is indicated by the triangle on the horizontal axis, while the average of the coefficient estimates is indicated by the round dot on the same axis. The grey area represents the overall distribution of the estimates (it is a kernel density estimate) and the solid curve represents the density of the asymptotic normal distribution centred on the true coefficient value and the inverse of the Fisher information matrix as variance. It is quite obvious that with a sample size of 2000 the average of the simulation-based estimates, which represents the expected value of the MLE, is close to the true coefficient value, that is, the estimator is approximately unbiased. Also the diagram shows that the asymptotic distribution approximates the actual distribution of the estimates quite well.

The right-hand diagram in Figure 2.3 shows the result of another run of the simulation study where this time estimates of the slope in the same logistic regression model are based on much smaller samples of size $n = 25$. Of course, with such a much smaller sample size, the estimates show a much higher variation, which is why a much wider range of the horizontal axis is needed in this diagram. Yet the higher variation is not the only consequence of the smaller sample size:

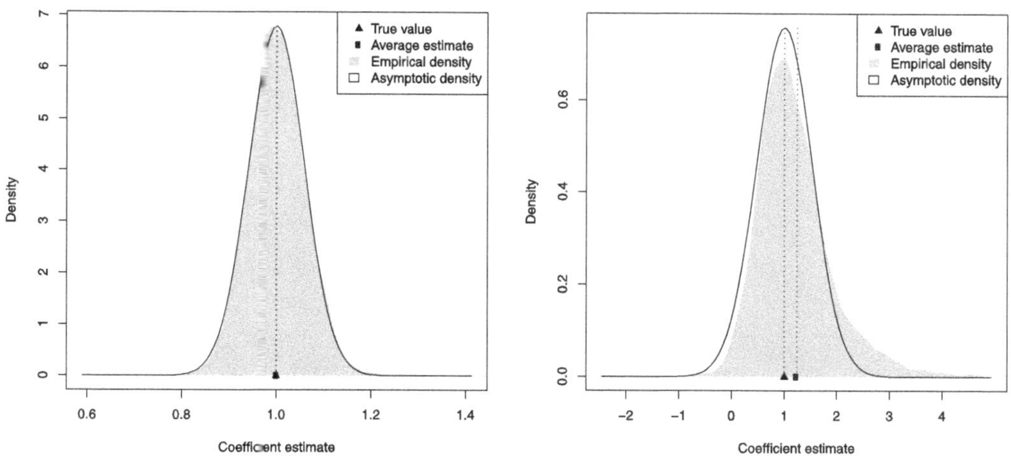

(a) Sample size 2000

(b) Sample size 25

Figure 2.3 The distribution of MLEs of a logistic regression coefficient based on different sample sizes. The grey areas represent kernel density estimates of the empirical distribution of the estimates, the solid lines represent the asymptotic normal distribution around the true parameter value with a variance based on the Fisher information matrix. The diagrams differ in scaling to facilitate the comparison between the empirical distribution of the estimates and their asymptotic distribution

the average of the estimates departs considerably from the true value of the regression coefficient (the average is 1.2487, 25 per cent larger than the true value). Also the empirical distribution of the estimates now differs from their asymptotic distribution, showing a skew to the right. So clearly in small samples, MLEs of logistic regression coefficients will not be very trustworthy and neither will be confidence intervals based on a normality assumption about the distribution of the estimator.

Fortunately, there are options for tackling such small-sample problems. In medium-sized samples the bias can be corrected to some degree using penalized likelihood methods (Firth, 1993). Resampling methods such as jackknifing and bootstrapping can be used to correct the bias and to obtain the empirical distribution of an estimator for a particular sample in order to obtain approximately correct confidence intervals and test statistics (Efron and Tibshirani, 1993; Davison and Hinkley, 1997).

It is not uncommon in small samples that an MLE does not even exist, for example in the case of *complete separation* in logistic regression. Here, the independent variables seem to completely determine the observed values of the dependent variable. In the bivariate case this means that there exists a value x_c of the independent variable such that $y_i = 0$ for all $x_i < x_c$ and $y_i = 1$ for all $x_i \geq x_c$. The consequence of complete separation is that no (finite) MLE for the coefficients of the logistic regression model of y_i on x_i exists. A Newton–Raphson algorithm will usually not converge. Software implementing this algorithm usually will stop after a predetermined maximum number of iterations, report very large estimates and extremely large standard errors and (if the software is transparent enough about this) indicate that the algorithm has not converged.

Figure 2.4 illustrates this phenomenon. The left-hand diagram shows a plot of a small data set with seven observations, an independent variable x ranging from -3 to 3 and a binary dependent variable y, which is completely separated by the independent variable: for all x-values smaller

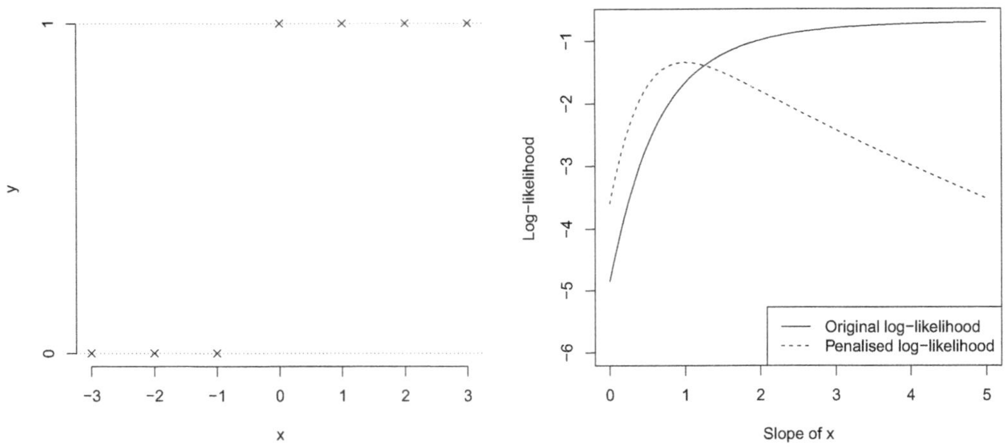

(a) Data with separation

(b) Log-likelihood function for the logistic regression slope

Figure 2.4 An illustration of complete separation

than or equal zero its value is 0, and for all x-values larger than or equal zero its value is 1, so that it is completely determined by the independent variable. As illustrated by the right-hand diagram, the log-likelihood function of the logistic regression slope has no maximum. Instead, it increases with the value of the slope and at the same time flattens out. If one runs a logistic-regression procedure on statistical software one will obtain a large estimate for the slope coefficient and an even larger standard error. For example, the function 'glm' of the software package R returns an estimate of 46.4 and a standard error of 93,010.3, reflecting the fact that the log-likelihood function is extremely flat for large values of the slope coefficient.

It is quite obvious that resampling based methods will not serve to overcome a case of complete separation. A safer method seems to be Firth's bias-correcting penalized ML estimator, as also illustrated by the right-hand diagram in Figure 2.4. In contrast to the original log-likelihood function, the penalized log-likelihood function employed by Firth's method does have a single maximum, suggesting that Firth's method will lead to finite estimates of logistic regression coefficients even in the presence of complete separation. Firth's penalized likelihood method was originally devised to reduce the finite-sample biased discussed earlier, and is constructed in such a way that the difference between a maximum likelihood estimate and a maximum penalized likelihood estimate gets smaller the larger the sample size is. That is, maximum likelihood and Firth's method are asymptotically equivalent.[6]

Small samples may even pose problems for the OLS estimator. Some observations may be outliers and some observations may be extremely influential because of their values on both dependent and independent variables. However, since these problems are covered by another chapter on regression diagnostics, they are not discussed here.

FURTHER READING

This chapter can only scratch the surface of the theory behind estimation techniques such as ordinary least squares and maximum likelihood. For readers who want to delve deeper into these matters there are two main avenues that can be taken. The first is the econometric literature. Here Greene's *Econometric Analysis* (2011) has become a contemporary classic. It provides a thorough discussion of linear models and their extensions as well as some theory of parameter estimation, which includes methods not discussed in this chapter, such as generalized method-of-moments estimators. In addition, it provides an introduction to probability theory necessary to understand the core concepts. Amemiya (1985) is a more advanced econometric text, which gives more room to a discussion of the general properties, such as efficiency, consistency and asymptotic normality, of a quite general set of estimators, such as extremum estimators. Gourieroux and Monfort (1995a,b) combine a discussion of the construction of econometric models with some thorough discussion of estimation theory. These advanced texts should however be read only after one has mastered texts such as Greene's *Econometric Analysis*.

The other strand of literature for potential further reading is that of theoretical statistics. This literature is less centred on models common in econometrics and more concerned with the principles of statistical inference in general. Chihara and Hesterberg (2011) has the advantage of providing a 'hands-on' approach to mathematical statistics in so far as concepts are explored and illustrated by simulation in a similar vein as done in this chapter for Figure 2.3. Casella and Berger (2002) give a nice introduction to modern probability theory and derive principles not only for constructing and evaluating estimators but also for tests of statistical hypotheses. Finally, a further discussion of various principles for the evaluation of estimators not addressed here, such as equivariance and minimaxity, is given in Lehmann and Casella (1998). Readers interested in the theory of hypothesis testing can supplement this with Lehmann and Romano (2005).

NOTES

1 Mathematics distinguishes between countable infinity – there are, for example, 'only' countably infinitely many integer numbers – and uncountable infinity – the size of the set of all real numbers is uncountably infinite.

2 For example, understanding the difference between almost sure convergence and convergence in the sense of traditional calculus requires some advanced concepts of probability theory that cannot be discussed here.

3 It should be noted that equation (2.26) is not a probability statement about the parameter value, but about the limits of the confidence interval. Probability statements about parameter values require a Bayesian framework, which is the topic of Chapter 3 of this volume.

4 Formally, a matrix $\mathbf{A}$ is positive definite if $\mathbf{b}'\mathbf{A}\mathbf{b} > 0$ for all vectors $\mathbf{b}$ that have a non-zero length, that is $\mathbf{b}'\mathbf{b} > 0$.

5 Linear dependence means here that there are numbers $a_0, a_1, \ldots, a_m$ such that $a_0\mathbf{x}_0 + a_1\mathbf{x}_1 + \ldots + a_m\mathbf{x}_m = 0$.

6 Note that for simplicity the statement of the theorem skips some of the more technical assumptions.

7 Note that there are also functions that are neither convex nor concave. Here a similar argument can be made based on the Taylor series expansion of the function.

8 It is noteworthy that Firth's penalized likelihood can also be interpreted as a posterior with Jeffrey's invariant prior. For Bayesian concepts in statistics, see Chapter 3 in this volume.

REFERENCES

Amemiya, T. (1985). *Advanced Econometrics*. Cambridge, MA: Harvard University Press.

Casella, G. and Berger, R. L. (2002). *Statistical Inference*, 2nd edn. Pacific Grove, CA: Duxbury.

Chihara, L. and Hesterberg, T. (2011). *Mathematical Statistics with Resampling and R*. Hoboken, NJ: Wiley.

Davison, A. C. and Hinkley, D. V. (1997). *Bootstrap Methods and Their Application*. Cambridge: Cambridge University Press.

Efron, B. and Tibshirani, R. J. (1993). *An Introduction to the Bootstrap*. New York: Chapman & Hall/CRC.

Firth, D. (1993). Bias reduction of maximum likelihood estimates. *Biometrika*, 80(1), 27–38.

Gourieroux, C. and Monfort, A. (1995a). *Statistics and Econometric Models, Volume 1: General Concepts, Estimation, Predicitions, and Algorithms*. Cambridge: Cambridge University Press.

Gourieroux, C. and Monfort, A. (1995b). *Statistics and Econometric Models, Volume 2: Testing, Confidence Regions, Model Selection, and Asymptotic Theory*. Cambridge: Cambridge University Press.
Greene, W. H. (2011). *Econometric Analysis*, 7th edn. Upper Saddle River, NJ: Prentice Hall.
Huber, P. J. (1967). The behavior of maximum likelihood estimates under nonstandard conditions. In L. M. L. Cam and J. Neyman Eds., *Proceedings of the Fifth Berkeley Symposium on Mathematical Statistics and Probability*, Volume 1 (pp. 221–233). Berkeley: University of California Press.
Lehmann, E. and Casella, G. (1998). *Theory of Point Estimation*, 2nd edn. New York: Springer.
Lehmann, E. L. and Romano, J. P. (2005). *Testing Statistical Hypotheses*, 3rd edn. New York: Springer.
White, H. (1982). Maximum likelihood estimation of misspecified models. *Econometrica*, 50(1), 1–25.

3

Bayesian estimation of regression models

Susumu Shikano

INTRODUCTION TO THE METHOD

Bayesian statistics provide an alternative to the maximum likelihood principle for the estimation of regression models. While maximum likelihood assumes a true value for a parameter and describes its estimator, including the standard error, the Bayesian statistics framework assumes a certain distribution as an inherent part of the parameter. This seemingly trivial difference has some important consequences in parameter estimation of statistical models, including regression models. To make this point clear, we explicate the difference between Bayesian statistics and maximum likelihood in what follows.

Basic idea of Bayesian estimation

In the maximum likelihood framework, a single data set is interpreted as one realization of potential data sets. This is very simple to understand if one has sampled data from a larger population. From the population, different samples can be drawn and the sample at hand is only one of them. Even if one has non-sampled data one can interpret a data set if one views its variables of interest as random variables. Based on this view of the data set, maximum likelihood focuses the probability of data (x) in hand given certain parameters (θ): $\Pr(x|\theta)$. This conditional probability is called likelihood. The goal of maximum likelihood estimation is to find the best possible parameter values which maximize $\Pr(x|\theta)$. For this purpose, one needs to have a function of the parameter given the data. This is obtained by $L(\theta|x) = \Pr(x|\theta)$ based on Bayes' theorem:

$$\Pr(\theta|x) = \frac{\Pr(\theta)\Pr(x|\theta)}{\Pr(x)}. \tag{3.1}$$

Inference based on maximum likelihood assumes that $\Pr(\theta)$ and $\Pr(x)$ are constant for all possible estimates of θ. Therefore, maximization of $L(\theta|x) = \Pr(x|\theta)$ and maximization of $\Pr(\theta|x)$ are equivalent to each other. Based on this logic, the equivalent likelihood function is constructed by simplifying $\Pr(x|\theta)$. Thus we search for a value for θ which maximizes $L(\theta|x) = \Pr(x|\theta)$.

In contrast to maximum likelihood, Bayesian statistics relaxes the assumption that $\Pr(\theta)$ is constant for all possible estimates of θ. Since $\Pr(x)$ is constant for all possible estimates of θ, equation (3.1) can be reformulated as

$$\Pr(\theta|x) \propto \Pr(\theta)\Pr(x|\theta). \tag{3.2}$$

Using this equation, Bayesian statistics calculates $\Pr(\theta|x)$. One might wonder what $\Pr(\theta)$ means. This is the probability of certain estimates for θ prior to data collection. Therefore, it is called prior probability. In contrast, $\Pr(\theta|x)$ is the probability of certain estimates for θ given the data collected, the so-called posterior probability. Accordingly, Bayes' theorem expresses the process in which a certain prior probability about parameter estimates is updated by the likelihood ($\Pr(x|\theta)$) to obtain the posterior probability of the parameter. Here, one can assume a constant prior probability for all parameter estimates as in maximum likelihood. However, it is not necessary, as one can also use different prior probabilities for different parameter estimates. The corresponding information can come from the existing literature, past data analysis or the researcher's subjective belief. The most important characteristic of Bayesian inference is the use of prior information. Bayesian inference can be understood as a process of updating the prior information to the posterior information using the data collected.

How to derive the posterior probability

To get an intuition, we begin with a very simple example model with a single parameter. Suppose we are interested in the unemployment rate and would like to obtain an estimate p. Furthermore, assume we found that one out of 10 persons investigated is currently unemployed ($x = 1, N = 10$). Which parameter is most likely to generate the data? In the maximum likelihood framework $p = 0.1$ is most likely to obtain data with one unemployed person out of 10. Using the binomial distribution, the likelihood of data for different parameter values is quite simple to calculate:

$$\Pr(x = 1|p = 0.1, N = 10) = \binom{10}{1} 0.1^1(1 - 0.1)^{10-1} \approx 0.387. \tag{3.3}$$

This means that, given $p = 0.1$, we would obtain the data ($x = 1, N = 10$) with probability 38.7%. If we use another value for p instead of 0.1 we obtain a smaller value, that is, a lower probability of this figure being the unemployment rate.

For Bayesian inference we first need to specify some prior information which represents our pre-existing knowledge, information or beliefs. Let us assume that we know for certain reasons the unemployment rate is either 0.05, 0.1, 0.15, 0.2 or 0.25. Furthermore, we also know with a probability of 10% that the unemployment rate equals 0.05. Analogously, we also know that the probability of the other values is 20%, 25%, 30% and 15%. This prior knowledge can be graphically displayed as in the left-hand panel of Figure 3.1. In the next step, we calculate the likelihood given our data using the binomial distribution for the possible values for π: $\{0.05, 0.1, 0.15, 0.2, 0.25\}$. The calculated likelihood values are presented in the middle panel of Figure 3.1. Note that the likelihood for $p = 0.1$ is identical with the result of equation (3.3). That is, the middle panel of Figure 3.1 verifies that $p = 0.1$ has the maximal value of the likelihood. Bayesian inference, however, proceeds further to calculate the posterior information using both prior and likelihood. The calculation is simply done based on Bayes' theorem (equation (3.1)). We can multiply the prior probability and likelihood for each value of p, which corresponds to

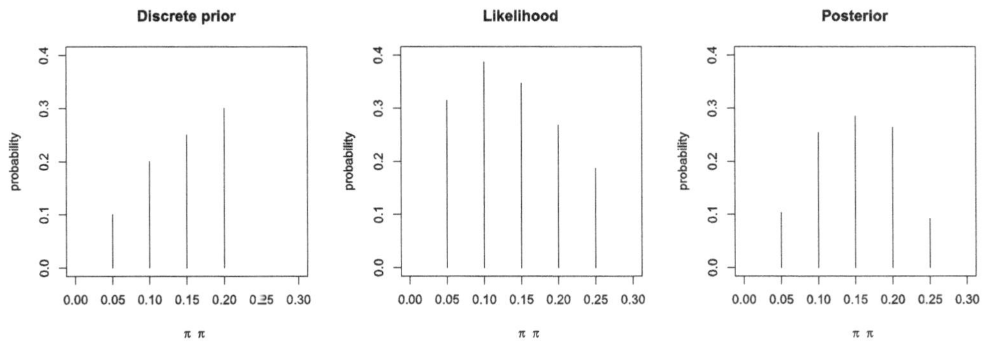

Figure 3.1 Discrete prior, likelihood and posterior (data with one person unemployed out of 10)

to the right-hand side of equation (3.2), $\Pr(\theta)\Pr(x|\theta)$. These probabilities are standardized to obtain the posterior probability whose sum is one. The posterior information thus obtained is shown in the right-hand panel of Figure 3.1.

We can compare the posterior information with the prior and the likelihood. In the prior information, we thought that the unemployment rate was most probably 0.2. According to the data collected, by contrast, the unemployment rate is most likely to be 0.1. Using this information from the data, we updated our prior information to the posterior according to which the unemployment rate is most probably 0.15. We can observe here that the posterior information is a kind of mixture of the prior and likelihood. The way to systematically mix both pieces of information is provided by the Bayesian framework of inference.

Let us now take another data set with three out of 30 persons unemployed. The unemployment rate in the data is 0.1, which is identical to the previous data (one unemployed out of 10). However, the new data set has more respondents. Using this data set, we can repeat the procedure introduced above. The corresponding prior, likelihood and posterior can be found in Figure 3.2. According to the new posterior, the unemployment rate of 0.1 is most probable. That is, the information provided by the data has a greater impact on the posterior than in the last analysis. This is because the new data has more information (30 observations instead of 10).

Finally, we use yet another kind of prior information. We have no idea which of the possible unemployment rates is more or less probable. Correspondingly, we assign the probability of 20% to each of five possible unemployment rates (left-hand panel of Figure 3.3). The impact of the data information is even stronger than in the previous analysis, so that the unemployment rate of 0.1 is much more probable in the posterior information. This is because the prior has less information than in the previous cases.

Note that the description of the posterior above was concentrated on the most likely unemployment rate, that is, the mode of the posterior distribution. This is, however, only one possible way to describe a posterior. It is also possible to use further point estimates (mean, median, etc.) as well as interval estimates (credible intervals). This might sound similar to the maximum likelihood estimator; however, one has to be careful about the difference between frequentist and Bayesian inference in interpretation. While maximum likelihood postulates a certain true value for a parameter, Bayesian inference views the parameter as random quantity. This difference can be illustrated by the frequentist confidence interval and the Bayesian credible interval. If one constructs a 95% confidence interval using a data set one postulates that the data is one of a

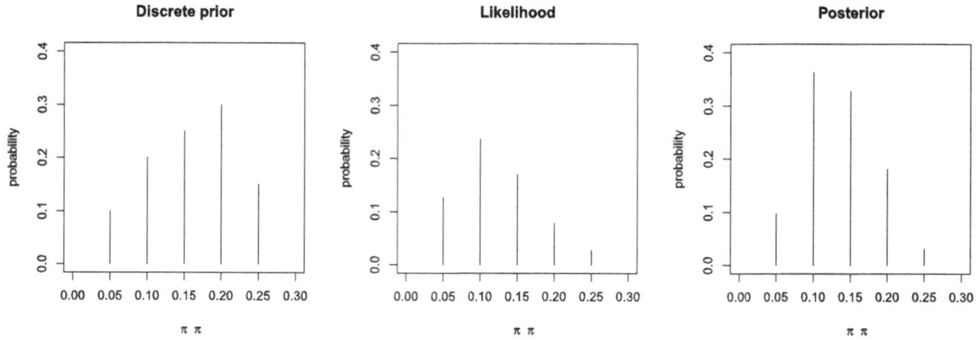

Figure 3.2 Discrete prior, likelihood and posterior (data with three persons unemployed out of 30)

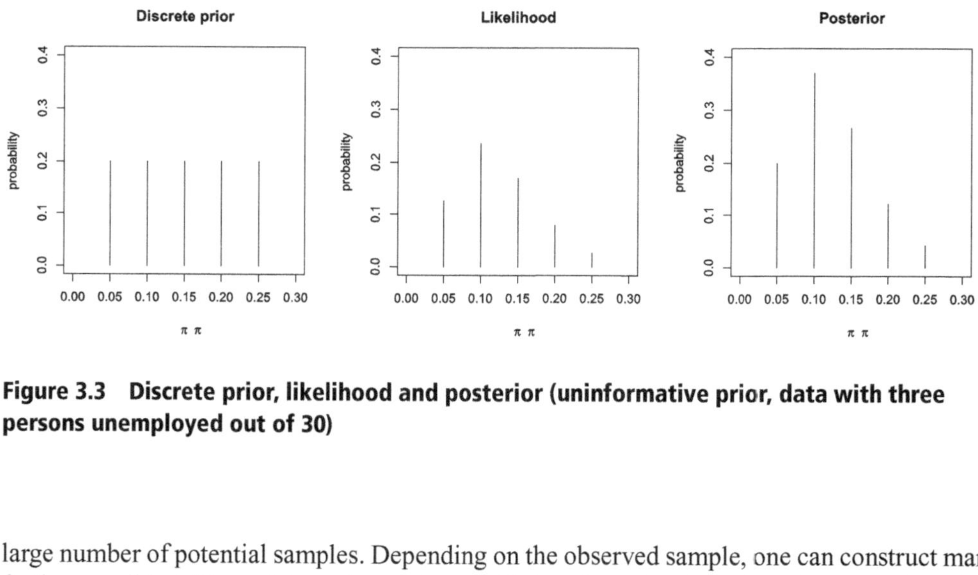

Figure 3.3 Discrete prior, likelihood and posterior (uninformative prior, data with three persons unemployed out of 30)

large number of potential samples. Depending on the observed sample, one can construct many further possible 95% confidence intervals, and 95% of them should include the true value of the parameter. In this frequentist view, the randomness comes from the observation process, including sampling and measurement. In contrast, Bayesian inference views the parameter itself as random entity and does not postulate a single true value behind the parameter. Thus, the Bayesian credible interval expresses the randomness of the parameter *per se*, and a 95% credible interval simply means that the parameter takes the value in the interval with 95% probability.

The posterior thus far has been quite simple to calculate since we have a prior information that π can take only five values: $\{0.05, 0.1, 0.15, 0.2, 0.25\}$. This kind of prior is called discrete prior. In contrast, we can also have a prior which can take all values on a certain scale: a continuous prior. In our example, we usually know that the unemployment rate can take all values on the scale between 0 and 1. To express this kind of prior, we can use, for example, a beta distribution. The beta distribution has two parameters, which enables us to express different shapes of a prior distribution. If we have no idea about the unemployment rate, all values between 0 and 1 are equally probable. This can be expressed as $f_{\beta}(p|1, 1)$, as shown in the upper left-hand panel of Figure 3.4. As can clearly be seen from the form of the distribution, we call this kind of prior a

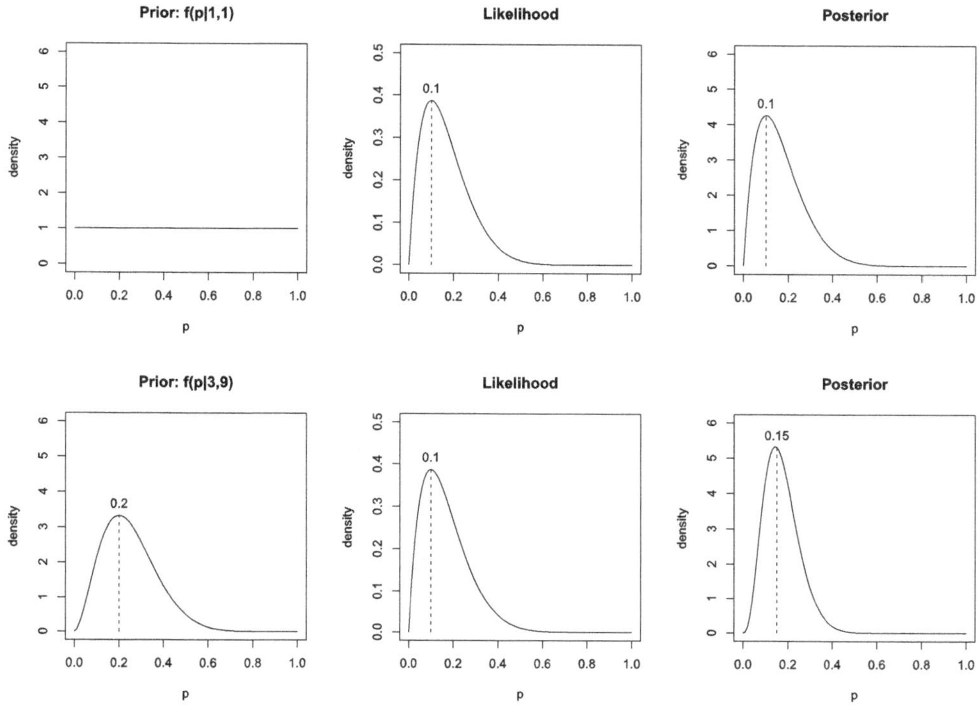

Figure 3.4 Continuous prior, likelihood and posterior (data with one person unemployed out of 10)

flat prior. If we have a continuous prior we also have to consider the likelihood for all possible values of π. This can be realized by using the density function of the binomial distribution. For the data with one person out of 10 unemployed, we have the likelihood function given in the upper middle panel of Figure 3.4. The posterior is calculated analogously to the examples above with the discrete prior. That is, we calculate the density for all possible p by multiplying the corresponding prior and likelihood. For the flat prior, this is quite simple. Since all values of p are equally probable in the prior, the posterior mode is identical to that of the likelihood (see the upper right-hand panel of Figure 3.4). We can now repeat this exercise using another prior. Assume that we have the prior given in the lower left-hand panel of Figure 3.4. That is, we believe that the unemployment rate of 0.2 is more probable than any other values. In contrast, we assume we have the same data as in the last exercise, so that the likelihood function is identical (the lower middle panel of Figure 3.4). In this case, too, the posterior can be calculated by multiplying the prior and likelihood for all π. Fortunately, it is known that the posterior is also a beta distribution and the parameters can simply be calculated using the prior parameter of the prior and some information from the data (lower right-hand panel of Figure 3.4).

The property that the prior and posterior have the same probability form is quite important since this offers the analytical way to derive the posterior. Note that this property depends on the form of the distribution used to calculate the likelihood. In the examples above, we used a binomial distribution to obtain the likelihood function. To this kind of distribution, it is known that a beta distribution is always *conjugate*. That is, if the prior has the form of a beta distribution and the likelihood comes from a binomial distribution, the posterior also takes the form of a beta distribution. This conjugacy is important in multiple senses. First, as we have seen in the example above, the calculation of the posterior is simplified through distribution parameters. If we have no conjugate prior we have to rely on numerical solution using Markov chain Monte Carlo (MCMC) techniques which will be introduced later. Second, the posterior obtained in a Bayesian inference can later be used as prior for a further inference using the same framework. Imagine that we obtained the posterior in the lower right-hand panel of Figure 3.4. Thereafter, we collected another data set. We do not use the same prior as in the previous analysis again, but its posterior as prior since we have updated our knowledge using the last data set. In this way, Bayesian inference enables us to combine information from different data sets.

To summarize, Bayesian inference provides a systematic way to integrate the prior information and data collected into the posterior information. In this process, data gains more impact on the posterior if the data contains more information (e.g. more observations) or if the prior has less information (e.g. larger dispersion). This applies not only to the simple examples presented here but also to regression analysis and further kinds of analysis with more complex statistical models.

Bayesian estimation of regression models

The examples in the previous subsection are quite simple in the sense that we had only one parameter. In estimation of regression models, we have more parameters to estimate. Even if a simple bivariate linear regression model with only one independent variable is estimated, we still have three parameters: two regression coefficients (intercept and slope) and error variance. That is, we need to build the joint posterior distribution using the prior and the data.

As in the previous subsection, we first need to specify the form of prior information for the coefficients and error variance. A conjugate prior is widely used for the reasons discussed in the previous subsection. In particular, the normal inverse gamma distribution is known to be conjugate to the likelihood based on a multivariate normal distribution. The normal inverse gamma distribution constitutes the joint distribution for the regression coefficients and the error variance. The marginal distribution of the coefficients corresponds to a multivariate t distribution, that of the error variance to an inverse gamma distribution.

By multiplying the prior with the likelihood which is known from the conventional likelihood-based methods, we can obtain the posterior distribution. Due to the conjugacy we again obtain a normal inverse gamma distribution for the posterior. As shown above, if we use a flat prior for the regression coefficients and variance error, the posterior corresponds to the likelihood so that we obtain the same result for the posterior mode as in conventional maximum likelihood estimation. If we use a specific prior, in contrast, the posterior becomes a mixture of the prior and the likelihood. Here, too, the rule is same as in the previous subsection. If the data has more observations or the prior is more widely dispersed the data has a greater impact on the posterior so that the likelihood and the posterior are similar, and vice versa.

If no conjugate prior is known for the likelihood function or if one wishes to specify a non-conjugate prior, obtaining the posterior is difficult for two reasons. First, to obtain certain posterior information (e.g. expectation) we need not only to multiply the likelihood and prior but also to divide it by $\Pr(y)$ (see equation (3.1)). Calculation of this denominator requires an integral which often has no analytical solution. Second, in most applications, including regression models, we have a statistical model with multiple parameters. That is, the posterior distribution is a multidimensional joint distribution. However, we are generally interested in the marginal distribution for individual parameters which can be obtained by integrating out the other parameters. Similarly, we often have no analytical solution for integrals of this kind. These are the reasons why Bayesian inference was for a long time limited mainly to conjugacy analysis and consequently had only limited impact on applied social sciences. For several decades, however, rapid growth of computational capacity has led to an alternative numerical toolkit for obtaining the posterior information: the Markov chain Monte Carlo techniques. These techniques enable us to randomly draw numbers from the target joint posterior distribution (Monte Carlo techniques). The joint posterior distribution is reached by a random process with the Markov property (Markov chain) which is specified by the available information. In particular, two specific classes of MCMC techniques are widely used: Gibbs sampling and the Metropolis–Hastings algorithm. For Gibbs sampling, one needs the conditional posterior distribution for individual parameters. By successively applying the conditional posteriors, one can obtain the joint posterior distribution. In contrast, the Metropolis–Hastings algorithm requires no conditional posteriors. Instead, by comparing the product of the likelihood and the prior among different sets of parameters, the Markov chains go through the parameter space and reach the joint distribution. More detailed descriptions of both methods can be found in the next section.

Independently of the methods to obtain the posterior distribution, its interpretation is quite straightforward. The posterior distribution obtained is simply described using for instance its mean, standard deviation or certain percentiles. If one has a hypothesis about the parameter, the posterior probability that the hypothesis is correct is simply calculated. For example, we have an alternative hypothesis that an independent variable should have a positive impact on the dependent variable. From the corresponding marginal posterior distribution, we can obtain the probability that the regression coefficient of an independent variable has positive sign.

MATHEMATICAL FOUNDATIONS

This section begins with a more formal presentation of Bayesian inference. In this context, it will also be clear why conjugacy is an important concept and when we need MCMC methods. The section then goes on to introduce the regression model in a formal way.

Bayesian inference

In terms of densities, Bayes' theorem states that

$$f(\theta|x) = \frac{f(x|\theta)f(\theta)}{f(x)} = \frac{f(x|\theta)f(\theta)}{\int f(x|\theta)f(\theta)\,\mathrm{d}\theta}, \tag{3.4}$$

where the denominator $f(x) = \int f(x|\theta)f(\theta)\,\mathrm{d}\theta$ is a normalizing constant which ensures that $\int f(\theta|x)\,\mathrm{d}\theta = 1$.

Based on $f(\theta|x)$, we can derive different kinds of point estimators. The posterior expectation is the expected value of the posterior distribution, which is given by

$$E(\theta|x) = \int \theta f(\theta|x)\,\mathrm{d}\theta. \tag{3.5}$$

The posterior median is given by

$$\hat{\theta} = \mathrm{Med}(\theta|x) \equiv \int_{-\infty}^{\hat{\theta}} f(\theta|x)\,\mathrm{d}\theta = 0.5. \tag{3.6}$$

The posterior mode is given by

$$\mathrm{Mod}(\theta|x) = \arg\max_{\theta} f(\theta|x). \tag{3.7}$$

Note that this posterior mode is in a close relationship to the point estimate based on maximum likelihood. Maximum likelihood finds an estimator which maximizes $L(\theta|x) = f(x|\theta)$. We would obtain the same estimator if $f(\theta)$ were constant, since $f(\theta|x) \propto f(x|\theta)f(\theta)$.

In addition to the point estimators, we can also obtain interval estimators. A credible interval at level $1-\alpha$ is defined by

$$\int_{t_a}^{t_b} f(\theta|x)\,\mathrm{d}\theta = 1-\alpha, \tag{3.8}$$

where $t_a, t_b \in \Theta$. According to this definition, θ lies in the interval between t_a and t_b with posterior probability $1-\alpha$. Note the difference between this and the frequentist confidence interval, which would include the true value of θ with probability $1-\alpha$ if one has a large number of repeated samples. In contrast, the Bayesian credible interval views θ as a random quantity.

Note, further, that equation (3.8) has no unique solution for t_a and t_b. Usually one uses the $\alpha/2$ and $1-\alpha/2$ quantile of the posterior distribution for t_a and t_b, respectively. Alternatively, one can calculate the highest posterior density (HPD) interval. Accordingly, an interval $C = [t_a, t_b] \subset \Theta$ is the $1-\alpha$ HPD interval for θ if equation (3.8) holds and

$$f(\theta|x) \geq f(\tilde{\theta}|x), \quad \forall\theta \in C \text{ and } \tilde{\theta} \notin C \tag{3.9}$$

That is, the HPD interval includes the parameter values with the highest posterior density.

Analytical solution with conjugate priors

In the previous subsection, we introduced different Bayesian estimators. Among them, the posterior mode is relatively easy to calculate since we can ignore the denominator of equation (3.4) and simply find the parameter values which maximize the numerator. In contrast, the posterior expectation requires information $f(x) = \int f(x|\theta)f(\theta)\,\mathrm{d}\theta$ as well as integration over θ as given by equation (3.5). In most cases the integrals in $f(x)$ and/or calculation of the posterior expectation have no solution in closed form. The exception is those instances where the prior distribution is conjugate to the distribution defining the likelihood.

A prior distribution is conjugate to the distribution defining the likelihood if the derived posterior density has the same functional form as the prior density with different parameter values. For example, a beta distribution is the conjugate to a binomial distribution. Using this property, we derived the posterior distribution in Figure 3.4. In general, the binomial distribution models the process in which N experiments yielding success with probability p are independently repeated. The probability that we obtain x successes can be calculated:

$$f(x|p,N) = \binom{N}{x} p^x(1-p)^{N-x}. \tag{3.10}$$

Note that here we have two kinds of parameter, p and N, and we are generally interested in p. For this p we can assume a beta distribution as the prior distribution, whose density is given by

$$f(p|\alpha,\beta) = \frac{p^{\alpha-1}(1-p)^{\beta-1}}{B(\alpha,\beta)}, \tag{3.11}$$

$$B(\alpha,\beta) = \int_0^1 p^{\alpha-1}(1-p)^{\beta-1}\,\mathrm{d}p = \frac{\Gamma(\alpha)\Gamma(\beta)}{\Gamma(\alpha+\beta)}. \tag{3.12}$$

$B(\alpha, \beta)$ is a normalizing constant which ensures that $\int f(p|\alpha, \beta)\,\mathrm{d}p = 1$. Further, the expected value of the beta distribution is also known:

$$E(p|\alpha, \beta) = \int_0^1 pf(p|\alpha, \beta)\,\mathrm{d}p = \frac{\alpha}{\alpha + \beta}. \tag{3.13}$$

The mode of the beta distribution is

$$\mathrm{Mod}(p|\alpha, \beta) = \arg\max_p f(p|\alpha, \beta) = \frac{\alpha - 1}{\alpha + \beta - 2}. \tag{3.14}$$

Both parameters of the beta distribution, α and β, are called hyperparameters since they are the parameters of a distribution which describes another distribution's parameter (here p).

By substituting both likelihood and prior into Bayes' theorem, we obtain:

$$\begin{aligned} f(p|x) &= \frac{f(x|p, N) f(p|\alpha, \beta)}{f(x)} \\ &= \frac{\binom{N}{x} p^x (1-p)^{N-x} \frac{p^{\alpha-1}(1-p)^{\beta-1}}{B(\alpha,\beta)}}{f(x)} \\ &= \frac{\binom{N}{x} B^{-1}(\alpha, \beta) p^{\alpha+x-1}(1-p)^{\beta+N-x-1}}{f(x)} \\ &= \frac{p^{\alpha+x-1}(1-p)^{\beta+N-x-1}}{B(x+\alpha, N-x+\beta)}. \end{aligned} \tag{3.15}$$

This means that the posterior distribution is also a beta distribution with parameters $\alpha + x$ and $\beta + N - x$. Therefore the expectation and mode can be easily obtained via equations (3.13) and (3.14):

$$E(p|x) = \frac{\alpha + x}{\alpha + \beta + N}, \tag{3.16}$$

$$\mathrm{Mod}(p|x) = \frac{\alpha + x - 1}{\alpha + \beta + N - 2}. \tag{3.17}$$

We can now return to the example in Figure 3.4. In the example we have data $x = 1$ and $N = 10$. In the upper panels, we had a beta prior with $\alpha = \beta = 1$. As is clear in the figure, the prior is completely flat between 0 and 1. The posterior expectation and mode can be calculated as follows:

$$E(p|x) = \frac{\alpha + x}{\alpha + \beta + N} = \frac{x+1}{N+2}, \tag{3.18}$$

$$\mathrm{Mod}(p|x) = \frac{\alpha + x - 1}{\alpha + \beta + N - 2} = \frac{1 + x - 1}{1 + 1 + N - 2} = \frac{x}{N}. \tag{3.19}$$

While the posterior mode coincides with the maximum likelihood estimates (x/N) the posterior expectation differs slightly and shrinks towards the prior expectation ($\frac{1}{2}$).

Table 3.1 Likelihood functions with conjugate priors

Likelihood	Model parameter	Conjugate prior
Bernoulli	p: probability	Beta
Binomial	p: probability	Beta
Negative binomial	p: probability	Beta
Poisson	γ: rate	Gamma
Multinomial	**p**: probability vector	Dirichlet
Normal	μ: mean	Normal
Normal	σ^2: variance	Inverse gamma
Normal	σ^2: variance	Scaled inverse chi-squared
Normal	μ and σ^2	Normal inverse gamma

Conjugate priors exist for other likelihood functions. In particular, it is known that a conjugate prior exists for likelihood functions which belong to the exponential family. Table 3.1 lists conjugate priors for likelihood functions frequently used in social science research. Among these conjugate distributions, the normal inverse gamma distribution is relevant for the linear regression model since the model generally has a likelihood function using the normal distribution with unknown mean and variance.

Linear regression model with conjugate priors

A linear regression model consists of a dependent variable $\mathbf{y}$ and some independent variables $\mathbf{X}$. Through a linear combination of $\mathbf{X}$ using $\boldsymbol{\beta}$ we can predict the dependent variable. The residual $\boldsymbol{\epsilon}$ is assumed to be distributed normal with mean zero and variance $\sigma^2\mathbf{I}$:

$$\mathbf{y} = \mathbf{X}\boldsymbol{\beta} + \boldsymbol{\epsilon}, \tag{3.20}$$

$$\boldsymbol{\epsilon} \sim N(\mathbf{0}, \sigma^2\mathbf{I}). \tag{3.21}$$

This can also be expressed as

$$(\mathbf{y}|\boldsymbol{\beta}, \sigma^2, \mathbf{X}) \sim N(\mathbf{X}\boldsymbol{\beta}, \sigma^2\mathbf{I}), \tag{3.22}$$

with $\mathbf{I}$ as the identity matrix. That is, our dependent variable is assumed to be a random variable from normal distribution with unknown mean and variance. The corresponding likelihood function is

$$f(\mathbf{y}|\boldsymbol{\beta}, \sigma^2) = \prod_i^n \frac{1}{\sqrt{2\pi\sigma^2}} \exp\left\{-\frac{(y_i - \mathbf{X}\boldsymbol{\beta})'(y_i - \mathbf{X}\boldsymbol{\beta})}{2\sigma^2}\right\}. \tag{3.23}$$

It is known that this likelihood function has several conjugate prior distributions. Among others, the normal inverse gamma distribution is typically used to derive the posterior. The normal inverse gamma distribution is a product of a normal distribution for $\boldsymbol{\beta}$ and an inverse gamma distribution for σ^2. An inverse gamma distribution has two parameters, the scale parameter a and the shape parameter d, and its probability density function is given by

$$f_{\Gamma^{-1}}(\sigma^2|a, d) = \frac{a^d}{\Gamma(d)} \sigma^{2(-d-1)} \exp\left(-\frac{a}{\sigma^2}\right). \tag{3.24}$$

Multiplication with a univariate normal distribution yields a normal inverse gamma distribution:

$$\begin{aligned} f_{N-\Gamma^{-1}}(\beta,\sigma^2|\mu,\lambda,a,d) &= f_N(\beta|\mu,\sigma^2/\lambda)f_{\Gamma^{-1}}(\sigma^2|a,d) \\ &= \frac{\sqrt{\lambda}}{\sqrt{2\pi\sigma^2}}\exp\left\{-\frac{\lambda(\beta-\mu)^2}{2\sigma^2}\right\}\frac{a^d}{\Gamma(d)}\sigma^{2(-d-1)}\exp\left(-\frac{a}{\sigma^2}\right) \\ &= \frac{\sqrt{\lambda}}{\sqrt{2\pi\sigma^2}}\frac{a^d}{\Gamma(d)}\left(\frac{1}{\sigma^2}\right)^{d+1}\exp\left\{-\frac{\lambda(\beta-\mu)^2+2a}{2\sigma_0^2}\right\}. \end{aligned} \tag{3.25}$$

In the linear regression with k independent variables (including a constant), $\boldsymbol{\beta}$ is a $k \times 1$ parameter vector and $\sigma^2\boldsymbol{\Sigma}$ is its variance–covariance matrix. For this reason, the prior is more precisely a *multivariate* normal inverse gamma distribution:

$$\begin{aligned} f_{N-\Gamma^{-1}}(\boldsymbol{\beta},\sigma^2|\boldsymbol{\mu},\boldsymbol{\Sigma},a,d) &= f_N(\boldsymbol{\beta}|\boldsymbol{\mu},\sigma^2\boldsymbol{\Sigma})f_{\Gamma^{-1}}(\sigma^2|a,d) \\ &= \frac{\sqrt{\lambda}}{\sqrt{2\pi\sigma^k|\boldsymbol{\Sigma}|}}\frac{a^d}{\Gamma(d)}\left(\frac{1}{\sigma^2}\right)^{d+1} \\ &\quad\times\exp\left\{-\frac{(\boldsymbol{\beta}-\boldsymbol{\mu})'\boldsymbol{\Sigma}^{-1}(\boldsymbol{\beta}-\boldsymbol{\mu})+2a}{2\sigma^2}\right\}. \end{aligned} \tag{3.26}$$

Now we specify the prior as follows:

$$f(\boldsymbol{\beta},\sigma^2)=f_{N-\Gamma^{-1}}(\boldsymbol{\beta}_0,\boldsymbol{\Sigma}_0,a_0,d_0). \tag{3.27}$$

It is known that multiplication of prior and likelihood yields another multivariate normal inverse gamma distribution:

$$f(\boldsymbol{\beta},\sigma^2|\mathbf{y})=f_{N-\Gamma^{-1}}(\boldsymbol{\beta}^*,\boldsymbol{\Sigma}^*,a^*,b^*), \tag{3.28}$$

with

$$\boldsymbol{\beta}^* = (\boldsymbol{\Sigma}_0^{-1}+\mathbf{X}'\mathbf{X})^{-1}(\boldsymbol{\Sigma}_0^{-1}\boldsymbol{\beta}_0+\mathbf{X}'\mathbf{y}), \tag{3.29}$$

$$\boldsymbol{\Sigma}^* = (\boldsymbol{\Sigma}_0^{-1}+\mathbf{X}'\mathbf{X})^{-1}, \tag{3.30}$$

$$a^* = a+\frac{1}{2}(\boldsymbol{\beta}_0'\boldsymbol{\Sigma}_0^{-1}\boldsymbol{\beta}_0+\mathbf{y}'\mathbf{y}-\boldsymbol{\beta}^{*\prime}\boldsymbol{\Sigma}^{*-1}\boldsymbol{\beta}^*), \tag{3.31}$$

$$d^* = d_0+\frac{n}{2}. \tag{3.32}$$

Note that this is a joint posterior distribution of $\boldsymbol{\beta}$ and σ^2. In general, σ^2 is a nuisance parameter and we are only interested in the marginal posterior distribution of $\boldsymbol{\beta}$, which can be obtained by integrating out σ^2. It is known that this marginal posterior follows a multivariate Student t distribution with $\nu = n - k$ degrees of freedom:

$$\begin{aligned} f(\boldsymbol{\beta}|\mathbf{y}) &= \int f(\boldsymbol{\beta},\sigma^2|\mathbf{y})\,\mathrm{d}\sigma^2 \\ &= f_t(\nu,\boldsymbol{\beta}^*,\boldsymbol{\Sigma}^*). \end{aligned} \tag{3.33}$$

Here, we can compare the posterior of $\boldsymbol{\beta}$ with the estimates via ordinary least squares and maximum likelihood. As can be seen in Chapter 2 of this volume, ordinary least squares and maximum likelihood give the same point estimate of $\boldsymbol{\beta}$:

$$\boldsymbol{\beta}=(\mathbf{X}'\mathbf{X})^{-1}\mathbf{X}'\mathbf{y}. \tag{3.34}$$

The posterior approximately yields this result if n approaches infinity:

$$\lim_{n\to\infty} \boldsymbol{\beta}^* = (\mathbf{X}'\mathbf{X})^{-1}\mathbf{X}'\mathbf{y}. \tag{3.35}$$

This is because $\boldsymbol{\Sigma}_0^{-1}$ has less weight as n increases. For the same reason, increasing $\boldsymbol{\Sigma}_0$ also leads to the same estimate as ordinary least squares and maximum likelihood.

Another important aspect is multicollinearity. Differently from ordinary least squares and maximum likelihood, it is also possible to obtain the posterior even in situations of perfect multicollinearity. In the case of perfect multicollinearity, that is, if there is an exact linear relationship among independent variables, $\mathbf{X}'\mathbf{X}$ does not have full rank and invertibility. Therefore, we can calculate neither $(\mathbf{X}'\mathbf{X})^{-1}$ nor $\boldsymbol{\beta}$. In calculating $\boldsymbol{\beta}^*$, by contrast, we do not have to invert $\mathbf{X}'\mathbf{X}$. Instead, if $\boldsymbol{\Sigma}_0^{-1} + \mathbf{X}'\mathbf{X}$ has full rank we can obtain $\boldsymbol{\beta}^*$.

Another approach: A numerical solution via Gibbs sampling

In the previous subsection, we derived the posterior distribution of $\boldsymbol{\beta}$ and σ^2 jointly using the conjugate normal inverse gamma prior. Alternatively, we can derive the posterior of $\boldsymbol{\beta}$ and σ^2 separately. That is, we specify a (multivariate) normal distribution for $\boldsymbol{\beta}$ and an inverse gamma for σ^2:

$$\begin{aligned} f(\boldsymbol{\beta}) &= f_{MN}(\boldsymbol{\beta}_0, \boldsymbol{\Sigma}_0), \\ f(\sigma^2) &= f_{\Gamma^{-1}}(a_0, d_0). \end{aligned}$$

Both priors are conjugate to the likelihood function from a normal distribution, and the posteriors are given by

$$f(\boldsymbol{\beta}|\sigma^2, \mathbf{y}) = f_{MN}(\boldsymbol{\beta}^*, \boldsymbol{\Sigma}^*), \tag{3.36}$$

$$f(\sigma^2|\boldsymbol{\beta}, \mathbf{y}) = f_{\Gamma^{-1}}(a^*, d^*), \tag{3.37}$$

with

$$\boldsymbol{\beta}^* = (\sigma^{-2}\mathbf{X}'\mathbf{X} + \boldsymbol{\Sigma}_0^{-1})^{-1}(\sigma^{-2}\mathbf{X}'\mathbf{y} + \boldsymbol{\Sigma}_0^{-1}\boldsymbol{\beta}_0), \tag{3.38}$$

$$\boldsymbol{\Sigma}^* = (\sigma^{-2}\mathbf{X}'\mathbf{X} + \boldsymbol{\Sigma}_0^{-1})^{-1}, \tag{3.39}$$

$$a^* = a_0 + \frac{1}{2}(\mathbf{y} - \mathbf{X}\boldsymbol{\beta})'(\mathbf{y} - \mathbf{X}\boldsymbol{\beta}), \tag{3.40}$$

$$d^* = d_0 + \frac{n}{2}. \tag{3.41}$$

Note that the derived posteriors of individual parameters are conditioned by the other parameter values. In contrast, we are generally interested in the marginal probability which gives an average picture of the posterior distribution over all possible parameter combinations. For this purpose, we utilize a numerical method called Gibbs sampling, one of the MCMC methods, instead of integrating out the other parameters.

Here, we wish to generate random draws from a joint posterior $f(\boldsymbol{\beta}, \sigma^2|\mathbf{y})$. Gibbs sampling can be described by the following iterative steps for $t = 1, \ldots, T$, with $\boldsymbol{\beta}^{(t)}$ and $\sigma^{2(t)}$ being the values generated in the tth iteration and $\boldsymbol{\beta}^{(0)}$ and $\sigma^{2(0)}$ arbitrary selected initial values:

1. Draw a random number from $f(\boldsymbol{\beta}|\sigma^{2(t-1)}, \mathbf{y})$ and save as $\boldsymbol{\beta}^{(t)}$.
2. Draw a random number from $f(\sigma^2|\boldsymbol{\beta}^{(t)}, \mathbf{y})$ and save as $\sigma^{2(t)}$.
3. Go to step 1 for the $(t + 1)$th iteration.

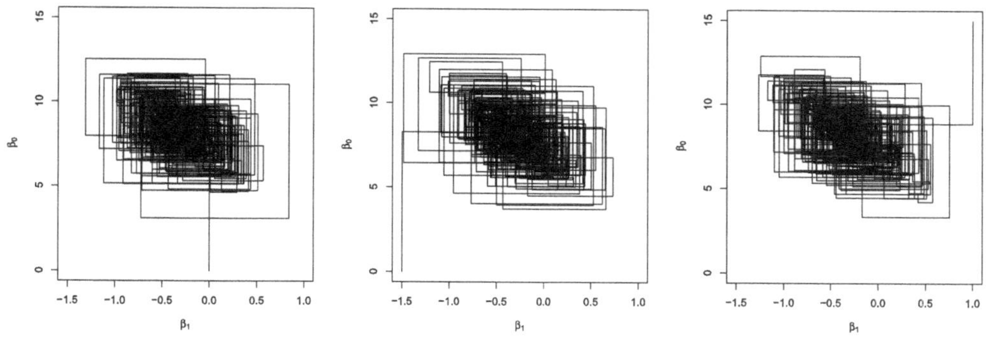

Figure 3.5 Three Markov chains with different initial values

By repeating these steps we can build a Markov chain. This kind of Markov chain (Gibbs chain) is known to have a unique invariant distribution. That is, after a large enough number of iterations chains converges to a stable state independently of their initial states. The invariant distribution obtained corresponds to the desired joint distribution. This is best illustrated by Figure 3.5 which shows three Markov chains for $\boldsymbol{\beta}$ with different initial values. These chains are set up by a simple bivariate regression model which will be described later in the example section. For simple statistical models of this kind, the Markov chain converges very quickly. In Figure 3.5 we can also see that all three chains reached a common region after just a few steps, that is, the posterior distribution. By summarizing the information for the individual parameters after convergence, we can also describe the individual marginal distribution (see Figure 3.7).

An important issue in practical Bayesian analysis is to evaluate whether the Markov chain reached convergence. While there are several methods and criteria suggested, it is most important to run not just one, but multiple chains with different initial values and to carry out a visual inspection. If we can confirm that multiple chains are converged in a certain region and well mixed we discard the part of Markov chain before convergence. This discarded phase of the Markov chain is called *burn-in*. After burn-in we should run the Markov chain further to obtain samples from the joint posterior. If the chain after the burn-in is not run long enough the joint posterior cannot be captured well enough. One simple way to evaluate whether one has run the Markov chain long enough is to observe whether the Markov chain changes the form of the captured posterior substantially. If it does not, we can stop running the Markov chain and begin summarizing the posterior.

One of the important advantages of Gibbs sampling and other MCMC techniques is their simplicity in describing posterior distributions. To calculate the posterior expectation one does not need to integrate, but simply take the average value of the random draws:

$$E(\theta|x) = \frac{1}{T}\sum_t \theta^{(t)}. \tag{3.42}$$

The posterior median is given by the median value. The credible interval can be constructed using quantiles. Only the posterior mode cannot be calculated in a straightforward way. For this purpose, one first needs a kernel density estimate based on which one can estimate the mode.

A more general numerical solution: Metropolis–Hastings algorithm

Gibbs sampling is a powerful technique which enables us to obtain the joint and marginal posterior in a very simple way. The algorithm, however, requires the full set of the conditional posterior for all parameters. This is not always the case. For example, a binary logistic regression with data of the form $y_i = 0$ or 1 has likelihood function

$$f(\mathbf{y}|\boldsymbol{\beta}, \mathbf{X}) = \prod_{i=1}^{n} (\operatorname{logit}^{-1}(\mathbf{x}_i'\boldsymbol{\beta}))^{y_i} (1 - \operatorname{logit}^{-1}(\mathbf{x}_i'\boldsymbol{\beta}))^{1-y_i}, \tag{3.43}$$

For this likelihood function, no conjugate prior distribution is known. If we take a multivariate normal prior for $\boldsymbol{\beta}$,

$$f(\boldsymbol{\beta}) = f(\boldsymbol{\beta}_0, \boldsymbol{\Sigma}_0), \tag{3.44}$$

we can write down the posterior:

$$\begin{aligned} f(\boldsymbol{\beta}|\mathbf{y}) \quad \propto \quad & \prod_{i=1}^{n} (\operatorname{logit}^{-1}(\mathbf{x}_i'\boldsymbol{\beta}))^{y_i} (1 - \operatorname{logit}^{-1}(\mathbf{x}_i'\boldsymbol{\beta}))^{1-y_i} \\ & \times \exp\left\{ -\frac{1}{2}(\boldsymbol{\beta} - \boldsymbol{\beta}_0)' \boldsymbol{\Sigma}_0^{-1}(\boldsymbol{\beta} - \boldsymbol{\beta}_0) \right\}. \end{aligned} \tag{3.45}$$

However, a closed form of the solution is known for neither the joint nor the conditional posterior. Therefore, neither conjugacy analysis nor Gibbs sampling proves useful for this statistical model. Fortunately, we can still use another class of MCMC algorithm and a general form of Gibbs sampling: the Metropolis–Hastings algorithm.

Let us assume that we wish to generate random draws from a posterior $f(\boldsymbol{\beta}|\mathbf{y})$. The Metropolis–Hastings algorithm can be described by the following iterative steps for $t = 1, \ldots, T$, with $\boldsymbol{\beta}^{(t)}$ being the vector of generated values in the tth iteration and $\boldsymbol{\beta}^{(0)}$ arbitrary selected initial values:

1. Set $\boldsymbol{\beta} = \boldsymbol{\beta}^{(t-1)}$.
2. Generate new candidate values $\boldsymbol{\beta}'$ from a proposal distribution $q(\boldsymbol{\beta}'|\boldsymbol{\beta})$.
3. Calculate $\alpha = \min\left(1, \frac{f(\boldsymbol{\beta}'|\mathbf{y})q(\boldsymbol{\beta}|\boldsymbol{\beta}')}{f(\boldsymbol{\beta}|\mathbf{y})q(\boldsymbol{\beta}'|\boldsymbol{\beta})}\right) = \min\left(1, \frac{f(\mathbf{y}|\boldsymbol{\beta}')f(\boldsymbol{\beta}')q(\boldsymbol{\beta}|\boldsymbol{\beta}')}{f(\mathbf{y}|\boldsymbol{\beta})f(\boldsymbol{\beta})q(\boldsymbol{\beta}'|\boldsymbol{\beta})}\right)$.
4. Update $\boldsymbol{\beta}^{(t)} = \boldsymbol{\beta}'$ with probability α. Otherwise set $\boldsymbol{\beta}^{(t)} = \boldsymbol{\beta}$.

The underlying idea is quite simple. If the candidate values have larger density in $f(\boldsymbol{\beta}|\mathbf{y})$ they are selected; if not, depending on the relationship of the density of the current and candidate values, one set of both is drawn. This process can be illustrated by Figure 3.6, which presents an example case of a binary logit model with one covariate. Therefore, we only need to estimate two parameters, β_0 and β_1. We specify $\{\beta_0, \beta_1\} = \{0, 0\}$ as initial values of the Markov chain. Furthermore, we use a normal distribution with variance 1 as the proposal distribution independently for β_0 and β_1:

$$\beta_0' \sim N(\beta_0, 1), \tag{3.46}$$

$$\beta_1' \sim N(\beta_1, 1). \tag{3.47}$$

The normal distribution has an advantage due to its symmetrical form which results in $q(\boldsymbol{\beta}'|\boldsymbol{\beta}) = q(\boldsymbol{\beta}|\boldsymbol{\beta}')$. Therefore, the calculation of α can be simplified:

$$\alpha = \min\left(1, \frac{f(\mathbf{y}|\boldsymbol{\beta}')f(\boldsymbol{\beta}')}{f(\mathbf{y}|\boldsymbol{\beta})f(\boldsymbol{\beta})}\right). \tag{3.48}$$

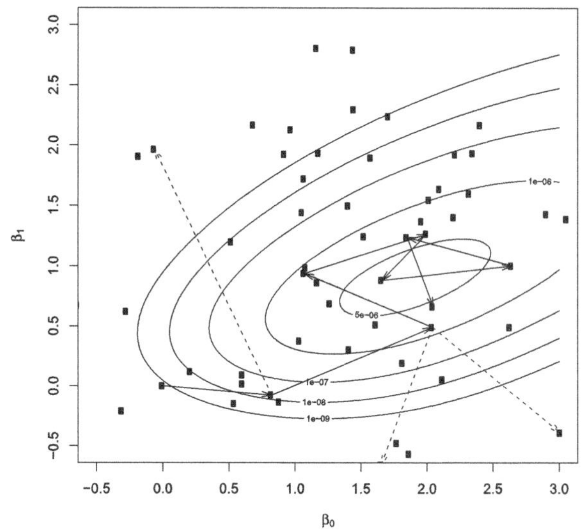

Figure 3.6 Graphical illustration of the Metropolis–Hastings algorithm. The solid arrows are the accepted moves of the Markov chain. The dashed arrows are (a part of) the rejected moves. The circles are rejected candidate proposals. The contour lines gives the product of the likelihood and the prior (equation (3.45))

The proposal distribution randomly generated a candidate vector $\boldsymbol{\beta}' =(0.823, -0.077)$. α can now be calculated based on equation (3.45). This candidate vector has a higher density, resulting in $\alpha = 1$. In this case, the Markov chain moves to this candidate vector with probability 1. In the next iteration, the normal distributions centred on the current status of the Markov chain, $(0.823, -0.077)$, generated a new candidate vector, $\boldsymbol{\beta}' =(-0.062, 1.963)$. As clearly seen in Figure 3.6, whose contour lines give the density level, the new candidate vector's density is much lower than that of the current status. Correspondingly, α is very low (2.662×10^{-6}). This does not exclude that the Markov chain moves to this candidate vector. However, in this example case, the Markov chain stayed in the current status. That is, $\boldsymbol{\beta}' =(0.823, -0.077)$ was drawn for the second time. In the next iteration, the proposal distributions generated a new candidate vector whose density is higher than that of the current status. Therefore, the Markov chain moved to the new vector. After the process described above is iterated a large number of times we obtain the samples from the joint posterior distribution. Figure 3.6 gives an example Markov chain with only 60 iterations. The solid dots are the selected $\boldsymbol{\beta}$ and the circles are the rejected candidate vectors.

The beauty of this algorithm is at least twofold. First, it is known that the Metropolis–Hastings algorithm will converge to its equilibrium distribution independently of the proposal distribution q being used. That is, we do not necessarily use the normal distribution as in the example above. We could also use a normal distribution with a different variance. However, a very small variance of the proposal distribution can slow down convergence. Second, one does not need to have the full density function with the normalizing constant since $\frac{f(\boldsymbol{\beta}'|\mathbf{y})}{f(\boldsymbol{\beta}|\mathbf{y})}$ cancels out the normalizing constant. Indeed, we had no full description of the posterior distribution in equation (3.45) which ignores the denominator of the posterior $(f(\mathbf{y}))$ and also some irrelevant component of the prior $f(\boldsymbol{\beta})$. Independently of these components, α can be calculated and the Markov chain can proceed.

Figure 3.6 may remind some readers of the iterative algorithm (see Chapter 2 in this volume) in the maximum likelihood estimation framework. Here, we have to note that the iterative algorithm in maximum likelihood and the MCMC techniques in the Bayesian inference have two different goals. The goal of maximum likelihood estimation is to find the set of parameter values which maximizes the likelihood. Therefore, the search of the iterative algorithm ends in the maximum of the likelihood surface. In contrast, MCMC techniques aim to draw samples from the target posterior distribution. Thus, there is no clear end-point of the Markov chain. Furthermore, a chain can also sometimes move away from the maximum of the posterior density surface. This can be also seen in Figure 3.6. This is because parameter values with lower density can be drawn so long as they have a certain density.

EXAMPLE ANALYSIS

In this section we apply the method sketched above to a bivariate regression model. As data we use the European Social Survey (ESS) which consists of multiple rounds of cross-sectional data. By using this data it is also demonstrated how results from the previous round can be integrated to the analysis of the current round.

In particular, we regress the opinion concerning European integration as a specific issue attitude on the left–right scale as a more general ideological orientation. The corresponding variables are measured by using the following questionnaire items:

- European integration (euftf) – 'Now thinking about the European Union, some say European unification should go further. Others say it has already gone too far. Using this card, what number on the scale best describes your position?' [0: Unification has already gone too far, …, 10: Unification should go further].
- Ideological orientation (lrscale) – 'In politics people sometimes talk of "left" and "right". Using this card, where would you place yourself on this scale, where 0 means the left and 10 means the right?'

The model parameters are estimated by using German respondents who neither are EU citizens nor were born in Germany. The reason for using this special group is to see whether the ideological orientation can have an impact on the specific issue attitude when the issue has less relevance. Thus, the analysis can provide an interesting test concerning the impact of ideology. However, the problem is that we do not have many respondents in the individual surveys: the second round has 57 respondents, third 43 respondents, and the fourth 40 respondents.[1] These figures reduced further to 38, 33 and 28 after listwise deletion due to missing values.

Before we proceed to the Bayesian inference, we first check the estimation results using conventional ordinary least squares in Table 3.2. Accordingly, all three estimates of the impact of ideological orientation on attitudes to the EU have negative sign, which means that more left-oriented respondents support further European integration, and vice versa. These effects have, however, relatively large standard errors, so that none of them is significant at the 5% level. Therefore, we cannot make any statements concerning the impact of ideology based on these data. If we look at the number of observations, however, we also realize that individual rounds each provide quite small pieces of information. Indeed, if we pooled all three rounds into one regression model the effect is significant. Another alternative is Bayesian inference which is useful for combining information from individual rounds.

What can we do if we apply Bayesian inference to the same data? One possible approach is that we begin with a flat prior belief and sequentially update our belief using the data from

Table 3.2 Estimates via OLS

	2nd round		3rd round		4th round		Pooled	
	Est.	SE	Est.	SE	Est.	SE	Est.	SE
β_0	8.216	1.402	8.754	2.097	7.594	2.372	8.633	1.025
β_1	−0.288	0.320	−0.618	0.390	−0.383	0.509	−0.528	0.213
R^2	0.022		0.075		0.021		0.059	
n	38		33		28		99	

individual rounds. In other words, with no idea at first about whether the ideological orientation has an impact on attitudes to the EU, we can collect data and update our belief about the existence of the effect.

Conjugacy analysis of the second round data with flat prior

At the beginning of the analysis, we assume we have no information about the impact of ideological orientation. Therefore, we specify a conjugate prior using a normal inverse gamma distribution $f_{N-\Gamma^{-1}}(\boldsymbol{\beta}_0, \boldsymbol{\Sigma}_0, a, d)$ with the following parameter values:

$$\boldsymbol{\beta}_0 = (0, 0), \tag{3.49}$$

$$\boldsymbol{\Sigma}_0 = \begin{pmatrix} 10000 & 0 \\ 0 & 10000 \end{pmatrix}, \tag{3.50}$$

$$a = 0.0001, \tag{3.51}$$

$$d = 0.0001. \tag{3.52}$$

Substituting these hyperparameter values and information from data into equations (3.29)–(3.32), we obtain the following parameters of the posterior:

$$\boldsymbol{\beta}^* = (8.216, -0.288), \tag{3.53}$$

$$\boldsymbol{\Sigma}^* = \begin{pmatrix} 0.239 & -0.051 \\ -0.051 & 0.012 \end{pmatrix}, \tag{3.54}$$

$$a^* = 148.160, \tag{3.55}$$

$$d^* = 19.000. \tag{3.56}$$

Note that $\boldsymbol{\beta}^*$ is very similar to the ordinary least squares point estimates. This is because we specified a very flat prior distribution.

An inverse gamma distribution with parameter values a and d has expectation $a/(d-1)$. Therefore:

$$E(\sigma^2|\mathbf{y}) = \frac{148.160}{19.000 - 1} = 8.231. \tag{3.57}$$

Thus, the posterior variance–covariance matrix of $\boldsymbol{\beta}$ has expectation

$$E(\sigma^2|\mathbf{y}) \cdot \boldsymbol{\Sigma}_0 = \begin{pmatrix} 1.965 & -0.423 \\ -0.423 & 0.102 \end{pmatrix}. \tag{3.58}$$

The square root of the diagonal elements is (1.402, 0.320). This corresponds to the standard error of the ordinary least squares estimates.

Drawing the posterior using Gibbs sampling

In this subsection, we construct the same posterior distribution using a numerical method, namely Gibbs sampling. As in the conjugacy analysis above, we again assume we have no information about the impact of ideological orientation. Here, however, we specify our prior using a multivariate normal and an inverse gamma distribution instead of the normal inverse gamma distribution. The parameters of both distributions are the same as in equations (3.49) to (3.52).

The conditioned posteriors for $\boldsymbol{\beta}$ and σ^2 correspond to equations (3.36) and (3.37), respectively. By using these posteriors we set up the Gibbs sampling. As initial values we use the following three sets of values:

$$\{\beta_0, \beta_1, \sigma^2\} = \{0, 0, 1\},\ \{0, -1.5, 2\},\ \{15, 1, 3\}.$$

That is, we run three Markov chains. Figure 3.5 traces the Markov chains of $\boldsymbol{\beta}$ (β_0 and β_1). The three chains started from different initial values; however, they converged after a few steps to a certain common region. This common region corresponds to the joint posterior distribution of β_0 and β_1. The individual points in the Markov chains correspond to the samples from the joint posterior distribution. Therefore, we can simply summarize the information of the sampled data from the posterior. Before summarizing the data, we discarded the first 1000 iterations as burn-in. Thereafter, we ran a further 2000 iterations to sample from the posterior distribution.

Figure 3.7 shows the density distributions for the individual parameters, which visualize the corresponding marginal posterior distribution. We can also use the mean, standard deviation or quantiles to obtain further information about the posterior which is presented in Table 3.3. The right-hand half of the same table presents the corresponding summary statistics of the posterior distributions which were derived in the conjugacy analysis in the previous subsection. The results are almost identical, which means that the Gibbs sampling worked very well. Slight differences are due to the random draws in the Gibbs sampling. In the previous subsection we have already found that the conjugate posterior is almost identical to the ordinary least squares estimates (Table 3.2), which is of course also the case for the posterior via Gibbs sampling. This does not mean, however, that we have to make exactly same statements using ordinary least squares and Bayesian estimation. To see this, we again look at Figure 3.7 which also presents the interval between the 5th and 95th percentiles. This is not called a 90% confidence interval, but a 90% credible interval, and its interpretation is not the same. The confidence interval gives information about how frequently the true parameter value is inside the interval in repeated experiments or equivalent (frequentist probability). In contrast, the posterior distribution expresses our belief about the parameter with specific uncertainty. Therefore, we can simply state that the parameter value is in the credible interval with probability 90%. Consequently, we can also calculate the probability that a certain parameter lies in a certain interval. The middle panel of Figure 3.7 also gives the probability of β_1 having a negative value. By using this information we can also state that right-wing ideology leads to an anti-European attitude with 81% probability. This stands in clear contrast to the conventional inference where we can only either accept or reject certain hypotheses.

Updating belief using data from further rounds

In the analysis of the second round data in the previous two subsections we derived the posterior using certain flat priors. That is, we assumed we had no clear belief about the parameter values of the statistical model of interest. Having done the first analysis, however, we formed certain

Table 3.3 Summary statistics of the posterior distributions (second round data)

	Gibbs sampling				Conjugacy analysis			
	Est.	SD	5%	95%	Est.	SD	5%	95%
β_0	8.223	1.456	5.819	10.568	8.216	1.402	5.849	10.583
β_1	–0.290	0.331	–0.834	0.264	–0.288	0.320	–0.828	0.253
σ^2	8.732	2.205	5.817	12.844	8.231	1.996	5.551	11.908

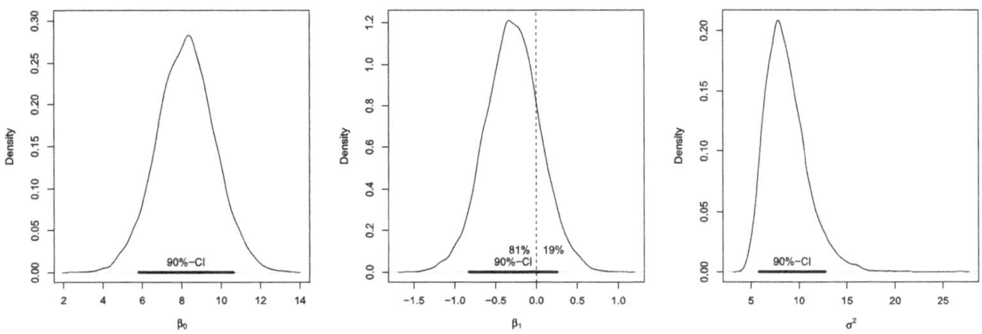

Figure 3.7 (Marginal) posterior distribution

beliefs which we can now use as priors in further analyses. In particular, the ESS data allows us to estimate the same statistical model using the third and fourth round data.

Here, we can exploit the advantage of conjugacy analysis. Due to the conjugacy of the normal inverse gamma distribution to the normal likelihood function, we can use the parameters of the posterior distribution as those of the prior in further analysis. In the analysis of the third round data, we can substitute the posterior's parameter values in equations (3.53)–(3.56) as prior parameter into equations (3.29)–(3.32). This results in the following parameter values of the new posterior distribution which again has the form of the normal inverse gamma distribution:

$$\boldsymbol{\beta}^* = (8.900, -0.553), \tag{3.59}$$

$$\boldsymbol{\Sigma}^* = \begin{pmatrix} 0.122 & -0.024 \\ -0.024 & 0.005 \end{pmatrix}, \tag{3.60}$$

$$a^* = 386.625, \tag{3.61}$$

$$d^* = 35.500. \tag{3.62}$$

Using these parameter values, we can calculate the summary statistics of the (marginal) posterior distribution (the left-hand half of Table 3.4). If we look at β_1, which is substantively of most interest to us, we find a smaller impact in the posterior mean (-0.553) than in the ordinary least squares estimate (-0.618). This is because we used a prior distribution with a mean of -0.288. While we find a smaller magnitude of the effect in the mean, its uncertainty is relatively small. While the standard error in the corresponding ordinary least squares estimation was 0.390, the standard deviation of the posterior is only 0.241. This is because we have the prior as an information source in addition to the third round data. Therefore, we can be more certain regarding our results. Consequently, the probability that β_1 has negative sign increases from the previous analysis to 99%.

Table 3.4 Summary statistics of the posterior distributions (third and fourth round data)

	3rd round				4th round			
	Est.	SD	5%	95%	Est.	SD	5%	95%
β_0	8.900	1.172	6.914	10.886	8.632	1.025	6.885	10.381
β_1	−0.554	0.241	−0.962	−0.145	−0.527	0.213	−0.892	−0.164
σ^2	11.206	1.936	8.435	14.700	10.192	1.479	8.023	12.831

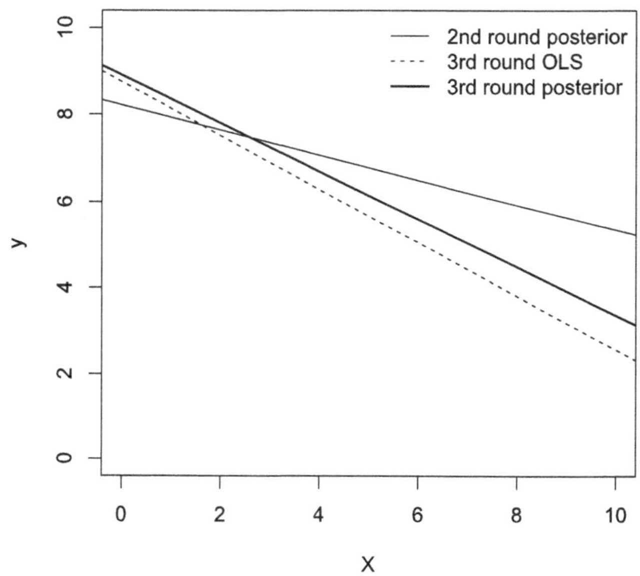

Figure 3.8 Estimated regression lines from second and third round data

Here one might wonder that the posterior mean of β_0 (8.900) is higher than that both of the prior (8.223) and the ordinary least squares estimate (8.754). Indeed, we learned at the beginning of this chapter that the posterior is a mixture of the prior and the likelihood. However, we are now not working on the posterior with a single parameter, but on the joint posterior with multiple parameters. That is, the posterior of a parameter depends on the prior and the likelihood as well as the other parameters. In this particular case, the posterior after the second round (i.e. the prior for the third round analysis) has a relatively flat regression line with a higher value for the constant (the thin solid line in Figure 3.8). That is, the dependent variable is in general at a higher level. The ordinary least squares result of the third round analysis shows, by contrast, a steeper regression line (the dashed line in Figure 3.8). Now the posterior based on the third round data is a mixture of this steeper regression line and a generally high level of the dependent variable (the thick solid line in Figure 3.8). And both factors raise the value of constant.

After updating our belief using the third round data, we can now further update our belief using the fourth round data. The procedure is analogous to the analysis of the third round data. The only difference is that we use the posterior from the third round data as prior. The results appear in the right-hand half of Table 3.4. If we compare the posterior of β_1 with the corresponding ordinary least squares estimate, it is clear that the prior has a stronger impact on the posterior than the data. This is reasonable if we consider the following points. First, the fourth round data provides

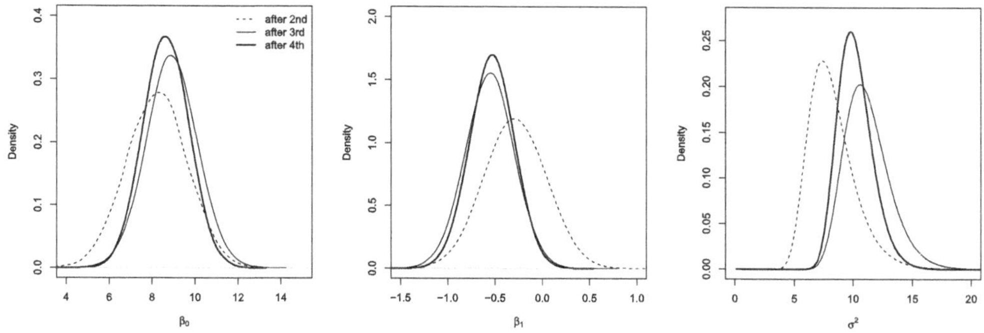

Figure 3.9 Sequential updating of the (marginal) posterior distributions

only 28 observations. Second, the standard error of β_1 in the ordinary least squares estimation is larger than those in the earlier rounds. Furthermore, the prior (posterior from the third round) is smaller than the standard errors from ordinary least squares estimation as discussed above. For these reasons, the prior information is much more weighted than the information from the fourth round data.

Figure 3.9 gives a graphical presentation of the development of the posteriors. In the posteriors of β_0 and β_1 we can clearly observe that the uncertainty about the parameter was reduced in the course of analysis. Concerning β_1 we can be quite sure after the analysis of fourth round data that the ideological orientation has an impact on the EU attitude. In contrast, σ^2 has a different development. After building the first posterior using the second round, the next posterior has a larger mean and also a larger variance. This suggests that the prior from the second round and the data from the third round data provided conflicting pieces of information concerning σ^2. Therefore, after the analysis we could not be as sure as after the first analysis. However, the further data from the fourth round provided more information supporting a smaller value of σ^2, so that our posterior became similar again to that after the first analysis.

Careful readers will surely have realized that the posterior from the fourth round data coincides with the pooled ordinary least squares result (Table 3.2). This is always the case if we conduct this kind of sequential updating of the posterior. To illustrate this relationship, we can return to the normal inverse gamma posterior (equations (3.29)–(3.32)). Here, we focus on $\boldsymbol{\beta}$ and $\boldsymbol{\Sigma}$, but the same holds also for the other parameters (a and d). Denote the data from the second, third and fourth round by $\mathbf{y}_2$, $\mathbf{y}_3$, $\mathbf{y}_4$, $\mathbf{X}_2$, $\mathbf{X}_3$ and $\mathbf{X}_4$. We first derive the posterior based on a flat prior and second round data, which corresponds to the ordinary least squares estimate:

$$\boldsymbol{\beta}_2^* = (\mathbf{X}_2'\mathbf{X}_2)^{-1}(\mathbf{X}_2'\mathbf{y}_2), \tag{3.63}$$

$$\boldsymbol{\Sigma}_2^* = (\mathbf{X}_2'\mathbf{X}_2)^{-1}. \tag{3.64}$$

We substitute these parameters into $\boldsymbol{\beta}_0$ and $\boldsymbol{\Sigma}_0$ in equations (3.29) and (3.30) to obtain the posterior after the third round data analysis:

$$\boldsymbol{\beta}_3^* = (\underbrace{\mathbf{X}_2'\mathbf{X}_2}_{\boldsymbol{\Sigma}_0^{-1}} + \mathbf{X}_3'\mathbf{X}_3)^{-1}(\underbrace{(\mathbf{X}_2'\mathbf{X}_2)}_{\boldsymbol{\Sigma}_0^{-1}}\underbrace{(\mathbf{X}_2'\mathbf{X}_2)^{-1}(\mathbf{X}_2'\mathbf{y}_2)}_{\boldsymbol{\beta}_0} + \mathbf{X}_3'\mathbf{y}_3) \tag{3.65}$$

$$= (\mathbf{X}_2'\mathbf{X}_2 + \mathbf{X}_3'\mathbf{X}_3)^{-1}(\mathbf{X}_2'\mathbf{y}_2 + \mathbf{X}_3'\mathbf{y}_3), \tag{3.66}$$

$$\boldsymbol{\Sigma}_3^* = (\underbrace{\mathbf{X}_2'\mathbf{X}_2}_{\boldsymbol{\Sigma}_0^{-1}} + \mathbf{X}_3'\mathbf{X}_3)^{-1}. \tag{3.67}$$

Analogously, we can obtain the posterior after the fourth round data analysis:

$$\boldsymbol{\beta}_4^* = (\mathbf{X}_2'\mathbf{X}_2 + \mathbf{X}_3'\mathbf{X}_3 + \mathbf{X}_4'\mathbf{X}_4)^{-1}(\mathbf{X}_2'\mathbf{y}_2 + \mathbf{X}_3'\mathbf{y}_3 + \mathbf{X}_4'\mathbf{y}_4), \tag{3.68}$$

$$\boldsymbol{\Sigma}_4^* = (\mathbf{X}_2'\mathbf{X}_2 + \mathbf{X}_3'\mathbf{X}_3 + \mathbf{X}_4'\mathbf{X}_4)^{-1}. \tag{3.69}$$

Here, it is clear that the sequential posterior updating corresponds exactly to the pooled analysis using the OLS or maximum likelihood framework. Of course, this is not the case if we use certain informative priors for the analysis of the second round data.

From this observation another advantage of the Bayesian inference is clear in that we can also estimate statistical models with a smaller number of observations. This is because the prior belief serves as a further observation in estimation. Furthermore, Bayesian inference calculates parameters' posterior distribution directly instead of constructing an estimator. Thus it is free from the conventional asymptotic theory. For these reasons, the Bayesian inference is also possible for smaller numbers of observations than would be sufficient for ordinary least squares and/or maximum likelihood.

CAVEATS AND FREQUENT ERRORS

In using a Bayesian inferential framework one always has to be aware of the underlying conception about prior beliefs, data, probability and the manner of inference. In particular, one should be careful in interpreting the estimation results. Bayesian inference, for example, does not know the null hypothesis significance test. Therefore, it is completely wrong to discuss the significance level, rejection or acceptance of the null hypothesis, etc. Further, we have already discussed the difference between the confidence interval in conventional inference and the Bayesian credible interval. The interpretation of the credible interval *per se* is quite straightforward, and one should be rather careful in the interpretation of the conventional confidence interval. However, one still has to be careful not to make significance-test-style statements using a credible interval.

The use of prior information and its systematic integration into the posterior is one of the most important features of Bayesian inference. Again, one should be aware of the conception of priors and the idea of Bayesian updating. The basic idea of Bayesian inference is to update our prior beliefs using new information from the data. A belief is prior here in the sense that we have the information prior to the data collection. Therefore, the choice of prior distribution cannot be reasoned by information from the data which is used to update the prior information. A prior can be also specified after data collection. This is the case even in many concrete contexts. However, the legitimation of the choice of certain priors can be never made using information from the data. If one is sufficiently conscious of this point the criticism against the Bayesian manner of inference for its use of prior information does not have to be taken seriously. It is obvious that the choice of priors is not grounded by the data. This kind of a priori decision is, however, also made by conventional inferential statistics. To conduct statistical analyses, we have to make a series of assumptions, such as the choice of independent variables and the distribution form of the stochastic term, in order to identify the statistical model. If these assumptions can be made we can also make assumptions about parameter values – of course, if we have good reasons for that.

Even if one is convinced of the use of prior information, it is difficult to bring the information into a certain probability form. As one possibility this chapter has presented an example which uses past estimates as prior. If one has no information from other comparable analyses one needs to build one's own prior distribution in a discrete or continuous form. In this case, one can have a certain choice between different probability forms. If this is the case and one cannot decide on a specific probability form with a strong argument, one should use multiple priors to check whether the results based on the posterior are strongly affected by the choice of prior. This kind of check is called a sensitivity test.

In our favorable case with prior information from the other analysis, one still has to be careful about the previous data used to obtain the prior. If a strong measurement and/or sampling error exists which systematically affects both past data and current data, the posterior result can contain strongly boosted errors. The example above might also suffer from this problem. We used only those respondents born outside Germany who do not possess EU citizenship. If there are any systematic sampling errors which are relevant to the relationship of interest to us, our posterior provides information biased towards the error. In contrast, if we can expect the errors to be balanced out in the course of the updating process we do not have to worry about this problem.

While one can never overemphasize the role of prior information, it is not the only advantage of the Bayesian inference. In particular, Bayesian inference can work with a wide range of statistical models, thanks in particular to the MCMC techniques. For example, some statistical models may cause difficulties in finding the maximum likelihood due to some local maxima or a very flat likelihood surface. This difficulty is less relevant for Bayesian inference due to the random walk property of MCMC. Another advantage concerns missing-data problems. In the framework of MCMC one can treat missing data as a random entity and a parameter to be estimated. Furthermore, hierarchical models which have been increasingly utilized in social sciences fit the basic idea of Bayesian inference. In hierarchical models parameters at one level are conceptualized as random variables and modelled by higher-level parameters, which corresponds to the idea of hyperparameters. And this idea is quite simple to implement in the MCMC framework.

In using MCMC, we can never be careful enough in evaluating the convergence of Markov chains. This topic is crucial since random draws in the burn-in phase can offer completely different pictures of the target posterior. For this reason running multiple chains with significantly different initial values is inevitable. At the same time, the phase after the burn-in is also important. In this phase, we can collect the information from the posterior distribution. If one runs the Markov chain not long enough the collected information may be biased due to random draws. In particular, such information as quantiles requires a sufficiently large number of iterations. In principle, the more iterations after burn-in the more accurate the information obtained about the posterior. To evaluate whether one has reached a certain precision through sampling after the burn-in phase one can utilize the *Monte Carlo error*, which is based on the idea of the variance in autocorrelated samples.

FURTHER READING

The aim of this chapter is to provide a first look at Bayesian inference for the readers who are unfamiliar with it and to make the relevant literature accessible. For this reason, the topics dealt with in this chapter are deliberately limited. In particular, this chapter only presents the basic idea of Bayesian inference and its application to simple linear regression using conjugacy analysis and Gibbs sampling. Beyond these topics, there are a large number of statistical models and their accompanying topics. For those who have just started with this chapter, some journal articles are recommended as further reading.

Gill (1999) elegantly summarizes the differences between conventional and Bayesian inference. Simon Jackman, a political scientist who helped disseminate Bayesian inference in the social sciences, has written several articles which provide a short introduction. Among them,

Jackman (2000) delves deeply into the relationship between maximum likelihood and Bayesian inference and gives a good introduction to the Markov chain Monte Carlo method. Western and Jackman (1994) is also a well-written introductory article whose focus is more on its application and advantages in the social sciences. Due to their limited length, journal articles always have deficits in terms of broad coverage of topics. Having gained a feel for Bayesian statistics, those who wish to extend their knowledge of Bayesian analysis would be well advised to read Gelman and Hill (2007). This book is about the regression model and multilevel modelling, but also covers further important and relevant topics such as missing values, and offers many practical tips on programming and computation.

For those who would like to extend their knowledge further, Gill (2002) and Gelman et al. (2003) deal extensively with diverse topics in Bayesian inference. These books are very technical and more appropriate for those who are familiar with Bayesian inference and have specific problems in mind.

NOTE

1 The first round is not considered here since this round does not include the questionnaire item concerning European integration.

REFERENCES

Gelman, A., Carlin, J. B., Stern, H. S., and Rubin, D. B. (2003). *Bayesian Data Analysis*, 2nd edn. Boca Raton, FL: Chapman & Hall/CRC.

Gelman, A. and Hill, J. (2007). *Data Analysis Using Regression and Multilevel/Hierarchical Models*. Cambridge: Cambridge University Press.

Gill, J. (1999). The insignificance of null hypothesis significance testing. *Political Research Quarterly*, 52(3), pp. 647–674.

Gill, J. (2002). *Bayesian Methods: A Social and Behavioral Sciences Approach*. Boca Raton, FL: Chapman & Hall/CRC.

Jackman, S. (2000). Estimation and inference via Bayesian simulation: An introduction to Markov chain Monte Carlo. *American Journal of Political Science*, 44(2), pp. 375–404.

Western, B. and Jackman, S. (1994). Bayesian inference for comparative research. *American Political Science Review*, 88(2), pp. 412–423.

PART II

Regression analysis for cross-sections

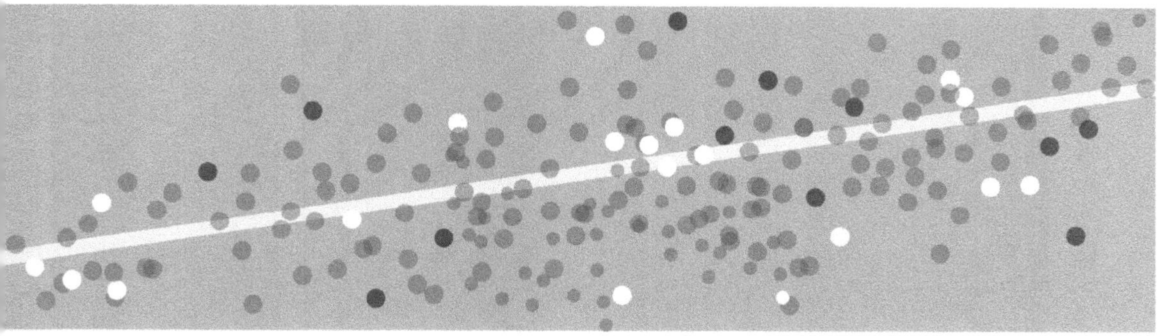

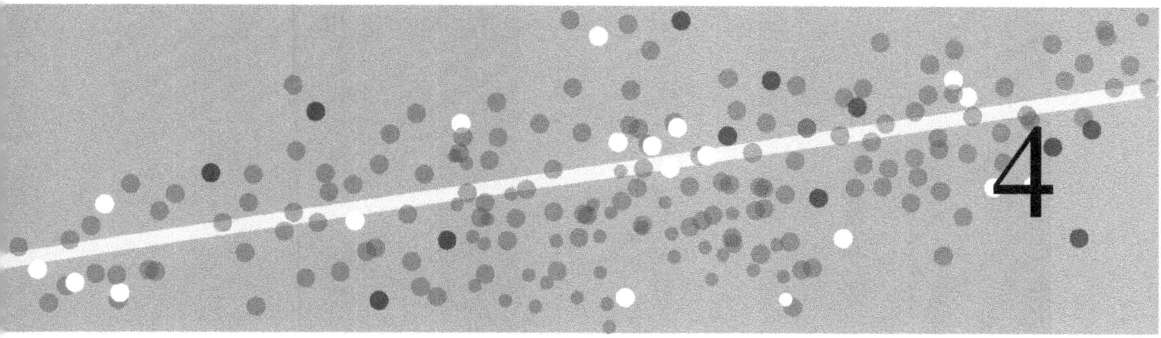

4 Linear regression

Christof Wolf and Henning Best

INTRODUCTION

In this chapter we first present the basic idea of linear regression and give a non-technical introduction. Next we cover the statistical basis of this method and discuss estimation, testing and interpretation of regression results. The third section is devoted to the presentation of an example analysis. In closing, we first discuss issues related to the causal interpretation of OLS regression coefficients and then mention some general problems encountered in linear regression and recommend further reading.

In science we often are interested in studying hypotheses of the form 'the more X, the more/less Y', for example 'the higher the education of a person is, the more willing s/he is to accept immigrants'. Thus, we assume that acceptance of immigrants is partly determined by education, or more technically that the acceptance of immigrants is a function of education. We can express this idea mathematically as

$$\text{Acceptance of Immigrants} = f(\text{Education}) \quad \text{or} \quad y = f(x).$$

If we choose a linear function $f(\cdot)$ to link y with x_1 the result is a linear regression model. Alternative link functions result in other regression models; some of which are discussed in Chapters 8 and 9 of this volume. Expressed as linear model, we get

$$y = \beta_0 + \beta_1 x + \varepsilon, \tag{4.1}$$

where y is referred to as the dependent (endogenous) variable and the x as an independent (exogenous) variable or predictor. This equation can be seen as a specification of our hypothesis 'the higher the education of a person is, the more willing s/he is to accept immigrants'. The specification consists of the assumption that we have a linear effect of education on accepting immigrants of size β_1. This means we assume that if education increases by one unit the acceptance of immigrants changes by β_1 units. It is also important to note that if we specify a linear effect as in equation (4.1) we assume that the effect of education is the same for any given level of education, that is, the effect is constant throughout the range of x.

Equation (4.1) contains an element that has not yet been introduced. The term ε is referred to as an error term or residual. It is equal to the difference between the observed values of y and the

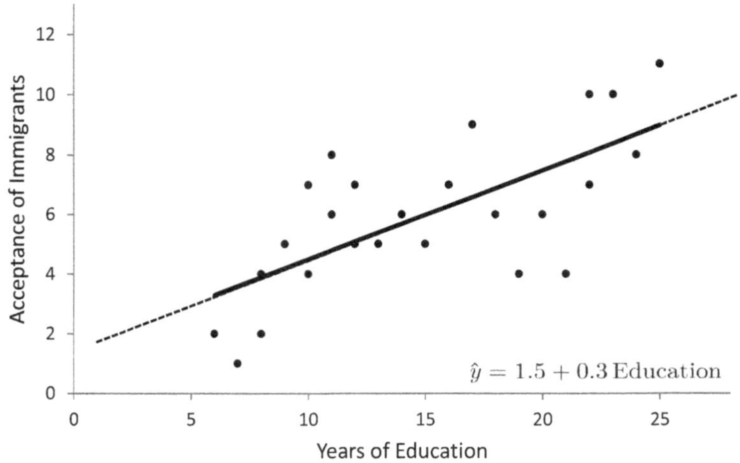

Figure 4.1 Scatter plot with linear fit line

values predicted by the independent variable(s). Subtracting ε from both sides of equation (4.1) yields

$$y - \varepsilon = \hat{y} = \beta_0 + \beta_1 x, \tag{4.2}$$

where $\hat{y}$ denotes the predicted values of y.

Figure 4.1 illustrates the relationship between education and acceptance of immigrants for 25 people, where education is measured in years of full-time education and acceptance is measured on an 11-point scale ranging from 1 (no acceptance) to 11 (full acceptance). Each dot in this scatter plot represents one person in this property space. The line drawn in Figure 4.1 represents the linear relationship between education and acceptance; it follows the equation $\hat{y} = 1.5+0.3x$. Here, $\beta_1 = 0.3$ implies that a person with one more year of education on average scores 0.3 points higher on the acceptance of immigration scale. Figure 4.1 also illustrates why β_1 is referred to as the slope: it determines how shallow or steep the regression line is. β_0, in this example equal to 0.5, is called the intercept, since it equals the value of y at which the regression line 'intercepts' or crosses the y-axis. In our little example the intercept of 0.5 can be interpreted as the predicted value for a person with zero years of education. The problem with this interpretation is that we have no data for this range of values of the independent variable and thus we should abstain from making any predictions. Of course, this is also true for the other end of the distribution. For example, we would predict a value of $\hat{y}\,(x=50) = 16.5$ for someone with 50 years of education, which is an impossible value given the current measurement of acceptance with a maximum value of 11. In general, we should restrict our analysis and interpretation of results to those areas of the property space for which we have empirical data. This range is implied by the solid regression line.

In a real application we would usually assume that the phenomenon of interest is affected by more than one factor. In the case of opinions towards immigrants such additional factors could be age, sex, employment status, income, etc. The idea that acceptance of immigrants is affected by all these factors again can be expressed in a linear model as

$$y = \beta_0 + \beta_1 x_1 + \beta_2 x_2 + \cdots + \beta_k x_k + \varepsilon. \tag{4.3}$$

In this equation we have refined our hypothesis that 'the higher the education of a person is, the more willing s/he is to accept immigrants' in an important way. Let us assume that education is represented by x_1 and consequently the effect of education is β_1. As in the earlier example, we assume that if education increases by one unit the acceptance of immigrants changes by β_1 units. But equation (4.3) extends our hypothesis by stating that the effect of education on acceptance of immigrants is estimated while 'holding all other variables constant'. This means that equation (4.3) allows us to observe the effect of education controlling for third variables such as age and sex. The same in turn is true for the effects of all other independent variables x_j in equation (4.3). Their effects are estimated under the assumption that all other independent variables are held constant. Thus, linear regression allows us to estimate the effect of an independent variable on a dependent one as if the units of analysis did not differ with respect to other characteristics contained in the model. For social science applications this is an enormous advantage because, unlike other sciences, we often cannot experimentally manipulate the variables we want to study.[1]

Now assume that instead of analyzing the effect of education we are interested in analyzing the effect of church membership on the sentiment towards immigrants. A variable like this with only two levels, member and non-member, is referred to as binary variable. In linear regression analysis the effects of such binary variables can be modeled straightforwardly. All we have to do is to decide how to code this variable. The standard approach is dummy coding, that is, assigning one of the two categories the value 0 (this category serves as the so-called 'reference category'), and the other the value 1. For example, we could create a variable having the value 0 for non-members and the value 1 for members of churches. Let us denote this variable by D_c and insert it in the regression equation. This gives

$$\hat{y} = \beta_0 + \beta_1 D_c. \tag{4.4}$$

To understand what this means we look at this model for non-members only, $D_c = 0$. Then equation (4.4) reduces to

$$\hat{y}(D_c = 0) = \beta_0.$$

For members, $D_c = 1$, equation (4.4) yields

$$\hat{y}(D_c = 1) = \beta_0 + \beta_1.$$

Thus, in equation (4.4) the intercept (β_0) is identical to the expectation of $\hat{y}$ for non-members, while the slope (β_1) equals the expectation of the difference between members and non-members with respect to $\hat{y}$. Figure 4.2 illustrates a regression result for a dummy variable. In this example non-members average 4.6 points and members 7.0 on the immigration acceptance scale (indicated by the vertical bars). From these figures we obtain the regression results

$$\hat{y} = 4.6 + (7 - 4.6)\ \text{Member} = 4.6 + 2.4\,\text{Member}.$$

The inclusion of binary predictors in regression models can easily be extended to categorical variables with several categories, for example marital status or nationality. In this case we need more than one dummy variable to represent these effects. More precisely, we need one variable fewer than we have groups. Assume we want to distinguish between single, married, divorced/separated and widowed persons. Then we would need three dummy variables.[2] One of the categories will be the reference category, that is, the category relative to which all the differences are expressed. If we take 'single' as the reference category, the regression of y on marital status will give us

$$\hat{y} = \beta_0 + \beta_1 D_m + \beta_2 D_d + \beta_3 D_w.$$

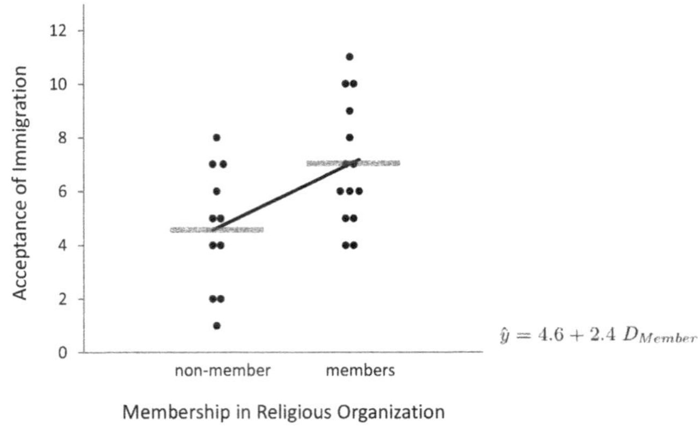

Figure 4.2 Scatter plot for binary predictor

Here β_0 is the expectation of $\hat{y}$ for single persons, the reference category, while β_1, β_2 and β_3 are the differences in expectation of married, divorced and widowed people, respectively.

Using dummies to represent a variable does not have to be restricted to non-metric variables. It can also be a means to test whether the relationship of an independent variable to the dependent variable is non-linear. Take again the example of the effect of education on sentiments towards immigrants. We may have doubts as to whether years of education are linearly related to our variable of interest. In this case we might group years of education into 3-year bands, for example under 8 years, 8 to 10 years, 11 to 12 years, 13 to 15 years, 16 to 18 years, 19 to 21 years, 22 to 24 years, 25 years and over. Inspecting the regression slopes of the respective dummy variables gives us an indication whether or not the assumption of a linear relationship between education and attitudes towards immigrants is warranted.

MATHEMATICAL FOUNDATIONS

The model

In the previous section we have already introduced the general model of multiple linear regression analysis as

$$y = \beta_0 + \beta_1 x_1 + \beta_2 x_2 + \cdots + \beta_k x_k + \varepsilon.$$

Using the summation sign, this expression can be written more compactly as

$$y = \sum_{j=0}^{k} \beta_j x_j + \varepsilon,$$

with $x_0 = 1$. If we use matrix notation this can be written as

$$\boldsymbol{y} = \boldsymbol{X\beta} + \boldsymbol{\varepsilon}, \tag{4.5}$$

which is identical to

$$\begin{bmatrix} y_1 \\ y_2 \\ \vdots \\ y_n \end{bmatrix} = \begin{bmatrix} 1 & x_{11} & \cdots & x_{1k} \\ 1 & x_{21} & \cdots & x_{2k} \\ \vdots & \vdots & \ddots & \vdots \\ 1 & x_{n1} & \cdots & x_{nk} \end{bmatrix} \begin{bmatrix} \beta_0 \\ \beta_1 \\ \vdots \\ \beta_k \end{bmatrix} + \begin{bmatrix} \varepsilon_1 \\ \varepsilon_2 \\ \vdots \\ \varepsilon_n \end{bmatrix}.$$

The predicted values $\hat{\boldsymbol{y}}$ can be written as $\boldsymbol{X\beta}$.

Identifying the regression coefficients

Once we have specified a regression model of the kind presented in equation (4.5), the next step is to identify the regression coefficients, that is, the β_j. Think back to the bivariate case for a moment and look again at Figure 4.1. We can easily imagine different lines representing the cloud of points in the scatter plot; the question is which one of these is the best. Obviously the line we want (i.e. the regression coefficients sought) should minimize the difference between observed and predicted values of the dependent variable. However, there are different ways to 'minimize' this difference. The most common approach is to minimize the *sum of the squared differences* between observed and predicted values (i.e. errors). That is, the regression coefficients are found by minimizing

$$\min \sum_{i=1}^{n} \varepsilon_i^2 = \min \sum_{i=1}^{n} (y_i - \hat{y}_i)^2 = \min \sum_{i=1}^{n} (y_i - (\beta_0 + \beta_1 x_{i1} + \cdots + \beta_k x_{ik}))^2 . \qquad (4.6)$$

Because this method minimizes the sum of squared errors it is usually referred to as *ordinary least squares* (OLS) regression. We can now find values of β_j that minimize equation (4.6) by partially differentiating (4.6) for each β_j, setting the resulting equation equal to zero and solving for β_j. To illustrate how this works, let us explain this procedure for β_1 in more detail. Because equation (4.6) is a composition of two functions we have to obey the chain rule, that is, we have to differentiate the outer and inner part and multiply the result. The derivative of $\sum(\cdot)^2$ equals $2\sum(\cdot)$; the derivative of $(\boldsymbol{y} - \boldsymbol{X\beta})$ for β_1 equals $-x_{i1}$. Multiplying both results and setting the equation to 0 yields

$$2\sum_{i=1}^{n} -x_{i1}(y_i - \beta_0 - \beta_1 x_{i1} - \beta_2 x_{i2} - \cdots - \beta_k x_{ik}) = 0$$

or

$$-2\sum_{i=1}^{n} x_{i1}(y_i - \beta_0 - \beta_1 x_{i1} - \beta_2 x_{i2} - \cdots - \beta_k x_{ik}) = 0. \qquad (4.7)$$

We can simplify this expression by dividing both sides of the equation by -2, giving

$$\sum_{i=1}^{n} x_{i1}(y_i - \beta_0 - \beta_1 x_{i1} - \beta_2 x_{i2} - \cdots - \beta_k x_{ik}) = 0. \qquad (4.8)$$

To complete the exercise, we have to repeat the differentiation for the other regression coefficients. The resulting system of equations is (cf. Wooldridge, 2009, p. 800)

$$\begin{aligned}
&\sum_{i=1}^{n} (y_i - \beta_0 - \beta_1 x_{i1} - \beta_2 x_{i2} - \cdots - \beta_k x_{ik}) = 0 \\
&\sum_{i=1}^{n} x_{i1}(y_i - \beta_0 - \beta_1 x_{i1} - \beta_2 x_{i2} - \cdots - \beta_k x_{ik}) = 0 \\
&\qquad\vdots \\
&\sum_{i=1}^{n} x_{ik}(y_i - \beta_0 - \beta_1 x_{i1} - \beta_2 x_{i2} - \cdots - \beta_k x_{ik}) = 0,
\end{aligned} \qquad (4.9)$$

where the first equation results from differentiation for β_0, the second for β_1, etc. In matrix notation this can be written as

$$\boldsymbol{X}'(\boldsymbol{y} - \boldsymbol{X\beta}) = \boldsymbol{0}. \tag{4.10}$$

Multiplying and rearranging terms gives

$$(\boldsymbol{X}'\boldsymbol{X}')\,\boldsymbol{\beta} = \boldsymbol{X}'\boldsymbol{y}. \tag{4.11}$$

Assuming $(X'X)$ has full rank, that is, none of the independent variables is a perfect linear combination of other independent variables, we can left-multiply both sides by the inverse matrix $(X'X)^{-1}$, resulting in

$$\boldsymbol{\beta} = (\boldsymbol{X}'\boldsymbol{X}')^{-1}\boldsymbol{X}'\boldsymbol{y}. \tag{4.12}$$

This equation gives regression coefficients that minimize the sum of squared errors. Thus, unlike maximum likelihood estimation used in logistic or probit regression, there exists a closed-form solution for finding regression coefficients in OLS regression (for more on OLS and maximum likelihood estimation see Chapter 2 of this volume).

Assessing model fit

As we have seen in the previous subsection, it is always possible to solve a linear regression problem using the OLS principle. And each model estimated with this principle minimizes the squared difference between observed and estimated values of the dependent variable. However, this does not imply that every regression model fits the data equally well. On the contrary, some models will have very poor fit while others will fit the data better. The degree of fit obviously depends on the degree to which the predicted values for the dependent variable $\hat{y}$ are similar to the observed values of y. Or, to put it slightly differently, the more differences in the observed variables a model can account for, the better its fit.

To operationalize this idea of lesser or better fit we make use of a basic concept from the analysis of variance, namely that the total variation of the dependent variable can be partitioned into a part explained by the regression model and a part not explained by the model. Put more formally,

$$\sum(y-\bar{y})^2 = \sum(\hat{y}-\bar{y})^2 + \sum(y-\hat{y})^2$$

or

$$\text{TSS} = \text{ESS} + \text{RSS},$$

where TSS, ESS and RSS stand for total sum of squares, explained sum of squares and residual sum of squares, respectively. We can now look at the ratio of explained variation relative to the total variation,

$$R^2 = \frac{\text{explained variation}}{\text{total variation}} = \frac{\text{ESS}}{\text{TSS}} = \frac{\sum(\hat{y}-\bar{y})^2}{\sum(y-\bar{y})^2}. \tag{4.13}$$

The ratio R^2, also called the coefficient of determination, can vary between 0 and 1 and reflects the proportion of variance explained by the regression model.

One problem with this measure of fit is that if we add more variables to a given model R^2 can only increase, even though the variables we add may be irrelevant with respect to the dependent variable of interest. This happens because R^2 and consequently also changes in R^2 can only be positive. Therefore, independent variables unrelated to the dependent variable can by chance produce an increase in R^2. To correct for this tendency an adjusted version of R^2 has been proposed which is given by

$$R^2_{\text{adj}} = 1 - \frac{n-1}{n-k-1}(1-R^2),$$

with n denoting the number of cases and k the number of independent variables. In contrast to R^2, R^2_{adj} can decrease when adding new variables to an equation that are irrelevant with respect to the dependent variable.

If using a regression approach to model data we would often start with a basic model to which we would add more variables step by step each time making it slightly more complex. For example, if we were interested in the effect political orientation has on attitudes towards immigrants we could first estimate a model containing only socio-demographic variables. This first model will give a fit of R_1^2. In a second model we then add political orientation to the independent variables and estimate a model which gives us R_2^2. The difference between the two measures of fit, $R_2^2 - R_1^2$, then indicates the effect of political orientation on attitudes towards immigrants net of the socio-demographic variables controlled for in the model. This strategy is particularly useful if we want to estimate the effect of several variables, an approach we can also use to determine the impact of a set of dummy variables on the explained variance.

If the models we are interested in are not nested, R^2 should not be used for comparisons. If we want to compare the same model in different populations (e.g. men and women or Switzerland and Germany), then we can apply the Chow test presented below on page 66.

Before closing this section, we would like to add a final word on the size of R^2. A question often asked is how large R^2 should be. The answer to this question depends on the purpose of our research. If we are aiming to explain a certain variable, such as attitudes towards immigrants, then we would like to maximize the R^2 of our model. If, on the other hand, we are interested in the effect of political orientation and class position on the attitudes of immigrants then we would not care about the overall fit of our model so much but focus on the effect sizes for the variables we are interested in.

Statistical inferences of regression results

Usually a regression model is estimated on the basis of data from a (random) sample of the target population of interest. For example, the short empirical analysis we present in the next section of this chapter is run on data from a Swiss and German sample of the European Social Survey. When we estimate the regression models we are not so much interested in studying the sample as such, but instead aim to learn something about the population from which the sample was drawn. So, in our illustrative analysis presented below, we aim to gain better knowledge of certain attitudes of the adult populations of Switzerland and Germany. As in other areas of statistics, we can apply technics of statistical inference to draw these conclusions for the populations based on data from random samples. For simplicity of the presentation the tests we discuss in this section assume that the data come from a simple random sample. In most cases this will be an oversimplification because our data typically stem from multistage (stratified) samples. If this is the case we should use the appropriate adaptations of the tests we present here. Some of these tests will be presented in Chapter 11 of this volume which discusses regression analysis for data from 'complex' samples. However, the logic of statistical inference remains unchanged.

To indicate that a regression model is estimated with sample data the standard notation is slightly modified by adding a circumflex (^ or 'hat') to the regression coefficients estimated by the model. The regression model then becomes

$$y = \hat{\beta}_0 + \hat{\beta}_1 x_1 + \hat{\beta}_2 x_2 + \cdots + \hat{\beta}_k x_k + \varepsilon = \sum_{j=0}^{k} \hat{\beta}_j x_j + \varepsilon \tag{4.14}$$

or, in matrix notation,

$$y = X\hat{\beta} + \varepsilon. \tag{4.14'}$$

The $\hat{\beta}_j$ are estimates of the β_j computed from data from a sample. Related to such a model we can make two different types of inferences: first, inferences about the model itself or a comparison between different models; and second, inferences about one or two regression coefficients. We begin by discussing this second type of inference about regression coefficients.

Inferences about one regression coefficient

A first question we may want to ask is whether it can be assumed with some reasonable level of certainty that a regression coefficient in the population (β_j) is equal to or different from some value a. The decision between the statistical hypotheses[3]

$$H_0\colon \beta_j = a,$$
$$H_1\colon \beta_j \neq a$$

is made based on the following test statistic:

$$t_{\hat{\beta}_j} = \frac{\hat{\beta}_j - a}{s_{\hat{\beta}_j}}, \tag{4.15}$$

with $s_{\hat{\beta}_j}$ as standard error of the regression coefficient. The standard error is given by

$$s_{\hat{\beta}_j} = \sqrt{\frac{\sum_{i=1}^{n}(y_i - \hat{y}_i)^2/(n-k-1)}{\sum_{i=1}^{n}\left(x_{ij} - \bar{x}_j\right)^2\left(1 - R_j^2\right)}}, \tag{4.16}$$

with R_j^2 denoting the amount of variance explained in x_j by the other independent variables (cf. Wooldridge, 2009, p. 89). If the assumptions behind OLS analysis are met, the t-statistic follows a t-distribution with $n-k-1$ degrees of freedom. Based on this test statistic, we can test $\hat{\beta}_j$ against any value a. Standard software usually gives results for the two-sided test for $a = 0$, that is, the hypotheses

$$H_0\colon \beta_j = 0,$$
$$H_1\colon \beta_j \neq 0.$$

The test tells us if we can assume with a given degree of certainty that the null hypothesis (H_0) can be rejected, meaning that we can assume that x_j has an influence on y in the target population. The degree of certainty we adopt is a convention which is often set to 95% or 99% in social science applications. However, depending on the research question, different levels of certainty will make sense.

Inferences about the relative size of two regression coefficients from the same population

Sometimes we may be interested in testing whether the effect of one variable is stronger than that of another variable from the same model. For example, we could ask if the effect of education (β_1) on acceptance of immigrants is stronger than the effect of age (β_2). The statistical hypotheses are

$$H_0\colon \beta_1 \leq \beta_2,$$
$$H_1\colon \beta_1 > \beta_2$$

and the test statistic is

$$t = \frac{\hat{\beta}_1 - \hat{\beta}_2}{\sqrt{s^2_{\hat{\beta}_1} + s^2_{\hat{\beta}_2} - 2s_{\hat{\beta}_1\hat{\beta}_2}}}, \tag{4.17}$$

with $s_{\hat{\beta}_1\hat{\beta}_2}$ being the covariance between the estimates $\hat{\beta}_1$ and $\hat{\beta}_2$; t has $df = n - k - 1$ degrees of freedom.

Inferences about the relative size of a regression coefficient in two different populations

In other situations we may be interested in learning whether a predictor has the same effect in different populations. We could be interested, for example, in testing whether the effect of education on acceptance of immigrants is the same in Switzerland and Germany. The statistical hypotheses would be

$$H_0: \beta_1|\text{Switzerland} = \beta_1|\text{Germany},$$
$$H_1: \beta_1|\text{Switzerland} \neq \beta_1|\text{Germany}.$$

One way to answer this question is to combine the samples of interest into one data set and add an indicator variable to the data set taking the value 0 for data from the first sample and 1 for data from the second sample. Finally, we would have to create an interaction term between the data set indicator and the independent variable we want to test. The resulting model can be expressed by the equation

$$y = \hat{\beta}_0 + \hat{\beta}_1 x_1 + \hat{\beta}_d D + \hat{\beta}_{1d} x_1 D + \sum_{j=2}^{k} \hat{\beta}_j x_j + \varepsilon. \tag{4.18}$$

Based on this equation, the statistical hypotheses are modified to

$$H_0: \beta_1|\text{Switzerland} - \beta_1|\text{Germany} = \beta_{1d} = 0,$$
$$H_1: \beta_1|\text{Germany} - \beta_1|\text{Germany} = \beta_{1d} \neq 0.$$

As can be seen, from this expression for the statistical hypotheses, the original question has been transformed into one asking if a regression coefficient is different from zero. This question can easily be answered by the test introduced above (see equation (4.15)) which allows us to test whether β_{1d} is significantly different from zero. If so, we would have to conclude with a given level of confidence that the effect of x_1 is different in the two populations.

Inferences about an entire model

Having covered some tests concerning regression coefficients, we now turn to testing entire models. The question we ask is whether the regression model we estimate can be expected with reasonable certainty to explain at least some of the variation of the dependent variable we study in the population. The statistical hypotheses can be formulated as

$$H_0: \beta_1 = \beta_2 = \cdots = \beta_k = 0,$$
$$H_1: \text{at least one } \beta_j \neq 0.$$

This pair of hypotheses can be tested by the statistic

$$F = \frac{\sum(\hat{y} - \bar{y})^2 / k}{\sum(y - \hat{y})^2 / (n - k - 1)} = \frac{\text{ESS}/k}{\text{RSS}/(n - k - 1)} = \frac{\text{EMS}}{\text{RMS}}, \tag{4.19}$$

which follows an F-distribution with $df_1 = k$ and $df_2 = n - k - 1$. This statistic is well known from the analysis of variance framework, in which the numerator is also known as explained mean squares (EMS) and the denominator as residual mean squares (RMS). If the F statistic is significant then we can conclude with the given level of certainty that at least one independent variable included in the model is (linearly) related to the dependent variable in the population of interest.

Inferences about two nested models

Often we take a stepwise approach to modeling a dependent variable with regression analysis. As mentioned before, we may, for example, first want to see how much attitudes towards immigrants depend on socio-demographic variables and then add political orientation in a second step. But we could also take the reverse route and first explore the effect of political orientation on attitudes towards immigrants and only then control for socio-demographics. In either analysis we may want to know if the additional variables added in the second step improve the model fit significantly. If we want to test the difference of such 'nested' models, that is, models where the parameters of the first model are a true subset of the parameters of the second model, we can use the following F-distributed test (Fox, 2008, p. 201):

$$F = \frac{(\mathrm{RSS}_1 - \mathrm{RSS}_2)/(k_2 - k_1)}{\mathrm{RSS}_2/(n - k_2)}, \tag{4.20}$$

with RSS_1 and RSS_2 referring to the residual sum of squares for model 1 and model 2, respectively, where model 1 is nested in model 2, so that $k_1 < k_2$. If the F-statistic is not significant then model 2, the model with more variables, does not predict the dependent variable better than model 1, the simpler model. If the F-statistic is significant we can be reasonably certain that model 2 fits the data better than model 1.

Inferences about two models for different populations

There may be cases in which we are interested in knowing whether the same model holds for different populations. Thus, we may want to know whether a given model intended to explain attitudes towards immigrants leads to identical conclusions for Switzerland and Germany. In this situation, instead of comparing two slopes we will have to compare the overall fit of the two models. The relevant test statistic, also known as the Chow statistic (see Wooldridge, 2009, p. 245), again follows an F-distribution and is defined by

$$F = \frac{\left(\mathrm{RSS}_p - (\mathrm{RSS}_1 + \mathrm{RSS}_2)\right)/(k+1)}{(\mathrm{RSS}_1 + \mathrm{RSS}_2)/\left(n - 2(k+1)\right)}, \quad df_1 = k+1, \quad df_2 = n - 2(k+1), \tag{4.21}$$

where RSS is the residual sum of squares of a pooled (RSS_p) analysis and the separate analyses (RSS_1 and RSS_2). A significant test result implies that the regression models in the two groups are not identical, that at least one slope or the intercept differs between the two populations.[4]

In closing this subsection on significance tests, we would like to remind readers that these tests only indicate whether a certain hypothesis holds or does not hold with a specified, predefined level of certainty. Even more importantly, statistical significance does not tell us anything about the substantive significance of an effect. If our samples are large, even very small effects will become statistically significant but often they would be not very meaningful from a substantive viewpoint. Take, for example, a literacy test with mean 250 and standard deviation 50 points. Assume that males score 3 points higher than females, and that this difference is statistically significant. Should we conclude that the gender difference is important and should we advise policy-makers to act on this? Most likely this would not be very sound advice. Given that the difference between males and females amounts to less than a tenth of a standard deviation of the literacy measure we should probably focus our attention on other factors, perhaps education. The bottom line of this is that substantive importance of regression results has to be judged based on substantive criteria.

Assumptions in ordinary least squares regression

Ideally OLS regression estimators are best linear unbiased estimates (BLUE). This is the case if the data meet the assumptions on which this method is based. Analysts should be aware of these assumptions and test whether they apply. We will briefly describe the most important assumptions underlying OLS regression; for a fuller discussion of this issue, see Berry (1993) or the next chapter of this volume:

- The dependent variable has to be metric; the independent variables may be metric or coded as dummy variables or other contrasts.
- If we want to draw inferences from our data it must come from a random sample of the population of interest.
- The independent variables have to be measured without measurement error.
- None of the independent variables must be a constant or a linear combination of the other independent variables; that is, there should be no perfect multicollinearity. Technically this means the matrix X must have full rank.
- The error terms (residuals) must follow a normal distribution.
- For each value of the independent variables the variance of the error term has to be identical, $\text{var}(\varepsilon|x) = \text{const.}$; this is also referred to as a situation of homoscedasticity.
- For each combination of independent variables the expectation of the error term has to be zero, $E(\varepsilon|x) = 0$. This assumption implies that no independent variable is correlated with the error term – a situation described in econometrics as strict exogeneity.
- The aforementioned assumption implies that the model is correctly specified, that is, all relevant variables are included in the model and the model does not contain irrelevant variables. In addition, the parametrization of the model has to be correct, that is, in the given operationalization and parametrization the independent variables have to be linearly associated with the dependent variable.

The OLS estimates of the regression coefficients and their standard errors are BLUE if these assumptions are met. However, in real-world applications of OLS the assumptions listed above will only be met to a certain degree, with the effect that the OLS estimates will deviate from the ideal of being unbiased and efficient (have minimum variance). To assess the quality of a regression model it is important to be aware of the consequences of deviations from the assumptions. Multicollinearity, heteroscedasticity and non-normal residuals lead to biased standard errors of regression estimates which lead to incorrect significance tests and confidence intervals. The estimates of the regression coefficients (intercept and slopes), however, remain unbiased. Deviations from the other assumptions have an even stronger effect on results. In this case not only the standard errors but also the regression coefficients are biased.

We would like to briefly show why this happens in the case of misspecification. Let us assume the true model is

$$y = \beta_0 + \beta_1 x_1 + \beta_2 x_2 + \cdots + \beta_k x_k + \beta_m x_m + \varepsilon.$$

Now let us assume we were not aware of the factor x_m and we specify the model as

$$y = \beta_0 + \beta_1 x_1 + \beta_2 x_2 + \cdots + \beta_k x_k + \varepsilon^*,$$

without x_m. The error term of this model will be identical with the error term of the true model plus the variable x_m, that is, $\varepsilon* = \beta_m x_m + \varepsilon$. If x_m is correlated with at least one of the other independent variables – and this will be the case in almost all real situations – the error term of the misspecified model is correlated with independent variables. Thus, the assumption of strict exogeneity is violated and the estimates for the regression coefficients will be biased. If we inspect equation (4.12) we see why this is the case: the estimation of regression coefficients takes the correlations between the independent variables into account. If important independent variables which are both related to the dependent and the independent variables are left out of the model the estimates are biased – a bias also referred to as *omitted variable bias*. The only way to avoid misspecification of regression models is to root them in a sound theoretical foundation and use adequate operationalizations of the concepts of interest.

Another often encountered problem is unreliable measurement of independent variables. Measurement error of independent variables, be it systematic or random measurement error, leads to biased estimates of the regression coefficients and their standard errors (Cohen et al., 2003, p. 119). The larger the measurement error of a variable x_k, the more the regression coefficient β_k underestimates the true effect of x_k on y, an effect also known as attenuation. Therefore, we should strive to improve our measurement instruments and scaling techniques. If we have several indicators for the concept of interest we could consider using structural equation modeling instead of OLS regression (see Kline, 2010). If we only have a single indicator for a concept, we might be able to obtain a reliability estimate through the web-based Survey Quality Prediction program maintained by the Research and Expertise Centre for Survey Methodology at Universitat Pompeu Fabra under the guidance of Willem Saris (see `http://sqp.upf.edu/`). The estimated reliability could be used to estimate the amount of attenuation of regression coefficients.

The next chapter of this volume presents a much more comprehensive discussion of the assumptions underlying OLS regression as well as ways to test to what extent they are met.

Interpretation of regression results

Once we have established that a regression model does 'explain' the dependent variable at least partly (i.e. is statistically significant), we still are faced with interpreting the regression results from a substantive point of view. Let us first focus on interpreting the regression coefficients or slopes β_j (these coefficients are often referred to as unstandardized regression coefficients, in contrast to standardized coefficients which we will discuss in the next section). Frequently one reads that these coefficients indicate the unit change in the dependent variable if the independent variable is increased by one unit. So if $\beta_1 = 0.5$ a one unit increase in x_1 would result in an increase of half a unit of y. In most practical instances this interpretation will be incorrect. In particular, if we use cross-sectional data to estimate a regression model we should abstain from interpreting the results in a dynamic way. A correct interpretation would be that the expectation for y is 0.5 units higher for those with $x_1 = a + 1$ compared to those with $x_1 = a$. Additionally, if β_j is estimated in a regression model with more than one independent variable then this coefficient is conditional on the other predictors. In other words, the estimate is an attempt to model a situation in which the other independent variables are held constant. If we can assume that all relevant variables are included in the model and parametrized correctly, x_i is conditionally uncorrelated with ε and β_i can be interpreted as causal effect (for a more thorough discussion see below).

Let us have a closer look at the following model (numbers in parentheses are standard errors of estimates):

$$\text{Control Immigration} = 4 + \underset{(0.01)}{0.05}\,\text{Age} - \underset{(0.5)}{0.5}\,\text{Female} - \underset{(0.00004)}{0.0001}\,\text{Income} + \varepsilon, \quad R^2 = 0.04. \tag{4.22}$$

Both age and income have significant effects on the attitudes towards immigration (the coefficients are more than twice their standard errors). In contrast, being female rather than male has no significant effect. For each age group the expected value on the attitude scale is 0.05 points higher than for the group one year younger, irrespective of sex and income. Similarly, each additional dollar decreases the opposition towards uncontrolled immigration by a small amount (0.0001 points) controlling for age and sex. This interpretation draws attention to three crucial characteristics of multiple regression. First, the regression coefficients reflect the estimated effect of one variable, controlling for all other variables in the model. In our case this means that the effect of age is estimated by taking sex and income into account. Second, only linear effects are modeled and correctly reflected in regression estimates. In our example this means we assume that there is the same difference in attitudes towards immigration between a 21- and a 20-year-old person as between an 81- and an 80-year-old person, namely 0.05 units. If we had reason to believe that the relationship between a predictor and a dependent variable is non-linear we could still model this in the framework of linear regression. However, we would have to transform the variable in question in such a way that the regression model is linear with respect to the transformed variable. For example, if we assume that the increase in opposition to immigration gets smaller with increasing age we could use the logarithm of age instead of age in our model. Third, our model implies that the predictors' effects are additive and do not depend on each other. Again, if we had reason to believe that the effect of one independent variable depends on levels of another independent variable we could model this in the framework of linear regression by incorporating interaction effects. Because Chapter 6 exclusively discusses non-linear and non-additive effects in linear regression we do not discuss these issues here any further.

The interpretation of the effects of 0–1 coded binary variables is similar to the interpretation of effects of continuous variables. The coefficient reported above for being female implies that the conditional expected value of y is 0.8 units higher for females than for males, controlling for age and income. But as we have seen, this difference is not statistically significant.

The interpretation of regression results can often be facilitated by changing the scale of the independent variable. Assume we had measured income not in dollars but in tens of thousands of dollars. Then the regression coefficient would change from 0.0001 to 1, implying that an income difference of \$10,000 is associated with an expected difference of one unit on the dependent variable, controlling for age and sex. Another example would be age. If we divide age in years by 10, thus measuring age in decades, the above age effect would change from 0.05 to 0.5, indicating that people who are 10 years apart are expected to be half a scale point apart on the immigration scale.

How can we interpret the intercept β_0? This is the expectation for the dependent variable if all of the independent variables x_j are zero. For the above model we could say that for men (female $= 0$) who are zero years old and who have zero income we expect a value of 4 on the attitude scale. Here and in most other cases this information is of no interest. It could even be misleading because $x_1 \cdots x_k = 0$ most often lies outside of the window we observe. Here, for example, we would assume that the observations were restricted to the adult population. Also, it is safe to assume that newborns do not have attitude towards immigrants. One way to avoid misleading interpretations of the intercept is to center all (metric) variables on their mean (or alternatively on some other meaningful value). Then the intercept reflects the expectation for the 'average' person, a figure which might be of substantive interest.

Standardized regression coefficients

The size of the regression coefficients we have reported and interpreted so far depends on the units used to measure the independent and dependent variable. As long as variables have 'natural' or intuitive measurement units (e.g. age in years or income in dollars) the interpretation of coefficients is straightforward. However, many measures in the social sciences are based on arbitrary units derived from answers to rating scales of various types and lengths. Because of the arbitrariness of the units of such measures, their regression coefficients are not very informative. A related problem arises if we are interested in the relative effect of variables measured on different scales. Reconsider the example given above where we found that the effect of age was 0.05 and the effect of income was 0.0001 (see equation (4.22)). Do these coefficients imply that attitudes towards immigrants are more affected by age than by income? Obviously not, because – as we have seen – the size of the unstandardized regression coefficient depends on the units used to measure the variables. As mentioned above, the slope for income would have been 1 if we had measured income in tens of thousands of dollars.

Arbitrary units of measurement and the assessment of relative importance of predictors are typically addressed by interpreting standardized regression coefficients. These coefficients are computed by multiplying the unstandardized coefficient by the ratio of the standard deviations of the independent and dependent variable,[5] that is,

$$\beta_j^s = \beta_j \frac{\sigma_{xj}}{\sigma_y}. \tag{4.23}$$

Expressed in standard deviations of y, these standardized coefficients tell us how much two groups are expected to differ with respect to y if they differ by one standard deviation on x_j. In this parametrization all effects are expressed in standard deviations of y and x_j. An increase of one standard deviation of x_j results in a change of y by β_j^s standard deviations.

In the social sciences it has long been common practice to report and interpret only standardized regression coefficients. However, their use has been criticized for several reasons (cf. Bring, 1994). One problem is that standardized regression coefficients reflect not only effect sizes but also variation of the variables. Therefore, standardized coefficients may vary between samples or populations just because the variables of interest have different variances, even though the effects of x_j on y are identical. Assume we want to compare the effect of income on attitudes for men and women. If the variation of income differs between men and women this will influence the standardized coefficients and so it will be impossible to know if the difference in standardized coefficients is due to differences in the effect of income or the variation of income in the two groups. Consequently, we should use unstandardized measures in comparative analysis.

But even the comparison of standardized coefficients within the same model has been criticized. To see why, let us inspect the following example. Let us assume that

$$\text{Control Immigration} = \beta_0 + \beta_1 \text{Age} + \beta_2 \text{Income} + \varepsilon$$

is the model of interest. As we have seen above, β_1 reflects the effect of age, holding income constant. According to equation (4.23) the standardized effect of age is determined by multiplying β_1 by the standard deviation of age (σ_{x_1}). This may be seen, as Bring (1994) has claimed, as inconsistent because β_1 is an estimate conditional on other variables (controlling for ...) in the model while σ_{x_1} is the unconditional standard deviation referring to the entire population unadjusted for other measures. In essence, the slope and standard deviation refer to different populations. As a solution Bring (1994, p. 211) suggests using the partial standard deviation of x_j averaged over the groups formed by the independent variables.

A further critique of standardized coefficients is that they only reflect the relative contribution of an independent variable to R^2, the explained variance, if the independent variables are all uncorrelated. In this very limited case R^2 is identical to the sum of squared correlation coefficients between independent variables and dependent variable which are in this special case identical to the standardized coefficients. Therefore, only if all independent variables are unrelated to each other do the standardized regression coefficients reflect the relative contribution of each variable to the explained variance. This special case, however, never occurs when working with real data. And if it did occur we would not need to use multiple regression because there would be no need to 'control' the effects of independent variables for other independent variables. With correlated independent variables R^2 can be decomposed as

$$R^2 = \sum_{j=1}^{p} \beta_j^{s\,2} + 2\sum_{j=1}^{p-1} \sum_{k=j+1}^{p} \beta_j^s \beta_k^s \rho_{jk}, \tag{4.24}$$

with β_j^s and β_k^s representing the standardized effect of x_j and x_k, respectively, and ρ_{jk} indicating the correlation between x_j and x_k (cf. Grömping, 2007, p. 140). Bring (1994) has suggested measuring the relative importance of independent variables by multiplying the correlation between independent and dependent variable by the unstandardized regression coefficient. This measure has the advantage of summing to R^2 over all independent variables. That is, R^2 can be partitioned as

$$R^2 = \sum_{j=1}^{k} \beta_j \rho_{jy}. \tag{4.25}$$

Though the products $\beta_j \rho_{jy}$ sum to the explained variance and thus can be considered as indicating relative importance of the predictors, there is a problem with this interpretation when the signs of the two factors differ. In this case the product has a negative sign implying that the independent variable contributes not to the explained but rather to the unexplained variance.

In the meantime several measures to reflect relative importance of predictors have been proposed in an attempt to overcome the shortcomings of standardized coefficients. Chao et al. (2008) compare six of these proposals to capture the relative importance of predictors in multiple linear regression. They base their comparison on three criteria: (a) (squared) coefficients of relative importance should sum to R^2; (b) coefficients of relative importance should never be negative; (c) coefficients of relative importance should not depend on the order in which predictors are entered into a regression equation. Only two of the six measures examined by Chao et al. (2008) meet all three criteria: a proposal by Budescu (1993) and one by Johnson (2000). Because the coefficient proposed by the former is very cumbersome to compute and because both approaches result in very similar estimates, Chao et al. (2008) recommend using the method introduced by Johnson (2000). Therefore, we will only discuss this latter measure briefly.

Suppose we have a regression model with k predictors. Using principal components analysis, Johnson (2000) suggests extracting k orthogonal factors z_m and rotating them so as to minimize the sum of squared differences between the rotated factors and the variables x_j. Then the dependent variable of interest is regressed on the rotated factors z_m. Because the factors are orthogonal the squared standardized coefficients $\beta^s_{z_m}$ sum to R^2. In a final step we have to determine the importance of the original variables x_j by

$$\beta^\dagger_{x_j} = \sum_{m=1}^{k} \lambda_{jm} \beta^s_{z_m},$$

with λ_m denoting the correlation or loading between x_j and z_m.

A more readily available alternative to determine relative importance of predictors is the t-value of the typically reported two-sided test of the slopes (see equation (4.15)). This test statistic can also be represented by

$$t_1 = \sqrt{\frac{R^2_{1,2,3,\ldots,k} - R^2_{2,3,\ldots,k}}{(1 - R^2_{1,2,3,\ldots,k})/(n-k-1)}}, \tag{4.26}$$

implying that the t-value is a direct function of the increase in R^2 produced by entering the variable of interest into a model containing all other independent variables (for more details, see Bring, 1994, p. 213). Hence, comparing t-values of the same model allows us to order the independent variables with respect to their importance relative to the dependent variables.

As we have seen, users of multiple regression have several choices to determine the relative importance of independent variables in a regression model. Standardized coefficients β^s are readily available in most statistical software packages but may be problematic. Alternatives like the one developed by Johnson (2000) and presented above seem to better capture relative importance as contributions to R^2 but are not easy to obtain. We also do not know how stable these coefficients are between samples. We do know, however, that comparing standardized coefficients between samples or populations can lead to incorrect conclusions because they do not only depend on the effect of independent variable on the outcome. Rather, they are also affected by the variance of both variables in the two groups. Our recommendation therefore is to always report unstandardized coefficients, the t-value which indicates if a predictor is statistically significant and also reflects relative importance. Additionally, standardized coefficients may be useful but should be interpreted with caution. In the end the 'importance' of a predictor has to be determined from a substantive point of view.

MODELING ATTITUDES TOWARDS IMMIGRATION: AN EXAMPLE ANALYSIS

In this section we present a sample analysis. In contrast to many textbooks we will use real data and we will partly replicate a research paper by Green (2009). In her paper Green studies the determinants of support for different criteria to restrict immigration. Drawing on data from the first round of the European Social Survey, she studies endorsement of ascribed and acquired characteristics as criteria for granting immigration. The central individual level predictors Green studies are perceived threat and social status of host country members. In brief, she hypothesizes that those perceiving negative consequences of immigration for their own life chances (i.e. feel threatened) will oppose immigration more and be more in favor of restricting immigration. Similarly, those in lower social strata will experience more competition for jobs or affordable housing by immigrants than people in high social positions. Therefore, social status should be negatively related to accepting unconditional immigration. To test her hypotheses Green focuses on nationals who do not claim to belong to a minority group within their country (Green, 2009, p. 47).

Our replication of Green's analysis is limited in a number of ways. First, we only study attitudes with respect to immigration criteria which can be acquired (e.g. education). Second, our analysis focuses on only two countries, Switzerland and Germany, and thus, we do not study country-level predictors. Third, we only use a subset of the individual level predictors employed by Green. Our analysis is based on version 6.1 of the data from the first round of the European Social Survey.

In the following analysis we study the importance given to education, proficiency in national language, having work skills and being committed to the way of life in the host country for deciding about immigration. The answers to these four items are combined into an additive index serving as our dependent variable (see the appendix to this chapter for a detailed description of items). Higher values reflect greater importance of these criteria and can be interpreted as being in favor of higher restrictions on immigration. Perceived threat was measured by seven items which we again combine into an additive index with higher values reflecting higher levels of threat (see appendix for items). Social status was measured by education in years. To capture political orientation we follow Green and use the answers to the 11-point left–right scale. Because this measure contains a substantial number of missing values, Green categorizes this variable into left orientation (values 0 to 3), middle orientation (values 4 to 6), right orientation (values 7 to 10) and missing information on political orientation. In the following analysis we use the middle category as reference. Additionally, we control for sex and age.

Our analysis is focused on a comparison between Switzerland and Germany, two countries differing substantially with respect to immigration. In 2002, the year round 1 of the European Social Survey was carried out, 20% of the population in Switzerland did not hold Swiss citizenship, while in Germany only 9% of the population were foreigners. In the same year Switzerland welcomed over 125,000 new long-term immigrants (1.7% of its population), and Germany welcomed almost 850,000 new immigrants (amounting to 1% of the population). With these figures Germany is slightly above the European average whereas Switzerland (together with Luxembourg) is the country having the largest non-national and foreign-born population.

Table 4.1 gives an overview of the variables we will use in our analysis and their distribution in Switzerland and Germany. According to this table, Swiss people place less importance on acquired characteristics for immigration and perceive slightly less threat than Germans. Also Swiss people are on average a little older, a little less educated and a little more oriented towards the political right than Germans. To model the attitudes towards immigration in Switzerland and Germany we proceed in two steps. First, we estimate a model including only sex, age, education and political orientation. In a second step we add perceived threat to the model.

From the left panel of Table 4.2 we see that, in accordance with the social status hypothesis, higher educated people in Switzerland and Germany place less importance on acquired immigration criteria than their less educated compatriots. Although both effects are statistically significant, the effect is much larger in Germany than in Switzerland. In Germany 10 more years of schooling are associated with almost one point less on the importance scale (−0.91), while in Switzerland the same educational difference is only associated with a quarter point change (−0.26). As one might expect, older persons and persons who identify themselves as politically right-wing have more reservations towards unconditional immigration in both countries. In neither country do we observe strong sex differences, so we may conclude that attitudes of men and women with regard to immigration do not differ.

Before we inspect the results of our second model, we would like to draw readers' attention to two issues related to political orientation. As pointed out above, we followed Green in treating this variable as categorical and in using a separate category for the missing values. Because this leads to three variables and three regression coefficients in the model they do not allow us to say anything about the size of the effect political orientation has. An alternative approach would be to estimate a model without the dummies for political orientation and compare it with the

Table 4.1 Descriptive statistics

	Switzerland ($n = 1516$)				Germany ($n = 2349$)			
	min	max	mean	sd	min	max	mean	sd
Restrict immigration	0	10	6.43	1.96	0	10	7.38	1.91
Female	0	1	0.51	0.50	0	1	0.50	0.50
Age	15	103	48.90	16.95	15	93	47.39	17.37
Education	0	31	10.85	3.36	0	30	13.09	3.27
Left	0	1	0.20	0.40	0	1	0.26	0.44
Middle	0	1	0.56	0.50	0	1	0.56	0.50
Right	0	1	0.19	0.39	0	1	0.13	0.34
LR missing	0	1	0.05	0.21	0	1	0.05	0.21
Perceived threat	0.06	0.97	0.47	0.14	0.07	1	0.52	0.16

Data: European Social Survey round 1, version 6.1. Only nationals not belonging to a minority; listwise deletion of missing cases, unweighted.

Table 4.2 Model 1 to explain attitude towards immigration

	Switzerland				Germany			
	$\hat{\beta}$	$s_{\hat{\beta}}$	t	β^s	$\hat{\beta}$	$s_{\hat{\beta}}$	t	β^s
Constant	6.419	0.083	77.2	–	7.522	0.061	123.0	–
Female	−0.019	0.100	−0.2	−0.005	−0.110	0.074	−1.5	−0.029
Age[a]	0.015	0.003	5.2	0.132	0.019	0.002	8.6	0.171
Education[a]	−0.026	0.016	−1.7	−0.044	−0.091	0.011	−8.0	−0.156
Left[b]	−0.619	0.130	−4.8	−0.127	−0.759	0.091	−8.4	−0.174
Right[b]	0.596	0.132	4.5	0.119	0.486	0.109	4.5	0.087
LR missing[b]	0.169	0.239	0.7	0.018	0.015	0.178	0.1	0.002
R^2		.065				.125		
R^2_{adj}		.062				.123		
$F(df_1; df_2)$		17.61	(6; 1509)			55.63	(6; 2342)	

[a] Centered on the country-specific mean.
[b] Reference category political orientation 'middle'.
Data: European Social Survey round 1, version 6.1. Only nationals not belonging to a minority; listwise deletion of missing cases, weighted by design weight.

model displayed in Table 4.2 in terms of the increase in R^2. If we compare two such models for Germany we see that adding political orientation to the model increases the explained variance by 4.2 percentage points. By applying the F test given in equation (4.20) to the two models just estimated, we can test whether political orientation is statistically significant – which it is.

A final remark on the variable political orientation: As the reported results show, there is no statistically significant difference between those without a valid response to the left–right question and those placing themselves in the middle of the political spectrum. Thus, at least with respect to the dependent variable studied here, we find no difference between these two groups and we might come to the conclusion that substituting the missing cases on this variable by its mean and then treating this variable as continuous would make our model more parsimonious without distorting the results for left–right placement.

This brings us to our second model, in which we add perceived threat as a predictor. Again, we report separate analyses for Switzerland and Germany in Table 4.3. In both countries perceived threat has a strong effect. In fact, based on the t-value and the standardized coefficient, perceived

Table 4.3 Model 2 to explain attitude towards immigration

	Switzerland				Germany			
	$\hat{\beta}$	$s_{\hat{\beta}}$	t	β^s	$\hat{\beta}$	$s_{\hat{\beta}}$	t	β^s
Constant	6.378	0.081	78.5		7.504	0.058	128.3	
Female	−0.012	0.097	−0.1	−0.003	−0.044	0.071	−0.6	−0.012
Age[a]	0.014	0.003	4.9	0.122	0.016	0.002	7.6	0.145
Education[a]	0.007	0.016	0.5	0.012	−0.033	0.012	−2.9	−0.057
Left[b]	−0.466	0.128	−3.6	−0.096	−0.555	0.088	−6.3	−0.127
Right[b]	0.558	0.129	4.3	0.112	0.323	0.105	3.1	0.058
LR missing[b]	0.017	0.234	0.1	0.002	−0.193	0.171	−1.1	−0.021
Perceived threat[a]	3.186	0.365	8.7	0.226	3.745	0.252	14.9	0.310
R^2			.110				.202	
R^2_{adj}			.106				.198	
$F(df_1; df_2)$		26.70	(7; 1508)			83.83	(7; 2341)	

[a] Centered on the country-specific mean.
[b] Reference category political orientation 'middle'.
Data: European Social Survey round 1, version 6.1. Only nationals not belonging to a minority; listwise deletion of missing cases, weighted by design weight.

threat is the most important predictor in our model in both countries. The difference between those perceiving no threat and those perceiving the maximum level of threat is 3.2 (Switzerland) and 3.7 (Germany) points on the importance scale used to measure attitudes towards immigration. Consequently, adding this perceived threat to the model substantially increases the amount of explained variance in both countries; in Switzerland by 4.5, in Germany by 7.7 points, leading to $R^2_{\text{CH}} = 0.11$ and $R^2_{\text{DE}} = 0.20$, respectively. Thus, it seems that the model fits the German data much better than the Swiss. Whether or not this difference is statistically significant can be tested with the Chow test given in equation (4.21). The result of this test clearly indicates that the model indeed does not fit the Swiss and German data equally well.

When comparing the effects of the other predictors in the model to their effects in model 1, we see that some of them are strongly affected when perceived threat is entered into the model. In particular, the regression coefficients for education and political orientation show a quite strong reduction in absolute size. This implies that some of the explained variance attributed to these variables in model 1 actually has to be attributed to perceived threat. Indeed, model 2 implies that education in Switzerland does not seem to be directly related to attitudes towards immigrants once we control for perceived threat. This implies that the social status hypothesis does not hold for Switzerland. One possible explanation for this result may be that many foreigners in Switzerland are highly qualified and so competition between the indigenous and migrant population for jobs, dwellings and so on may not be concentrated in lower status groups in Switzerland.

Finally, we may be interested in testing whether the predictors we examined differ in their effects between the two countries. To do so we applied the model given in equation (4.18). That is, we estimated a model based on a combined sample, including in the equation a dummy variable indicating the country and interaction terms for this variable with all predictors. This analysis shows that the only independent variable with significantly different effects in both countries is education. All the other variables seem to have identical effects in Switzerland and Germany.

Here we end our example analysis. If we wanted to publish the results of our analysis we should go further and test whether the assumptions of OLS regression are reasonably well met by our data. We do not have the space to do this here, but refer readers to the next two chapters in which these assumptions, diagnostic tools and possible remedies are extensively discussed.

PROBLEMS AND REMEDIES FOR CAUSAL INFERENCE BASED ON OLS REGRESSION

In recent years, causal analysis in the social sciences has increasingly developed consensus on applying the potential outcome model, also called the counterfactual model of causality (e.g. Rosenbaum and Rubin, 1985). This framework explicates an intra-individual concept of causal analysis. The simplest setup assumes a binary causal state with treatment and control (untreated) conditions. The causal effect of the treatment D on an outcome y then can be defined as

$$\Delta_i = y_i^1 - y_i^0 , \tag{4.27}$$

with i as a person index and 0/1 denoting control and treatment state. Identification of the treatment effect is complicated by the fact that a person never will be in both states at the same time, and Δ_i hence cannot be observed. In presence of panel data, identification of the average treatment effect oftentimes is attempted using fixed-effects panel regression or related methods (see Chapter 15 of this volume). With only cross-sections being available, groups of persons which are observed under different causal states have to be compared. This, of course, imposes massive problems on the researcher if a controlled and fully randomized experiment cannot be conducted. Notably, unbiased identification of the treatment effect is possible only if the potential outcome is unconditionally or at least conditionally independent of treatment assignment:

$$(y^0, y^1) \perp\!\!\!\perp D \tag{4.28}$$

or

$$(y^0, y^1) \perp\!\!\!\perp D|z . \tag{4.29}$$

The former is a very strong assumption, and valid only in absence of selection into treatment groups, i. e. in randomized trials. The conditional independence assumption (equation 4.29) is somewhat weaker as independence only needs to hold after controlling a number of covariates z.

OLS regression has developed a bad reputation in causal inference (see e. g. Morgan and Winship, 2007, section 1.1.2). This is mostly due to the misuse of explorative regression models for causal conclusions and, more generally, the focus on "fully" explaining the variance of a dependent variable rather than identifying treatment effects of a specific manipulation (e. g. Blalock, 1964).[6] However, as proponents of the potential outcome model have pointed out, OLS regression can play a role in estimating causal effects if applied sensibly. To show why and in how far this is the case we will briefly review the main obstacles of drawing causal inference from observational data and how these are addressed by OLS regression for cross-sectional data (but see Chapter 15 of this volume for causal inference based on longitudinal data).

In our opinion, the most severe problem in using OLS regression for causal inference with non-experimental data stems from self-selection or policy endogeneity, both of which result in violating the unconditional independence assumption. For example, if we study the effect of job interview training on earnings, a self-selection bias may occur due to higher-skilled persons participating in the training classes with a higher probability. Policy endogeneity occurs when the organizer of a program targets a specific sub-group that is expected to show the strongest effects (e. g. persons with an academic education). As mentioned before, selection bias is avoided in experimental research by randomly assigning individuals to control and treatment groups thereby ensuring that the state of treatment is the only systematic difference between the two groups; the two groups are unconditionally independent. However, if we know all relevant factors in which the "treated" and "non-treated" differ, we can adjust statistically for these factors to achieve conditional independence. In an OLS regression we can assume conditional independence provided that all selection variables are included in the regression and the parameterization of the model is correct.[7] Under these circumstances the regression coefficient of interest can be interpreted as a causal effect (cf. Angrist and Pischke, 2009, pp. 51–59; Gelman and Hill, 2007, p. 169). We must also check if there is sufficient overlap of covariates across treatment groups. This means that confounders should not be highly correlated with the treatment variable (for a more in-depth discussion see Gelman and Hill, 2007, Chapter 10.1). Furthermore, we must assume treatment homogeneity or monotonicity for a meaningful interpretation of the regression coefficient as average treatment effect (see Humphreys, 2009).

In a way, the reservations against regression may be seen as stemming from a different epistemological background of those having these reservations and a more stringent focus on research design in the potential outcomes framework – rather than from inherent statistical shortcomings of the regression approach. Applied in a careful and well-conceived way, regression can be used as a statistical tool for the estimation of treatment effects (see also Angrist and Pischke, 2009, Chapter 3; Morgan and Winship, 2007, pp. 123ff). The fundamental difference to the more direct matching approach is that the conditional independence problem is tackled by adjusting for covariates rather than by balancing, and that a different set of assumptions – such as correct parameterization – has to be met for the identification of causal effects.

That said, there will be situations in which cross-sectional OLS models are simply not suited for causal inference. In a situation with weak overlap between treatment- and control group or when the functional form of effects is unknown, propensity score matching may be the method of choice. Matching estimators have been suggested as the most direct approach to solving the problem of balancing treatment and control groups and meeting the assumption of conditional independence. As a semi-parametric model, matching rests on less assumptions than the regression approach. Additionally it allows to at least estimating a local treatment effect when overlap is weak, provided that the dataset is large enough to identify a sufficient number of matches (for a more detailed exposition of propensity score matching and its identifying assumptions see Chapter 12 of this volume). However, there may be a total lack of overlap of covariates that originates from a variable that was the basis for assigning cases to control and treatment group. Imagine we are interested in studying the long-term career effects of scholarships. Imagine further that scholarships are awarded to all students scoring in the top 80 % of an aptitude test. In this case aptitude and being granted a scholarship are perfectly related and neither OLS nor matching are suitable methods for creating conditional independence. Alternatively, we could concentrate on only those students just below and above the critical threshold for obtaining a scholarship. Taking into account measurement uncertainty we can assume that these students have equal aptitude and only differ with respect to the scholarship. This is the basic idea behind the regression discontinuity approach presented together with a more elaborate description of the example in Chapter 14 of this volume. In many cases we will also be unable to observe all relevant covariates. As we have seen in above omitting a relevant variable from the regression equation leads to biased estimates, the so-called omitted variable bias. In this case an approach known as instrumental variable regression (IV regression) may help if we have a good proxy or instrument for the omitted variable (see Chapter 13 of this volume). In the presence of panel data, we could use fixed effects regression to control for unobserved time-constant covariates and obtain unbiased estimates (see Chapter 15 of this volume).

CAVEATS AND FREQUENT ERRORS

We can only apply statistical methods and interpret their results correctly if we have at least some basic understanding of their general purpose and the assumptions on which they are built. Linear regression is no exception. One of the major questions we should ask is whether we have specified our model correctly. Were we able to include all relevant predictors? Are the predictors linearly related to the outcome variable? Are the effects of the independent variables additive or does the effect of one variable depend on the level of another variable? To answer these questions satisfactorily we first have to rely on sound theory about the substantive matter we are investigating. Second, as Chapters 5, 6 and 10 of this volume show, we can and should test the assumptions of linearity and additivity. Applying sound theory and rigorous testing of assumptions are important because – as we have seen – misspecification of a model leads to the violation of strict exogeneity, that is, violates the assumption that residuals and predictors should not be correlated. This in turn leads to biased estimates of regression coefficients and their standard errors.

Another practical issue we must look at in every analysis is the sample size. Often we would run a regression analysis with many independent variables using listwise deletion of missing values – in many statistical programs this is the default and we might not even make a deliberate decision about this choice. When our model contains many predictors or at least one predictor with many missing values, we can end up estimating our model on a relatively small, selective subsample. Therefore, we should monitor sample size on which our results are based at all stages of the analysis.

Further, we should be aware of the difference between statistical significance and substantive importance. The assertion that a given coefficient is statistically significant does not tell us anything about its substantive importance. As we have seen, it might even be problematic to rely on standardized regression coefficients for this matter. Instead, we have to make well-founded judgments based, for example, on a comparison with other effects or on the benefits/cost of (changing) the effect.

In the examples we presented in this chapter we relied on cross-sectional data. As we have pointed out, the usual interpretation of regression slopes as reflecting the changes in the dependent variable if the independent variable is increased by one unit is not valid in this situation. Instead, we should say that for someone having a value of $x_j = a+1$ the conditional expectation for the dependent variable is β_j units higher than for someone with $x_j = a$. If, for example, the effect of one additional year of education on monthly earnings is $50, then those having 15 years of education are expected to earn $150 more than those with 12 years of education, all else being equal – that is, controlling for the other independent variables in the model.

We also should remind ourselves that results from cross-sectional analysis should not be used for making predictions. Suppose we decided to increase the earning potential of a person by giving him or her one more year of education. Can we hope that this person's earnings will increase by $50? Probably not, because many factors which we might not have controlled in our model may lead to higher earnings and may also be responsible for staying in education longer, for example, cognitive abilities and endurance. In contrast to cross-sectional data analysis, these unobserved, time-constant factors can be controlled for in the framework of panel regression, a method discussed in Chapter 15 of this volume (see also Gelman and Hill, 2007, Chapter 9).

Another danger when interpreting regression results from cross-sectional surveys is what may be called the individualistic fallacy. Suppose again that we have found earnings to rise with increased education. Suppose further that we publicize this result widely and encourage people to obtain more education so as to be able to earn more. However, if everyone increases their level of education, gains in earnings will most likely diminish strongly because the overall situation has completely changed (see Boudon, 1974, for a theoretical and empirical analysis of this phenomenon).

FURTHER READING

Linear regression is covered by almost every introductory textbook in statistics. In addition, there are countless monographs dealing with regression techniques. Therefore, it is neither easy nor particularly important to give advice on further reading. That said, we want to recommend some of the books we like and have profited from. Gelman and Hill (2007) give an excellent introduction to regression analysis. An easy-to-understand introduction to the assumptions underlying linear regression is presented by Berry (1993). Fox (2008) offers a comprehensive overview of linear regression and more general regression models, and Fox and Weisberg (2011) show how these models can be estimated with the R package. A mathematically precise and in-depth coverage of regression models can be found in Wooldridge (2009).

APPENDIX

This appendix lists the items we used to construct the indices reflecting support for restrictive immigration and perceived threat.

Criteria for immigration

Please tell me how important you think each of these things should be in deciding whether someone born, brought up and living outside [country] should be able to come and live here. Please use this card. Firstly, how important should it be for them to:

... have good educational qualifications? (D10)
... be able to speak [country's official language(s)]? (D12)
... have work skills that [country] needs? (D16)
... be committed to the way of life in [country]? (D17)

extremely unimportant (0) ... extremely important (10)

Perceived threat

- Average wages and salaries are generally brought down by people coming to live and work here. (D18)

  agree strongly (1) ... disagree strongly (5)

- People who come to live and work here generally harm the economic prospects of the poor more than the rich. (D19)

  agree strongly (1) ... disagree strongly (5)

- Using this card, would you say that people who come to live here generally take jobs away from workers in [country], or generally help to create new jobs? (D25)

  take away jobs (0) ... create new jobs (10)

- Would you say it is generally bad or good for [country]'s economy that people come to live here from other countries? Please use this card. (D27)

  bad for the economy (0) . . . good for the economy (10)

- And, using this card, would you say that [country]'s cultural life is generally undermined or enriched by people coming to live here from other countries? (D28)

  cultural life undermined (0) . . . cultural life enriched (10)

- It is better for a country if almost everyone shares the same customs and traditions. (D40)

  agree strongly (1) . . . disagree strongly (5)

- It is better for a country if there are a variety of different religions. (D41)

  agree strongly (1) . . . disagree strongly (5) (reversed)

NOTES

1 In the recent discourse on causal analysis many methods for the identification of a causal effect of x on y have been proposed (see Chapters 12–15 of this volume). Nonetheless, linear regression models may as well be used for causal inference under certain circumstances (Angrist and Pischke, 2009). Some important assumptions that need to be met are described in below and in the next below and in the next chapter.

2 Alternative coding schemes are discussed at `http://statsmodels.sourceforge.net/devel/contrasts.html`.

3 The given test statistic can also be used to test one-sided hypotheses.

4 If we wanted to test for slope differences only we could include an indicator variable for the samples in the pooled analysis (see equation (4.18)).

5 We would obtain the same result when performing a regression analysis on z-standardized variables, that is, in this case the regression coefficients β_j are standardized coefficients. Because the regression always passes through the centroid of the data ($\bar{y}$, $\bar{x}_1,\ldots,\bar{x}_k$) and because the centroid of z-standardized measures is zero the 'standardized' intercept is also zero.

6 In particular, controlling for irrelevant variables and those that can be considered themselves outcomes of the outcome variable of interest (i.e. endogenous variables) do not belong into a regression equation (cf. Angrist and Pischke, 2009, pp. 64–68). This is why strong theories and carefully constructed models are of paramount importance.

7 We have to assume that we specify the functional form of the relation between covariates and outcome correctly. We can, however, fulfill this assumption by including covariates and their interactions as indicator variables; an approach also dubbed the saturated regression model.

REFERENCES

Angrist, J. D. and Pischke, J.-S. (2009). *Mostly Harmless Econometrics: An empiricist's companion*. Princeton: Princeton University Press.

Berry, W. D. (1993). *Understanding Regression Assumptions*. Newbury Park, CA: Sage.

Blalock, H. M. (1964). *Causal Inferences in Nonexperimental Research*. Chapel Hill: The University of North Carolina Press.

Boudon, R. (1974). *Education, Opportunity and Social Inequality: Changing Prospects in Western Society*. New York: Wiley.

Bring, J. (1994). How to standardize regression coefficients. *The American Statistician*, 48(3), 209–213.

Budescu, D. V. (1993). Dominance analysis: A new approach to the problem of relative importance of predictors in multiple regression. *Psychological Bulletin*, 114(3), 542–551.

Chao, Y.-C. E., Zhao, Y., Kupper, L. L. and Nylander-French, L. A. (2008). Quantifying the relative importance of predictors in multiple linear regression analyses for public health studies. *Journal of Occupational and Environmental Hygiene*, 5(8), 519–529.

Cohen, J., Cohen, P., West, S. and Aiken, L. S. (2003). *Applied Multiple Regression/Correlation Analysis for the Behavioral Sciences*. Hillsdale: Lawrence Erlbaum.
Fox, J. (2008). *Applied Regression Analysis and Generalized Linear Models*. Thousand Oaks: Sage.
Fox, J. and Weisberg, S. (2011). *An R Companion to Applied Regression*. Los Angeles: Sage.
Gelman, A. and Hill, J. (2007). *Data Analysis Using Regression and Multilevel/Hierarchical Models*. Cambridge: Cambridge University Press.
Green, E. G. T. (2009). Who can enter? a multilevel analysis on public support for immigration criteria across 20 european countries. *Group Processes and Intergroup Relations*, 12(1), 41–60.
Grömping, U. (2007). Estimators of relative importance in linear regression based on variance decomposition. *The American Statistican*, 61, 139–147.
Humphreys, M. (2009). *Bounds on least squares estimates of causal effects in the presence of heterogeneous assignment probabilities*. Working Paper, Columbia University. Last accessed 31.03.2014: `http://www.columbia.edu/~mh2245/papers1/monotonicity7.pdf`.
Johnson, J. W. (2000). A heuristic method for estimating the relative weight of predictor variables in multiple regression. *Multivariate Behavioral Research*, 35(1), 1–19.
Kline, R. B. (2010). *Principles and Practice of Structural Equation Modeling*. New York: Guilford Press.
Morgan, S. L. and Winship, C. (2007). *Counterfactuals and Causal Inference. Methods and Principles for Social Research*. Cambridge: Cambridge University Press.
Rosenbaum, P. R. and Rubin, D. B. (1985). Constructing a control group using multivariate matched sampling methods that incorporate the propensity score. *American Statistician*, 39, 33–38.
Wooldridge, J. M. (2009). *Introductory Econometrics. A Modern Approach*. o.O.: South-Western.

5

Regression analysis: Assumptions and diagnostics

Bart Meuleman, Geert Loosveldt and Viktor Emonds

INTRODUCTION

As shown in the previous chapter, ordinary least squares (OLS) regression links the values of dependent variable Y_i ($i = 1, 2, \ldots, n$) to the values of a set of independent variables X_{ik} by means of a linear function and an error term ϵ_i:

$$Y_i = \beta_0 + \beta_1 X_{i1} + \beta_2 X_{i2} + \ldots + \beta_k X_{ik} + \epsilon_i = \sum_k \beta_k X_{ik} + \epsilon_i, \qquad (5.1)$$

where k ranges from 0 to $p-1$. This model thus contains p regression parameters (namely effects of $p-1$ predictors and one intercept: X_0 equals 1 for all cases). The linear function is called the linear predictor or the structural part of the model, while the error term is the random or stochastic component of the model. In general, regression analysis can be used for two purposes: (1) to describe the data structure or (2) to make inferences about the population parameters of the regression model.

Regression analysis can only perform these functions optimally, however, if certain conditions are fulfilled. This chapter deals with the assumptions on which the OLS regression model as described above is built.[1] The exact number of assumptions (and the way in which they are categorized) varies considerably across regression textbooks. In this account, we will limit ourselves to six assumptions we believe to be the most important and widely cited ones. A first group of four classical assumptions follows from the statistical theory underlying regression analysis. The relationships between dependent and independent variables are assumed to be linear (1). Furthermore, it is required that error terms are homoscedastic (2), independent (3) and normally distributed (4). If these assumptions are met, the Gauss–Markov theorem guarantees that the OLS coefficients are the best linear unbiased estimators (BLUE). Here, 'best' means that these estimators have the smallest mean squared error. Violations of these four classical assumptions, however, are not the only factors that can hamper regression analysis. In addition,

Table 5.1 Regression output

Variable	Parameter	Standard error
Intercept	3.819***	(0.287)
Age	–0.007**	(0.003)
Female	0.003	(0.092)
Hincfel	–0.268***	(0.056)
Eduyrs	0.052***	(0.013)
Plcpvcr	0.434***	(0.025)
Adjusted R^2	.177	

* $p < 0.05$** $p < 0.01$ *** $p < 0.001$

accurate regression analysis requires that predictors are not multicollinear (5) and that influential observations are absent (6).

In the discussion of the six assumptions, we follow a similar structure. First, we indicate the role the respective assumption plays in the regression machinery, and how violations can affect conclusions. Subsequently, we discuss how violations of the assumptions can be diagnosed, and how possible violations can be remedied.

To make the diagnostics and remedies accessible for applied researchers, we provide illustrations using the fifth round of the European Social Survey (this data set can be downloaded from `http://ess.nsd.uib.no/ess/round5/download.html`). For the Belgian subsample, we examine a linear regression model with trust in the police (trstplc, ranging from 0 (no trust at all) to 10 (complete trust)) as dependent variable. Predictors in the model are age in years (age), gender (female), number of years of full-time education completed (eduyrs), subjective income in four categories (hincfel – higher scores indicate a lower subjective income), and an evaluation (from 0 to 10) of how successful the police are at preventing criminality in the respondent's country (plcpvcr). The results of this regression (see the output in Table 5.1) show that older people and those with a lower subjective income tend to have a lower trust in the police. Conversely, higher trust levels are found among the higher educated and persons who positively evaluate the ability of the police to prevent crime.

The online appendix to this chapter provides Stata code that can be used to reproduce these examples.

ASSUMPTION 1: LINEARITY

What is it?

The assumption of linearity is essential for using regression analysis as a descriptive tool. When we use the OLS regression model to describe the relationship between the dependent variable Y and a set of $p - 1$ independent variables X, we assume that Y is a linear function of X. The basic regression model is fully or completely linear, meaning that the model is linear in both its parameters and its variables. The qualifying characteristic of a model that is linear in the parameters is that a unit change in any parameter value leads to the same change in the dependent variable whatever the values of the parameters. So linearity is a characteristic of the parameters of the model (Krzanowski, 1998). The basic regression model is also linear in the variables: a unit change in one of the variables produces a constant change in the dependent variables whatever the value of the variable.

The linearity assumption can also be looked at from the perspective of the error terms. There is a perfect linear relationship between Y and X when all error terms in the model equal zero. In this situation, all observations characterized by their coordinates $(x_1, x_2, \ldots, x_k, y)$ lie on a hyperplane in a $(k+1)$-dimensional space. In bivariate regression, for example, perfect linearity means that all observations are positioned on a straight line in a two-dimensional space. When there are two independent variables, all observations should be situated on a two-dimensional surface in a three-dimensional space. In realistic research settings, however, perfect linear relationships do not occur, and the errors represent the deviations from the perfect linear model. For each covariate pattern (i.e. a specific combination of X-values), there will be several error terms ϵ_i which can be considered as the difference between the observed Y_i-values and the predicted value for each unit based on the specific values of X_{ik} in the structural part of the model. Now linearity implies that, for each covariate pattern, positive and negative errors balance each other out. Conditional on the X's, the expected value of the errors should equal zero. Or in other words, the linearity assumption stipulates that the conditional means of the dependent variable Y (i.e. given the values of X) equal the predicted values of Y.

Consequences of non-linearity

The parameters in the regression model we discuss are the least squares (OLS) estimators of the population regression parameters. These parameters are stochastic variables with a distribution. The assumption of linearity is used to determine the mean of the distribution of these stochastic variables. Only when the assumption of linearity holds will the expected value of the parameter equal the population value of the parameter. When linearity is violated, estimates of the regression parameters can be biased.

When we specify a linear regression model in which the relationship between the dependent variable Y and the independent variables X is not linear, a specification error is made. In that case, the model is not appropriate to describe the dependency between Y and X. Notice that it is always possible to calculate the regression parameters of a model when the assumptions are not fulfilled. However, in that case the estimated parameters of the regression model will be biased.

Diagnostics

To check the assumption of linearity, a statistical test and a few graphical methods can be used.

Statistical test: the lack-of-fit test

To test the linearity assumption one can partition the error sum of squares (SSE) of the estimated regression model into a 'pure' error sum of squares (SSE′) and lack-of fit sum of squares (SSLF). The 'pure' error sum of square is the variation of the observed values of Y around their conditional mean (i.e. the mean given a particular value of X). The lack-of-fit sum of squares is the variation of the conditional mean values around the prediction based on the linear model. As such, the SSLF represents deviations from linearity. The basic structure of this partitioning is: SSE $=$ SSE′ $+$ SSLF. Or, more fully elaborated:

$$\sum_i \sum_{j=1}^{n_i} (Y_{ij} - \widehat{Y}_i)^2 = \sum_i \sum_{j=1}^{n_i} (Y_{ij} - \overline{Y}_i)^2 + \sum_i n_i(\overline{Y}_i - \widehat{Y}_i)^2 . \tag{5.2}$$

Figure 5.1 illustrates this idea by means of a simple example. Imagine that the relation between X and Y is modelled as a linear function (i.e. the dashed line). In reality, however, the conditional means do not lie on a straight line, but follow a quadratic function (the solid line). Now, the distance between observed value Y_i and predicted score $\widehat{Y}_i$ (total error) can be divided into the

Table 5.2 Lack-of-fit test output

Statistic	Value	df
pN	1648	
SSLF (df)	5366.5905	(1544)
SSE′ (df)	275.2833	(98)
$F\ (df_n, df_d)$	1.2374	(1544, 98)
$p > F$	0.0873	

* $p < 0.05$ ** $p < 0.01$ *** $p < 0.001$

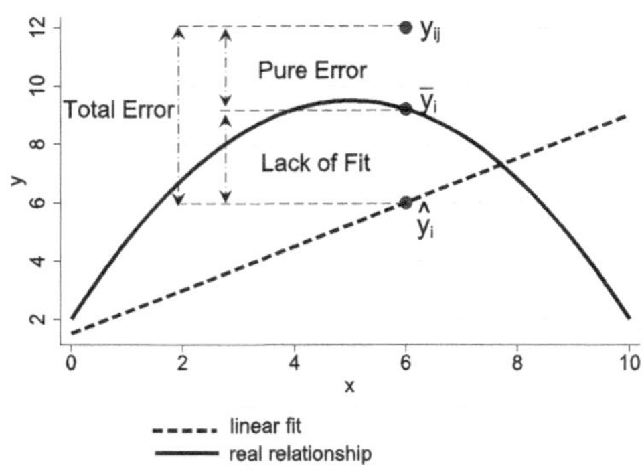

Figure 5.1 Partitioning error sum of square into 'pure' error sum of square and lack-of-fit sum of squares

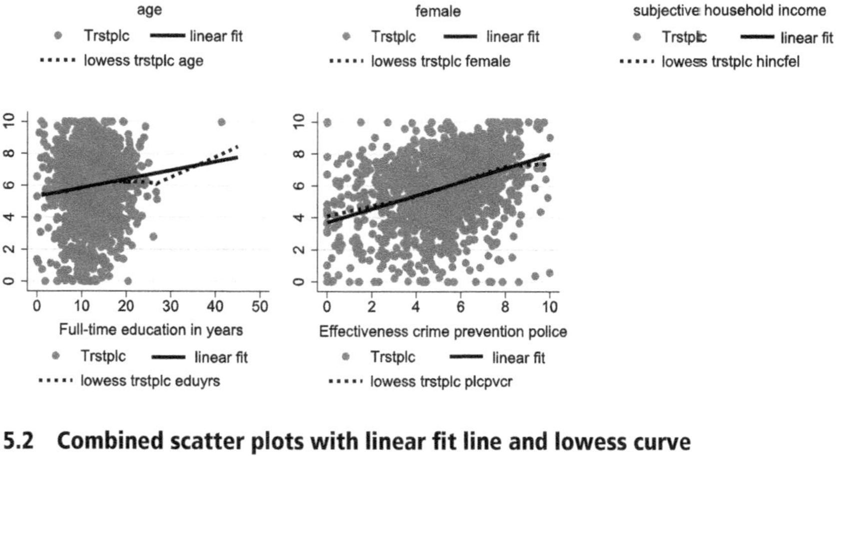

Figure 5.2 Combined scatter plots with linear fit line and lowess curve

distance between the conditional mean $\overline{Y}_i$ and the observed score $\widehat{Y}_i$ (i.e. the pure error) and the distance between the conditional mean $\overline{Y}_i$ and predicted score $\widehat{Y}_i$ (lack of fit). The larger the lack-of-fit sum of squares, the stronger the deviation from linearity.

To test the linearity assumption formally, the amount of pure and lack-of-fit error can be compared by calculating the ratio of SSLF and SSE'. The resulting test statistic follows an F-distribution.[2]

$$F = \frac{\text{SSLF}/(c-p)}{\text{SSE}'/(n-c)} \tag{5.3}$$

SSLF has $c-p$ degrees of freedom, where c is the number of actually existing covariate patterns (i.e. the number of combinations of categories of independent variables for which we have observations in the data); p is the number of regression parameters. SSE$'$ has $n-c$ degrees of freedom (with n equal to the sample size). The null hypothesis of this F-test states that SSLF is zero, implying that all error is pure error and that the relationship between X and Y is linear. The alternative hypothesis is that the relationship is not linear.

By way of illustration, we perform a lack-of-fit test for the model presented in Table 5.1. The test gives an F-value of 1.2374 (see Table 5.2). The associated p-value (0.0873) is greater than $\alpha = 0.05$, meaning that the assumption of linearity is not violated. Note that, because our model includes several variables with a large number of categories (e.g. age and eduyrs), the number of observed covariate patterns (c) is very high (1550).

It should be stressed that the F-test is an overall test. When the null hypothesis is rejected, there is evidence that the assumption of linearity is not tenable for at least one independent variable. However, we cannot precisely locate the problem. The graphical methods discussed below are useful tools for identifying the variables that are problematic in this regard.

Graphical method 1: Scatter plots with lowess curve

Graphical methods can be used as an exploratory tool to get a first idea of the relationship between the dependent and the independent variables and also to evaluate the linearity assumption. In a simple scatter plot with the dependent variable and one independent variable, the data points must show a negative or a positive linear relationship between both variables. To get a better visualization of the trend in the scatter plot, one can superimpose a lowess (locally weighted scatter plot smoother) fit line. The lowess method makes no assumption about the form of the relationship between Y and X (e.g. linear model) and produces a smooth line that follows the trend in the data. The lowess method successively calculates a predicted value for Y using a subset of cases (smoothing window) surrounding each value of X. If the lowess curve approximately follows a straight line, the linearity assumption is supported.

Figure 5.2 presents scatter plots with lowess curves for each of the independent variables in the regression model explaining trust in the police. Deviations between the lowess curve and the linear fit are minimal for all variables. As such, this graphical method confirms the lack-of-fit test. A very detailed look reveals that the association between age and trust in the police is slightly curvilinear, with a reversal of the negative age effect around age 70. However, this deviation from linearity is too small to be substantial. Furthermore, we see an outlying observation for education that strongly bends the lowess curve, and could influence the linear fit. We return to this issue when discussing outliers and influential observations later in this chapter.

Graphical method 2: Residual and partial residual plots

Partial residual plots are a second useful graphical tool to evaluate the assumption of linearity, complementing the information from the scatter plots. In a scatter plot, we get an overview of the marginal relationships between Y and X. In a multiple regression model, however, our interest lies in the partial relationship between Y and X, that is, controlling for the other independent variables. Partial residual plots can tackle this problem. Yet before we discuss the partial residual plot we briefly introduce the simple residual plot.

The residual values $e_i = Y_i - \widehat{Y}_i$ can be plotted against each independent variable X. This results in a simple residual plot for a particular independent variable. The residuals are represented on the vertical axis, while the values of the independent variable are plotted on the horizontal axis of the graph. In the graph a horizontal line is drawn where the residuals equal zero (the 0-line). This line represents the situation where there is no difference between observed and predicted values of Y. When the relationship between Y and X is properly specified, the points should be scattered randomly around this line, not showing a systematic pattern. This means that the values of X are not systematically related with positive or negative residuals; predictive values are not systematically higher or lower than the observed values for particular values of X. Once again, a lowess line can be used to visualize the trend. Linearity implies that a 'lowess line' approximately follows the 0-line.

Notice that in a multiple regression model the predicted values and, as a consequence, the residuals are determined by several independent variables. As a result residual plots do not make it possible to link deviations from linearity to a particular independent variable. A partial residual plot can solve this problem. In a partial residual plot an adjusted dependent variable is used:

$$Y_i - \sum_{k,k \neq j} \beta_k X_{ik} = \beta_j X_{ij} + \epsilon_i. \tag{5.4}$$

In the adjusted dependent variable the linear effects of all independent variables except one (X_j) are subtracted. This means that the dependent variable is corrected for the linear effects of the dependent variables, except X_j. The values of the adjusted dependent variables are called the partial residuals. These values contain two components: the linear effect of the independent variable X_j, and the residual values. For this reason, the partial residual plot is sometimes also called a 'component-plus-residual plot'. It is the partial residuals that should be linearly related to X_j. A partial residual plot is appropriate to evaluate this relationship. In this plot the y-axis is identified by the partial residuals and the x-axis by the independent variable X_j. If linearity holds, the data points in this plot should follow a straight line. Once again, one can plot a lowess curve and a linear fit line to get a clearer visualization. Deviations between the lowess curve and fit line are indicative of deviations from linearity in the partial relationship between X and Y.

Partial residual plots for the model explaining trust in the police are shown in Figure 5.3. These plots are similar to the scatter plots presented in Figure 5.2, but whereas the plots in Figure 5.2 show bivariate associations, those in Figure 5.3 show partial associations. None of the plots give evidence for violations of the linearity assumption, as there are no substantial differences between the lowess curve (the dashed line) and the linear fit line (the solid line). In this case, the conclusions from the scatter plots and partial residual plots are identical, but this need not always be the case.

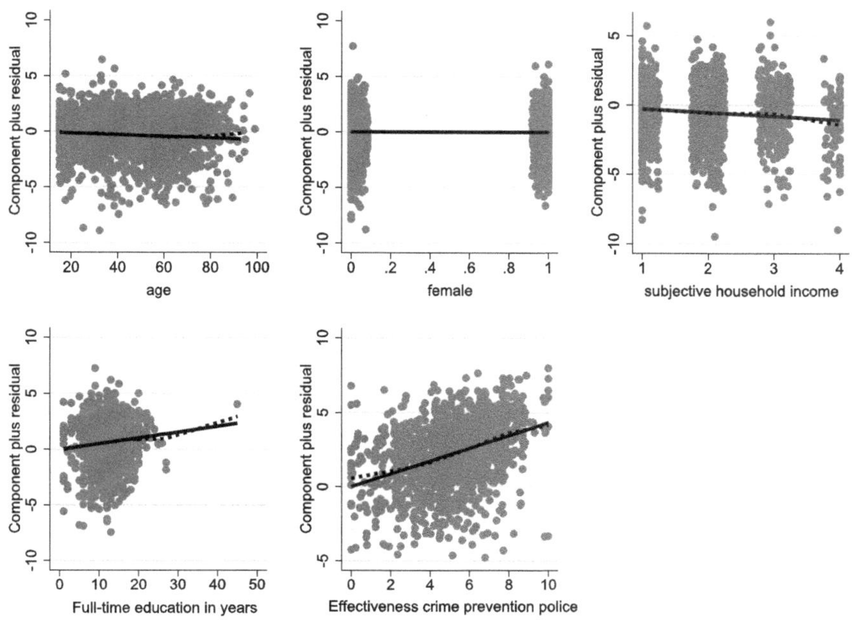

Figure 5.3 Partial residual (component-plus-residual) plots x

Remedies

As we mentioned before, violations of the assumptions of linearity can have severe consequences and may lead to biased estimates. When non-linear relationships between Y and X are detected, it is recommended to use procedures that account for the non-linearity. The literature on modelling non-linearity is extensive. In this presentation, we restrict ourselves to a brief explanation of two popular approaches: polynomial regression and piecewise regression. A more detailed account of polynomial regression can be found in Chapter 6 of this volume, while piecewise regression is discussed extensively in Chapter 14.

Both procedures can be considered as an adjustment or manipulation of the independent variable(s). The independent variable(s) are transformed in such a way that the relationship between the transformed independent variable and the dependent variable follows the structure in the data.

Polynomial regression

In a polynomial regression model, second- or higher-order terms are entered into the model. These terms are powers of the X variable and they serve as additional predictors. When the relationship between X and Y is not linear, several higher-order terms can solve the non-linearity. When, for example, the relationship between X and Y is curvilinear with one maximum, the appropriate model is a quadratic regression model (i.e. a second-order polynomial in X_1): $\widehat{Y}_i = \beta_0 + \beta_1 X_{i1} + \beta_2 X_{i1}^2$. In this model, there is only one substantive independent variable: X_1. A positive β_2 in the quadratic model indicates a model that is U-shaped (concave upwards); a negative indicates a curve that is inverted U-shaped (concave downward). One can elaborate the polynomial model with several higher-order terms. In a cubic model, for example, we introduce a third-order polynomial in X_1 and we get $\widehat{Y}_i = \beta_0 + \beta_1 X_{i1} + \beta_2 X_{i1}^2 + \beta_3 X_{i1}^3$. The number of higher-order terms needed in the model depends on the strength of the non-linear relationship between X and Y. Notice that polynomial regression models are still linear models, because they are linear in their parameters.

Piecewise regression

Sometimes the non-linear curve is not smooth, but characterized by one or several breaking points. In that case, the relationship between dependent and independent variables is different for different segments of the range of the independent variable. The segments are delimited by breaking points where the effect of the independent variable substantially changes.

Take, for example, a simple regression model with one breaking point and where the relationship between the independent and the dependent is linear before and after the breaking point. In a piecewise regression model for this situation, the range of X is divided into two segments, $X < b$ and $X \geq b$, where b is the value of X at the breaking point in the regression line. To solve the problem of non-linearity we need a separate regression equation for each segment. Therefore, we define and use a new independent variable: $X_B = 0$ if $X < b$, and $X_B = X - b$ if $X \geq b$. This new variable is entered into the following regression equation:

$$Y_i = \beta_0 + \beta_1 X_i + \beta_b X_{iB} + \epsilon_i. \tag{5.5}$$

The regression coefficient of the new independent variable X_B is now the change in the slope of the regression line used before the breaking point.

ASSUMPTION 2: HOMOSCEDASTICITY OR CONSTANT VARIANCE ASSUMPTION

What is it?

The assumption about homoscedasticity is related to the dispersion of error terms or residuals of the model. The assumption is that for each covariate pattern of X, the variance of the residuals is constant, $\text{Var}(\epsilon_i | x_{i1}, \ldots, x_{ik}) = \sigma^2$. This means that, when the residuals are uncorrelated (see the independence assumption), the variance–covariance matrix of the residuals can be written as

$$\text{Var}(\boldsymbol{\epsilon}) = \sigma^2 \mathbf{I}. \tag{5.6}$$

The conditional variances of the residuals represent the variability of the residuals around the predicted value based on a specific combination of values of independent variables X. So homoscedasticity means that all the conditional residual variances are equal: residual variances are constant regardless of the values of the independent variables X.

Heteroscedasticity conversely refers to the situation of non-constant variance of the residuals. The variance of the residuals changes as the value of X changes. Heteroscedasticity or the dependency of the variances of the residuals and the values of X can occur in different ways. It is possible, for example, that the variance of residual values increases when the predicted value of Y increases. This means that the predictions based on the model are better for low predicted values of the dependent variable than for high predicted values. The reverse situation is also possible: a decrease in the variances when the predicted values of Y increase. More complex patterns of heteroscedasticity can also occur.

Consequences of heteroscedasticity

The assumption of constant error variance is used to determine the variance of the distribution of the parameters in the standard OLS estimation procedure of a regression model. Faulty inferences can be made when this assumption does not hold. Remember that the OLS procedure is used to estimate the parameters **b** (vector of the estimated regression parameters) and **V**(**b**) (the variance–covariance matrix of the regression parameters). In matrix notation, this can be written as

$$\mathbf{b} = (\mathbf{X}'\mathbf{X})^{-1}\mathbf{X}'\mathbf{y}, \tag{5.7}$$

$$\mathbf{V}(\mathbf{b}) = \mathbf{V}\left[(\mathbf{X}'\mathbf{X})^{-1}\mathbf{X}'\mathbf{y}\right] = (\mathbf{X}'\mathbf{X})^{-1}\mathbf{X}'\mathbf{V}(\mathbf{y})\mathbf{X}(\mathbf{X}'\mathbf{X})^{-1} \tag{5.8}$$

Now, when $\mathbf{V}(\mathbf{y})$ is assumed to equal $\sigma^2\mathbf{I}$ (thereby implying constant error variance), the expression for $\mathbf{V}(\mathbf{b})$ simplifies greatly to:

$$\mathbf{V}(\mathbf{b}) = \sigma^2(\mathbf{X}'\mathbf{X})^{-1}. \tag{5.9}$$

Yet, when the error variance is not constant in the data, the calculation of the estimated standard errors of the regression parameters is no longer valid. In this situation, the least squares estimators are still unbiased, but they are inefficient. This means that when non-constant error variance occurs, there exist other estimation procedures (weighted least squares; see the subsection below on remedies) that generally produce estimators with smaller standard errors. It is clear that this has an impact on the significance tests and confidence intervals.

Diagnostics

Graphical method: studentized residual plots

Once again a graphical tool is useful to evaluate the homoscedasticity assumption. The most appropriate graph is a plot of studentized residuals against the predicted values of Y or against certain predictors. Studentized residuals are residuals divided by an estimate of their variance (this variance is unknown, and therefore it is impossible to standardize). Concretely, residuals are studentized by dividing them by $s_i^2\sqrt{1-h_i}$ (where s_i^2 is the estimate of σ^2 obtained after deleting the ith observation and h_i is the leverage of i – see the section on influential observations for more information on leverage). Plotting studentized rather than raw residuals makes it easier to observe patterns of changing spread (Fox, 2008, p. 272). When the studentized residuals are randomly spread around the mean of zero and contained within a horizontal band of ± 2 standard deviations of the mean, the homoscedasticity assumption can be considered valid. A pattern of changing dispersion in the studentized residuals (e.g. increase or decrease of the spread with the level of the predicted values of Y or one of the Xs) is indicative of heteroscedasticity.

Figure 5.4 displays studentized residuals for the regression model explaining trust in the police. Studentized residuals are plotted against fitted values (upper left-hand corner) as well as against the various predictor variables in the model. The plot clearly reveals some anomalies. High studentized residuals (> 2) occur more often for low fitted values, while low studentized residuals are mostly present when fitted values are high. This pattern suggests that heteroscedasticity is present. A similar pattern can be seen for the plot for predictor plcpvcr (effectiveness of crime prevention).

Statistical test: White's test

White's test is a formal procedure for detecting heteroscedasticity (White, 1980). Although this test is a more general test for model misspecification, it is also appropriate for testing

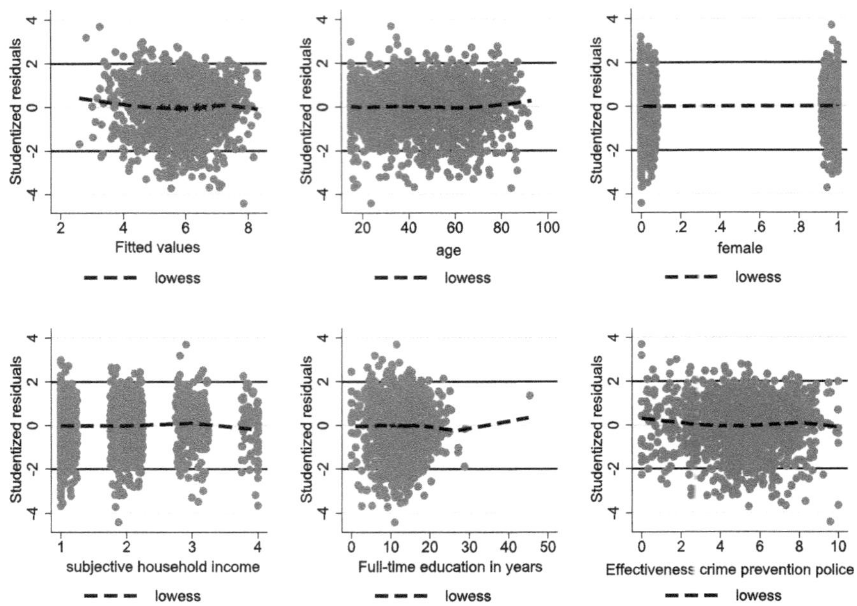

Figure 5.4 Residual versus fitted plot (upper left) and residual versus predictor plots

homoscedasticity. The test does not make any assumption about the pattern of non-constant error variance. The null hypothesis is that the residuals are homoscedastic:

$$H_0 : \sigma^2_{\epsilon_i} = \sigma^2_{\epsilon}. \tag{5.10}$$

The rejection of this null hypothesis is evidence of heteroscedasticity. In White's test, the squares of the residuals of a substantive model with $p-1$ predictors (step 1) are regressed on all predictors in the model plus all cross products among the predictors (step 2). The test statistic in White's test equals nR^2 of the last model (step 2). Under the null hypothesis of homoscedasticity, the test statistic is distributed as chi-squared with degrees of freedom the number of predictors of the model in step 2.

In our example, White's test renders a χ^2-value of 116.13 (for 19 degrees of freedom). This value is strongly significant ($p < 0.0001$) and confirms that a significant amount of heteroscedasticity is present in the data.

Remedies

Several strategies can be used to tackle the problem of heteroscedasticity. A first remedy is a transformation of the dependent variable. When, for example, the spread of the residuals is an increasing linear function of the predicted values one can use the square root of the dependent variable. Alternatively, a log or inverse transformation can be used.

Another frequently used strategy to deal with heteroscedastic data is to perform a weighted least squares (WLS) estimation procedure instead of ordinary least squares. Remember that in an OLS procedure the values of the parameters of the model are estimated by minimizing the value of the sum of the squared residuals: $\min(\sum e_i^2)$. In the OLS procedure, all units have the same weight, $w_i = 1$. In a WLS estimation procedure, each unit is given a different weight w_i and the sum of the weighted squared residuals is minimized: $\min(\sum w_i e_i^2)$.

The generalized least squares procedure is the starting point for tackling heteroscedasticity. In this generalization of OLS, the inverse of the diagonal matrix $\mathrm{Var}(\epsilon_i) = \mathbf{V}$ is used. We get

$$\mathbf{b}_w = (\mathbf{X}'\mathbf{V}^{-1}\mathbf{X})^{-1}\mathbf{X}'\mathbf{V}^{-1}\mathbf{y}. \tag{5.11}$$

We obtain $\mathbf{V}^{-1}$ by inverting the diagonal elements of $\mathbf{V}$. This results in $\mathbf{D}_{wi}$, a diagonal matrix with $w_i = 1/\sigma_i^2$ on the diagonal and $\mathbf{b}_w = (\mathbf{X}'\mathbf{D}_{wi}\mathbf{X})^{-1}\mathbf{X}'\mathbf{D}_{wi}\mathbf{y}$. So, to treat the heteroscedasticity problem, the inverse of the conditional variance of the residuals is used as weight for each unit: $w_i = 1/\sigma_i^2$. In this way, units with larger residual variance are given a smaller weight than units with less variance. One can also consider the conditional variance of the residuals as an indicator of the precision of the estimate in that condition. Units with less precision (large variance) are given smaller weights than units with more precision.

It can be proven that when $\mathbf{X}$ and $\mathbf{y}$ are multiplied by $\sqrt{w_i}$, an OLS regression with these transformed variables results in $\mathbf{b}_w$, and that this process is equivalent to minimizing the weighted sum of squares: $\min(\sum e_i^2/\sigma_i^2)$.

To illustrate the functioning of this weighting procedure, take a regression model with one independent variable X and heteroscedastic residuals (Panik, 2009, p. 157). We assume that the other assumptions of the regression model are correct. The basic expression for the model is

$$Y_i = \beta_0 + \beta_1 X_{i1} + \epsilon_i, \quad \mathrm{Var}(\epsilon_i) = \sigma_i^2,\ i = 1, \ldots n. \tag{5.12}$$

After transforming this model by multiplying both sides by $1/\sigma_i$, we get the expression

$$\frac{Y_i}{\sigma_i} = \beta_0 \frac{1}{\sigma_i} + \beta_1 \frac{X_{i1}}{\sigma_i} + \frac{\epsilon_i}{\sigma_i}, \qquad \sigma_i \neq 0,$$

or

$$Y_i^* = \beta_0 W_i^* + \beta_1 X_{i1}^* + \epsilon_i^*, \tag{5.13}$$

with

$$Y_i^* = \frac{Y_i}{\sigma_i}, \quad W_i^* = \frac{1}{\sigma_i}, \quad X_{i1}^* = \frac{X_{i1}}{\sigma_i}, \quad \epsilon_i^* = \frac{\epsilon_i}{\sigma_i}.$$

For the residuals of this model, we have

$$\begin{aligned} E(\epsilon_i^*) &= \frac{E(\epsilon_i)}{\sigma_i} = 0, \\ \mathrm{Var}(\epsilon_i^*) &= \frac{\mathrm{Var}(\epsilon_i)}{\sigma_i^2} = 1. \end{aligned} \tag{5.14}$$

This means that in the transformed model, the conditional variance of the residuals is constant. Notice that we obtain this result by multiplying the dependent and independent variable by the square root of the weights and these weights are the reciprocals of the residual variances. When it is possible to assume that the conditional variance of the residuals is a function of an independent variable X, we get other weights and another transformation of the variables in the model. Suppose, for example, $\sigma_i^2 = \sigma^2 X_i^2$; then $w_i = 1/\sigma^2 X_i^2$ and we transform the model by multiplying both sides by $1/X_i$. Then the residual equals

$$\epsilon_i^* = \frac{\sigma\epsilon_i}{\sigma_i}, \quad \text{and} \quad \mathrm{Var}(\epsilon_i^*) = \frac{\sigma\ \mathrm{Var}(\epsilon_i)}{\sigma_i^2} = \sigma. \tag{5.15}$$

Once again we obtain constant error variance.

In practice, the variance of the residuals is not known and must be estimated from the data. It can be shown that squared observed residuals (or the absolute values of the residuals) are unbiased estimates of the population variance. However, the reciprocal of the squared observed residuals cannot directly be used as weight in the WLS procedure. After all, units with the same covariate pattern (same values on the independent variable) do not always have the same value for the dependent variable Y. As a consequence, they have a different (squared or absolute) residual value and this results in a different weight. However, we need the same weight for units with the same covariate pattern. This can be realized using the following procedure. We first specify a regression model with a substantive dependent ($\mathbf{y}$) and independent variables ($\mathbf{X}$). This regression analysis produces a residual value for each unit. Second, to get the same weight for each unit with the same covariate pattern, the squared residuals (or the absolute values of the residuals) are used as the dependent variable in a model with the relevant substantive variables which sufficiently explain the residuals. To avoid negative weights it is better to use the log of the squared residuals ($\log e_i^2$). This regression model results in the log of predicted squared errors ($\log \widehat{e_i^2}$) which are equal for all the units with the same covariate pattern. To recover the estimated squared residuals ($\widehat{e_i^2}$), the inverse log is taken. In the last step the WLS estimators are produced by an OLS regression analysis with the transformed variables. The weights used to transform the variables are $w_i = 1/\widehat{e_i^2}$. The stepwise summary of the procedure following DeMaris (2004, p. 206) is:

1. Regress Y on X and save e_i.

2. Transform e_i into $\log e_i^2$.

3. Regress $\log e_i^2$ on X and save the predicted values $\log \widehat{e_i^2}$.

4. Take the inverse of $\log \widehat{e_i^2}$ to get $\widehat{e_i^2}$.

5. Regress Y on X using $w_i = 1/\widehat{e_i^2}$.

It is important to notice that the reported R^2 on the output of WLS regression analysis is calculated for the transformed data and not for the original data. Because of this, it is not valid to compare the R^2 of both analyses. To obtain a comparable value of R^2 we must use the original data and the WLS estimates and calculate the predicted values and the WLS residuals. Then the sum of squared WLS residuals is used to calculate R^2_{WLS}. Usually this value is not reported by the software.

A final small comment is related to the use of sampling weights. Sampling weights are used to correct for different probabilities of selection into the sample and to make sample and population distributions comparable. Sampling weights are different from the weights used in the WLS procedure. These weights are a function of the error variances and used to produce correct estimates of the standard error. It can be shown that using sampling weights produces heteroscedasticity even when the unweighted data are homoscedastic.

Table 5.3 compares the OLS estimates from Table 5.1 with results obtained through WLS estimation with $w_i = 1/\widehat{e_i^2}$ as weights. As expected, the WLS procedure generally leads to slightly decreased standard errors, while parameter estimates do not differ substantially. Yet the differences between OLS and WLS are not dramatic. This is not surprising, since the studentized residual plots revealed relatively small amounts of heteroscedasticity.

Table 5.3 OLS and WLS regression output

	OLS		WLS	
Variable	*b*	*SE*	*b*	*SE*
Constant	3.819	(0.287)	3.398	(0.246)
Age	−0.007	(0.003)	−0.005	(0.002)
Female	0.003	(0.092)	−0.051	(0.086)
Hincfel	−0.268	(0.056)	−0.249	(0.054)
Eduyrs	0.052	(0.013)	0.067	(0.010)
Plcpvcr	0.434	(0.025)	0.460	(0.026)
Adjusted R^2	0.177		0.190	

Note: Analytic weights = 1/exp (*g*).

ASSUMPTION 3: INDEPENDENCE OF RESIDUALS

What is it?

As a third assumption, the regression model requires independence of the error terms. The residuals should be patternless, meaning that the residual value for one observation cannot depend on the residual for other data points. Formally, this means that the residual values should not be correlated:

$$E(\epsilon_i, \epsilon_j) = 0 \quad \text{for all } i \neq j. \tag{5.16}$$

The independence assumption can also be looked at from the perspective of the dependent variable. It implies that, conditional on the independent variables, the values Y_i (with $i = 1, \ldots, n$) should be independent draws from the population distribution of Y.

Violations of independence occur when important explanatory factors are neglected in the model. Think of a dependent variable, such as personal income, for which important gender differences exist (males earning more than females). If gender is omitted from the regression model explaining income, we tend to underestimate male incomes systematically, and overestimate female incomes. Males, on average, will have positive residuals and females negative ones. As result, the residuals are patterned: for a gender-matched couple of respondents, a positive correlation between the error terms is expected; for a male and a female respondent, the correlation between residuals will be negative. Note, however, that this pattern will be hard to detect, since in this example gender is not a variable in the model.

In practice, non-independence mainly occurs in two situations. The first situation is time series data, where some dependent variable is measured several times on different occasions. Since social phenomena tend to change only gradually over time, the observation at time point t will probably depend to some extent on how the situation was at time point $t-1$. In this case, the data is said to be autocorrelated: the correlation for residual values of consecutive observations is not zero. Clustering of observations is a second common violation of non-independence. In many instances, social scientists encounter hierarchically structured data: observations are nested in higher-level units. This clustering is often a result of the fact that social reality is layered. Persons are not atomized individuals, but instead grow up in families, work together in organizations and live together in neighborhoods. But clustering can also arise from the research design (e.g. when groups of respondents are interviewed by the same interviewer). If cluster membership is related to the dependent variable but not taken into account into the model, observations belonging to the same cluster will have more similar residuals as observations belonging to other clusters. Again, the assumption of independent error terms is violated.

Consequences

Violations of the independence assumption affect the standard errors of the regression parameters rather than the parameter estimates themselves. As a result, non-independence can affect statistical inference, but does not invalidate regression analysis as a descriptive tool.

The consequences of non-independence for statistical inference can be illustrated in an intuitive way using the following example. Imagine a study in which 100 monozygotic twins are investigated. Obviously, the independence assumption does not hold here. Twins share their genetic material as well as important life experiences. As a result, twin siblings are more similar than two respondents who are not siblings, and will consequently have similar residuals. Yet because of this similarity, a pair of twins contains less information than two independent individuals do. Including someone's twin brother or sister in the study adds relatively little new empirical evidence, because in large part the same information is replicated. Including a completely unrelated individual instead contributes a larger amount new information. Thus, although our hypothetical twin study contains 200 respondents, the data is less informative than a study containing 200 unrelated persons. Nevertheless, the regression model assuming independence proceeds as if 200 independent observations are present, overestimating the amount of available information and consequently also the reliability of the estimates. As a result, the estimated standard errors tend to be smaller than they are in reality. This can result in pseudo-significant effects.

The same point can be made in a more formal manner as well. The independence assumption is used in the derivation of the variance of the regression coefficients. To simplify matters, take the example of simple linear regression with one independent variable. In this case, the point estimator of the regression coefficient can be written as (Neter et al., 1996, p. 46):

$$b_1 = \frac{\sum_{i=1}^{n}(X_i - \overline{X})\, Y_i}{\sum_{i=1}^{n}(X_i - \overline{X})^2} \tag{5.17}$$

Consequently, the variance of this point estimator equals

$$\text{Var}(b_1) = \text{Var}\left[\frac{\sum_{i=1}^{n}(X_i - \overline{X})\, Y_i}{\sum_{i=1}^{n}(X_i - \overline{X})^2}\right]. \tag{5.18}$$

Regression theory considers the values of the predictor X as known constants. As a result, the observations Y_i (keeping X constant) are the only random variables in this expression, and $\text{Var}(b_1)$ can be rewritten as

$$\text{Var}(b_1) = \frac{\text{Var} \sum_{i=1}^{n}(X_i - \overline{X})\, Y_i}{\left[\sum_{i=1}^{n}(X_i - \overline{X})^2\right]^2}. \tag{5.19}$$

The formula above expresses the variance of estimator b_1 as the variance of a (weighted) sum of random variables Y_i. A fundamental law of variances and covariances states that the variance of a sum of two random variables equals the sum of the variances of the random variables plus two times its covariance:

$$\text{Var}(A + B) = \text{Var}(A) + \text{Var}(B) + 2 \times \text{Cov}(A, B). \tag{5.20}$$

The fact that covariances between all the Y_i need to be taken into account makes the further elaboration of $\text{Var}(b_1)$ a very complex operation. Assuming that the covariances of the random variable Y_i (conditional on X) equal 0 simplifies calculation considerably. Under this condition, the variance of a sum of the random variables can be written as the sum of its variances:

$$\frac{\text{Var}\sum_{i=1}^{n}(X_i-\overline{X})\,Y_i}{\left[\sum_{i=1}^{n}(X_i-\overline{X})^2\right]^2}=\frac{\sum_{i=1}^{n}\text{Var}\left[(X_i-\overline{X})\,Y_i\right]}{\left[\sum_{i=1}^{n}(X_i-\overline{X})^2\right]^2}. \tag{5.21}$$

Now that the covariances have been removed as obstacles, and assuming that all the Y_i have the same variance σ^2 (i.e. homoscedasticity – see assumption 2), the variance of the regression coefficient can be easily obtained:

$$\text{Var}(b_1)=\frac{\sum_{i=1}^{n}(X_i-\overline{X})^2}{\left[\sum_{i=1}^{n}(X_i-\overline{X})^2\right]^2}\text{Var}(Y_i)=\frac{\sigma^2}{\sum_{i=1}^{n}(X_i-\overline{X})^2}. \tag{5.22}$$

If in reality the covariances of the Y_i (conditional on X) are not zero while we assume them to be, the variance of the regression coefficient will be incorrectly calculated. The harmful effects for statistical inference can be serious. Scariano and Davenport (1987), for example, show by means of a simulation study that even mild clustering can produce pseudo-significant effects, since probabilities of falsely rejecting a null hypothesis of an F-test for equal means are inflated.

Diagnostics and remedies

Earlier, we mentioned that non-independence most often occurs in the case of time series or nested data. For both situations, diagnostic tools as well as specific models taking the non-independence into account have been developed. Because of their importance, these models are discussed in greater detail elsewhere in this volume. Chapter 7 explains how clustered data can be properly analyzed using multilevel models. Models for time series data are discussed in Chapter 17.

ASSUMPTION 4: NORMALITY

What is it?

The standard regression model presupposes that the distribution of the residuals or error terms ϵ_i has a particular form; the error terms are assumed to follow a normal distribution. Concretely, the density function of residuals should have the famous bell shape of a Gaussian curve described by the following function:

$$f(\epsilon_i;\mu,\sigma^2)=\frac{1}{\sigma\sqrt{2\pi}}e^{-\frac{1}{2}\left(\frac{x-\mu}{\sigma}\right)^2}, \tag{5.23}$$

where μ is the mean of the residuals (which equals 0 by definition in regression analysis) and σ^2 is the variance of the residuals. This implies, among others, that the distribution of the residuals should have a single peak (i.e. be unimodal), be symmetric instead of skewed (skewness = 0), and that the tails of the distribution should be neither too light nor too heavy (kurtosis = 0).

The normality assumption has repercussions for the distribution of the dependent variable as well. If we consider values of the predictor variables that are considered as fixed (as regression theory does), the observations Y_i and residuals ϵ_i are communicating vessels. The assumption of the normal residuals implies normality for observations of the dependent variable Y_i conditional on the independent variables X (i.e. for specific values of the X's). This is, however, not necessarily the same as saying that the marginal distribution of dependent variable Y should be normal.

One can easily imagine a dependent variable Y following a normal distribution for males and females separately. If these gender-specific normal distributions have a different mean, however, the distribution of Y in the population of men and women together can be bimodal and thus not normal. Likewise, a normally distributed dependent variable does not guarantee normally distributed residuals. It is the dependent variable Y, controlling for the independent variables, that needs to fulfill the requirement of normality (Lumley et al., 2002). Nevertheless, the normality assumption bears some implications for the distribution of the dependent variable Y as well. The residuals can only fully match the normal density function if Y is continuous (rather than categorical) and not truncated (i.e. not having a lower or upper limit).

Consequences of non-normality

The normality assumption places rather strict restrictions on the residuals, and will very often be violated in social science research. Fortunately, the consequences of non-normality range from quite mild to even non-existent. The normality assumption is used to determine the functional form of the sampling distribution of the regression estimates. To illustrate this, consider the case of a linear regression model for Y_i with only one predictor variable X. We have seen before that the point estimator of the regression slope (b_1) can be written as

$$b_1 = \frac{\sum_{i=1}^{n}(X_i - \overline{X})\,Y_i}{\sum_{i=1}^{n}(X_i - \overline{X})^2}. \tag{5.24}$$

Regression theory considers the values of the predictor X as fixed. As a result, the observations Y_i (keeping X constant) are the only random variables in the expression for regression estimate b_1; b_1 is a linear combination of observations Y_i. If we assume that these observations Y_i, conditional on X, are normally distributed – in addition to being independent (see assumption 3) – the sampling distribution of b_1 is also normal (Neter et al., 1996, p. 1320). The mean of this normal distribution is the population regression slope β_1 and the variance equals $\sigma_{b_1}^2$:

$$b_1 \sim N(\beta_1, \sigma_{b_1}^2). \tag{5.25}$$

Or the standardized regression slope is a standard normal variable (i.e. with mean 0 and standard deviation 1):

$$\frac{b_1 - \beta_1}{\sigma_{b_1}} \sim N(0, 1). \tag{5.26}$$

Because the population standard deviation σ_{b_1} is unknown, however, it needs to be estimated by S_{b_1} (see Chapters 2 and 3 in this volume). If we divide by the estimate S_{b_1} instead of by the population value σ_{b_1}, we get a so-called studentized statistic, following a t-distribution with $n - p$ degrees of freedom:

$$\frac{b_1 - \beta_1}{S_{b_1}} \sim t_{(n-p)}, \tag{5.27}$$

where p refers to the number of regression parameters (i.e. the number of predictors plus one intercept).

Thus, the normality assumption allows us to derive how the regression estimates are distributed. This information is used to perform significance tests for regression coefficients and to constructed confidence intervals. Consequently, statistical inference regarding the regression coefficient could go off the rails if normality of the residuals does not hold. In that case, the regression parameters (divided by their estimated standard error) are not guaranteed to follow a

t-distribution. The reported t-values and their corresponding p-values, as well as the constructed confidence intervals, can be biased.

In practice, however, the consequences of non-normality are often limited. First, the normality assumption is not necessary for point estimates of the regression parameters to be unbiased, and thus has no consequences for regression as a descriptive tool. Second, statistical tests for regression parameters are quite robust against deviations from normality. If sample sizes are sufficiently large, the regression coefficients will approach a t-distribution even if the observations Y_i are not normally distributed. This can be explained by the central limit theorem, which asserts that the sum of a large number of random variables (here, the observations Y_i) will be approximately normally distributed irrespective of the distribution of these random variables. The data sets used in applied research are usually sufficiently large to provide a reasonable approximation (Lumley et al., 2002).

Therefore, some authors argue that practical researchers can neglect the normality assumption (see Gelman and Hill, 2007, p. 46). To a certain extent, we agree with this relaxed position on the normality assumption. This does not mean that normality is completely irrelevant, though. In small data sets with strong violations, the normality assumption can still be an issue. Furthermore, non-normal data often contain influential observations that can distort the regression analysis (see assumption 5). For that reason, we offer a brief discussion on diagnostics and remedies for non-normality below.

Diagnostics

Graphical method: The quantile–quantile plot

Normality of the residuals can be assessed by drawing a so-called normal quantile–quantile (QQ) plot or normal probability plot. The QQ plot is a standard graphical tool used to compare observed values to a theoretical distribution. In the normal QQ plot, the quantiles of the observed residuals and the quantiles of the standard normal distribution are plotted against each other. To obtain a QQ plot, the residuals obtained from a regression model are ranked from low to high. These values are called the observed quantiles of the residuals, and are plotted on the vertical axis of the QQ plot. Then, every residual is given a rank order i, going from 1 to n (given that there are n observations). These rank orders are then divided by n. The resulting proportional rank (i/n) of a residual indicates the proportion of the observations that have a smaller or equal residual value. Subsequently, the z-score that would cut off exactly the same proportion in a cumulative standard normal distribution is identified. These z-values are the quantiles of the standard normal distribution, and are plotted against the quantiles of the observed residuals (on the horizontal axis of the QQ plot). If the dots in the resulting scatter plot form a straight line, the residuals are normally distributed. Deviations from a straight line are indicative of non-normality. A drawback of the QQ plots is that the decision whether the plot displays a straight line is always arbitrary to a certain extent.

Figure 5.5 shows a normal QQ plot (right) for the regression model explaining trust in the police, as well as a kernel density plot of the residuals (left). Both plots indicate that the residuals follow a normal distribution reasonably well. The kernel density curve maps rather well onto the Gaussian curve, and the dots on the QQ plot more or less follow a straight line. The most important deviation can be found in the left tail of the distribution. The dots below the line on the left-hand side of the QQ plot mean that the negative residuals are somewhat more extreme than expected. In other words, the left tail is slightly heavier than expected. Yet these are only minor deviations and the main conclusion from the plots is that the normality assumption is sufficiently well approximated.

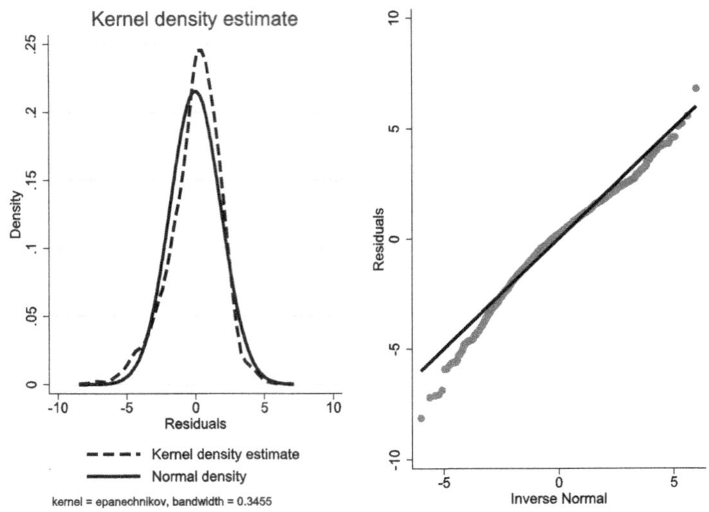

Figure 5.5 Kernel density and QQ plots

Statistical tests for normality: Shapiro–Wilk and Kolmogorov–Smirnov

Alternatively, formal statistical tests for normality are available as well. The reader should be aware, however, that the practical use of these tests is often limited. When sample sizes are small (the only condition under which non-normality substantially affects the conclusions), the tests lack power to detect deviations from normality. When sample sizes are large (thus when the non-normality hardly has perceivable consequences), the tests are sometimes overly sensitive, and even detect insubstantial violations of normality. Nevertheless, we briefly discuss two tests that are readily available in most statistical software packages.

Shapiro and Wilk (1965) developed a statistic, W, that quantifies the straightness of the line formed by the dots in a QQ plot. W ranges from 0 to 1, where higher values are indicative of a closer correspondence to a normal distribution. If W falls below a certain critical value, the null hypothesis of normality is rejected.

The Kolmogorov–Smirnov test evaluates the discrepancy between the cumulative density function of the observed residuals and the cumulative density function of a normal distribution with the same mean and variance. If the residuals are normally distributed, both functions are expected to be highly similar. The Kolmogorov–Smirnov test is based on test statistic D, the maximum distance between both functions. A p-value is given, representing the probability that an even larger distance is found in the sample, assuming both density functions are equal in the population. If the p-value is smaller than 0.05, then the null hypothesis that the cumulative density functions are equal at all points is rejected, and we conclude that normality does not hold.

Of these two tests, the Shapiro–Wilk test has the larger statistical power (Razali and Wah, 2011). Therefore, it is preferable to use Shapiro–Wilk when sample sizes are small ($n < 100$). For larger sample sizes, Kolmogorov–Smirnov is to be preferred.

Table 5.4 shows both tests for the example regression model. Despite the minor deviations from normality we saw in the kernel density and QQ plot, both tests have a p-value smaller than 0.05 and thus indicate that the assumption of normality is significantly violated. This

Table 5.4 Results for the statistical tests for normality

Test	Test statistic and p-value
Shapiro–Wilk	0.977 (0.000)
Kolmogorov–Smirnov	0.068 (0.000)

illustrates that these normality tests are very sensitive to deviations from normality when sample sizes are large.

Remedies

Two main strategies exist if non-normality is deemed to be a problem. First, one can adapt the regression model to the data, and model the non-normality. The generalized linear model is a generalization of the basic regression model that makes it possible to relax the assumption of normality and assume other error distributions instead. This makes it possible to model variables following binomial (logistic regression – cf. Chapter 8 in this volume), multinomial (cf. Chapter 9), count, Poisson and numerous other distributions. The alternative strategy consists of adapting the data to the model by transforming the dependent variable Y so that its distribution becomes (approximately) normal. Commonly used transformations include the family of Box–Cox transformations or the logarithmic transformation. Because the ability of these transformations to deal with non-normality is of limited use for applied researchers, we do not discuss them in detail here and refer to Carroll and Ruppert (1988) for more details instead.

ASSUMPTION 5: ABSENCE OF INFLUENTIAL OBSERVATIONS

The nature of the problem and its consequences

In contrast to the other assumptions, which refer to relations between variables or error distributions in the complete data set, this assumption deals with the position of specific observations. Data sets sometimes contain a small number of observations that are separated from the rest of the data, in the sense that they have values that deviate strongly from the other observations. Potentially, such 'extreme cases' can become influential observations that affect the results of the regression analysis. An influential observation can be defined as a case 'that alters the value of a regression coefficient whenever it is deleted from an analysis' (Allen, 1997, p. 177). Obviously, the presence of influential observations is not desirable, as this means that the regression results are distorted by a couple of extreme observations.

To understand under what conditions extreme cases can become influential observations, two types of extreme observations need to be distinguished. First, outliers are observations with a deviating score on the dependent variable. Often (but this does not always have to be the case, as we shall see shortly) outliers are characterized by a high residual value. The residual plots introduced earlier in this chapter can be useful tools to identify outliers.

Second, cases that have extreme scores on the predictor variables are called *leverage points*. Exceptional values on an independent variable can function as a lever, tilting the regression line towards them. The more extreme the X-values, the more powerful the lever is. More formally speaking, the degree of leverage of an observation i can be defined as the impact that the X-values of observation i have on the predicted value of that same observation i. Leverage values can be calculated as follows. From the previous chapter we know that, using matrix notation, the predicted values of a regression model can be expressed as:

$$\widehat{\mathbf{y}} = \mathbf{X}(\mathbf{X}'\mathbf{X})^{-1}\mathbf{X}'\mathbf{y} \tag{5.28}$$

where $\widehat{\mathbf{y}}$ is an $n \times 1$ vector containing the predicted values, $\mathbf{y}$ is an $n \times 1$ vector of the observed values for the dependent variable, and $\mathbf{X}$ is an $n \times p$ matrix of containing the scores on the predictor variables. The impact of the X-values on the predicted scores is determined by the so-called hat matrix $\mathbf{H}$:

$$\mathbf{H} = \mathbf{X}(\mathbf{X}'\mathbf{X})^{-1}\mathbf{X}', \tag{5.29}$$

which is an $n \times n$ matrix. The elements h_{ij} capture the influence that the X-values of observation j have on the predicted value for observation i. In consequence, the diagonal elements (h_{ii}) of the hat matrix are the leverage values as defined above. The higher the value of h_{ii}, the more extreme observation i is with respect to the X-values. The leverage values h_{ii} have some useful properties. It can be shown that they range between 0 and 1, and that their sum equals the number of regression coefficients to be estimated (p). As a result, the average leverage value equals p/n. A leverage value is considered as large if it is more than twice as large as the average leverage value.

Outliers and leverage points are not necessarily influential cases. Only under certain conditions will they affect the regression estimates. This is illustrated in Figure 5.6, which shows a scatter plot for hypothetical data. Point A has an exceptionally high value on the dependent variables and is thus an outlier. Point B is a leverage point due to its extreme value on the predictor. C and D are at the same time outliers and leverage points. Not all of these points are influential observations. Point A will not influence the regression slope strongly, because there are many other observations with similar values for the predictor. The only influence A might exercise is a slight increase in the intercept of the regression line. Although C is an outlier and has strong leverage, it will not substantially influence the regression line either. The reason is that C is positioned in exactly in the direction of the cloud of dots. B and D, on the other hand, will tilt the regression line severely in a downward direction. Summarizing, observations are especially influential if they possess leverage and are inconsistent with the regression relation for the other observations.

Diagnostics

Influential observations can be identified by evaluating the impact that single cases have on the regression outcomes. We will discuss three commonly used measures to quantify this impact, namely the DFFITS, Cook's distance and DFBETA. These measures are built on the same general principle. The regression analysis is repeated leaving out a single observation. Subsequently, one assesses how leaving out that observation changes the regression outcomes. This procedure is repeated for every observation in the data set. The three measures differ, however, in the specific outcomes that are evaluated.

The DFFITS (a mnemonic for 'difference in the fitted value, standardized') measures the change in the predicted value for an observation when the very same observation is left out of the analysis. The DFFITS for observation i can be written as

$$DFFITS_i = \frac{\widehat{Y}_i - \widehat{Y_{i(i)}}}{\sqrt{MSE_{(i)}h_{ii}}} \tag{5.30}$$

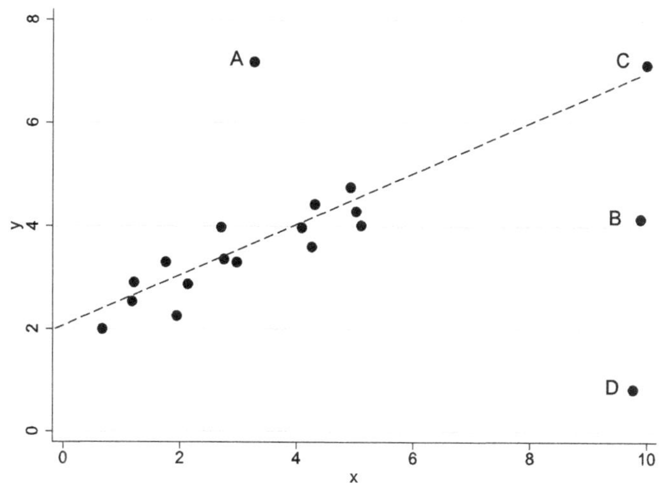

Figure 5.6 Scatter plot for hypothetical data, showing outliers and leverage points

where $\widehat{Y_i}$ and $\widehat{Y_{i(i)}}$ are the predicted values for observation i for a regression model respectively including and excluding observation i, $MSE_{(i)}$ is the mean squared error for the model excluding observation i, and h_{ii} is the leverage value of observation i. Alternatively, the formula for DFFITS can be written as

$$DFFITS_i = t_i\sqrt{\frac{h_{ii}}{1-h_{ii}}} \tag{5.31}$$

with t_i equal to the studentized deleted residual, that is, the difference between the observed score Y_i and $\widehat{Y_{i(i)}}$ divided by its estimated standard error. The studentized deleted residual expresses how inconsistent an observation is with the remainder of the cases. Thus, the more inconsistent an observation and the larger its leverage, the larger the DFFITS will be. Observations with a DFFITS larger than $2\sqrt{\frac{p}{(n-p)}}$ are considered influential observations.

While the DFFITS measures the influence of observation i on its own predicted value, Cook's distance (D_i) summarizes the impact of observation i on predicted values for all other observations:

$$D_i = \frac{\sum_{k=1}^{n}(\widehat{Y_k} - \widehat{Y_{k(i)}})^2}{p\,MSE}, \tag{5.32}$$

which is equivalent to

$$D_i = \frac{e_i^2}{p\,MSE}\frac{h_{ii}}{(1-h_{ii})^2}. \tag{5.33}$$

As with DFFITS, higher residuals and leverage values will lead to a larger Cook's distance. D_i has the advantage, however, that it is less sensitive than DFFITS. Furthermore, strongly influential observations will stand out more clearly since residuals as well as leverage values are squared in the calculation of Cook's distance (Freund and Wilson, 1998, p. 130).

To determine whether observations are influential, the D_i obtained are sometimes compared with the 50th percentile of an F distribution with p and $n - p$ degrees of freedom (Neter et al., 1996, p. 409). If D_i is larger than this value, observation i is considered as influential. Others have argued that $4/n$ should be used as a cut-off point instead (Bollen and Jackman, 1990).

The DFBETAs, finally, quantify how strongly a single observation influences the estimated regression parameters rather than predicted values. For each observation i and independent variable j, a DFBETA can be calculated as follows:

$$DFBETA_{j(i)} = \frac{b_j - b_{j(i)}}{\sqrt{MSE_{(i)}c_{jj}}}, \tag{5.34}$$

where b_j is the parameter estimate for independent variable j calculated on the complete data set and $b_{j(i)}$ is the same parameter estimate but then calculated excluding observation i. C_{jj} refers to the jth diagonal element of the matrix $(\mathbf{X}'\mathbf{X})^{-1}$.

A positive (negative) DFBETA indicates that excluding a particular observation would lead to a decrease (increase) in the respective regression parameters. A DFBETA with an absolute value greater than $2/\sqrt{n}$ is considered large, and points in the direction of an influential observation.

Alternative methods for detecting influential observations are discussed in Freund and Wilson (1998, Chapter 4).

To illustrate the detection of influential observations, we rerun the regression analysis, save the leverage measures and calculate how many cases exceed the critical value (for Cook's distance, the $4/n$ rule is used). Based on Cook's distance and the DFBETAs, 9–11% of the observations can be considered influential. The DFFITS gives a more conservative estimate of 5.9%. Table 5.5 lists the cases with the ten highest values on each of those measures (in the case of the DFBETAs, the highest absolute values/deviations from 0).

There are some very extreme values, and it is instructive to look at the associated case numbers. Two cases appear most frequently in the top ten lists (1509 and 965). Are there any logical explanations for their influence on the parameter estimates of our regression model? Listing their values on the model variables reveals that both have the maximum distance between trust in the police (our dependent variable) and rating police effectiveness in crime prevention. Since there is a significantly positive relationship between those variables in our model, the extremely negative relationship between them for these observations strongly influences the parameter estimates. Not surprisingly, case 965 and 1509 are both in the top three of highest DFBETAs for the crime prevention predictor.

In our previous diagnostic graphs, one could identify a clear leverage point on educational level. This case, observation 1091, also appears three times in the list of extreme values and has an especially high DFBETA for education. A list of its values on the model variables shows that this observation concerns a 78-year-old woman shown as having completed no less than 45 years of full-time education!

Remedies

How influential observations should be remedied is, strictly speaking, not a statistical problem, and as such no straightforward statistical solutions exist (Freund and Wilson, 1998, p. 143). Influential observations represent an anomaly in the regression model. The reasons for this anomaly can be manifold. Further investigation of the nature of the influential observation is required before remedying action can be undertaken.

First, influential observations might be caused by errors made during the data collection process. For example, respondents might have provided an incorrect (and implausible) score, or a mistake might have occurred during coding or inputting (this is probably the reason for the exceptional value of observation 1091 on eduyrs). If there is convincing evidence that a data collection error is present, the incorrect value should be corrected, or alternatively the influential observation can be excluded from the analysis.

Table 5.5 Ten highest values on Cook's distance, DFFITS and DFBETAs (case numbers in square brackets)

Rank	Cook's D		DFFITS		DFBETA age		DFBETA female		DFBETA hincfel		DFBETA eduyrs		DFBETA plcpvcr	
	$Cv^* = 0.0024$		$Cv = 0.0061$		$Cv = 0.0493$		$Cv = 0.0493$		$Cv = 0.0493$		$Cv = 0.0493$		$Cv = 0.0493$	
1	0.0286	[1509]	0.3397	[965]	0.1837	[1535]	0.1271	[1509]	0.2385	[1509]	0.2417	[1504]	0.2843	[121]
2	0.0198	[121]	0.2620	[64]	0.1625	[862]	0.1229	[121]	0.2131	[1119]	0.2294	[1091]	0.2668	[965]
3	0.0191	[965]	0.2391	[1091]	0.1533	[560]	0.1124	[306]	0.1692	[376]	0.2103	[862]	0.2399	[1509]
4	0.0175	[862]	0.2341	[1369]	0.1365	[1426]	0.1045	[965]	0.1582	[8]	0.1963	[1573]	0.2142	[64]
5	0.0116	[1504]	0.2330	[1573]	0.1320	[121]	0.0999	[1119]	0.1441	[306]	0.1698	[756]	0.1783	[1028]
6	0.0114	[64]	0.2177	[1028]	0.1284	[1573]	0.0934	[560]	0.1359	[407]	0.1472	[1509]	0.1594	[87]
7	0.0102	[1119]	0.2007	[927]	0.1172	[1369]	0.0922	[862]	0.1354	[777]	0.1546	[1580]	0.1498	[1270]
8	0.0098	[306]	0.1809	[1581]	0.1151	[149]	0.0903	[376]	0.1282	[1614]	0.1387	[977]	0.1496	[927]
9	0.0095	[1091]	0.1803	[8]	0.1138	[911]	0.0830	[336]	0.1196	[965]	0.1385	[534]	0.1450	[911]
10	0.0095	[560]	0.1753	[185]	0.1124	[650]	0.0812	[1535]	0.1150	[1379]	0.1252	[196]	0.1397	[335]

Associated case numbers in square brackets.
* Cv = critical value

Second, some observations might behave differently compared to the rest of the data set because they are subject to a factor not accounted for in the model. In this case, influential observations are informative, because they point in the direction of model misspecification. Here, in-depth study of the influential observations is warranted to identify the factor causing the influential observations to behave differently. When successful, the regression model can be respecified by including this factor in the model.

In practice, however, it is often not easy to determine the reasons for the exceptional behavior of some observations. When uncertain about the nature of the influential observations, it is not good practice to simply remove these observations from the data set. Obtaining a better model fit is not a valid argument for modifying the data. In this case, robust regression estimation (a technique that is less sensitive to single observations having exceptional scores) might be considered. A discussion of robust regression estimation can be found in Carroll and Ruppert (1988, Chapter 6).

ASSUMPTION 6: ABSENCE OF MULTICOLLINEARITY

What is it?

In multiple regression analysis, independent variables are not only often related to the dependent variable, but also regularly correlated among themselves. Multivariate regression analysis was specifically designed as a statistical tool to deal with the situation of several correlated variables predicting a single outcome variable. However, if the correlations between the independent variables become too strong, problems can arise during the analysis. Multicollinearity refers to this situation where (a set of) predictor variables show very strong intercorrelations. We speak of perfect multicollinearity when one independent variable can be perfectly predicted by the other independent variables.

Consequences

A first consequence of multicollinearity is that the interpretability of the results of the regression analysis is made more difficult. As a multivariate analytical tool, the aim of regression analysis is to disentangle the effects of several predictors and to estimate net effects of one independent variable, keeping other predictors constant. Yet when strong multicollinearity is present, this task becomes meaningless. If predictors are almost perfectly intertwined, attempts to disentangle them are not fruitful. And the idea of increasing one predictor while keeping all other variables constant is not meaningful if predictors covary almost perfectly. Two strongly correlated predictors might turn out to have very small and even insignificant effects, even if separately they have strong predictive power. In sum, multicollinear data makes regression coefficients hard to interpret.

The statistical consequences of multicollinearity can be illustrated by looking at the formula for the point estimators for the regression coefficients:

$$\mathbf{b} = (\mathbf{X}'\mathbf{X})^{-1}\mathbf{X}'\mathbf{y}. \tag{5.35}$$

In this formula, the inverse of $\mathbf{X}'\mathbf{X}$ is used. This inverse can be calculated by dividing the cofactor matrix of $\mathbf{X}'\mathbf{X}$ by the determinant of $\mathbf{X}'\mathbf{X}$:

$$(\mathbf{X}'\mathbf{X})^{-1} = \frac{(\mathbf{X}'\mathbf{X})_{\mathrm{cof}}}{|\mathbf{X}'\mathbf{X}|}. \tag{5.36}$$

If a perfect linear dependency between two or more predictors exists, then matrix **X** is not of full rank. As a result, $(\mathbf{X}'\mathbf{X})$ will also not be of full rank, and the determinant $|\mathbf{X}'\mathbf{X}|$ will equal 0. The consequence is that the inverse of $(\mathbf{X}'\mathbf{X})$ is not defined, and that the regression coefficients cannot be estimated.

In practice, however, perfect multicollinearity is rather exceptional. But also when the intercorrelations between predictors are very strong instead of perfect, statistical problems are encountered. In this case, the determinant $|\mathbf{X}'\mathbf{X}|$ will not equal zero, but be very close to zero. Because this determinant figures in the denominator, $(\mathbf{X}'\mathbf{X})^{-1}$ will be inflated. In the first place, this results in very unstable estimates for the regression coefficients. After all, the vector of regression coefficients **b** is the product of $(\mathbf{X}'\mathbf{X})^{-1}$ and $\mathbf{X}'\mathbf{y}$. Small changes in $\mathbf{X}'\mathbf{y}$ will lead to substantial differences in the regression coefficients. Furthermore, the inverse of $\mathbf{X}'\mathbf{X}$ is used in the estimation of the variance–covariance matrix of the regression coefficients (cf. Chapter 4 of this volume):

$$\sigma_b^2 = \sigma^2(\mathbf{X}'\mathbf{X})^{-1}. \tag{5.37}$$

Multicollinearity thus also increases the standard errors of regression coefficients, and renders it more difficult to find statistically significant effects.

The consequences of multicollinearity are thus potentially very severe. In social science research (where perfect relationships between variables are exceptional), however, this phenomenon mainly occurs in very specific situations. There is a risk of multicollinearity, for example, when interaction effects or higher-order polynomial terms are introduced in the model (in these cases, centering the predictors prior to calculating interactions or polynomial terms can remedy the problem; cf. Brambor et al., 2006).

Diagnostics

The presence of multicollinearity can be tested formally by making use of the so-called tolerance and *variance inflation factor* statistics. Both measures quantify the extent to which a certain predictor j depends on the set of other predictors.

The tolerance of a predictor j equals 1 minus the proportion of explained variance of a regression model explaining predictor j by means of the other independent variables:

$$TOL_j = 1 - R_j^2. \tag{5.38}$$

In other words, the tolerance expresses the amount of unique variance in a predictor. Tolerance values range from 0 to 1. Small tolerance values are indicative of multicollinearity, as they imply that a predictor depends strongly on the other independent variables. As a rule of thumb, tolerance values smaller than 0.1 are considered problematic.

The variance inflation factor is defined as the inverse of the tolerance:

$$VIF_j = \frac{1}{TOL_j} = \frac{1}{1 - R_j^2}. \tag{5.39}$$

Intuitively, VIF_j can be interpreted as the factor by which the variance of independent variable j increases due to the intercorrelations with the other variables. A variance inflation factor of 2, for example, means that the variance of X_j in the multivariate model has doubled compared to a model where X_j would be the only predictor variable, thereby leading to less stable estimates and larger standard errors. The variance inflation factor ranges between 1 and $+\infty$. High values indicate that the respective predictor depends more strongly on the other independent variables. Variance inflation factor values larger than 10 indicate that potentially harmful multicollinearity is present.

Table 5.6 Variance inflation factors

Variable	VIF
Age	1.07
Female	1.06
Hincfel	1.05
Eduyrs	1.01
Plcpvcr	1.01

Table 5.6 shows the variance inflation factors for the regression model explaining trust in the police. In this example, all VIFs are close to 1, and they come nowhere close to the cut-off value of 10. We conclude that no multicollinearity is present.

Remedies

The most straightforward remedy for multicollinearity is to remove one or more predictors with strong intercorrelations. Alternatively, data reduction techniques, such as principal components or factor analysis, can be used to summarize the information contained by a set of interrelated predictors by a more limited number of factors.

Sometimes, removing variables or data reduction is not an option for theoretical reasons. In that case, ridge regression can be applied. The idea behind ridge regression is that the linear dependency between the columns of matrix $\mathbf{X'X}$ can be reduced by adding a constant value to the diagonal elements of this matrix. As a result, the multicollinearity problem will be attenuated, but small biases in the parameter estimates can be introduced. Because multicollinearity is encountered only rarely in applied social research, we do not discuss ridge regression in detail here, but refer to Neter et al. (1996, pp. 394ff.).

CONCLUDING REMARKS

This chapter has provided an overview of six often-mentioned assumptions underlying OLS regression. Three of these conditions concern the error terms of the regression equation: error terms are assumed to be homoscedastic, independent and normally distributed. Other assumptions regard the functional form of relations between independent and dependent variables (linearity), interrelations among predictors (absence of multicollinearity) or the position of individual data points (absence of influential observations).

If these assumptions are not fulfilled, regression results can be distorted and conclusions might be misleading. The nature and severity of the consequences of violations, however, vary greatly from one assumption to the other. Generally speaking, regression analysis has two main functions. First, it is employed as an analytical tool to describe the structure of the observed data. This first purpose requires accurate point estimates of regression coefficients. Violations of the assumption of linearity and the presence of influential observations, however, can cause bias in the regression parameters. Furthermore, multicollinearity among the predictors can cause a large amount of unreliability in regression estimates. As such, violations of these three assumptions can pose a threat to the descriptive function of regression analysis. Second, regression is also often used to make statistical inferences, and generalize findings to a wider population. For this second purpose, not only point estimates but also estimated standard errors of regression parameters are of great importance. Heteroscedasticity, non-normality and non-independence of residuals each potentially create bias in the estimates of standard errors. Additionally, multicollinearity can lead to unreliable estimates of standard errors. In consequence, violations of these assumptions can hamper statistical inference.

Yet, it has to be mentioned that, from the perspective of applied researchers, not all assumptions are equally important. When the size of the sample analyzed is sufficiently large, for example, deviations from normality usually do not have harmful consequences. Also violations of the assumption of homoscedasticity are in many cases relatively minor (Gelman and Hill, 2007, p. 46).

The list of assumptions presented in this chapter is not exhaustive, however, and compliance with these six assumptions is not a sufficient condition for drawing valid and reliable conclusions from the regression model. As is the case in any statistical model, the strength of conclusions crucially depends on the measurement quality of the variables (Carmines and Zeller, 1979). As such, regression analysis assumes that measurements are valid, that predictor variables are not contaminated by random measurement error,[3] and that the dependent variable is measured on an interval scale. Furthermore, regression assumes that no model misspecifications are present and that no important causal factors are left out of the model (so-called omitted variable bias) – refer to Berry (1993) for a more detailed treatment of these issues.

Finally, it has to be repeated that this chapter reviews assumptions for the OLS regression model specifically. Assumptions for other regression models are discussed elsewhere – see Harrell (2001) for logistic regression and survival analysis; and Snijders and Bosker (2012) for multilevel models.

NOTES

1 This chapter focuses on the assumptions of OLS regression, and does not deal with the assumptions underlying other regression models (e.g. logistic regression and multilevel modelling). A thorough discussion of assumptions for these models can be found elsewhere – see Harrell (2001) for logistic regression and Snijders and Bosker (2012) for multilevel models.

2 Various alternatives to perform the lack-of-fit test are possible, such as an ANOVA decomposition of error or a likelihood ratio test comparing a regression where predictors are specified as categorical and one where predictors are specified as continuous. These alternatives lead to identical results.

3 Random measurement error in the dependent variable is less problematic, since this source of error is accommodated in the model by means of the residual term.

REFERENCES

Allen, M. P. J. (1997). *Understanding Regression Analysis*. New York: Plenum Press.

Berry, W. D. (1993). *Understanding Regression Assumptions*, volume 07–092 of *Quantitative Applications in the Social Sciences*. Newbury Park, CA: Sage.

Bollen, K. A. and Jackman, R. W. (1990). Regression diagnostics: An expository treatment of outliers and influential cases. In J. Fox and J. S. Long (Eds), *Modern Methods of Data Analysis*. Newbury Park, CA: Sage.

Brambor, T., Clark, W. R. and Golder, M. (2006). Understanding interaction models: Improving empirical analyses. *Political Analysis*, 14(1), 63–82.

Carmines, E. G. and Zeller, R. A. (1979). *Reliability and Validity Assessment*. Newbury Park, CA: Sage.

Carroll, R. J. and Ruppert, D. (1988). *Transformation and Weighting in Regression*. New York: Chapman & Hall.

DeMaris, A. (2004). *Regression with Social Data: Modeling Continuous and Limited Response Variables*. Hoboken, NJ: Wiley.

Fox, J. (2008). *Applied Regression Analysis and Generalized Linear Models*. Thousand Oaks, CA: Sage.

Freund, R. J. and Wilson, W. J. (1998). *Regression Analysis: Statistical Modeling of a Response Variable*. San Diego, CA: Academic Press.

Gelman, A. and Hill, J. (2007). *Data Analysis Using Regression and Multilevel/Hierarchical Models*. Cambridge: Cambridge University Press.

Harrell, F. E. (2001). *Regression Modelling Strategies*. New York: Springer.

Krzanowski, W. (1998). *An Introduction to Statistical Modelling*. London: Arnold.

Lumley, T., Diehr, P., Emerson, S. and Chen, L. (2002). The importance of the normality assumption in large public health data sets. *Annual Review of Public Health*, 23, 151–169.

Neter, J., Kutner, M. H., Nachtsheim, C. J. and Wasserman, W. (1996). *Applied Linear Statistical Models*. Chicago: Irwin.

Panik, M. (2009). *Regression Modeling. Methods, Theory and Computation with SAS*. Boca Raton, FL: Taylor & Francis.

Razali, N. and Wah, Y. B. (2011). Power comparisons of Shapiro-Wilk, Kolmogorov-Smirnov, Lilliefors and Anderson-Darling tests. *Journal of Statistical Modeling and Analytics*, 2(1), 21–33.

Scariano, S. M. and Davenport, J. M. (1987). The effects of violations of independence assumptions in the one-way anova. *American Statistician*, 41(2), 123–129.

Shapiro, S. S. and Wilk, M. B. (1965). An analysis of variance test for normality (complete samples). *Biometrika*, 52(3–4), 591–611.

Snijders, T. A. B. and Bosker, R. J. (2012). *An Introduction to Basic and Advanced Multilevel Modeling*, 2nd edn. London: Sage.

White, H. (1980). A heteroscedasticity-consistent covariance matrix estimator and a direct test for heteroscedasticity. *Econometrica*, 48(4), 817–838.

6

Non-linear and non-additive effects in linear regression

Henning Lohmann

INTRODUCTION

This chapter discusses the approaches to modeling non-additive and non-linear relationships within the framework of multiple regression. When using multiple regression in its simplest form we have to assume that relationships are linear and additive (see Chapter 5 in this volume). Linearity means that the strength of a relationship between a variable X and a variable Y does not differ across the range of the variable X. However, many research questions in the social sciences address non-linear relationships. We speak of additivity if the relationship between a variable X_1 and a variable Y is not dependent on the value of a variable X_2. Yet often we observe interactions between two variables: the relationship between X_1 and Y differs according to the value of a variable X_2.

Take, for example, the relationship between happiness, individual income and societal affluence (Delhey, 2010). With the proliferation of post-materialist self-expression values in more affluent societies we might expect happiness to depend less on the income domain. In other words, we assume an interaction between income and the level of affluence on happiness. If we ignore this interaction we will overestimate the effect of income for some groups and underestimate it for other groups. Many research questions consider interactions with categorical variables such as gender, race or ethnicity. For instance, labor market research suggests that the influence of work experience on wages differs according to gender (Fernandez-Mateo, 2009). To establish the gender-specific influence of job experience we can estimate two separate models, one for women and one for men. Specifying an interaction term instead gives us the same results in one model. The advantage of this is that we can see easily – using the coefficients' test statistics – if there are significant interactions. Social science literature is full of examples where interaction effects are used to analyze the differences between two groups. In addition to interactions, many research questions focus on non-linear relationships. For instance, the intensity of social contacts changes across the life course (McDonald and Mair, 2010). However, there is no linear relationship as the intensity of contacts does not just increase or decrease when people get older. There is a non-linear relationship between age and social contacts. If we ignore the non-linearity we systematically underestimate the intensity of contacts at some ages while we overestimate it at other ages. As a consequence we would obtain a biased estimate of the age effect or would even conclude that there is no effect, as a positive and a negative effect might cancel each other out.

The chapter addresses the approaches to dealing with non-linearity and non-additivity. It is structured as follows. The second section discusses frequently used options for dealing with non-additivity (interaction effects) and non-linearity (quadratic terms and higher-order polynomials, spline regression). Having presented the statistical foundations of the approaches, the chapter goes on, in the third section, to give example analyses using survey data. Readers who prefer an illustrated non-technical introduction can begin with the third section without thoroughly reading the second. The fourth section discusses some caveats and frequent errors. The chapter concludes with an overview of further reading options.

INTERACTION EFFECTS, POLYNOMIALS AND SPLINES

The specification of interaction terms, polynomials and splines are frequently used approaches to incorporating non-additive and non-linear relationships within the framework of multiple regression. This section provides a general introduction which will be illustrated by the example analyses presented in the next section.

Non-additivity: Interaction effects

We speak of non-additivity when the strength of a relationship between a variable X_1 and a variable Y is dependent on the value of a variable X_2. There exists an interaction between the variables X_1 and X_2. In other words, the relationship between X_1 and Y is moderated by X_2. Therefore, we often label X_2 as a moderator variable. Looking at the example mentioned in the first section, societal affluence is interpreted as a moderator variable as it moderates the relationship between individual income and happiness. However, statistically it would be reasonable to interpret income as a moderator variable, assuming that the relationship between affluence and happiness is dependent on the level of individual income. The question which variable is the moderator variable cannot be answered by statistical means, but only by theoretical means. A model containing an interaction between X_1 and X_2 technically will provide the same results, irrespective of which variable we regard as the moderator. Therefore, before we run a model we need to decide at a theoretical level not only whether we expect an interaction between two variables but also whether the relationship between X_1 and Y is moderated by X_2 or if the relationship between X_2 and Y is moderated by X_1.

Theoretical reasoning is often the starting point that leads us to a model specification which includes an interaction term. However, the violation of the ordinary least squares (OLS) regression assumption of variance homogeneity of the residuals may also be the reason for the respecification of a model. Systematic patterns in the distribution of residuals (e.g. differences in residual variance according to the values of a variable X) may indicate the need for the specification of an interaction. I will discuss a simple example below. Yet, in more complex regression models it is often difficult or impossible to derive hints for model specification just from the analysis of residuals. Therefore, in most cases theory or previous research motivates the specification of a model with interaction effects. We can specify interactions between variables with different measurement levels. In this section I discuss the interactions between two continuous variables and the interactions between a continuous and a dichotomous variable. The latter applies when we compare the effect of a variable across one or more groups.

We multiply the two variables for which we hypothesize an interaction to generate a new variable. The product of the two variables $X_1 \cdot X_2$, the interaction term, can be added to a model like any other variable. Including the interaction results in the regression equation

$$Y = b_0 + b_1 \cdot X_1 + b_2 \cdot X_2 + b_3 \cdot (X_1 \cdot X_2) + e. \tag{6.1}$$

When the equation is rearranged into

$$Y = b_0 + (b_1 + b_3 \cdot X_2) \cdot X_1 + b_2 \cdot X_2 + e, \tag{6.2}$$

we can see why the inclusion of the multiplicative term reflects the idea of an interaction between two variables. The effect of X_1 on Y is dependent on the value of X_2. The effect of X_1 is a function of b_1 and the product $b_3 \cdot X_2$. Suppose that b_1 is positive. When b_3 is negative the effect of X_1 on Y diminishes with higher values of X_2. When b_3 is positive the effect increases. For each value of X_2 the effect of X_1 on Y differs, that is, X_2 moderates the effect of X_1 on Y. As a consequence of the inclusion of the interaction term we cannot interpret b_1 and b_2 in the same way as in a model without an interaction term. b_1 depicts the relationship between X_1 and Y only if $X_2 = 0$. We have to interpret b_2 in an analogous manner. It refers to the influence of X_2 on Y only under the condition that $X_1 = 0$. The same is true for the test statistics of b_1 and b_2. They provide information on the significance of an effect of one variable only for the case where the other variable equals zero. The test statistic refers only to persons where this condition applies, usually a very restricted part of the population. We therefore talk about a conditional effect because it is the effect of X_1 on Y for one specific condition, namely a specific value of X_2. This conditional effect is often labeled as the main effect, in contrast to the interaction effect. As the term 'main effect' is somewhat misleading since it implies that we can interpret it independently of the value of the interacting variable, I will use the term 'conditional effect' in the following section.

A problem arises when the variable X_2 cannot take the value zero. In this case, the conditional effect of X_1 cannot be interpreted in a meaningful way. This problem can be circumvented when we use mean-centered versions of X_1 and X_2 to compute the interaction term (for a detailed discussion of this issue, see Aiken and West, 1991). Why is mean-centering a solution for this problem? After the transformation the zero refers to the mean of the original variable and, assuming that the mean of a variable has a meaningful interpretation, we can interpret the conditional effect in a meaningful way. We proceed step by step as follows. First, we mean-center both variables ($X_1' = X_1 - \overline{X}_1, X_2' = X_2 - \overline{X}_2$). Second, we use the transformed variables to compute the interaction term ($X_1' \cdot X_2'$). Third, we regress Y on the mean-centered variables X_1', X_2' and the interaction term $X_1' \cdot X_2'$. It is not necessary to transform the dependent variable and we therefore retain the same metric as in a model with non-mean-centered coefficients. However, we have to bear in mind that the positive and negative values of the X_1' and X_2' reflect deviations from the mean and not deviations from zero. One aim of mean-centering is to give the conditional effect a meaningful interpretation. A positive side-effect is that mean-centering most often reduces the problem of multicollinearity which is inherent to the use of interaction terms (Cronbach, 1987). However, this approach is restricted to the estimation of unstandardized regression coefficients. Aiken and West (1991, pp. 40ff.) discuss an application with standardized regression coefficients.

As mentioned above, we can use variables of different measurement levels to generate interaction terms. Very often dichotomous variables are used. The following example contains an interaction between a dichotomous variable and a continuous variable. However, in an analogous manner we can also interact two dichotomous variables or categorical variables with more than two values (Jaccard and Turrisi, 2003, pp. 57ff.). Dichotomous variables are frequently used to model differences in the relationship between a variable X_1 on Y between two groups (e.g. men and women). Usually, the dichotomous variables are dummy-coded (i.e. 0 and 1). In the following equation X_1 is a continuous variable and X_2 a dummy variable:

$$Y = b_0 + b_1 \cdot X_1 + b_2 \cdot X_2 + b_3 \cdot (X_1 \cdot X_2) + e, \tag{6.3}$$

The model in equation (6.3) is often called a joint model, as it contains the coefficients for two (or more) groups. However, rearranging the equation, we can also obtain the equations for two separate models. If X_2 equals zero all terms which include X_2 drop out of the equation and the constant b_0, $b_1 \cdot X_1$ and the error term remain. The simplified equation describes the model for the group for which X_2 equals zero:

$$Y = b_0 + b_1 \cdot X_1 + e, \quad \text{with } X_2 = 0. \tag{6.4}$$

If X_2 were a gender variable coded 0 = male and 1 = female the reduced equation would describe the model for men. X_1 can be interpreted as in a model without an interaction term under the condition that $X_2 = 0$ (e.g. that a person is male and not female).

Rearranging the equation, we can also show the model for persons of the second group ($X_2 = 1$, i.e. women):

$$Y = \underbrace{(b_0 + b_2)}_{b'_0} + \underbrace{(b_1 + b_3)}_{b'_1} \cdot X_1 + e, \quad \text{with } X_2 = 1. \tag{6.5}$$

As $b_2 \cdot 1$ is a constant we interpret it as part of the intercept $(b_0 + b_2)$. The regression coefficient of X_1 is $b_1 + b_3$ (given $X_2 = 1$). Including interaction terms with a dichotomous variable leads to the same results as the estimation of two separate models for the two groups. Equation (6.4) shows the model of the first group ($x_2 = 0$). Equation (6.5) with its regression coefficients b'_0, b'_1 shows the model of the second group ($x_2 = 1$). If the intercept and slope do not differ, $b_0 = b'_0$ and $b_1 = b'_1$. In the joint model (equation (6.3)) b_2 depicts the differences between b_0 and b'_0 and the interaction effect b_3 depicts the differences between b_1 and b'_1. Therefore, we can use b_2 and b_3 for testing on group differences.

It should be noted that we can easily expand the model and include more than one variable of interest. The following equation shows a model with two variables (X_1, X_2) and a dichotomous group variable (X_3):

$$Y = b_0 + b_1 \cdot X_1 + b_2 \cdot X_2 + b_3 \cdot X_3 + b_4 \cdot (X_1 \cdot X_3) + b_5 \cdot (X_2 \cdot X_3) + e. \tag{6.6}$$

If $X_3 = 0$ we achieve a reduced equation where all terms which include X_3 drop out and the constant $b_0, b_1 \cdot X_1, b_2 \cdot X_2$ and the error term remain. As in the example above, we can transform the equation if $X_3 = 1$. An example using survey data is given in the third section of this chapter.

Omission of interactions as a form of misspecification

When interaction terms are omitted in the specification of a model, this results, as with the omission of other independent variables, in a violation of the assumption of variance homogeneity of the residuals. This is illustrated in the examples in Figure 6.1, which shows the relationship between years of education, motivation and earnings. Let us assume that education and motivation have a positive effect on earnings. In addition, we assume that education pays off in particular for the highly motivated, that is, there is a positive interaction between motivation and education. Figures 6.1a and 6.1c contain observed values of simulated data on this relationship. The different types of markers refer to different levels of motivation. As assumed, education pays

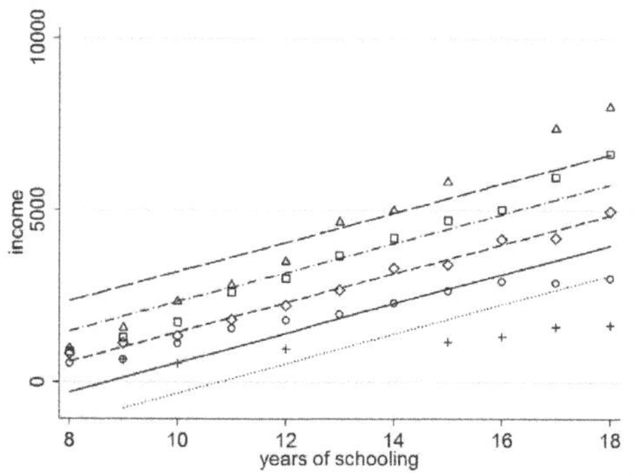

(a) no interaction effect (observed and estimated values)

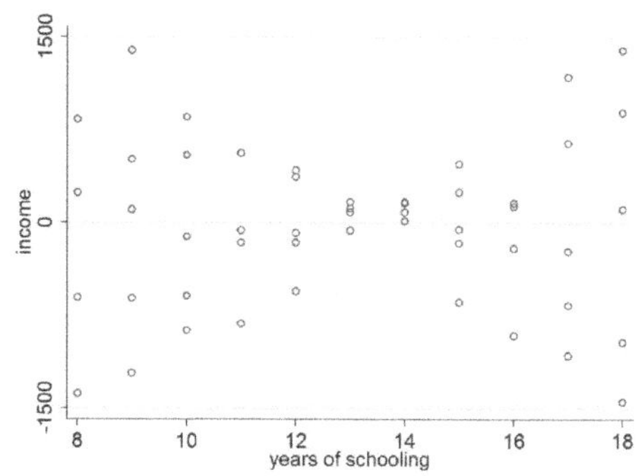

(b) no interaction effect (residuals)

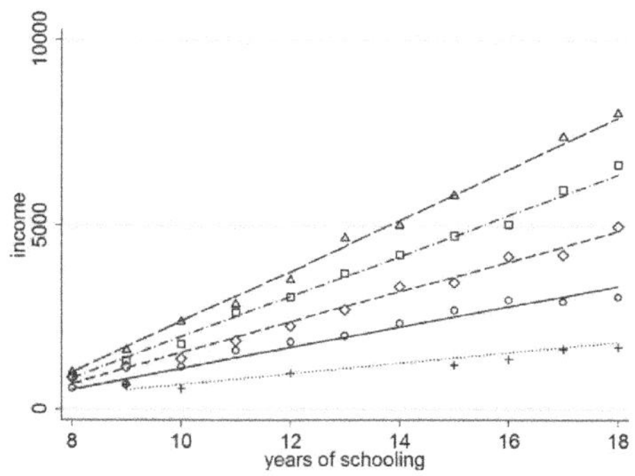

(c) interaction effect included (observed and estimated values)

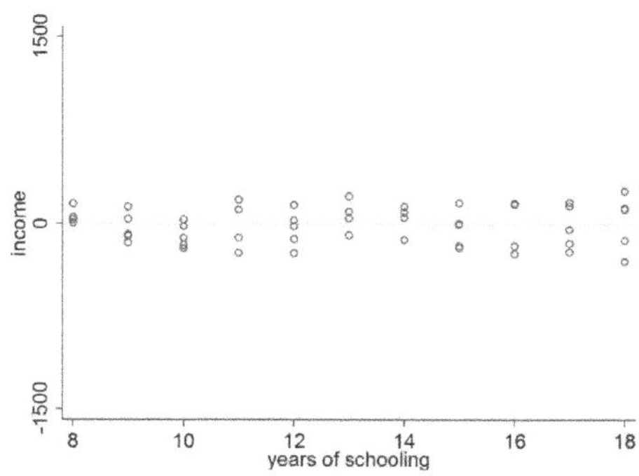

(d) interaction effect included (residuals)

Figure 6.1 Model specifications with and without interaction effect

off most for the group with the highest level of motivation (triangles) and least for the group with the lowest level of motivation (crosses). In addition, Figure 6.1a contains the predicted values of a model without interaction effects in the form of regression lines. The slope of the regression lines shows the positive effect of education. Due to the positive effect of motivation, we see a number of parallel regression lines with different intercepts. The effect is additive, that is, the effect of education is independent of the effect of motivation (and vice versa). The slope is the same for each level of motivation.

As we know that there is an interaction between motivation and education, the model is misspecified. In Figure 6.1b we can see that the model only fits for the average values of the independent variable 'years of education', rather than for high or low values. As a consequence, we observe a clearly patterned plot of residuals. The variance of the residuals is much higher for high or low values of education. Thus, the assumption of variance homogeneity of the residuals is violated as a consequence of the omission of the interaction term. Figure 6.1c contains the predicted values of a model with interaction term. The regression lines are no longer parallel but the slope differs according to the level of motivation. The higher the level of motivation, the steeper the slope of the regression line. Education pays off more for the highly motivated. In contrast to the additive model, the model with interaction term fits well for all levels of education. We can also see this in the plot of residuals (Figure 6.1d). Due to the better fit of the model the total variance of the residuals is reduced. Above all, however, it does not differ systematically according to years of education (variance homogeneity of the residuals).

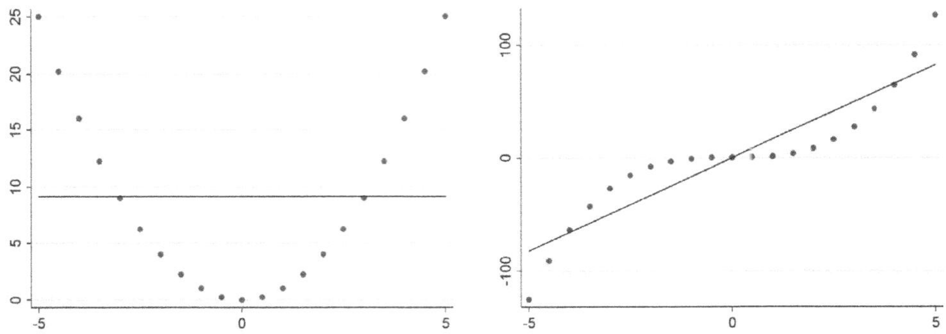

(a) U-shaped relationship (b) S-shaped relationship

Figure 6.2 Non-linear relationships

Non-linearity: Polynomials and splines

While the inclusion of interaction terms in a model offers a standard option for the specification of non-additive models, there are various options for the specification of non-linearity. In this subsection I discuss two options: polynomials and splines. The basic idea of polynomial regression is to transform the independent variable using a non-linear function. A simple and often used transformation is to square a variable (second-order polynomial) to model a U-shaped relationship. More complex relationships can be modeled using higher-order polynomials. There are other transformations in widespread use (e.g. the logarithmic transformation) which I do not discuss (but see Wooldridge, 2009). While transformations of the independent variable are one approach, spline regression follows a different route. The range of the independent variable is split into different intervals for which separate slopes are estimated.

Figure 6.2 shows two examples of non-linear relationships which can be modeled using polynomials or splines. The pattern of observed values in Figure 6.2a shows a clearly U-shaped relationship between a variable X and a variable Y. At first Y decreases with increasing values of X, and then the effect turns positive with further increasing values of X. The regression line in the figure shows that a linear model does not fit the data. The model erroneously assumes that there is no relationship between the two variables. The regression coefficient equals zero. The reason is that the model just minimizes the deviations of the regression line from the observed values across the full range of the variable X. Therefore, the negative effect for lower values of X and the positive effect for higher values of X cancel each other out. Figure 6.2b shows an S-shaped relationship. At first Y increases with increasing values of X, then the relationship weakens for average values of X, before the strength increases again for higher values of X. The linear model ignores the change in the strength of the relationship and indicates a positive effect for the full range of the variable X. A plot of the residuals would show that the variance of the residuals follows a systematic pattern and that the assumption of variance homogeneity is violated.

Polynomials

We can use simple transformations of the independent variable to transform the non-linear relationships shown in the two examples above into a linear relationship. As introduced in the first section above, we speak of linearity when the strength of a relationship between a variable

X and a variable Y does not differ across the range of the variable X. When we observe a U-shaped relationship as in Figure 6.2a, the square of the independent variable is a suitable transformation ($X' = X^2$). Instead of the square of the variable we may also speak of a second-order polynomial. Accordingly, the term 'first-order polynomial' refers to the non-squared, linear variable ($X = X^1$). What do we gain from using the squared variable? If we regress Y on X^2 using the example data from Figure 6.2a we can see that the effect of X^2 on Y is constant for all values of X, although the relationship is non-linear. The regression equation of this example is $Y = 0 + 1 \cdot X^2 + e$, which gives us predicted values of 100 for $X = -10$, 0 for $X = 0$ and 25 for $X = 5$. In the case of an S-shaped relationship we can use a third-order polynomial to linearize the relationship. A fourth-order polynomial would be adequate to model a U-shaped relationship with a rather flat bottom, where the slope for values in the medium range is not as steep as with a second-order polynomial.

In practice, we will seldom find pure U-shaped or S-shaped relationships. Therefore, polynomials are usually specified in addition to a linear term in order to pick up deviations from a linear relationship. Most often just a squared term is introduced into the model:

$$Y = b_0 + b_1 \cdot X + b_2 \cdot X^2 + e. \tag{6.7}$$

The predicted values of different models of this type are depicted as examples in Figure 6.3. The linear part in each of the equations (b_1) equals 5. The squared part (b_2) differs. Regression curves for four different squared coefficients are shown ($-3, -0.3, 0.3, 3$). Depending on the size and direction of b_2, the regression curves differ. When b_2 is positive the curve is U-shaped, if it is negative it follows an inverted U-shape. If b_2 equals 3 (or -3) the slope of the curve is steeper. If b_2 equals 0.3 (or -0.3) the regression line is only slightly bent and it is not easy to see that it is bent at all. This example shows that a coefficient of a squared variable which is different from zero does not necessarily indicate a strongly bent curve or even that the direction of the effect is turning from negative to positive (or vice versa). It often just means that the strength of an effect is increasing or decreasing with higher values of an independent variable X. This is always the case when the sign of b_1 and b_2 point in the same direction and the range of X is limited to positive or negative values.

We can use models with a squared term for a simple test of the basic assumption of linear regression, namely whether or not a relationship is linear. This is accomplished using the test statistic of b_2. If b_2 is significantly different from zero then we must assume that the relationship between Y and X is non-linear. However, the opposite is not always true. If b_2 is not significantly different from zero it could be evidence in favor of a linear relationship, but it could also mean that the relationship is non-linear but in a way which is not reflected in a model with a second-order polynomial. However, higher-order polynomials might pick up the non-linearity. For instance, the relationship between X and Y shown in Figure 6.2b can be modeled by using a third-order polynomial. In the following example analysis I will use models which contain polynomials up to fourth order:

$$Y = b_0 + b_1 \cdot X + b_2 \cdot X^2 + b_3 \cdot X^3 + b_4 \cdot X^4 + e. \tag{6.8}$$

However, in applications with micro data in the social sciences often not more than a squared term is required. Complex non-linearities are more frequent in time-series analysis or similar applications. In particular, in the econometrics literature we find formalized tests such as Ramsey's regression specification error test (see Ramsey, 1969) which relies on models with higher-order polynomials. It provides the F statistic for testing the hypothesis that any of the coefficients of the higher-order polynomials is different from zero. As we can draw similar conclusions from the single regression coefficients of such a model I will not discuss the regression specification error test in detail (but see Wooldridge, 2009).

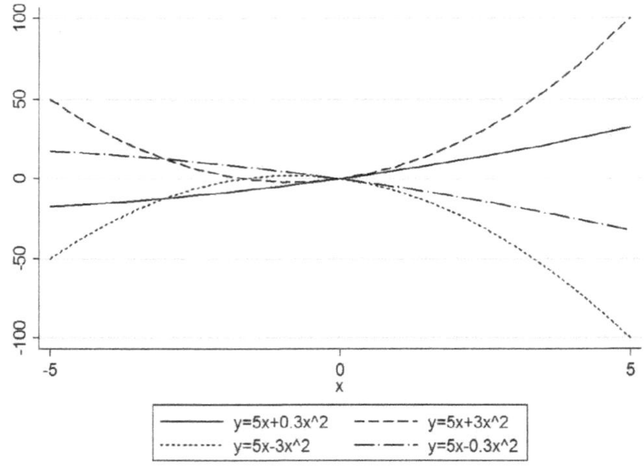

Figure 6.3 Model specifications with squared term (second-order polynomial)

If we find significant coefficients of polynomials we conclude that the relationship is non-linear. However, as in the case of a non-significant squared term discussed above, the opposite need not be true. Two additional caveats should be mentioned. Models with higher-order polynomials often face problems of multicollinearity. As in the case of interaction terms, this problem is reduced when the respective variable is mean-centered before the polynomials are generated. While multicollinearity renders your estimates inefficient, further problems affect the model predictions (Magee, 1998). Regression models with polynomials result in regression curves with – depending on the order of the polynomials – a given number of inflection points. With an inadequate number of inflection points peaks and valleys in the curve may be overemphasized. Related to this, polynomial regression often results in an inadequate model fit near the minimum or the maximum of a variable, (i.e. in the tails of the curve). In the context of the example analysis in the next section I will discuss the interpretation and problems of models with polynomials in more detail.

Splines

Spline regression is an alternative to polynomial regression which is less likely to generate the above mentioned problems (Harrell, 2001; Ruppert et al., 2003). The approach follows the general idea that the range of the independent variable is split into different intervals for which separate slopes are estimated. A similar result can be obtained by using dummy variables for each interval and interaction effects (Marsh and Cormier, 2001, p. 7). Figure 6.4 contains an example where both approaches are compared. We can see a clearly non-linear relationship where the effect changes twice across the range of the independent variable. There is a strong positive effect for values below -5, it is rather weak in the range -5 to 5, and strongly positive again for higher values. Figure 6.4a shows the predictions of a dummy variable regression (dashed lines) and of a dummy variable regression with interaction effects (solid lines). The dummy variable regression predicts an average value for each of the three groups. The use of interaction terms produces a much better fit with strongly and less strongly sloping segments of the regression line. As discussed above in the subsection on interaction effects, this approach equals the estimation

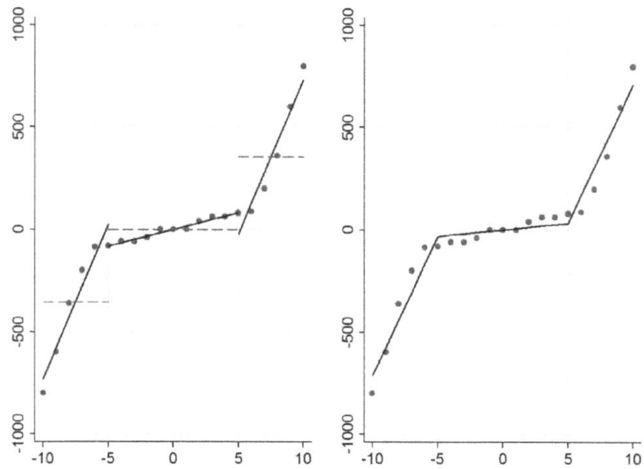

Figure 6.4 Dummy variable regression and regression with linear splines

of a separate model for each of the three groups. However, at the margins of each group we can observe jumps or gaps, i.e. the predicted value increases or decreases rather strongly with a small change in the independent variable only. The regression lines in different intervals are not connected to each other (see Figure 6.4a). Compare, for instance, the predicted values when the independent variable is -5 instead of -4.99. The model predicts a slightly positive value for the former and around -80 for the latter. It is rather improbable that these gaps reflect a pattern in the data when we use a continuous independent variable. (a) (b)

Figure 6.4b shows the same example using spline regression. The main difference compared to the dummy variable approach with interactions is that it produces a prediction without jumps across the full range of the independent variable, although the steepness of the slope changes twice. The values of the variable X for which a change in the slope is assumed (and modeled) are called *knots*. In this example two knots are defined: -5 and 5. The range of X is divided into three intervals. For each interval a separate regression coefficient is estimated. In its simplest form, which is illustrated in the example, we assume that the effect of the variable X on the variable Y is constant within each interval (i.e. linear). Therefore, we speak of *linear spline* or *piecewise constant* regression. However, as we assume different slopes for each of the intervals the relationship between X and Y across the full range of the variable X is non-linear. As the number of knots is unrestricted, we can model rather complex functional forms using linear splines. Below I will discuss how to choose the number and location of knots. But first suppose that the number and location of the knots are known. Then a regression model with three knots k_1, k_2 and k_3 is defined as

$$Y = b_0 + b_1 \cdot X + b_2 \cdot (X - k_1)_+ + b_3 \cdot (X - k_2) + b_4 \cdot (X - k_3)_+ + e, \tag{6.9}$$

where

$$(u)_+ = \begin{cases} u & \text{if } u > 0 \\ 0 & \text{if } u \leq 0. \end{cases}$$

The three knots divide the range of the variable X into four intervals. For each of these intervals we estimate a separate slope using $(u)_+$, that is, transformations of the original variable X with

$u = X - k$. If $(u)_+ = 0$ the respective terms are left out of the equation. As a consequence, we can simplify the equation for given intervals of the variable X:

$$Y = \begin{cases} b_0 + b_1 \cdot X + e & \text{if } x \leq k_1 \\ b_0 + b_1 \cdot X + b_2 \cdot (X - k_1) + e & \text{if } k_1 < x \leq k_2 \\ b_0 + b_1 \cdot X + b_2 \cdot (X - k_1) + b_3 \cdot (X - k_2) + e & \text{if } k_2 < x \leq k_3 \\ b_0 + b_1 \cdot X + b_2 \cdot (X - k_1) + b_3 \cdot (X - k_2) + b_4 \cdot (X - k_3) + e & \text{if } k_3 < x. \end{cases} \tag{6.10}$$

The model allows for a different slope for each of the four intervals. In the case of a non-linear relationship between X and Y at least one of the regression coefficients b_2, b_3 or b_4 differs significantly from zero. Although linear splines are suited to modeling rather complex non-linear relationships between X and Y, we may also use a polynomial instead of a linear function to define the splines. In particular, with highly curved functions polynomial splines are smoother and will render a better fit (Harrell, 2001, pp. 19ff.). Most common is the use of third-order polynomials (*cubic splines*). However, as with polynomials in general, cubic splines may result in a loose fit at the tails of the curve. *Natural splines* (also *restricted cubic splines*) offer a solution to this problem. In contrast to cubic splines, with natural splines we assume a linear relationship at the margins of the range of the independent variable (i.e. below the first knot and above the last knot). A model with natural splines and m knots is given by

$$Y = b_0 + b_1 \cdot X_1 + b_2 \cdot X_2 + \ldots + b_{m-1} \cdot X_{m-1} + e, \tag{6.11}$$

where $X_1 = X$ and, for $j = 1, \ldots, m-2$,

$$X_{j+1} = (X - k_m)_+^3 - (X - k_{m-1})_+^3 \cdot \frac{(k_m - k_j)}{(k_m - k_{m-1})} + (x - k_m)_+^3 \cdot \frac{(k_{m-1} - k_j)}{(k_m - k_{m-1})}.$$

Harrell (2001, p. 23) proposes a normalization where the cubic terms are divided by the square of the difference in the outer knots $(k_m - k_1)$. This renders an interpretation of all terms in units of X.[1]

So far we have assumed that the number and location of the knots are known. Usually, theory or previous research will provide a basis for the selection of knots. However, there are also approaches to defining the number and location of knots analytically (Marsh and Cormier, 2001). Most of these techniques are not convincing, but as they are covered in the literature I provide a brief discussion. If the number of knots is known it is possible to estimate the location of knots jointly with all other parameters (e.g. using non-linear least squares). However, in situations with more than one knot the approach often yields unstable results, strongly reducing its feasibility. Selecting starting values works as a fix as it increases the robustness of the estimation results. However, the outcome strongly depends on the choice of starting values, which again requires at least some prior knowledge about the location of the knots. If neither the number nor the location of knots is known Marsh and Cormier (2001, pp. 49ff.) propose an approach using stepwise regression. First, the independent variable is split into a large number of small intervals and a spline is constructed for each of these intervals. Second, the variable Y is regressed stepwise on all splines. This approach identifies all significant splines, that is, intervals which have a significantly different slope compared to the prior interval. This approach results in a smooth fit also with highly curved functions. Nevertheless, the approach does not seem to be suited for most applications as it exhibits all problems of stepwise regression. It is data-driven and not based on hypotheses on the form of the relationship. Nevertheless it extensively uses the model's test statistics which would require a prior specification of hypotheses (for a critical discussion, see Harrell, 2001, pp. 56ff.). In addition, a number of other problems arise. For instance, using a stepwise forward or backward selection yields different results.

Table 6.1 Suggested quantiles for the specification of knots in natural splines

k	quantiles						
3			0.1	0.5	0.9		
4			0.05	0.35	0.65	0.95	
5		0.05	0.275	0.50	0.725	0.95	
6	0.05	0.023	0.41	0.59	0.77	0.95	
7	0.025	0.1833	0.3417	0.5	0.6583	0.8167	0.975

Source: Harrell (2001, p. 23)

Apart from the automated but problematic approaches, some 'rules of thumb' have been established. In many situations five knots or fewer have proven sufficient (Stone, 1986, p. 313). A larger number of knots provides a better fit but may also result in highly curved functions which pick up every small, potentially random variation in the data. The distribution of the independent variable is often used as a starting point for choosing the location of knots – if prior evidence is not available. Choosing equally spaced intervals (e.g. quartile borders) is one approach. With natural splines this does not apply to the selection of the outer knots, which need to be placed near the minimum and maximum of the independent variable. Harrell (2001, p. 23) provides examples of the selection of knots using natural splines (according to the number of knots). As Table 6.1 shows, the outer knots are located near the minimum and maximum of X. These knots are positioned in a way that we can assume that the effect within the outer intervals is more or less constant. If this is not the case the fit in the tails will be rough – similar to polynomial regression – since extreme values are picked up.

EXAMPLE ANALYSIS

In the previous sections I have discussed different strategies to model non-additive and non-linear relationships. This section contains applications of these methods using data from the European Social Survey (ESS, Round 5).

Interaction effects

This example addresses the question of whether the relationship between age, earnings and satisfaction with the state of health services differs between East and West Germany. Satisfaction is measured using a standard 11-digit scale, where 0 represents the most negative and 10 the most positive value. The sample is restricted to working persons with positive earnings. Earnings are measured as monthly gross pay in euros.[2] The example analysis first looks at the influence of age and earnings on satisfaction with health services. In a second step, East–West differences are addressed.

Table 6.2 contains the coefficients of several regression models. Model 1 shows that age negatively affects an individual's evaluation of health services, while earnings affect this positively. The model predicts that persons become less satisfied the older they get. Higher earnings have the opposite effect.

Model 2 contains an age–earnings interaction term to test the assumption that the effect of age on the satisfaction with health services differs by level of earnings. We estimate the following model:

$$\widehat{Y} = b_0 + b_1 \cdot \text{age} + b_2 \cdot \text{earn} + b_3 \cdot (\text{age} \cdot \text{earn})\,. \tag{6.12}$$

Table 6.2 shows that the coefficient of the interaction term is positive. In Model 1 we see that there is a negative effect of age on satisfaction with health services. A positive interaction

Table 6.2 OLS regression with interaction effect (Example 1: satisfaction with health services)

	Non-mean-centered		Mean-centered
	M1	M2	M2c
age (in years)	−0.025***	−0.043***	−0.019**
	(0.007)	(0.012)	(0.007)
gross pay (in € 1000)	0.120**	−0.335	0.073
	(0.038)	(0.207)	(0.042)
age · gross pay		0.010*	0.010*
		(0.005)	(0.005)
intercept	5.383***	6.231***	4.581***
	(0.289)	(0.490)	(0.079)
R^2	0.021	0.028	0.028
RSS	5769.4	5724.4	5724.4
n	1057	1057	1057

Source: ESS Round 5 (own calculations, weighted), Subsample: Germany, monthly gross pay up to € 25,000.
Notes: Unstandardized coefficients and standard errors in parentheses. Significance levels: *** 0.001, **0.01, *0.05.

indicates that there is a less negative effect of age with higher levels of earnings. However, the coefficients of age and earnings have changed. These coefficients must not be compared to those in Model 1. In Model 2 we estimate conditional effects which refer to the effect of age for an individual who has no earnings (i.e. € 0 of gross pay). Given that we are considering a sample of working persons, this is an unlikely case. Hence, the conditional effects alone lack a meaningful interpretation and should not be interpreted without taking into account the interaction effect.

A better way to obtain meaningful results is to mean-center the variables age and earnings before we compute the interaction term (Model 2c). Now the effect of age refers to individuals with average earnings, which allows for a meaningful interpretation. Comparing Models 1 and 2c, we see that the coefficients of age and earnings have not changed fundamentally. Yet, at first glance the results of Models 2 and 2c seem to differ strongly. However, this is only due to the different scale of the variables (original values versus mean-centered values). Taking this difference into account, we obtain exactly the same predictions from both models.

Let us assume that the effect of age on satisfaction with health services differs between East and West Germany due to the differences in their twentieth-century histories. As discussed above, we use interaction terms of a categorical variable to model group differences (here: East and West Germans). We also use separate models for both East and West Germany (Table 6.3). In these models we can see that the coefficient of age is negative in West Germany as well as in East Germany. However, the latter coefficient is very small and insignificant. In West Germany older individuals are less satisfied with the health services, while in East Germany we observe no such relationship. The coefficients of earnings point in the same direction but have different values (0.101 versus 0.440). The difference in the intercepts shows that on average East Germans are slightly less satisfied with the state of the health services compared to their West German counterparts. The two models indicate differences between East and West Germany, but there is no direct way to test whether these differences are significant. This can be accomplished by estimating a joint model with interaction terms for each of the independent variables and the group variable as shown in the following equation:

$$\widehat{Y} = b_0 + b_1 \cdot \text{age} + b_2 \cdot \text{earn} + b_3 \cdot \text{east} + b_4 \cdot \text{age} \cdot \text{east} + b_5 \cdot \text{earn} \cdot \text{east}. \tag{6.13}$$

Table 6.3 OLS regression with interaction effects, group differences (Example 1: satisfaction with health services)

	separate models		East and West Germany	
	M1 (West)	M2 (East)	M1	M2
age (in years)	−0.028***	−0.002	−0.028***	−0.021**
	(0.007)	(0.012)	(0.007)	(0.008)
gross pay (in € 1000)	0.101*	0.440***	0.101*	0.048
	(0.039)	(0.096)	(0.039)	(0.046)
age · gross pay				0.011*
				(0.005)
East (Ref.: West)			−0.089	−0.041
			(0.163)	(0.171)
age · East			0.026	0.022
			(0.014)	(0.015)
gross pay · East			0.340	0.382***
			(0.103)	(0.113)
age · gross pay · East				−0.008
				(0.009)
intercept	4.670***	4.581***	4.670***	4.608***
	(0.086)	(0.139)	(0.086)	(0.089)
R^2	0.022	0.060	0.029	0.038
RSS	4383.5	1222.3	5720.3	5668.5
n	792	265	1057	1057

Source: ESS Round 5 (own calculations, weighted), Subsample: Germany, monthly gross pay up to 25,000 euros.
Notes: Unstandardized coefficients and standard errors in parentheses. Significance levels: ***0.001, **0.01, *0.05.

Table 6.3 contains the results (Model 1). Again we have to take into account that the coefficients of age and earnings refer to individuals for whom the group variable 'East' equals zero (i.e. West Germans). The same applies to the intercept. This is easy to see when we compare the coefficient of the separate model for West Germany (M1 West) with these coefficients in the joint model. Point estimates and standard errors are exactly the same. The coefficient of the group variable and the interaction refer to differences between East and West Germany. By adding the conditional age coefficient which refers to West Germany (−0.028) and the coefficient of the respective interaction term (0.026) we obtain a value of −0.002 which equals the coefficient in the separate East German model. Along these lines we obtain the East German intercept (4.670−0.089) and the East German earnings coefficient (0.101+0.340). The test statistics of the earnings' interaction coefficients provide evidence on significant group differences. Accordingly, we conclude that the relationship between earnings and satisfaction with the health services differs significantly between East and West Germany. However, the coefficients of the joint model do not directly provide information on the question of whether age and earnings have a significant effect in East Germany (only with reference to West Germany). But this is easily obtained by testing the linear combination of the conditional and the interaction effect:

$$H_0 : \beta_1 + \beta_4 = 0,$$

$$H_0 : \beta_2 + \beta_5 = 0,$$

$$H_0 : \beta_0 + \beta_3 = 0.$$

Testing these hypotheses gives us the same results as seen before in the separate model for East Germany. The model predicts that the intercept and the earnings coefficient differ significantly

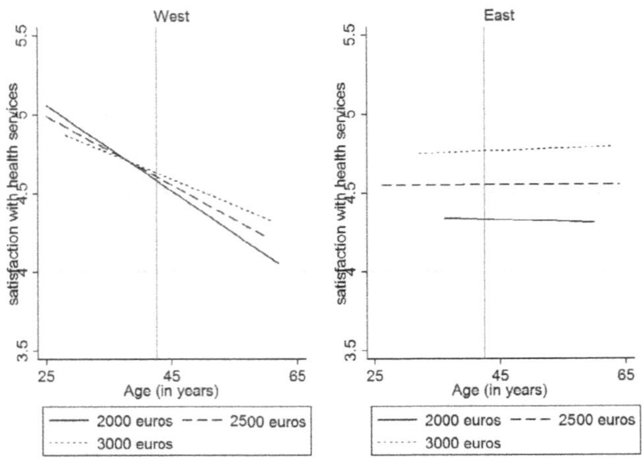

Figure 6.5 Age–gross pay interaction and group differences (Example 1: satisfaction with health services)

from zero. Thus, the model with group variable interactions provides the same information as separate models but also allows testing for group differences using the test statistic of the interaction terms and the group variable. As a summary test of group differences we can carry out a joint test of the null hypothesis that none of the coefficients β_3, β_4 and β_5 differs from zero. In this example the F statistic is 6.41. As this value is larger than the critical value of the 95th percentile of the F distribution, we reject the null hypothesis that East and West Germany do not differ. There is a significant difference in the constant or in at least one of the slope coefficients. The econometrics literature usually refers to the Chow statistic as a test for group differences (Wooldridge, 2009, p. 245). It is calculated differently but provides the information that the joint coefficient test has just described.

So far, I have not considered the interaction between age and earnings that I discussed above, but this is included in Model 2 (Table 6.3). Again, the conditional effect refers to West Germany (0.011), the coefficient of the interaction term to the difference between East and West (−0.008). The model indicates that there is an interaction between age and earnings in West Germany. For East Germany the separate Model 1 indicated that there is no such interaction. However, the coefficient of the three-way interaction term (age · earnings · East) in the joint model is non-significant. Despite this result that there is no significant difference to West Germany, we cannot conclude that the interaction between age and earnings is significant in East Germany. To check if there is a significant effect in East Germany we need to test the linear combination of the coefficients (0.011 − 0.008) which gives us the same insignificant result as in a separate model.

Graphical displays are helpful for interpreting three-way interactions such as in this example. Figure 6.5 shows the predicted values of the satisfaction with the state of the health services for three different earnings groups (gross monthly pay of €2,000, €2,500 and €3,000) according to age. The second value is near mean earnings. At first glance we see that higher age is related to lower satisfaction in West Germany, while there is no such relationship in East Germany. Furthermore, the almost parallel lines in the East German graph indicate a lack of interaction between age and earnings, while in West Germany the impact of age on satisfaction is strongest for those with lower earnings. These differences are clearly visible in the graphical display. In addition, the figures offer insights into the interpretation of the intercept. The intercept indicates the predicted value for an individual with mean age and earnings. The vertical line in the graphs shows the mean age in the sample (42.5 years). The regression line for individuals with earnings of €2,500 (almost the mean) intersects the vertical line near the value of the intercept (West, 4.670; East, 4.581).

Polynomials and splines

This example uses a subsample of ESS data on Danish men aged 19–74 years. The respondents were asked how many hours they would like their partners to work. The example analysis addresses the question of how the preferred working hours of the partner differ according to the age of the respondent. Suppose that this preference changes more than once over the life course. Thus, it is unlikely to observe a linear effect of age on the preferred working hours of the partner. Apart from theoretical reasoning on the impact of events such as having children in younger and middle age or leaving the labor force in older age, a visual test of the linearity assumption is used as a starting point for the analysis. Figure 6.6 shows the relationship between respondent's age and the preferred working hours of the partner in the form of a scatter plot. The bubble size represents the number of cases. Due to the large number of cases no pattern would be detectable from a simple scatter plot. Most pronounced are the bubbles around 30 working hours, slightly below 40 hours and at zero hours when the respondent is aged 60 or older. The figure contains conditional means (for each year of age) which show that preferred working hours are highest on average in younger years, fall slightly in the early thirties and go up again before a decrease starts which is related to leaving the labor force. In addition, the figure contains the result of a lowess smoother (Cleveland, 1979). The curve represents the results of a larger number of regression models where the dependent variable is regressed on the independent variable, taking into account only a restricted interval of the latter. The conditional means as well as the lowess smoother also provide evidence that the linearity assumption is violated. Despite this fact, we begin with a linear regression for the preferred working hours of the partner on respondent's age. This model will be compared against models which take non-linearity into account. Table 6.4 shows that Model 1 predicts a linear decrease of preferred working hours with increasing age. Figure 6.7 contains a graphical display of this and other estimation results. It also contains conditional means which were also provided in Figure 6.6. We see that the negative slope of the regression line is mainly driven by the decline in preferred working hours with higher age. The model does not pick up the decrease in hours in the respondents' thirties and underestimates the hours for respondents in the age range of about 40–60 years. It clearly overestimates the working time preference in higher age groups. Still, the model explains 15.4% of the variance of the preferred working hours of the partner.

As discussed in the previous section, we can use higher-order polynomials to model non-linearity. Model 2 contains a second-order polynomial of the independent variable. Additional models (Models 3 and 4) contain third- and fourth-order polynomials. The range of these variables has been restricted by dividing them by 10, 100 or 10,000. Without this transformation we would obtain very small coefficients for higher-order polynomials which would require tables with more decimal places than usual. The coefficient of the squared term in Model 2 is negative. In contrast to Model 1, the coefficient of the linear term is positive. In broad terms these results suggest that with increasing age preferred working hours go up, but go down with even higher age. Yet, it is not always easy to derive this result directly from the coefficients as the metric of the variable is different (years versus squared years divided by 10). However, using some

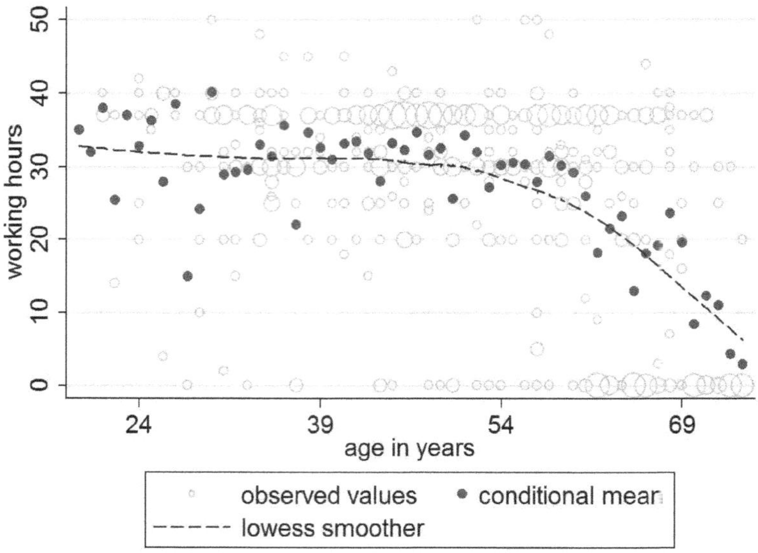

Figure 6.6 Respondent's age and partner's preferred working hours (Example 2: Working hours)

Table 6.4 OLS regression with polynomials (Example 2: Working hours)

	Non-mean-centered		Mean-centered		
	M1	M2	M2c	M3	M4
Age:					
in years	−0.409***	1.597***	−0.469***	−0.226**	−0.240*
	(0.048)	(0.284)	(0.046)	(0.079)	(0.096)
in years^2/10		−0.204***	−0.204***	−0.250***	−0.279***
		(0.029)	(0.029)	(0.032)	(0.079)
in years^3/100				−0.677***	−0.620*
				(0.173)	(0.247)
in years^4/10,0000					0.053
					0.126
intercept	47.759***	2.095	30.625***	31.080***	31.263***
	(2.284)	(6.675)	(0.684)	(0.709)	(0.794)
R^2	0.154	0.229	0.229	0.247	0.247
RSS	75416	68689	68689	67113	67092
n	472	472	472	472	472

Source: ESS Round 5 (own calculations, weighted), Subsample: Denmark, men, 19–74 years, two outliers excluded.

Notes: Unstandardized coefficients and standard errors in parentheses. Significance levels: ***0.001, **0.01, *0.05.

example values clearly shows the interplay between the two coefficients. The model predicts the following preferred working hours according to respondent's age:

$$\begin{aligned}
&20 \text{ years: } 25.9 \text{ hours } (= 2.095 + 1.597 \times 20 - 0.204 \times 20^2/10),\\
&40 \text{ years: } 33.3 \text{ hours } (= 2.095 + 1.597 \times 40 - 0.204 \times 40^2/10),\\
&60 \text{ years: } 24.5 \text{ hours } (= 2.095 + 1.597 \times 60 - 0.204 \times 60^2/10).
\end{aligned}$$

The positive coefficient of the linear term dominates within the lower age range before the negative squared coefficient gains in importance. From the example values above we can see that the curve turns around the age of 40 years. At the turning point the slope equals zero. Using calculus we can analytically determine the turning point of the curve. At this point the first derivative of the model equation which refers to the curve's slope equals zero. Hence, we set the first derivative to zero and solve the equation

$$\begin{aligned}
1.597 - 0.204 \times 2 \times \text{age}/10 &= 0\\
\Leftrightarrow \text{age} = \frac{1.597}{2 \times 0.204} \times 10 &= 39 \text{ years.}
\end{aligned}$$

According to the model, men aged 39 years prefer their partners to work the largest number of hours while younger and older men have a preference for a smaller number of hours. The predicted values of Model 2 underline this result. The model fits better than the linear model (with an R-squared value of 0.229 versus 0.154). However, Figure 6.7 also shows that the model is ill-fitting for younger men who prefer higher working hours. As discussed in the previous section, polynomial regression often results in an inadequate model fit in the tails of the curve. Apparently neither Model 1 nor Model 2 fully captures the complex relationship between age and the preferred working hours. Models with higher-order polynomials may offer a solution but – as discussed above – often face problems of multicollinearity. Therefore, we use a mean-centered age variable before we compute the polynomials. Mean-centering changes the scale of the variables, which we can clearly see when comparing Model 2 and Model 2c. In the latter model the intercept refers to the mean of the variable. Values of age below the mean turned into negative values. This explains the negative coefficient of the linear term, although the curve is upward sloping. Even when bearing in mind the mean-centering it is difficult to interpret the coefficients without calculating or plotting predicted values. It is important to note that although the size and the direction of the coefficients of Models 2 and 2c differ, the predicted values are exactly the same. Test statistics differ as the coefficients refer to different reference values and as mean-centering reduces multicollinearity. In this example, the bivariate correlation of age and the interaction term is $r = 0.10$, compared to $r = 0.50$ using the variables on the original scale. However, the change in the test statistic is hard to see as in both models the linear age effect is significant at the 0.1% level. Looking more closely at Models 2 and 2c reveals a difference in the age coefficient by a factor of 3.5, while the standard error differs by a factor of 6. Hence, compared to the size of the coefficient the standard error is smaller, indicating a more efficient estimation. A lack of efficiency is primarily a problem when using higher-order polynomials, which makes mean-centered variables generally advisable. Model 3 contains a third-order polynomial. The inclusion of a cubic term again increases the explanatory power of the model (R-squared is now 0.247). As Figure 6.7 clearly shows, this is due to the better fit in the lower age range. Model 4 additionally contains a fourth-order polynomial which does not increase the fit. Here we see that more complex models do not necessarily capture a given relationship better. Yet taken together, the analyses show that there is a non-linear relationship between age and the preferred working hours of the partner. Including higher-order polynomials

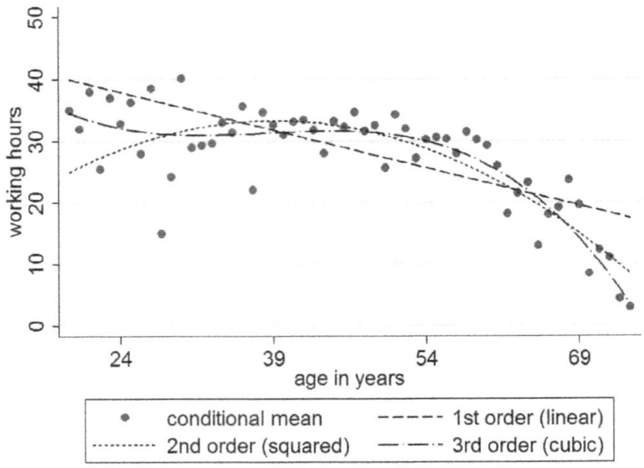

Figure 6.7 Polynomials – regression estimates and conditional means (Example 2: working hours)

is a simple way to relax the linearity assumption. Many research questions in the social sciences address non-linear relationships. In the example analysis the inclusion of polynomials up to third order offers an adequate solution.

An alternative to the use of polynomials is spline regression. As discussed in the previous section, it follows the basic idea of estimating different slopes for intervals of the independent variable – separated by so-called knots. In a first step I assume that the number and location of the knots are known.[3] As argued above, with events such as having children or the advent of retirement, working time preferences are likely to change. Therefore, I define two knots at the ages of 35 and 55 years and estimate the equation

$$Y = b_0 + b_1 X + b_2(X - 35)_+ + b_3(X - 55)_+ \qquad (6.14)$$

where

$$(u)_+ = \begin{cases} u & \text{if } u > 0 \\ 0 & \text{if } u \leq 0. \end{cases}$$

Table 6.5 shows the estimation results (Model 1). The three regression coefficients (Splines 1, 2 and 3) depict the influence of age on the preferred working time of the partner within the intervals separated by the two knots. The model predicts a decrease of preferred working hours up to the age of 35 years (−0.198). However, as the coefficient is insignificant at the 5% level we must assume that there is a constant effect of age with this interval. The model predicts an increase above age 35 (+0.234), a strong decrease from 55 years onwards (−0.990) and thus captures the non-linearity well. Accordingly, the model fits better than the linear model (R-squared 0.231 versus 0.154). However, we cannot rule out that more complex models might fit even better. Model 2 offers a solution with four splines (knots at 25, 45 and 60 years). In fact, the fit increased (R-squared increasing to 0.247). The model does not show an increase in preferred working hours from the mid-thirties onwards but predicts that reduced working hours are preferred, not just at age 55 but even earlier. The comparison to Model 1 shows that a slightly different choice of knots may change the substantial interpretation of the model.

Table 6.5 OLS regression with linear and natural splines (Example 2: Working hours)

	Linear splines		Natural splines		
	M1	M2	M3	M4	M5
age:					
Spline 1	−0.198	−0.881	0.229*	−0.043	−0.088
	(0.193)	(0.682)	(0.091)	(0.134)	(0.165)
Spline 2	0.234*	0.082	−0.853***	0.148	0.485
	(0.116)	(0.123)	(0.118)	(0.383)	(0.795)
Spline 3	−0.990***	−0.292*		−3.436*	−4.503
	(0.103)	(0.124)		(1.477)	(4.782)
Spline 4		−1.594***			3.304
		(0.207)			(7.925)
intercept	37.185***	52.500**	35.638***	30.688***	29.731***
	(6.243)	(15.889)	(1.215)	(2.214)	(3.026)
R^2	0.231	0.247	0.234	0.247	0.246
RSS	68528	67117	68253	67162	67171
N	472	472	472	472	472

Source: ESS Round 5 (own calculations, weighted), Subsample: Denmark, men, 19–74 years, two outliers excluded.
Notes: Unstandardized coefficients and standard errors in parentheses. Significance levels: ***0.001, **0.01, *0.05.

Models 3–5 use natural splines which combine the characteristics of spline and polynomial regression. The aim is to fit smooth regression curves with a minimum of predicting variables. Natural splines assume a linear function in the tails of the curve (values below the first knot and above the last knot) and a cubic function in the intervals between the knots. The incorporation of cubic terms makes the interpretation of the coefficients without a graphical display more difficult. Knots are selected using the suggested values provided in Table 6.1. Regarding the overall fit, Models 3–5 hardly differ from the models with linear splines. Model 3 predicts an inverted U-shaped curve. Model 4 is similar to Model 2. Model 5 does not seem to capture the relationship between the respondent's age and preferred working hours very well. All its coefficients are insignificant. The number of parameters is higher than in Model 4, but the overall fit does not increase.

Figure 6.8 contains the results of some selected models (Models 1, 3 and 4). Although the distances between the curves are quite small, we can see some relevant differences. Model 3 (natural splines, three knots) is rather ill-fitting within the lower age range. As discussed in the previous section, with natural splines we assume a linear relationship at the margins of the range of the independent variable. In the case of Model 3, too small a number of knots seemingly leaves wide margins which results in a far from perfect fit in the lower tail of the curve. Models 1 (natural splines, two knots) and 5 (natural splines, four knots) perform quite similarly within this age range. The main difference is that Model 1 predicts increasing working hours above age 35, while according to Model 5 there is hardly any change in the slope up to age 50. It is certainly hard to tell which model to choose. Following the theoretical considerations, one might prefer Model 1 over Model 5. However, we cannot completely rule out that the second spline in Model 1 just picks up a random increase in preferred working hours which proves to be irrelevant when using a slightly different model. Comparing the predictions of the spline regression models with those of the polynomial regression models, we see the main differences in the curve's tails. At the margins of the variable's range the predictions of the spline regression models are slightly less extreme.

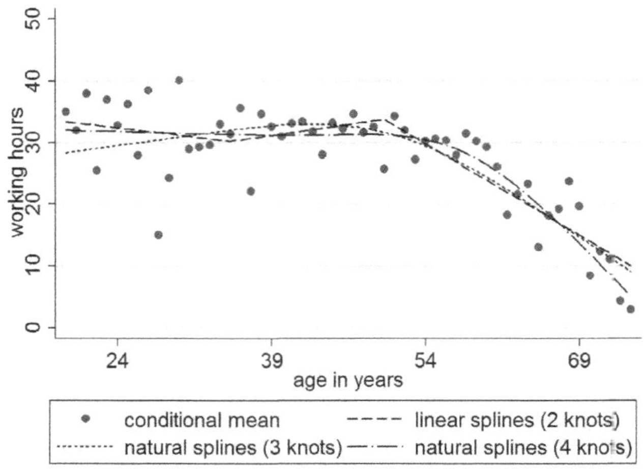

Figure 6.8 Linear and natural splines – Regression estimates and conditional means (Example 2: Working hours)

CAVEATS AND FREQUENT ERRORS

This chapter has discussed frequently used approaches towards non-additivity and non-linearity which are easily applied in the framework of multiple regression analysis. Concluding I discuss some caveats and frequent errors. *Interaction* terms are easily included in a regression model but not as easily interpreted, and graphical displays are highly advisable. Above all, it must be acknowledged that the inclusion of an interaction term $X_1 \cdot X_2$ changes the meaning of the coefficients of the variables X_1 and X_2 into conditional coefficients. The conditional coefficient of X_1 only refers to the case of $X_2 = 0$. Related to this, a further problem may occur. The conditional effect has no meaningful interpretation if the range of the variable X_2 does not include 0. Using mean-centered variables solves this problem. As the mean of a centered variable equals zero, the conditional coefficient of X_1 refers to an average case with respect to X_2 which is provided with a meaningful interpretation. However, it should be noted that this solution cannot be easily transferred to standardized regression coefficients (Aiken and West, 1991, pp. 40ff.). In general, mean-centering reduces the degree of multicollinearity in models with interaction terms.

The use of higher-order *polynomials* for modeling non-linearity is even more prone to the problem of multicollinearity. Again, mean-centering helps to deal with this problem. However, mean-centering affects the interpretability of the results. In particular, if more than a squared term is used graphical displays are advisable. The application of polynomials is often difficult to justify on a theoretical basis as exact turning points cannot be specified in advance but are data-driven. Furthermore, the fit in the tails of the curve is often rough.

In contrast, *spline regression* allows for an a priori specification of turning points. However, following the 'rules of thumb', splines can be specified without detailed theoretical reasoning. However, you should bear in mind that the fit of spline regression models strongly depends on the choice of the number and location of knots. Sensitivity analyses may help to rule out misspecified models. I have not discussed in detail the approaches that 'automatically' define the number and location of knots (but see Ruppert et al., 2003). One reason for this omission is that the resulting complex functions are often hard to justify from a substantive perspective. Such approaches and the use of test statistics can certainly help to detect violations of standard regression assumptions but should not fully replace the specification of models starting off from theoretical considerations. If the choice is between a theoretically grounded parsimonious model and a more complex model which fits only slightly better, the former model is certainly preferable.

FURTHER READING

The topics of this chapter are broadly discussed in the encompassing literature on regression analysis. Jaccard and Turrisi (2003) provide a short and illustrative overview of interaction effects in multiple regression. Aiken and West (1991) provide an in-depth discussion on the topic (including a discussion of interaction terms in the context of standardized regression coefficients). Ai and Norton (2003) give a brief but focused discussion of interaction effects in logistic regression. Harrell (2001) provides a general treatise on regression model specification which includes an advanced discussion of non-linear modeling approaches. Ruppert et al. (2003) discuss spline regression in detail, including approaches not covered in this chapter such as penalized spline regression which is often used for scatter plot smoothing.

NOTES

1 This normalization is implemented in the spline regression commands of standard statistical software packages (e.g. mkspline in Stata: see StataCorp, 2011, pp. 1174ff.).
2 For some respondents the variable seems to contain information on annual pay. As such outliers have a strong impact on the results, as an easy fix 36 observations with earnings higher than €25,000 have been excluded.
3 The examples in this chapter do not require a complex construction of splines and can easily be generated manually. However, standard statistical software packages often have commands which provide for spline construction. The splines of this example analysis were generated using Stata's mkspline.

REFERENCES

Ai, C. and Norton, E. C. (2003). Interaction terms in logit and probit models. *Economics Letters*, 80(1), 123–129.
Aiken, L. S. and West, S. G. (1991). *Testing and Interpreting Interactions*. Thousand Oaks, CA: Sage.
Cleveland, W. S. (1979). Robust locally weighted regression and smoothing scatterplots. *Journal of the American Statistical Association*, 74, 829–836.
Cronbach, L. (1987). Statistical tests for moderator variables: Flaws in analysis recently proposed. *Psychological Bulletin*, 102, 414–417.
Delhey, J. (2010). From materialist to post-materialist happiness? National affluence and determinants of life satisfaction in cross-national perspective. *Social Indicators Research*, 97, 65–84.
Fernandez-Mateo, I. (2009). Cumulative gender disadvantage in contract employment. *American Journal of Sociology*, 114(4), 871–923.
Harrell, F. E. (2001). *Regression Modelling Strategies*. New York: Springer.
Jaccard, J. and Turrisi, R. (2003). *Interaction Effects in Multiple Regression*. Thousand Oaks, CA: Sage.
Magee, M. (1998). Nonlocal behavior in polynomial regressions. *American Statistician*, 52(1), 20–22.
Marsh, L. C. and Cormier, D. R. (2001). *Spline Regression Models*. Thousand Oaks, CA: Sage.
McDonald, S. and Mair, C. A. (2010). Social capital across the life course: Age and gendered patterns of network resources. *Sociological Forum*, 25(2), 335–359.
Ramsey, J. B. (1969). Tests for specification errors in classical linear least-squares analysis. *Journal of the Royal Statistical Association, Series B*, 71, 350–371.
Ruppert, D., Wand, M. P. and Carroll, R. J. (2003). *Semiparametric Regression*. Cambridge: Cambridge University Press.
StataCorp (2011). *Stata Base Reference Manual, Release 12*. College Station, TX: StataCorp.
Stone, C. J. (1986). Comment: Generalized additive models. *Statistical Science*, 1, 312–314.
Wooldridge, J. M. (2009). *Introductory Econometrics: A Modern Approach*. Mason, OH: Thomson/South-Western.

7

The multilevel regression model

Joop Hox and Leoniek Wijngaards-de Meij

INTRODUCTION

Social and behavioral research often concerns data that have a hierarchical structure, with individuals nested within groups. In multilevel analysis, such data structures are viewed as a multistage sample from a hierarchical population. For example, in educational research we may have a sample of schools, and within each school a sample of pupils. This results in a data set consisting of pupil data (e.g. socioeconomic status, intelligence, school career) and school data (e.g. school size, denomination, but also aggregated pupil variables such as mean socioeconomic status). In this chapter, the generic term 'multilevel' is used to refer to analysis models for hierarchically structured data, with variables defined at all levels of the hierarchy. Typically, the research problem includes hypotheses of relationships between variables defined at different levels of the hierarchy.

A well-known multilevel model is the hierarchical linear regression model, which is essentially an extension of the familiar multiple regression model. It is known in the literature under a variety of names, such as 'hierarchical linear model' (Bryk and Raudenbush, 1992; Raudenbush and Bryk, 1986), 'variance component model' (Longford, 1989), and 'random coefficient model' (Longford, 1993; De Leeuw and Kreft, 1986). It has become so popular that 'multilevel modeling' has become almost synonymous with 'applying a multilevel regression model'. However, since we also have multilevel extensions of other models, such as factor analysis or structural equation models, we reserve the term 'multilevel model' for the general case, and refer to specific classes of models as 'multilevel regression analysis' and 'multilevel structural equation models'.

The multilevel regression model is a hierarchical linear regression model, with a dependent variable defined at the lowest (usually the individual) level, and explanatory variables at all existing levels. Using dummy coding for categorical variables, it can be used for analysis of variance models. The model has been extended to include dependent variables that are binary, categorical, or other non-normal data, and generalized to include multivariate response models and cross-classified data (Bryk and Raudenbush, 1992; Longford, 1989; Goldstein, 1995).

We begin by introducing a two-level regression model. We illustrate what kind of questions can be answered, and how these questions will be answered with a multilevel analysis. Next we explain the multilevel model in more detail, with equations for the separate levels, and the mathematics behind the multilevel model. We then present two examples: first a three-level analysis with data from the European Social Survey (ESS); and second a repeated measures multilevel analysis. We conclude with some caveats and frequent errors of multilevel analyses.

THE BASIC MULTILEVEL REGRESSION MODEL

In most applications of multilevel regression analysis, the first (lowest) level consists of individuals, the second level of groups of individuals, and there might be higher levels of sets of groups. For the example of this section data of the third round of the ESS is used (for more information, see the examples section below). On the individual level, we have the dependent variable, satisfaction with the state of the economy (stfeco) (Y_{ij}), and the explanatory variables, gender (gndr) and trust in politicians (trstplt) (X_{ij}). On the country level, we have the explanatory variable, average working hours a week (wkhtot, aggregated by country) (Z_j). Missing data on the first-level variables were handled using listwise deletion. The hierarchical data set and the availability of predictor variables at different levels give rise to several possible research questions.

First, we want to know whether satisfaction with the state of the economy can be predicted by gender and trust in politicians, just as in single-level regression analysis. We will add the first-level predictors gender and trust in politicians to the model as a predictor of the individual differences in satisfaction with the state of the economy. Second, we shift our attention to the country level and want to know whether there are differences between the countries in their average satisfaction scores. By doing this we are checking the assumption of dependency in the data: do individuals from the same country look more alike than individuals from different countries? When individuals from one country look more alike, this will result in differences between the average satisfaction with the state of the economy of countries. When there are no differences between the countries in satisfaction, it is neither necessary nor possible to do a multilevel regression analysis on this data set. On the other hand, when there are differences between the countries on average satisfaction, the question can be asked whether the differences between the countries can be predicted by the average working hours a week. Is it true that with a higher average of working hours a week the average satisfaction with the state of the economy of a country is higher? While looking at the country level another question arises, a question that is related to predicting the satisfaction of the individual by a first-level predictor, such as gender. In addition to looking at the average regression coefficient of gender for the total sample, we might wonder whether this relationship between gender and satisfaction is actually the same for all the countries in the population. It could be that males are more satisfied in general, but that the strength of this relationship between gender and satisfaction varies between the countries. It could also be that the relationship is alike for some countries, but that the relationship is opposite for other countries (females are more satisfied). The hierarchical structure in our data enables us to check these differences between the countries. When there are indeed these differences between the countries, the question can be asked why this relationship between gender and satisfaction differs: can the average working hours of people in the country be the moderator for this relation? Do countries with a higher average number of working hours have a stronger relationship between gender and satisfaction than countries with a lower number of average working hours, or perhaps it is the other way around?

These questions will now be addressed, and the multilevel regression model will be built step-by-step. To be able to build and understand a multilevel regression model, you should be well acquainted with the basic terms of single-level regression analysis because they play a crucial role in multilevel regression analysis (see Chapter 4 of this volume).

We will work with the example of predicting the satisfaction with the state of the economy of individuals who are nested within countries. The predictors of the first level are gender and trust in politicians, and the predictor of the second level is the average number of working hours. The sample consists of 40,526 individuals in 29 European countries.

Intercept-only model

The individuals are nested within countries, and therefore we expect a dependency in the satisfaction scores of the individuals within countries. To check whether this is correct we must ask: do the satisfaction scores of individuals within a country look more alike than the satisfaction scores of individuals from different countries? Or in other words, do the means of the countries on average satisfaction differ between the countries? To answer this question, we must compare the means of the countries. To be able to compare the means of the countries in a multilevel analysis we start, in contrast to a single-level regression analysis, with a model without any predictors. Just as in a single-level regression analysis, the intercept in a model without predictors is equal to the average satisfaction in the sample. To test whether the means of the groups differ, we do not want to estimate just one intercept for the whole sample as in a normal regression analysis. Instead, we want an intercept for each group estimated separately. Within the multilevel regression model this is accommodated by adding a variance term to the average intercept, so allowing the intercepts to differ between countries. When the means of satisfaction with the state of the economy of the countries are the same, this will result in a variance component around the intercept of (almost) zero. In contrast, when there is dependency in the data, the means (and therefore the intercepts) will differ between the countries and this will be represented in the multilevel model as a variance component of the intercept that is larger than zero. In this first step, the total variance of the satisfaction scores in the sample is allocated to two different places, variance of the individuals representing the individual differences within the countries (on the first level), and variance of the intercept representing the differences between the countries (on the second level). To give an indication of the level of dependency on the dependent variable in the data, the intra-class correlation (ICC) can be calculated. To calculate the ICC, the second-level variance of the intercept is divided by the total variance (individual variance plus intercept variance). The ICC can be interpreted in two ways. The first interpretation is that the ICC is the expected correlation between two randomly chosen units within one class. The second interpretation is that the ICC is the proportion of the total variance that is located at the second level. The higher the value of the ICC, the stronger the dependency in the data will be. In our example a high ICC would mean that individuals within a country resemble each other more in satisfaction than when there is a low ICC.

When we turn to the data, the results show that the variance at the individual level is 4.24 and the variance at the class level is 2.25. From this we can calculate the ICC: $2.25/(2.25+4.25) = 0.35$. This indicates that 35% of the total variance of satisfaction is located at the country level, and the remaining 65% of the variance of satisfaction is therefore located at the individual level. To put it another way, the expected correlation of satisfaction between two people from the same country is expected to be 0.35. We conclude that in the sample there are differences between the countries, so the next step is a statistical test to establish the significance of the results.

To test the significance of the variance term at the country level a chi-square test is used on the difference between the deviance between two models: the model with only the individual variance (model 0, the normal regression model) and the model with both the individual variance and the country-level variance (model 1, the two-level regression model). The difference between the two deviance terms has a chi-square distribution. This model has one degree of freedom because, compared to model 0, only one extra parameter is estimated in model 1. In our example we find $\chi^2(1) = 15{,}917.65, p < 0.001$ (see below for a more detailed description of this test). We conclude that the satisfaction of individuals within countries looks more alike than the satisfaction of individuals between countries, therefore it is necessary to analyze the satisfaction in this data set by doing a multilevel regression analysis.

First-level predictors

Having established that there is dependency in the data (and that a multilevel analysis is therefore needed!), predictors are added to the model. First, we try to predict the individual differences by adding the variables of the lowest level, the individual level, to the equation (see model 2 in Table 7.1). This is actually the same as in a normal regression analysis. So, to predict differences between the individuals, the variables gender and trust in politicians are added to the model. Note that these effects are estimated for the whole sample, and that at this point in the analysis the grouping structure is not yet incorporated in the estimation of the effect of these first-level explanatory variables. In general, males appear to be more satisfied than females (females coded as 1; $b = -0.35, t(40523) = -18.30, p < 0.001$) (see the next section for a more detailed description of this test). Furthermore, the higher the trust in politicians, the higher the satisfaction ($b = 0.35, t(40523) = 78.30, p < 0.001$).

To give an indication of the relevance of the predictor, the explained variances (R^2) at the different levels are calculated (also called *pseudo* R^2; see the fourth section of this chapter for more information). By adding the predictors of the first level, not surprisingly, the variance component of the first level (the individual differences) has decreased. A part of the individual differences is explained by gender and trust in politicians. But not only the variance component of the first level has decreased; the variance term on the second level has also become smaller. Although the individual predictors are measured at the first level, there can be differences between the countries on the aggregated score of these predictors on the second level. For example, when males are more satisfied than females, the average satisfaction of a country with more males will be higher than the average satisfaction of a country with more females. By adding gender to the model, part of the differences on average satisfaction between countries will be explained.

To calculate the explained variance from the first level, we use the procedure described by Bryk and Raudenbush (1992) and we subtract the variance from model 2 from the variance from the baseline model (model 1), and divide it by the variance of the baseline model, $(4.25 - 3.66)/4.25 = 0.14$. Of the variance of the individual level 14% is explained by the variables gender and trust in politicians. When describing the intercept-only model (model 1) we saw that 65% of the total variance was at the first level. Of the 65% of the individual level 14% (so 9.1% of the total variance) is explained by the variables gender and trust in politicians. To calculate the explained variance of the second level, we use the same calculation but now with the variance terms of the second level, $(2.25 - 1.39)/2.25 = 0.38$. Of the variance of the second level 38% is explained by the two first-level predictors. In the baseline model 35% of the variance was on the second level and 38% of that (so 13.3% of the total variance) is explained by the first-level variables trust and gender. Combining this information, we can say that gender and trust in politicians explain 22.4% of the total variance.

Second-level predictors

Next, the differences between (the intercepts of) the countries are predicted by entering predictors of the country level (second level) into the model (see model 3 in Table 7.1). The principle is the same as predicting individual differences at the first level by individual variables, only now

Table 7.1 Results of multilevel regression models for satisfaction with the state of the economy

	Model 0	Model 1	Model 2	Model 3	Model 4	Model 5
Fixed part (B)						
Intercept	5.02	5.05	5.05	5.10	5.10	5.10
	(0.012)	(0.313)	(0.246)	(0.209)	(0.209)	(0.209)
Gender			−0.35	−0.35	−0.35	−0.35
			(0.019)	(0.019)	(0.039)	(0.032)
Trust in politicians			0.35	0.35	0.35	0.35
			(0.004)	(0.004)	(0.004)	(0.004)
Average working hours				−0.26	−0.26	−0.26
				(0.083)	(0.083)	(0.083)
Gender × AWH						0.04
						(0.013)
Random part						
σ_e^2	6.310	4.245	3.663	3.663	3.663	3.663
σ_{u0}^2		2.253	1.393	0.992	0.997	0.997
σ_{u1}^2					0.026	0.015
Fit						
Deviance	189,673.83	173,756.18	167,786.44	167,781.14	167,744.33	167,740.94

characteristics of the countries are used to predict the average score of the dependent variable of the country. When a country scores one unit higher on the predictor, the average satisfaction of the country is predicted to be higher with the value of the slope. In our example, the average satisfaction of the countries is predicted by average working hours, $b = -0.26$, $t(27) = -3.138$, $p = 0.005$, so for each average working hour the predicted average satisfaction of the country will go down by 0.16. The satisfaction of the individual is now predicted by both individual characteristics and country characteristics from the country he or she lives in.

In the previous subsection 38% of the intercept variance (the differences between the countries on satisfaction) was explained by the variables gender and trust in politicians. By adding the variable average working hours, the variance of the second level decreases to 0.99. To calculate the variance explained by gender, trust in politicians and average working hours we use the same procedure, $(2.25 - 0.99)/2.25 = 0.56$. Gender, trust in politicians and working hours explain 56% of the 35% of variance that was located at the second level in the intercept-only model. The first-level variables gender and trust in politicians explained 38%, so the unique contribution of average working hours is 18%.

Until now we have calculated the explained variance at each level separately. To give an indication of the total variance that is explained, we can say that 14% of the initial 65% of variance of the first level (=9.1%), plus 56% of the initial 35% of variance of the second level (=19.6%) is explained. In total 28.7% of the variance is explained by the variables gender, trust in politicians and working hours.

Random slope variance

The hierarchical structure of the data gives us the opportunity to study not only individual differences, but also difference between countries. In the previous subsection we looked at the difference in the average satisfaction between countries. Now we go one step further and look at the relationship between two individual variables within each country, and whether this relationship differs between the countries. In models 2 and 3 the relationship between gender and satisfaction is established based on the total sample and therefore an average slope over all the countries and all the individuals is established. We now want to study the relationship between gender and satisfaction not only for the entire sample, but also for each country separately. It could be that in one country there is a strong positive relationship between gender and satisfaction, while in another country this relationship is weak or even negative. To study whether the regression coefficients for gender differ between countries, a new variance term is added to the multilevel model (see model 4 in Table 7.1). When the slope for gender is equal for all countries, there is no random slope and the variance around the slope of gender will be zero. When the variance around the slope is significant, this means the slopes for gender differ between the countries. Be aware that the slope of average working hours from the second level cannot be made random at the second level. Each class has only one value for average working hours, and therefore we cannot calculate a slope for average working hours for each country separately. If we had a three-level model, with individuals nested in countries and countries nested in continents, we could make the average working hours random at the country level to see whether the relationship between average working hours and average satisfaction of the country differed between the continents.

When the variance around the slope of gender is added to the model, it is tested by a chi-square test (analogous to the chi-square test of intercept variance) and the variance is significant ($\sigma^2_{u1} = 0.026$, $\chi^2(1) = 36.81$, $p < 0.001$). In other words, the relationship between gender and satisfaction differs between countries. The difference between males and females scores on satisfaction ($b = -0.35$) is not the same in all the countries.

Cross-level interaction

The last step for this multilevel analysis depends on the presence of random slope variance in the model. If the relationship between gender and satisfaction does indeed differ between countries, the question is whether (part of) these differences can be explained by some characteristic of these countries. Perhaps in countries with higher average working hours the relationship between gender and satisfaction is less strong than in countries with lower average working hours. The effect of gender on satisfaction would in this case depend on the average working hours in the country, and therefore average working hours is a moderating variable (see Chapter 6 in this volume for moderation in standard one-level regression analysis). This can also be seen as an interaction between the two predictors on satisfaction. Because one predictor from the interaction is from the first level and the other predictor is from the second level, this type of interaction is a called a cross-level interaction.

In our example we have found a significant variance component around the average slope of gender and concluded that the relationship between gender and satisfaction is not the same for all the countries (see model 4 in Table 7.1). Now we want to predict part of these differences by adding the cross-level interaction between gender and average working hours to our model (see model 5 in Table 7.1). The interaction is significant ($b = 0.043$, $t(27) = 3.326$, $p = 0.003$), indicating that at least part of these differences in the slope of gender between the countries is explained by average working hours. To understand the interaction, the regression coefficient has to be interpreted together with the main effects of trust in politicians and working hours. The predictors were centered, and this makes the interpretation of the interaction easier. In model 5 we can see that in a country with average working hours (working hours = 0) the effect of gender on the predicted satisfaction is -0.35, indicating that females are predicted to be 0.35 points lower in satisfaction than males. When the average working hours is higher (for example, working hours above average=5) the effect of gender will be the main effect (-0.35) plus the effect of the interaction effect ($5 \times 0.043 = 0.22$ for gender), that is, $-0.35 + 0.22 = -0.13$.

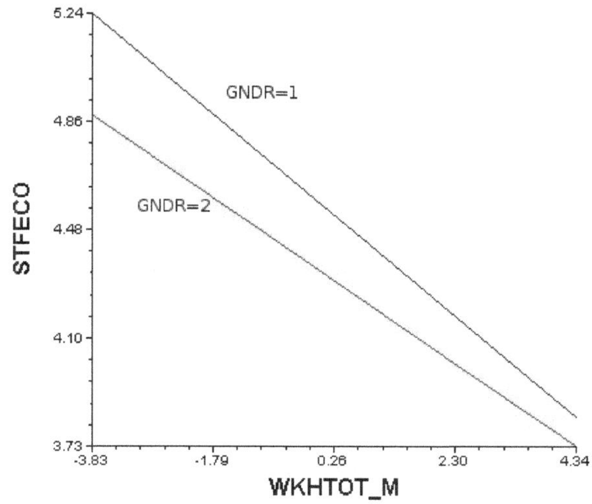

Figure 7.1 Regression lines for satisfaction on average working hours, for males and females

For an individual from a country with a higher average amount of working hours the effect of gender on satisfaction is less strong, and therefore the difference between males and females will be smaller. The effect of the cross-level interaction can also be seen in the interaction graph (Figure 7.1).

MATHEMATICAL FOUNDATIONS AND ADVANCED ASPECTS

There are two major applications of multilevel regression: analysis of subjects nested in larger units or contexts, as described in the previous section; and analysis of repeated measures nested in subjects, who may or may not be nested in groups. We will first describe the multilevel model for grouped data, and later describe how this applies to repeated measures, and what additional issues arise in modeling time-dependent data.

Multilevel analysis of grouped data

Multilevel regression can be conveniently described as sets of linked regression equations defined at each separate level. Assume that we have nested data with subjects nested in groups, an outcome variable y at the individual (lowest) level, and predictor variables at both the individual and the group level. At the individual level, we predict the outcome variable y using the explanatory variables x as follows:

$$y_{ij} = \beta_{0j} + \beta_{1j}x_{1ij} + \ldots + \beta_{pj}x_{pij} + e_{ij}. \tag{7.1}$$

In this regression equation, β_{0j} is the intercept, $\beta_{1j}, \ldots, \beta_{pj}$ are the regression coefficients (slopes) for the P explanatory variables $x_1, \ldots, x_p$ of the first level, and e_{ij} is the residual error term. Subscript i is for subjects and j is for groups. The regression coefficients β carry subscripts j for groups. This indicates that we assume that each group has a different intercept coefficient

β_{0j}, and different slope coefficients $\beta_{1j}, \ldots, \beta_{pj}$. The residual errors e_{ij} are assumed to have a mean of zero and a variance of σ_e^2 to be estimated. Most multilevel software assumes that the variance of the residual errors is the same in all groups. This is equivalent to the assumption in analysis of variance of homogeneity of variance. Some software allows estimation of separate variances in different groups, which is not a parsimonious solution, or modeling the distribution of variances across groups. For a discussion of heterogeneous variances, we refer to Raudenbush and Bryk (2002, Chapters 5 and 9).

The next step in the multilevel regression model is to model the variation of the regression coefficients j, introducing Q explanatory variables z at the group (second) level. For the intercept ($_{0j}$) this results in

$$\beta_{0j} = \gamma_{00} + \gamma_{01}Z_{1j} + \ldots + \gamma_{0q}Z_{qj} + u_{0j}. \tag{7.2}$$

In this regression equation, γ_{00} is the intercept in the second-level regression equation representing the average intercept over all the groups, while $\gamma_{01}, \ldots, \gamma_{0q}$ are the regression coefficients of the second-level variables predicting variation in the intercepts of groups.

For the regression coefficients (slopes) of the first level ($\beta_{1j}, \ldots, \beta_{pj}$) the variation across groups is modeled by

$$\beta_{pj} = \gamma_{p0} + \gamma_{p1}Z_{1j} + \ldots + \gamma_{pq}Z_{qj} + u_p. \tag{7.3}$$

In this regression equation, γ_{p0} is the intercept in the second-level regression equation representing the average slope value for the total group, and $\gamma_{p1}, \ldots, \gamma_{pq}$ are the regression coefficients predicting variation in the first-level regression slopes β_{pj}. These can also be seen as the moderator variables (part of the cross-level interactions). This particular representation is often referred to as the 'slopes-as-outcomes' model (Raudenbush and Bryk, 2002, p. 80). All regression coefficients β_j are assumed to have a multivariate normal distribution.

The u-terms $u_{0j}, u_{1j}, \ldots, u_{pj}$ are random residual error terms at the second (group) level. They are assumed to have a multivariate normal distribution with means of zero, and to be independent of the residual errors e_{ij} at the first level. The variance of the residual errors u_{0j} is specified as σ_{u0}^2 representing the intercept variance component, and the variances of the residual errors $u_{1j}, \ldots, u_{pj}$ are specified as $\sigma_{u1}^2, \ldots, \sigma_{up}^2$ representing the random slope variance components. The *covariances* between the residual error terms are generally *not* assumed to be zero.

Our model with separate first- and second-level regression equations can be written as a single regression equation. Substitution of equations (7.2) and (7.3) into equation (7.1) and rearranging terms gives the following mixed model equation:

$$\begin{aligned} Y_{ij} =& \gamma_{00} + \gamma_{10}X_{1ij} + \ldots + \gamma_{p0}X_{pij} + \gamma_{01}Z_{1j} \ldots + \gamma_{0q}Z_{qj} \\ &+ \gamma_{11}X_{1ij}Z_{1j} + \ldots + \gamma_{1q}X_{1ij}Z_{qj} + \ldots + \gamma_{p1}X_{pij}Z_{1j} + \ldots \gamma_{pq}X_{pij}Zqj \\ &+ u_{1j}X_{1ij} + \ldots + u_{pj}X_{pij} + u_{0j} + e_{ij}. \end{aligned} \tag{7.4}$$

Using summation notation, we can express the same equation as

$$Y_{ij} = \gamma_{00} + \sum_p \gamma_{p0}Xpij + \sum_q \gamma_{0q}Z_{qj} + \sum_p \sum_q \gamma_{pq}X_{pij}Z_{qj} + \sum_p u_{pj}X_{pij} + u_{0j} + e_{ij}. \tag{7.5}$$

If we view the intercept as a regression coefficient that is multiplied by a variable $x_0 = 1$, we can write the combined equation as

$$Y_{ij} = \sum_p \gamma_{p0}X_{pij} + \sum_q \gamma_{0q}Z_{qj} + \sum_p \sum_q \gamma_{pq}X_{pij}Z_{qj} + \sum_p u_{pj}X_{pij} + e_{ij}, \tag{7.6}$$

with $p = 0, \ldots, P$. This can be written very concisely in matrix format if we let X be the matrix of all explanatory variables in the fixed part, symbolize the residual errors at all levels by residual error matrix $\mathbf{u}^{(l)}$ with l denoting the level, and associate all error components with predictor variables $\mathbf{Z}$, which may or may not be equal to $\mathbf{X}$. This produces the very general matrix formula $\mathbf{Y} = \mathbf{X}\boldsymbol{\beta} + \mathbf{Z}^{(l)}\mathbf{u}^{(l)}$ (see Goldstein, 2011, Appendix 2.1). In this chapter, we will mostly use algebraic notation for the regression equations.

Equation (7.4) shows that the multilevel model equation can be decomposed into two parts: a fixed part that contains the linear regression model with predictor variables and fixed regression coefficients; and a random part that specifies a complex structure for the residual errors. For the purposes of interpretation, the focus is often on the fixed part. However, the structure of the random part is sometimes also of interest, which in this chapter will come up in the section on longitudinal modeling.

The assumptions of the multilevel regression model are firstly the usual assumptions of linear regression analysis, excluding the assumption of independent errors. So we assume predictors measured without error, linear relationships, homoscedasticity and normally distributed residual errors. In addition, we assume that the error terms at separate levels are uncorrelated, and that within each level residual errors are independent across cases. We also assume absence of outliers, which in multilevel regression is a complicated concept, because we can have outliers at several different levels. The diagnostics discussed in Chapter 5 are also useful in multilevel regression; some multilevel software produces multilevel diagnostics, such as indicating extreme residuals at the distinct levels.

Estimation methods

Maximum likelihood (ML) is the most commonly used estimation method in multilevel modeling. An advantage of the ML estimation method is that it is generally robust, and produces estimates that are asymptotically efficient and consistent. ML estimation proceeds by maximizing a function called the likelihood. Two different likelihood functions are used in multilevel regression modeling. One is full maximum likelihood (FML); in this method, both the regression coefficients and the variance components are part of the likelihood function. The second estimation method is restricted or residualized maximum likelihood (RML); here only the variance components are included in the likelihood function, and the regression coefficients are estimated in a second estimation step. Both methods produce parameter estimates with associated standard errors and an overall model deviance, which is equal to minus twice the log-likelihood (some software denotes this by −2LL). The difference is that RML takes into account the degrees of freedom lost by estimating the fixed effects. FML estimates of the variance components are biased downwards; RML estimates have in general less bias. We refer to De Leeuw and Meijer (2008) for a discussion of these and other estimation methods. In practice, the differences between the two methods are usually small (cf. Hox, 1998; Kreft and De Leeuw, 1998).

Some software has the option of using Bayesian estimation. In Bayesian statistics, we express uncertainty about the values of the model parameters by assigning to them a distribution of possible values. This distribution is called the *prior* distribution, because it is specified independently of the data. The prior distribution is combined with the likelihood of the data to produce a *posterior* distribution, which describes our uncertainty about the population values after observing the data. Typically, the variance of the posterior distribution is smaller than the variance of the prior distribution, which means that observing the data has reduced our uncertainty about the possible population values. This posterior is generally a complicated multivariate distribution, and simulation techniques are used to generate random samples from the posterior distribution. The simulated posterior distribution is then used to provide a point estimate (typically the mode or median of the simulated values) and a confidence interval.

Bayesian methods can provide accurate estimates of the parameters and the uncertainty associated with them (Goldstein, 2011). However, they are computationally demanding, and the simulation procedure must be monitored to ensure that it is working properly. The advantage of using Bayesian estimation is that it provides accurate estimates, even when the sample size is small (see Hox et al., 2012, for an example). A discussion of Bayesian methods is beyond the scope of this chapter, and we refer the reader to Hox (2010).

Statistical tests

Statistical tests are usually based on the information matrix, which indicates the precision of the parameter estimates. The diagonal of the inverse of the information matrix provides the sampling variances, and the square root of the sampling variance is the standard error. Several technical issues arise here.

Most software uses the standard errors to carry out a Wald test:

$$z = \frac{\widehat{\theta}}{se(\theta)}.$$

In the Wald test the parameter is divided by its standard error, and the result is compared against the standard normal distribution. Regression coefficients of the different levels are generally tested in this way. The Wald test is a large-sample method that assumes that the sample is large enough that it is not necessary to take the number of degrees of freedom for the test into account. In smaller samples, it is desirable to take the degrees of freedom into account. In multiple regression, this is straightforward: the degrees of freedom are simply $n - p - 1$. In multilevel data, the sample size is difficult to appraise since we have distinct sample sizes at different levels, and the effective sample size depends on the amount of dependency in the data. Therefore, the effective degrees of freedom have to be estimated. Two methods are often used to estimate the degrees of freedom for the regression coefficients: the Satterthwaite (Satterthwaite, 1946) or the Kenward–Roger (Kenward and Roger, 1997) estimate. The Satterthwaite correction corrects the degrees of freedom, taking the multilevel structure into account; the Kenward–Roger method corrects both the degrees of freedom and the standard error used in the Wald test. In small samples, using either of these corrections is better than relying on the asymptotic Wald test.

The issues that arise when the Wald test is used to test the significance of a variance component are more treacherous. The main issue is that the Wald test assumes that the parameter tested has a normal distribution. Variances do not have a normal distribution; for normally distributed data they have a chi-square distribution, which is skewed, especially in small samples. It is known that the Wald test performs rather badly for testing variances (Berkhof and Snijders, 2001). A generally good test for variance components is comparing the deviances of a model with and a model without that specific variance component. The difference between those two deviances is a chi-square variate, with degrees of freedom equal to the difference in degrees of freedom between the two models. If we just remove one variance component, the degrees of freedom are equal to one. We still have to take account of the matter that the null hypothesis for this test is on the boundary of the parameter space. To put it another way, the null hypothesis is of the form $H_0 : \sigma^2 = 0$. If the null hypothesis is true, an unbiased estimator of the variance would produce a sampling distribution that consists of 50% positive and 50% negative values. However, negative variances cannot exist, and all negative estimates are truncated to zero. As a result, the sampling distribution for this test is actually a mixture of two distributions, one distribution that consists only of zeros, and one distribution that corresponds to the right-hand tail (only positive values) of the sampling distribution of the variance. This means that the p-value from the deviance difference test for a variance must be halved (Berkhof and Snijders, 2001). The situation becomes more complicated when we test for the significance of a variance component of a regression slope. In general, there will also be a covariance of that slope with the intercept, and possibly also with other slopes. If the variance component for a slope is removed from the model, the corresponding covariance terms are also removed. As a result, the chi-square from the deviance difference test is an omnibus test for all effects removed simultaneously. This complicates the significance test considerably. For details we refer to Hox (2010). A practical solution is to test the significance of slope variances in a variance components model, which is a model where all covariances at the higher levels are constrained to be zero.

MULTILEVEL ANALYSIS OF DICHOTOMOUS DATA

When the outcome variable is dichotomous, or ordinal with few categories, treating it as a normally distributed dependent variable is not appropriate. What we need is a multilevel version of the logistic regression model (see Chapter 8 of this volume for a treatment of single-level logistic regression). For a 0–1 scored dichotomous dependent variable y, the probability π of observing the value 1 is modeled by an exponential function. The two-level regression equation for a logistic regression can conveniently be written as follows:

$$\text{Prob}(y_{ij} = 1) = \mu_{ij},$$

$$\text{logit}(\mu_{ij}) = \log\left(\frac{\mu_{ij}}{1-\mu_{ij}}\right) = \beta_{0j} + \beta_{1j}x1 + \ldots,$$

and

$$\beta_{0j} = \gamma_{00} + \gamma_{01}z_1 + u_0,$$

$$\beta_{1j} = \gamma_{10} + \gamma_{11}z_1 + u_1.$$

The single-equation version of the multilevel logistic model then becomes

$$\text{logit}(\mu_{ij}) = \gamma_{00} + \gamma_{10}x1 + \gamma_{01}z + \gamma_{11}x1z + u_0 + x1u_1. \qquad (7.7)$$

This looks much like an ordinary multilevel model equation, such as equation (7.4), with some alterations. The most important change is of course that we do not model the outcome variable directly, but via a logit transformation of the unobserved latent variable μ. As equation (7.7) is written, it might seem that we take the logit of the observed outcome, but this is clearly not possible with dichotomous data, since the logit of 0 or 1 is not defined. The latent variable μ_{ij} is the expected outcome for unit i in group j. The second change is that we have no error term e at the lowest level. We assume that the dependent variable has a binomial distribution, and in the binomial distribution the variance is a function of the mean. In other words, if we have an estimate of the probability μ_{ij}, we know the variance, and do not need to estimate it by adding an error term.

Multilevel logistic models are more difficult to estimate than multilevel models for continuous normal data. Two approaches have commonly been used. The first is a *Taylor series linearization*, a method that approximates the complicated likelihood function with a linear approximation. The second method relies on numerical methods to approximate the non-linear likelihood. The long and short of it is that the numerical estimation methods are generally more accurate, but may not work with some data and models, or may need an extremely long computation time. Except for small data sets with a high level-two variance, Taylor series linearization performs well enough (see Hox, 2010, for details). There is one drawback: Taylor series linearization only approximates the likelihood, and as a result fit indices such as the Akaike and Bayesian information criteria, which are based on the value of the likelihood function, cannot be used.

The interpretation of the results of a multilevel logistic regression is more complicated than interpreting a linear regression model, because of the logit transformation that is part of the model. However, the issues here are not different from the issues in interpreting the results of a single-level logistic regression, and we refer to Chapter 8 in this book. In the next section, we present a logistic two-level model, and discuss the interpretation of the results.

EXAMPLE ANALYSES

This section describes two empirical examples. The first example is a two-level analysis of data from the European Social Survey, which illustrates a number of issues that arise in multilevel modeling of individuals in groups. In this section, we also provide an example of multilevel logistic regression. The second example is a multilevel analysis of data collected in an experience sampling design, which is a modern version of a diary method.

Example 1: Increase in Internet access in different countries

This example uses data from the first four rounds of the European Social Survey, augmented with country-level information. The dependent variable is NetUse at the individual level. This variable records personal usage of Internet/web/email, and is coded from 0 = no access to Internet at home or work to 7 = every day. The ESS is a repeated survey, so there are different respondents at each data collection. For each wave new individuals are contacted for participation. The individuals are nested within countries. Across all countries, the mean score on the variable NetUse increased from 2.6 (less than once a month) in the first round in 2001 to 3.6 (more than once a month) in the fourth round in 2007. However, one may assume that the rate of Internet use is different in different countries, and it is a reasonable assumption that the rate of increase is also not the same in all available countries. We have four predictor variables at the individual level: the ESS round of data collection, recoded to 0 (2001), 2 (2003), 4 (2005) and 6 (2007), gender (0 = female, 1 = male), age (in years, removing some outliers, namely all ages over 100 years), and educational level (in five categories). At the country level we have one predictor variable, the gross national income per capita (in thousands of dollars, 2002; data from the World Bank database).

We have 29 countries, which is a rather small sample size at the country level. Preliminary analyses showed that the year of data collection had a random slope at the country level, but that the covariance between that slope and the intercept was not significant in any of the models. This implies that there is no relation between the starting point and rate of increase between the countries. To simplify the random part, this covariance was excluded from the model. The model estimated is thus a variance component model, and there are no covariances between the random terms at the country level. Also, the age variable was centered on its overall mean of 47 years.

Table 7.2 shows a sequence of models. The first model includes the year of data collection with a random slope. The regression coefficient for the year variable is 0.26. The Internet use variable has eight categories, and after six years the average Internet use in all countries increases to 1.56 of a category higher on the scale. The variance for the year slope is small but significant. Expressed as a standard deviation (and carrying more decimals than shown in the table), it is

Table 7.2 Sequence of multilevel models for Internet use in 29 countries

	Base model		+ individual predictors		+ country-level predictor	
Fixed part	Regression Coefficient (s.e.)		Regression Coefficient (s.e.)		Regression Coefficient (s.e.)	
Intercept	2.09	(0.21)	−0.03	(0.20)	−1.41	(0.22)
Year	0.26	(0.01)	0.25	(0.02)	0.25	(0.02)
Gender			0.36	(0.01)	0.36	(0.01)
Age			−0.06	(0.00)	−0.06	(0.00)
Education			0.65	(0.00)	0.65	(0.00)
GNI per capita					0.08	(0.01)
Random part	Variance (s.e.)					
Intercept	1.31	(0.35)	1.26	(0.34)	0.40	(0.11)
Year slope	0.01	(0.00)	0.01	(0.00)	0.01	(0.00)
Residual	7.61	(0.03)	5.16	(0.01)	5.16	(0.02)
Fit						
AIC	858,223		789,498		789,466	
BIC	858,259		789,528		789,503	

0.08. The range of 2 standard deviations below and above the mean is 0.10–0.42, so even this small variance term implies visible differences in the rate of increase in Internet use across countries. The second model shows the effects of the respondent-level individual predictors gender, age and educational level. They all have significant effects in the expected direction: Internet use is higher for males, younger respondents and respondents with a higher educational level. These individual predictors explain about 32% of the individual residual variance, and about 4% of the country-level variance. A small amount of the variance between countries can therefore be explained by differences between countries in the composition of the population in terms of gender, age and educational level. In the final model the country-level predictor, gross national income per capita, is added. This shows that in more affluent countries the rate of Internet use is higher. In an additional model (not shown in the table) we added the cross-level interaction between GNI per capital and year, but this was not significant, and adding this term did not visibly decrease the variance of the slope for year. This means that the difference in GNI per capita cannot explain the differences in the relation between year and rate of internet use. In addition, both the AIC and the BIC for this model were higher than the last values in Table 7.2, which also indicates that including this interaction does not improve the model.

A few methodological issues deserve mentioning here. Firstly, the sample size at the country level is small, and therefore we decided to keep the random part simple by leaving out the insignificant covariance between the intercept and the year slope. Secondly, although the table shows the standard errors of the variance components, we used the chi-square deviance difference test to test for the significance of the variances. Using this test, all variances in the models shown are significant at $p < 0.001$.

Finally, the dependent variable is measured in eight categories. We have treated it as an interval scale in this analysis, but it could be argued that a model that treats this variable as an ordinal dependent variable is more appropriate, especially since the distribution of Internet use is decidedly non-normal. These issues are discussed at length in Hox (2010), and we will not go into them here. A final remark concerns the design of the ESS. We include a year variable in the model, which suggests that we are analyzing longitudinal data. This is not the case: the ESS is a repeated survey, where in each year of data collection a new sample is taken. So there are

Table 7.3 Sequence of multilevel models for Internet use in 29 countries

	Base model		+ individual predictors		+ country-level predictor	
Fixed part	Regression Coefficient (s.e.)		Regression Coefficient (s.e.)		Regression Coefficient (s.e.)	
Intercept	–0.70	(0.19)	–0.13	(0.36)	–1.71	(0.22)
Year	0.16	(0.01)	0.24	(0.00)	0.24	(0.00)
Gender			0.35	(0.01)	0.35	(0.01)
Age			–0.07	(0.00)	–0.07	(0.00)
Education			0.76	(0.00)	0.76	(0.00)
GNI per capita					0.09	(0.01)
Random part	Variance (s.e.)					
Intercept	0.76	(0.28)	1.61	(0.71)	0.51	(0.19)
Year slope	0.01	(0.001)	0.02	(0.001)	0.003	(0.001)

no repeated measures, and there is no need to have an extra level for time, or to think about the structure of the measurements across time.

In general, we do not recommend dichotomizing multi-category variables, because it wastes information. In this particular case, however, the data are dichotomized into 0 (no Internet access or use) versus 1 (Internet). For policy reasons, or for scientific reasons such as deciding if a web survey is feasible, it may be necessary to have estimates of the proportion of respondents who actually use the Internet in different countries and across the years. Therefore, we recoded the outcome variable, NetUse, into a dichotomous variable that indicates whether a respondent is actually using the Internet (1) or not (0). For this large data set, numerical estimation methods turned out to be impossible, and the estimates were attained in HLM 7 using Taylor series linearization. Table 7.3 shows the results for the same series of models as presented in Table 7.2, but now for the dichotomous outcome variable.

The results are close to those obtained in the analysis of the continuous dependent variable. That is, however, just a coincidence. In the analysis presented in Table 7.3, the results must be interpreted on the underlying logit scale. For each year, Internet access increases by 0.16. To find out what this means for the actual proportions, we have to use the inverse transformation which is the logistic transformation of the predicted value, $e^{\text{predicted value}}/(1+e^{\text{predicted value}})$. If we use the first model in Table 7.3 to generate these predictions, we find that, averaged across all countries, Internet access has increased from 33% $(e^{-0.7}/(1+e^{-0.7}))$ in the first round to 56% $(e^{-0.7+(0.16\times3)}/(1+e^{-0.7+(0.16\times3)}))$ in the fourth round of the ESS. On average this means an increase of 4.6 percentage points per year. The slope variance is small: expressed as a standard deviation (and calculated with more decimals that are presented in the table), it is 0.043. Two standard deviations below and above the regression coefficient for year which is 0.16 in the first column of Table 7.3, yields a range of 0.07–0.25, which indicates a considerable variation in the rate of increase across the 29 countries.

Example 2: Experience sampling data

The data for this example are from a pilot study conducted on the Longitudinal Internet Studies for the Social Sciences (LISS) panel in the Netherlands, which is an Internet panel based on a true probability sample of households.[1] In March 2012, a small sample of panel members was asked to participate in an experience sampling study. In this study, the selected panel members received an app on their smartphone, or received a smartphone containing the app if they did not possess a smartphone. On two selected days, one weekday (Wednesday) and one weekend

Table 7.4 Sequence of multilevel models for Internet use in 29 countries

	Base model		+ time varying predictors		+ individual level predictor	
Fixed part	Regression Coefficient (s.e.)		Regression Coefficient (s.e.)		Regression Coefficient (s.e.)	
Intercept	6.90	(0.24)	5.84	(0.28)	6.56	(0.44)
Weekend			1.00	(0.17)	0.98	(0.17)
Afternoon			0.69	(0.19)	0.69	(0.19)
Evening			1.06	(0.20)	1.04	(0.20)
Relaxed					−0.27	(0.13)
Random part						
Intercept	2.67	(0.61)	2.56	(0.58)	2.38	(0.55)
Residual	2.29	(0.18)	1.96	(0.16)	1.96	(0.16)
FIT						
AIC	1480		1431		1429	
BIC	1488		1438		1437	

day (Saturday) the app would sound an alarm at random moments of the day, and pose a small number of questions. Respondents were asked to rate their mood on three 10-point scales: happy, relaxed, and awake. It was possible to ignore the alarm, and it was also possible to skip a rating. As a result, the number of measurement occasions varies from 3 to 14, and there are missing values, especially on the scales for relaxed and awake.

This type of data is related to diary studies, but by using a smartphone or similar device it is possible to sample random moments in time to observe changes in mood. These data can be viewed as multilevel data, with repeated measures nested within individual respondents. Analyzing such data with a fixed occasion model, for example repeated measures analysis of variance, would imply dealing with a huge missing-data problem, since the maximum number of measurements in the data is 14, but only 28 out of a total 455 respondents have data for all 14 measurements. In multilevel analysis this is not a particular problem.

For the multilevel analysis, we analyze happiness. The variables relaxed and awake have quite a number of missing data. If we included these as time-varying predictors, listwise deletion would be applied to our data, and we would have very few measurement occasions and respondents left. To solve this problem, these measures are aggregated to the respondent level. Thus, respondents were assigned the mean score on relaxed and awake on those occasions when they provided these data. This still leaves us with some incomplete data, because some respondents apparently never answered these questions. The final number of respondents in the analysis is 372. On the measurement occasion level we have dummy-coded the day as being a weekday (0) or the weekend (1). The time of day has been recoded into two dummies for the afternoon (after 12:00) and the evening (after 18:00), with the morning being the reference category. Thus, we have repeated measures with three time-varying predictors (weekend, afternoon, and evening) and two time-invariant predictors (mean score on relaxed and awake).

Table 7.4 presents the results of a sequence of analyses. The first model is an empty model with just an intercept and variances at the measurement and respondent level. The regression coefficients in model 2 show that happiness is higher at the weekend, and increases as the day proceeds from morning to afternoon and evening. In model 3 we can see that respondents who report a generally high level of relaxation are less happy. Being generally awake had no significant effects and is therefore removed from the analysis. Comparing variance terms across the three models, we can calculate that adding the dummies for weekend, afternoon and evening to the model explains 14% of the measurement occasion residual variance, and 4% of the individual-level variation. Adding the respondent-level predictor Relaxed increases the explained variance at the individual level to 11%.

The models in Table 7.4 are all variance component models. The structure over time is therefore very simple: compound symmetry is assumed. If we test the variances of the slopes of the three occasion-level predictors for significance, we find that only Weekend has a significant slope variation. The variance of the slope is estimated as 1.46 (s.e. 0.57), which is considerable. The chi-square test on the deviances is highly significant at $p < 0.001$. The AIC and BIC for this model are respectively 1408 and 1424, which means that this model is an improvement on the last model in Table 7.4.

To improve the model further, we can extend the random part in a different way, by including the measurement occasion (coded 0, 1, . . . , 13) in the model. Since measurement occasions are random moments spread over the day we only include it in the random part, we do not expect a fixed effect for this variable, nor could we interpret that if it were found significant. The first model we attempt models only variances over time, no covariances. This model is not nested in the model in the last column of Table 7.4, so a formal statistical test is not possible. But the AIC for this model is 1586 and the BIC is 1640, considerably higher than the AIC and BIC for the last model in Table 7.4, and therefore we may conclude that modeling only variances over time is not an improvement. Adding a first-order autoregression to the model, however, is a vast improvement: the AIC for this model is 1438 and the BIC is 1496. Removing the heterogeneous variances, which assumes that the variances are the same across time, results in an AIC of 1430 and a BIC of 1438. This model fits about as well as the last model in Table 7.4. The variance at each occasion is estimated as 4.21 (s.e. 0.45) and the autocorrelation is estimated as 0.64 (s.e. 0.04). The regression coefficients for afternoon (0.48, s.e. 0.21) and evening (0.90, s.e. 0.24) become a bit smaller, but the interpretation is still the same: respondents feel happier in the afternoon and still happier in the evening.

The experience sampling data illustrate two issues in longitudinal modeling. Firstly, we have choices for the structure of the random part. We can assume variation for the slopes of predictors at the measurement occasion level. If we do this in our example data, we find that Weekend has significant slope variation. We can also model the covariance structure over time directly. For our example data, the best structure turns out to be an autoregression structure. In fact, we can combine the two approaches by adding a random slope for Weekend to the autoregression structure. In our example in the combined model the slope variation for Weekend is not significant, while the autocorrelation coefficient remains high and significant. The second issue is the model comparison. Comparing structures across time often results in models that are not nested, and we have to rely on model fit indicators such as AIC and BIC. In our case, these indicate that the model with a varying slope for Weekend fits better than the autoregression model. Nevertheless, the autoregression model indicates a strong stability of happiness across time, which is an interesting result in itself. A discussion of choices in modeling structures over time can be found in Hox (2010).

CAVEATS AND FREQUENT ERRORS

When doing a multilevel analysis the researcher should be aware that the notion of 'amount of variance explained at a specific level' is not a simple concept. As suggested by Bryk and Raudenbush (1992), we have taken the intercept-only model as a benchmark model to calculate the explained variances at the separate levels, by examining how much the residual variance goes down when explanatory variables are added to the model. In certain cases this strategy leads to inconsistencies, because the intercept variance can in fact go up when a predictor is added to the model, leading to negative explained variance at the second level! A well-known example is the longitudinal model with fixed measurement moments which has the variable time as one of the predictors. By adding the variable time to the intercept-only model, the variance component of the measurement occasion level (first level) will decrease, but the variance of the individual level (second level) will in general increase. As Snijders and Bosker (1994) explain in detail, this problem arises because the statistical model behind multilevel models is a hierarchical sampling model: groups are sampled at the higher level, and at the lower level individuals are sampled within groups. This sampling process creates some variability between the groups, even if there are in fact no real group differences. In time series, the lowest level is a series of measurements, which in many cases are (almost) the same for all individuals in the sample. Thus, the variability between persons in the time series variable is in fact much lower than the hierarchical sampling model assumes. Although several statisticians have come up with procedures to correct the problem (e.g. Snijders and Bosker, 1994), there is to our knowledge no procedure that completely solves the problem. Therefore we recommend calculating the explained variance by the procedure of Bryk and Raudenbush (1992). When negative explained variances can be expected based on the hierarchical structure of the data, instead of using the intercept-only model as a benchmark model, the model with the predictor that causes the problem can be used as the benchmark model. Our analysis of the Internet use ESS data follows this pattern; the baseline model in this analysis is not the intercept-only model, but a model that already incorporates time as a predictor.

The multilevel regression model often contains interactions, especially cross-level interactions. For each of the two explanatory variables involved in an interaction, the interpretation of its slope is that it is the expected value of the slope when the other variable has the value zero. If the value zero does not occur in the data, or it may not even be a possible value, the value of the slope may not be interpretable at all. This poses a serious interpretation problem. When both variables in the interaction are centered on their grand mean, the problem disappears, because the value zero now refers to the average in the sample. Thus, it is good practice to grand-mean-center predictor variables that have random slopes or are involved in an interaction.

Some multilevel analysts advocate a totally different way of centering, called group mean centering. Group mean centering means that the group means are subtracted from the corresponding individual scores. Since different values are subtracted from different scores, this is not the same as centering on some overall value, such as the grand mean.

In the end, the choice of centering method depends on its link to the substantive research question, as grand mean centering and group mean centering address different research questions. Enders and Tofighi (2007) discuss the various centering options in detail and conclude that this choice should be driven by the research question at hand. They suggest that group mean centering (which they refer to as within-cluster centering) is most valuable when the research hypothesis is about the relationship between two level-one variables (within-group centering removes confounding with between-group effects), and when a hypothesis involves interactions among level-one variables.

The maximum likelihood estimation method used commonly in multilevel analysis is asymptotic, which translates to the assumption that the sample size is large. This raises questions about the accuracy of the estimation with relatively small sample sizes. The estimates for the regression coefficients are generally unbiased (Maas and Hox, 2004a,b). A large simulation by Maas and Hox (2004b) finds that the standard errors for the fixed parameters are slightly biased downward if the number of groups is less than 50. With 30 groups, they report an operative alpha level of 6.4% while the nominal significance level is 5%. Estimates of the residual error at the lowest level are generally very accurate. The group-level variance components are sometimes underestimated. Browne and Draper (2000) show that with as few as 6–12 groups, RML estimation can provide reasonable variance estimates. With 48 groups, FML estimation also produces good variance estimates. Maas and Hox (2004b) report that with as few as 20 groups, RML estimation produces accurate variance estimates. When the number of groups is around 10, the variance estimates are much too small. In the simulations by Maas and Hox (2004b), with 100 groups the operating alpha level was 6%, which is close to the nominal 5%. Summarizing: the estimates and standard errors of the regression coefficients tend to be quite accurate with as few as 20 groups, but for accurate estimates and standard errors of the variance components typically at least 50 and preferably 100 groups are needed.

Multilevel regression modeling is considerably more complex than classical regression modeling. As a result, a report of the results of a multilevel regression analysis must give more details than a report of a standard regression analysis. Dedrick et al. (2009) analyzed the reporting practices of 99 articles in 13 well-known journals in education and the social sciences. They found many articles that did not give enough information to enable readers to judge the decisions made in the analyses. Articles often did not report whether there was any checking of assumptions, whether centering was used and if so what kind of centering, or how the parameters were estimated. Especially the reporting practices on the variance components were seriously lacking. For example, 23% of the articles did not mention the random part of the multilevel model at all. Only 49% of the articles reported the actual variance estimates, and only 63% mentioned significance tests of the variance components. Of these 63 studies, 38 (60%) reported the test statistic used to test the variance components. Dedrick et al. (2009) end their article with guidelines for reporting multilevel regression analyses. These are well worth reading, even for readers who are not intending to do a multilevel regression analysis, but who need to read and understand other authors' results. As an example, we strongly suspect that until this point in our chapter, most readers have not realized that in our example analyses, we have not reported whether we used FML or RML estimation (it was the former). Although the differences are generally small, as Dedrick et al. point out, such details should be reported.

NOTE

1 See `http://www.centerdata.nl` for details.

REFERENCES

Berkhof, J. and Snijders, T. A. B. (2001). Variance component testing in multilevel models. *Journal of Educational and Behavioral Statistics*, 26, 133–152.

Browne, W. J. and Draper, D. (2000). Implementation and performance issues in the Bayesian and likelihood fitting of multilevel models. *Computational Statistics*, 15, 391–420.

Bryk, A. S. and Raudenbush, S. (1992). *Hierarchical Linear Models. Applications and Data Analysis Methods*. Newbury Park, CA: Sage.

De Leeuw, J. and Kreft, I. (1986). Random coefficient models for multilevel analysis. *Journal of Educational Statistics*, 11, 57–85.

De Leeuw, J. and Meijer, E. (2008). Introduction to multilevel analysis. In J. De Leeuw and E. Meijer (eds), *Handbook of Multilevel Analysis* (pp. 1–76). New York: Springer.

Dedrick, R. F., Ferron, J. M., Hess, M. R., Hogarty, K. Y., Kromrey, J. D., Lang, T. R., Niles, J. D. and Lee, R. S. (2009). Multilevel modeling: A review of methodological issues and applications. *Review of Educational Research*, 79, 69–102.

Enders, C. K. and Tofighi, D. (2007). Centering predictor variables in cross-sectional multilevel models: a new look at an old issue. *Psychological Methods*, 12, 121–138.

Goldstein, H. (1995). *Multilevel Statistical Models*. Kendall's Library of Statistics 3. London: Arnold.

Goldstein, H. (2011). *Multilevel Statistical Models*, 4th edn. Chichester: Wiley.
Hox, J. J. (1998). Multilevel modeling: When and why. In I. Balderjahn, R. Mathar and M. Schader (eds), *Classification, Data Analysis, and Data Highways* (pp. 147–154). New York: Springer.
Hox, J. J. (2010). *Multilevel Analysis. Techniques and Applications*. New York: Routledge.
Hox, J. J., van de Schoot, R. and Maththijsse, S. (2012). How few countries will do? Comparative survey analysis from a Bayesian perspective. *Survey Research Methods*, 6, 87–93.
Kenward, M. and Roger, J. (1997). Small sample inference for fixed effects from restricted maximum likelihood. *Biometrics*, 53, 983–997.
Kreft, I. and De Leeuw, J. (1998). *Introducing Multilevel Modeling*. Newbury Park, CA: Sage.
Longford, N. T. (1989). Fisher scoring algorithm for variance component analysis of data with multilevel structure. In R. Bock (ed.), *Multilevel Analysis of Educational Data*. San Diego, CA: Academic Press.
Longford, N. T. (1993). *Random Coefficient Models*. New York: Oxford University Press.
Maas, C. J. M. and Hox, J. J. (2004a). The influence of violations of assumptions on multilevel paramater estimates and their standard errors. *Computational Statistics and Data Analysis*, 46, 427–440.
Maas, C. J. M. and Hox, J. J. (2004b). Robustness issues in multilevel regression analysis. *Statistica Neerlandica*, 58, 127–137.
Raudenbush, S. and Bryk, A. S. (1986). A hierarchical model for studying school effects. *Sociology of Education*, 59(1), 1–17.
Raudenbush, S. W. and Bryk, A. S. (2002). *Hierarchical Linear Models*. Thousand Oaks, CA: Sage.
Satterthwaite, F. E. (1946). An approximate distribution of estimates of variance components. *Biometrics Bulletin*, 2, 110–114.
Snijders, T. A. and Bosker, R. (1994). Modeled variance in two-level models. *Sociological Methods & Research*, 22, 342–363.

8

Logistic regression

Henning Best and Christof Wolf

INTRODUCTION

Many questions raised in the social sciences involve analyzing binary variables. For example, Best (2009) uses logistic regression to study farmers' decisions whether to adopt organic farming, and Cornwell and Laumann (2011) study sexual dysfunction. Other topics studied with logistic regression include educational attainment (e.g. university degrees) and unemployment.

A dichotomous variable can take two distinct values. As in most regression-based methods, it is convenient to code the variable as a binary variable with the values 0 and 1. Let us assume a variable should indicate whether a person has xenophobic attitudes or not. In this case it would make sense to code xenophobic persons with '1' and non-xenophobic persons with '0'.[1] How can we analyze this variable?

The linear probability model

One rather simple way to analyze the determinants of xenophobia with regression methods is to estimate a linear regression model using 'xenophobia yes/no' as the dependent variable. However, the resulting estimate of the dependent variable $\hat{y}$ will not be dichotomous as the observed y was, but rather continuous. To make sense of this continuous variable, it is convenient to interpret it as the probability that a respondent is xenophobic ($y = 1$). Hence, the regression model assumes that the independent variables have linear effects on the probability $\Pr(y = 1) = \hat{y}$:

$$\Pr(y = 1) = \beta_0 + \beta_1 x_1 + \beta_2 x_2 + \cdots + \beta_k x_k + \varepsilon. \tag{8.1}$$

Due to the interpretation of $\hat{y}$ as the probability $\Pr(y = 1)$ and the linear model, this regression model is called a *linear probability model*. If we use the European Social Survey round 1 as a database for estimating the relationship between perceived group threat and the probability that a respondent does not want to allow any migrants with a different ethnic background, we get the regression line shown in Figure 8.1.

It is easy to see that, according to this model, the probability of xenophobia increases with group threat. However, some serious concerns have been raised against using the linear probability model for analyzing dichotomous outcomes (see Menard, 1995, pp. 1–11 for a more detailed discussion):

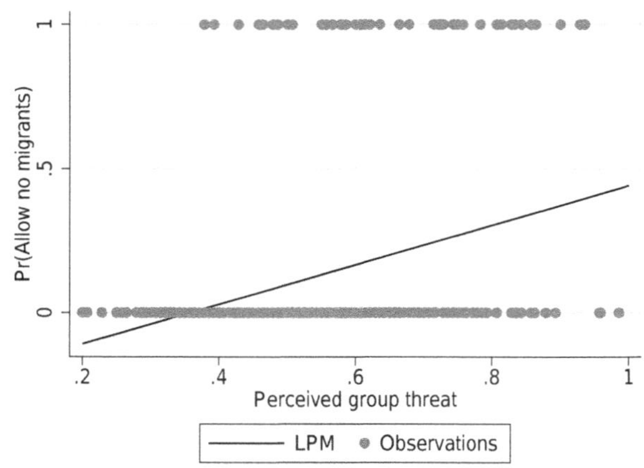

Figure 8.1 Linear probability model

- Some values of x can lead to predictions outside the defined value range of probabilities ($0 \leq \Pr(y = 1) \leq 1$). In the present example this would be the case for a subjective group threat lower than 0.35.
- As the observed dependent variable can only take the values 0 and 1, the variance of the error term necessarily correlates with y (*heteroscedasticity*). In ordinary least squares (OLS) models like the linear probability model, this leads to an inefficient estimation and biased standard errors.
- For given values of x, the residuals can only take two values, which leads to *non-normally distributed residuals*.
- The linear parametrization of the model is not adequate, because one can assume the probability of $y = 1$ to approach 0 and 1 gradually (see also Figure 8.3).

Due to these concerns we are well advised to use superior methods for analyzing dichotomous outcomes, such as logistic regression.

Logistic regression

In logistic regression we estimate a non-linear instead of a linear probability model. By switching to a non-linear parametrization and maximum likelihood estimation, it is possible to avoid violating the assumptions outlined above and at the same time to find a functional form better suited to modeling probabilities.

The basic idea of logistic regression is to assume a latent, unobserved variable y^* which causes the dichotomous outcomes $y = 1$ or $y = 0$ that we actually observe. y^* could be the unobserved subjective expected utility of an action alternative when analyzing binary choice, a latent personality trait such as attitudes, in our example of xenophobia, or – in the most general formulation – a latent propensity. As this unobserved y^* is defined for the interval of $-\infty$ to $+\infty$, it can be modeled using a linear equation:

$$y^* = \beta_0 + \beta_1 x_1 + \beta_2 x_2 + \cdots + \beta_k x_k + \varepsilon. \tag{8.2}$$

However, as y^* is unobserved, it must be linked to the observed dichotomous variable y to become meaningful to us. This link is provided by the assumption that $y = 1$ is observed once y^* exceeds a certain threshold as shown in Figure 8.2.

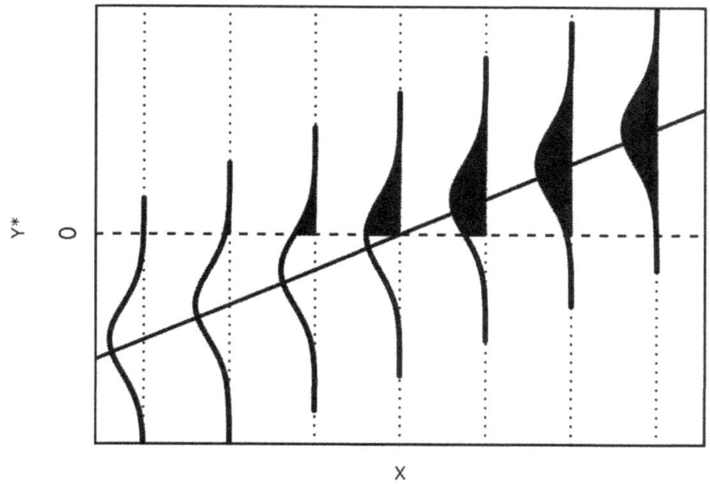

Figure 8.2 Latent and observed variables

Figure 8.2 displays the relationship between the linear prediction of y^* and y. As can easily be seen, $y = 1$ can be observed when y^* exceeds the threshold of 0. This threshold makes sense when thinking, for example, of expected utility as a latent variable: values below 0 would imply the costs to be higher than the benefits and it would be unwise to choose such an alternative. However, the figure also shows that y is not entirely determined by the linear prediction. Rather, there is some uncertainty, expressed by the distribution around $E(y^*)$. This uncertainty can be due to random error, due to variables not in the regression equation, etc. In equation (8.2) it is captured by the error term ε. The assumption of unobserved errors in the linear prediction leads to a probabilistic model: based on our knowledge of $\widehat{y^*}$ we cannot tell whether $y = 1$ or $y = 0$ will be observed, but under certain assumptions we can give a probability. The most important assumption concerns the distribution of the errors ε. As we will show at greater detail in the next section, assuming the errors to follow a logistic distribution leads to the logistic regression model:

$$\Pr(y = 1) = \frac{e^{\beta_0+\beta_1x_1+\beta_2x_2+\cdots+\beta_kx_k}}{1 + e^{\beta_0+\beta_1x_1+\beta_2x_2+\cdots+\beta_kx_k}}. \tag{8.3}$$

Consequently, logistic regression is linear regarding the latent dependent variable y^*, but non-linear regarding the probabilities. This may seem unfortunate at first sight, but it is necessary to avoid the problems with crossing the boundaries of the defined interval of probabilities and to provide an adequate parametrization for modeling the empirical curve of probabilities. Figure 8.3 plots the regression lines of a linear probability model and logistic regression. As a reference line the figure includes a non-parametric curve (lowess regression), fitted to the relative frequency of xenophobia and perceived group threat. It can be seen that both parametrizations are suited to modeling expected probabilities *in the middle of the range* of x. At the margins, however, the linear probability models fail to provide a good fit, while the logistic curve models the gradual approximation of the relative frequencies to 0 or 1 very well.

Interpretation of coefficients

Equation (8.2) shows that the coefficients of logistic regression refer to a linear model of the latent dependent variable. Hence, the interpretation regarding this latent outcome is the same

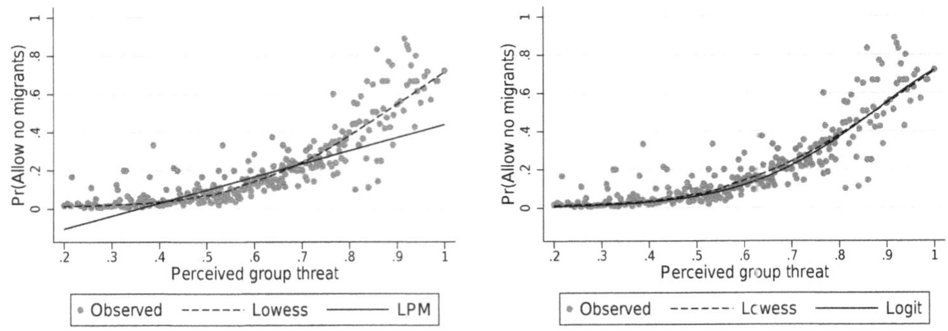

(a) Linear probability model (b) Logistic regression

Figure 8.3 Relative frequency of xenophobia by perceived group threat

as that with OLS regression (see also Chapter 4 of this volume): the constant term β_0 is the intercept of the y-axis and indicates the expected value of y^* when all $\mathbf{x}$ are 0. The slope of the regression line is given by the coefficients β_i, indicating a change by β_i in $\widehat{y^*}$ when x_i increases by one unit (holding the other $\mathbf{x}$ constant). Therefore a negative logit coefficient points to a negative effect of the independent variable on the latent dependent variable y^* (the larger x, the smaller y^*), and a positive coefficient for a positive effect (the larger x, the smaller y^*). This may seem a very straightforward interpretation at first, and in fact it is. However, as y^* is unobserved and unknown and the link to the probability of the observed y being 1 is non-linear (recall equation (8.3)), interpreting the scale of the coefficient is not meaningful in most cases. That said, the interpretation of the direction of the effect is meaningful, as the relationship between y^* and $\Pr(y = 1)$ is monotonic. Hence, the sign of β indicates the direction of the effect of x on $\Pr(y = 1)$.

Some researchers have argued that exponentiating logit coefficients leads to a meaningful and intuitive interpretation beyond the sheer direction of the effect (so-called 'odds ratios'). In fact, equation (8.3) can easily be rearranged to give

$$\frac{\Pr(y=1)}{1-\Pr(y=1)} = e^{\beta_0+\beta_1 x_1+\beta_2 x_2+\cdots+\beta_k x_k} = e^{\beta_0}\, e^{\beta_1 x_1}\, e^{\beta_2 x_2}\, \ldots\, e^{\beta_k x_k}, \tag{8.4}$$

with $\Pr(y = 1)/(1 - \Pr(y = 1))$ being the *odds* of $y = 1$. This leads to a multiplicative rather than an additive model and drastically changes the interpretation of the coefficients ($\exp(\beta)$ compared to β). Firstly, $\exp(\beta)$ indicates effects in the odds of $y = 1$ rather than in the latent dependent variable y^*. Secondly, the effect is multiplicative rather than additive – hence the name 'odds ratio'. For example, $\exp(\beta_i) = 2$ means that the odds of $y = 1$ double (multiplication by 2) when x_i increases by one unit, and $\exp(\beta_i) = 0.33$ indicates a decrease in the odds by two-thirds (multiplication by 0.33). More generally, $\exp(\beta) < 1$ indicates a negative relationship and $\exp(\beta) > 1$ a positive one between the two variables. The neutral value is 1, instead of 0. $\exp(\beta)$ coefficients therefore seem to offer a simple and intuitive interpretation as odds ratios, that is, multiplicative effects on the odds. The major problem with this interpretation, however, is that odds are quite complex and unintuitive, and the odds ratios therefore prone to misinterpretation.

Odds are ratios of probabilities ($\Pr(y = 1)/(1 - \Pr(y = 1))$), and *odds ratios are ratios of ratios of probabilities*. Moreover, the relationship between odds and probabilities is non-linear,

Table 8.1 The problem with odds ratios

		P	odds	OR	RR
A	Group 1	0.10	0.11	2.10	2.00
	Group 2	0.05	0.05		
B	Group 1	0.40	0.67	2.70	2.00
	Group 2	0.20	0.25		
C	Group 1	0.80	4.00	6.00	2.00
	Group 2	0.40	0.67		
D	Group 1	0.60	1.50	6.00	3.00
	Group 2	0.20	0.25		
E	Group 1	0.40	0.67	6.00	4.00
	Group 2	0.10	0.11		

that is, it varies with the level of probability. If odds ratios are misinterpreted as probability ratios (and this is unfortunately a common misinterpretation), the direction of the relationship will be intact, but the strength of the effect is biased downwards. Table 8.1 displays the complex relationship between odds, probabilities, odds ratios, and relative risks for multiple settings. It can easily be seen that the interpretation of the odds ratio depends on $\Pr(\cdot)$, and hence the interpretation is meaningless without knowledge of the base probability.[2]

We strictly advise against using odds and odds ratios when interpreting the results of logistic regression. An interpretation over and above the direction of the relationship is impossible based on the odds ratio alone, yet still attempted all too often (usually using a diffuse probabilistic terminology). If a researcher desires an interpretation that is more detailed than only direction and statistical significance, we recommend drawing on predicted probabilities, average marginal effects and graphical tools such as profile plots (see below and Chapter 10 of this volume).

MATHEMATICAL FOUNDATIONS AND ADVANCED ASPECTS

In this section we first discuss the derivation of logistic regression in detail. We then briefly present maximum likelihood estimation of the coefficients and statistical inference. Finally, we discuss problems that arise when estimating interaction effects or comparing logit coefficients between nested models and/or groups.

Derivation of the non-linear probability model

As briefly outlined above, logistic regression is based on the assumption of a latent, unobserved variable y^* that causes individuals to make decisions or more generally show a specific outcome that can be observed empirically as a dichotomous variable y.[3] The observed variable y takes the value 1 if y^* exceeds a threshold value τ. In logistic regression $\tau = 0$ by assumption. The latent variable can be expressed in a linear model:

$$y^* = \mathbf{x}'\boldsymbol{\beta} + \varepsilon. \tag{8.5}$$

It can be shown that we can estimate the probability of $y = 1$ when the distribution of the errors is known (see Long, 1997, p. 44; Wooldridge, 2002, p. 457). In order to derive the probability we start by assuming the threshold mentioned above and set up the equation

$$\Pr(y = 1|\mathbf{x}) = \Pr(y^* > \tau) = \Pr(y^* > 0)\,. \tag{8.6}$$

Substituting equation (8.5) leads to

$$\Pr(y=1|\mathbf{x}) = \Pr(\mathbf{x}'\boldsymbol{\beta} + \varepsilon > 0), \tag{8.7}$$

which expresses nothing more than the formalized assumption that the probability of $y = 1$ equals the probability of the predicted values of the regression model exceeding the threshold value $\tau = 0$. Rearranging the right-hand side of the equation leads to

$$\Pr(y=1|\mathbf{x}) = \Pr(\varepsilon > -\mathbf{x}'\boldsymbol{\beta}). \tag{8.8}$$

The right-hand side of the equation now expresses the probability that the errors ε are larger than a specific value (namely $-\mathbf{x}'\boldsymbol{\beta}$). We have already mentioned that the distribution of errors must be known. When assuming a continuous symmetric distribution with $E(\varepsilon) = 0$, $\Pr(\varepsilon > -a) = \Pr(\varepsilon < +a)$ due to symmetry. In other words, the area below the probability density function to the right of the negative value $-a$ is identical to the area to the left of the positive value $+a$. If we change the sign of $-\mathbf{x}'\boldsymbol{\beta}$ in the right-hand side of the equation we get

$$\Pr(y=1|\mathbf{x}) = \Pr(\varepsilon \leq \mathbf{x}'\boldsymbol{\beta}). \tag{8.9}$$

Now the right-hand side of the equation describes the probability of ε being smaller or equal to a specific value. Exactly this kind of probability is given by a cumulative probability density function (CDF). Using the shorthand $G(\cdot)$ for a CDF, we can write

$$\Pr(\varepsilon \leq \mathbf{x}'\boldsymbol{\beta}) = G(\mathbf{x}'\boldsymbol{\beta}). \tag{8.10}$$

Hence, equation (8.9) can be restated as

$$\Pr(y=1|\mathbf{x}) = G(\mathbf{x}'\boldsymbol{\beta}). \tag{8.11}$$

We have now arrived at our starting point: when we know the distribution of the errors – as described by $G(\cdot)$ – we can estimate the probability of $y = 1$. In logistic regression it is assumed that the errors follow a logistic distribution with expectation $E(\varepsilon|\mathbf{x}) = 0$ and variance $\mathrm{Var}(\varepsilon|\mathbf{x}) = \pi^2/3$. The assumption of a fixed variance is necessary as the error term is not observable given the unobservable y^*, as $\varepsilon = y^* - \widehat{y^*}$. Choosing the arbitrary value of $\mathrm{Var}(\varepsilon|\mathbf{x}) = \pi^2/3$ leads to a relatively simple formulation of $G(\cdot)$:

$$G(\mathbf{x}'\boldsymbol{\beta}) = \frac{e^{\mathbf{x}'\boldsymbol{\beta}}}{1+e^{\mathbf{x}'\boldsymbol{\beta}}}. \tag{8.12}$$

By substituting $G(\cdot)$ from equation (8.12) into equation (8.11), we get the base equation of logistic regression, as already known from the previous section:

$$\Pr(y=1|\mathbf{x}) = \frac{e^{\mathbf{x}'\boldsymbol{\beta}}}{1+e^{\mathbf{x}'\boldsymbol{\beta}}}. \tag{8.13}$$

In summary, a number of assumptions are necessary to identify and derive logistic regression: first, on the threshold value τ that needs to be exceeded to observe $y = 1$; second, on the distribution function of errors; third, on its expectation; and fourth, on the error variance. When the assumptions are made as described above, the model results in logistic regression.

Different assumptions on the error distribution lead to different models for $\Pr(y = 1)$. When assuming normally distributed errors with $\mathrm{Var}(\varepsilon|\mathbf{x}) = 1$ and $E(\varepsilon|\mathbf{x}) = 0$ we get the *probit model*. Due to the distribution assumption,

$$G(a) = \frac{1}{\sqrt{2\pi}} \int_{-\infty}^{a} e^{-z^2/2} \mathrm{d}z. \tag{8.14}$$

As a standard normal CDF differs only slightly from a logistic CDF with $\sigma = 1$, logit and probit models lead to almost identical results regarding the probability of $y = 1$. The β coefficients, however, are scale-dependent and vary with the variance of ε. Coefficients of logit models are usually about 1.7–1.8 times larger than probit coefficients (see Long, 1997, p. 48).

Estimation

As with regression methods, the major aim of the estimation process is to find values for the parameters (here $\boldsymbol{\beta}$) that allow us to reproduce the observed data as closely as possible – or, in other words, to provide a good fit. In logistic regression, estimation of the coefficients is done by *maximum likelihood estimation*.

The fit of the regression curve to empirical data is especially good in logistic regression when a high probability $\Pr(y = 1|\mathbf{x})$ is predicted for cases with $y_i = 1$, and a low probability $\Pr(y = 1|\mathbf{x})$ for cases showing $y_i = 0$. The latter condition is equivalent to $1 - \Pr(y = 1|\mathbf{x})$ being as large as possible. As the predicted probability is given by $\Pr(y = 1|\mathbf{x}) = \exp(\mathbf{x}'\boldsymbol{\beta})/(1 + \exp(\mathbf{x}'\boldsymbol{\beta}))$, for a single case the two conditions can be combined in the equations

$$f(y_i) = \Pr(y_i = 1)^{y_i}\,(1 - \Pr(y_i = 1))^{1-y_i} \tag{8.15}$$

and

$$f(y_i|\mathbf{x}_i;\boldsymbol{\beta}) = \left(\frac{e^{\mathbf{x}_i'\boldsymbol{\beta}}}{1 + e^{\mathbf{x}_i'\boldsymbol{\beta}}}\right)^{y_i}\left(1 - \frac{e^{\mathbf{x}_i'\boldsymbol{\beta}}}{1 + e^{\mathbf{x}_i'\boldsymbol{\beta}}}\right)^{1-y_i}. \tag{8.16}$$

Given that y can only take the values 0 and 1, and that the terms are exponentiated with y_i or $1 - y_i$ respectively, for given observations only part of the equation is of interest: the first part of the right-hand side when $y_i = 1$, and the second part of the right-hand side when $y_i = 0$. The equation can take values between 0 and 1, with larger values indicating a better fit. Maximum likelihood assumes $\ell(\boldsymbol{\beta}|\mathbf{x}_i;y_i) = f(y_i|\mathbf{x}_i;\boldsymbol{\beta})$. Consequently, the likelihood of a single case can be written

$$\ell(\boldsymbol{\beta}|\mathbf{x}_i;y_i) = \left(\frac{e^{\mathbf{x}_i'\boldsymbol{\beta}}}{1 + e^{\mathbf{x}_i'\boldsymbol{\beta}}}\right)^{y_i}\left(1 - \frac{e^{\mathbf{x}_i'\boldsymbol{\beta}}}{1 + e^{\mathbf{x}_i'\boldsymbol{\beta}}}\right)^{1-y_i}. \tag{8.17}$$

For estimating the parameters, all complete observations are taken into account by taking the product of all individual likelihoods. Therefore,

$$\mathcal{L}(\boldsymbol{\beta}|\mathbf{y};\mathbf{X}) = \prod_{i=1}^{n}\left(\frac{e^{\mathbf{x}_i'\boldsymbol{\beta}}}{1 + e^{\mathbf{x}_i'\boldsymbol{\beta}}}\right)^{y_i}\prod_{i=1}^{n}\left(1 - \frac{e^{\mathbf{x}_i'\boldsymbol{\beta}}}{1 + e^{\mathbf{x}_i'\boldsymbol{\beta}}}\right)^{1-y_i} \tag{8.18}$$

gives the maximization problem to be solved. This equation may be somewhat more complex than the individual likelihoods, but it follows the very same logic described above. In order to facilitate maximization, usually the logged form of the likelihood is used, called the log-likelihood:

$$\ln\mathcal{L}(\boldsymbol{\beta}|\mathbf{y};\mathbf{X}) = \sum_{i=1}^{n} y_i \ln\left(\frac{e^{\mathbf{x}_i'\boldsymbol{\beta}}}{1 + e^{\mathbf{x}_i'\boldsymbol{\beta}}}\right) + \sum_{i=1}^{n}(1 - y_i)\ln\left(1 - \frac{e^{\mathbf{x}_i'\boldsymbol{\beta}}}{1 + e^{\mathbf{x}_i'\boldsymbol{\beta}}}\right). \tag{8.19}$$

By taking the logarithm, products are transformed to sums, but the point where the equation has its maximum is unchanged. Computationally, maximization is achieved using an iterative numerical procedure, usually using the results of OLS as starting values for $\boldsymbol{\beta}$ (the Newton-Raphson method is a commonly used algorithm).

It is important to note that maximum likelihood is asymptotically unbiased only. Therefore, larger samples are advised than for OLS. Long (1997, p. 54) recommends a minimum of 100 cases and 10 cases per parameter, but the present authors would recommend larger samples.

Statistical inference

Tests of significance in logistic regression are rather straightforward. As in linear regression, the estimated sample coefficients $\hat{\beta}_j$ can be conceived as realizations of an asymptotically normal random variable with $E(\beta_j) = \hat{\beta}_j$ and $\sigma_{\beta_j} = se(\hat{\beta}_j)$.[4] The test statistic z is given by

$$z_j = \frac{\hat{\beta}_j - \beta_{H_0}}{se(\hat{\beta}_j)}. \tag{8.20}$$

The z-test can be used to test hypotheses of the form $H_0 : \beta_j = \beta_{H_0}$, including the traditional null hypothesis test of significance with $H_0 : \beta_j = 0$. In the latter case the z-value can easily be calculated by dividing the logit coefficient by its standard error.

More complex hypotheses may require the use of a *likelihood ratio test*, a *Wald test* or the *Lagrange multiplier test*. Such a complex hypothesis could result when using multiple variables to operationalize a construct, for example by using dummy variables for a multinomial construct such as religion. Another application might be comparing more and less constrained models.

In this chapter we present the likelihood ratio (LR) test that can be used to compare two nested models.[5] The LR test gives an answer to the question whether adding parameters to a model significantly improves model fit. More precisely, the LR test compares the likelihood of the less restricted model (with more parameters) with the likelihood of the more restricted model (with fewer parameters):

$$LR = -2\ln\frac{\mathcal{L}_R}{\mathcal{L}_U} = -2(\ln\mathcal{L}_R - \ln\mathcal{L}_U). \tag{8.21}$$

In this equation $\mathcal{L}_U$ denotes the likelihood of the unrestricted model and $\mathcal{L}_R$ the likelihood of the restricted model.

The test statistic LR follows a χ^2 distribution with $df_{LR} = df_U - df_R$. Here, the degrees of freedom correspond to the number of additional parameters in the unrestricted model. As already mentioned, the test requires nested models.

Goodness of fit and model comparison

The tests discussed above provide information on the statistical significance of differences between models. In many cases, however, researchers are also interested in a descriptive statistic of model fit – similar to the rate of explained variance R^2 in linear models. In logistic regression, such a measure of explained variance does not exist. However, a number of likelihood-based measures have been suggested, some of them mimicking the logic of R^2. These *pseudo-R^2* coefficients typically vary between 0 and 1, with 0 indicating a model with no explanatory power and 1 indicating a perfect fit.

The log-likelihood (LL or $\ln\mathcal{L}$) as given in equation (8.19) always takes negative values and is smaller in absolute value the better the model fits the data. Hence, the value of $-2LL$ provided by many statistical packages can be seen as an indicator of model fit (the smaller, the better), but its absolute scale depends on the number of observations. McFadden (1973) suggests one interpret the log-likelihood of an empty model as analogous to the total variance in R^2, and the log-likelihood of the fitted model as analogous to the explained variance. Accordingly, McFadden's pseudo-R^2 is given by

$$R^2_{\mathrm{MF}} = 1 - \frac{\ln \mathcal{L}(M_1)}{\ln \mathcal{L}(M_0)}, \tag{8.22}$$

with M_0 as the empty model and M_1 the fitted model with explanatory variables. R^2_{MF} is widely used (e.g. by Stata), but has the disadvantage of failing to reach 1. An alternative has been suggested by Cox and Snell (1989). They propose a correction based on the number of observations N:

$$R^2_{\mathrm{CS}} = 1 - \left(\frac{\mathcal{L}(M_0)}{\mathcal{L}(M_1)}\right)^{2/N}. \tag{8.23}$$

However, R^2_{CS} is also smaller than 1 in all cases. A further correction has been proposed by Cragg and Uhler (1970). Some statistical packages (e.g. SPSS) provide this statistic as under the name Nagelkerke R^2:

$$R^2_{\mathrm{NK}} = \frac{R^2_{\mathrm{CS}}}{\max(R^2_{\mathrm{CS}})} = \frac{R^2_{\mathrm{CS}}}{1 - \mathcal{L}(M_0)^{2/N}}. \tag{8.24}$$

In general, R^2_{NK} gives larger values than all other varieties of pseudo-R^2. We suggest one should interpret these likelihood-based measures with great caution as they do not relate to an easy-to-comprehend statistic like explained variance, and as there is no agreement as to which of the different variants to use.

All goodness-of-fit measures we have discussed additionally suffer from the problem of taking larger fit values the more explanatory variables the model includes. Many researchers, however, aim for models that provide a good fit and are parsimonious. In logistic regression, the Akaike information criterion (AIC) and the Bayesian information criterion (BIC) can be used as measures of model fit that take into account the number of parameters. The AIC and BIC even allow the comparison of non-nested models.

The AIC (see Akaike, 1973) is a likelihood-based measure that punishes each additional parameter used. The coefficient is given by

$$AIC = -2 \ln \mathcal{L}(M_1) + 2(k+1), \tag{8.25}$$

with k as the number of explanatory variables. The AIC can vary from 0 to $+\infty$, with smaller values indicating a better model.

The BIC (see Raftery, 1995) is also estimated using the likelihood of a model, but logically draws on a Bayesian comparison. Additional parameters are punished more strongly than with the AIC. The BIC can be computed as

$$BIC = -2 \ln \mathcal{L}(M_1) + \ln N(k+1), \tag{8.26}$$

where N denotes the number of observations, k the number of explanatory variables and M_1 the model specified. Lower values of the BIC indicate a better model fit.

Comparison between models

In the previous section we discussed the basic interpretation of logit coefficients and argued that β should be interpreted with regard to the direction of the effect only. A more detailed interpretation should be based on predicted probabilities and graphical tools such as profile plots (see also Chapter 10 of this volume). However, many researchers are interested in comparing the coefficients between nested models or between groups.

Such a comparison may be of great interest in OLS regression, but unfortunately is much more complicated and potentially misleading in logit models. This is due to fact that the variance of y^* is unobservable, and the error variance has to be fixed for the model to be identified ($\text{Var}[\varepsilon] = \pi^2/3$). The coefficients of logistic regression are therefore identified only to scale: the scale of the latent variables and hence the coefficients varies with model fit, and total variance increases with decreasing unobserved heterogeneity. This simply follows from fixing the error variance (see Karlson et al., 2012; Winship and Mare, 1984; Allison, 1999). Consequently, the logit coefficients of two nested models as well as those of two models estimated for subgroups refer to differently scaled dependent variables and their size cannot be compared in a meaningful way. It is not at all clear whether differences between the coefficients are due to different scaling or whether these differences reflect differential effects.

In summary, unobserved heterogeneity affects the size of logit coefficients even if $\mathbf{x}$ and ε are uncorrelated. This property of logistic regression models unfortunately has received insufficient attention in the past, and generations of researchers have erroneously compared results of logistic regression models between models. Best and Wolf (2012) use Monte Carlo simulations to estimate the magnitude of the scaling problem. They find that the bias is present in all cases, but increases with model fit. When the model fit is low (say, R^2_{MF} is around 5%), the bias due to scaling is not dramatic. It should be noted that the scaling problem is present in odds ratios as well (as these are simply $\exp[\beta]$), but does not affect predicted probabilities.

A number of solutions to the scaling problem have been proposed. Firstly, it is possible to standardize coefficients using the estimated variance of y^*. Secondly, average marginal effects are average effects on the probabilities and therefore not only provide a convenient interpretation but can also be compared between models. Thirdly, a newly proposed solution keeps unobserved heterogeneity constant between nested models and thus provides coefficients unbiased by scaling. We will briefly review these options below.

Standardized coefficients

Thirty years ago, Winship and Mare (1984) suggested standardizing coefficients using the estimated variance of the latent dependent variable to circumvent scaling and produce comparable coefficients. The fully standardized coefficient is defined as

$$\beta_j^s = \beta_j \frac{\sigma_{xj}}{\sigma_{y^*}}, \tag{8.27}$$

with σ_{y^*} being the latent standard deviation estimated from the data. This can be done quite easily as

$$\widehat{\text{Var}}(y^*) = \hat{\boldsymbol{\beta}}' \widehat{\text{Cov}}(\mathbf{x}) \hat{\boldsymbol{\beta}} + \text{Var}(\varepsilon). \tag{8.28}$$

If the independent variables are categorical, it may make sense to use a partial standardization of y^* only ($\beta^{s_{y^*}} = \beta/\sigma_{y^*}$). In cases where the statistical package does not provide an automatic routine for estimating standardized logit coefficients, the latent variance can easily be estimated by calculating the variance of the predicted values and adding $\pi^2/3$. Monte Carlo simulations by the authors show that standardized logit coefficients help in reducing the problems introduced by scaling, although they cannot completely avoid bias.

Average marginal effects

A good and simple alternative to standardized coefficients is provided by average marginal effects (AMEs). In a non-linear model such as logistic regression the effects due to an explanatory variable on the probability of $y = 1$ not only are non-linear, but also depend on the values of all other $\mathbf{x}$. More formally, the partial derivative of the function is not β, but rather

$$\frac{\partial \Pr(y = 1|\mathbf{x})}{\partial x_j} = g(\mathbf{x}'\boldsymbol{\beta})\,\beta_j,$$

with $g(\mathbf{x}'\boldsymbol{\beta})$ as density function of the logistic distribution. AMEs avoid the non-linearity of the model by averaging the effect of a variable and expressing the additive effect on $\Pr(y = 1|\mathbf{x})$ in a single coefficient.

An average effect can be defined in two ways: by calculating the marginal effect either with all variables held constant at their means, or for all cases in the sample and then averaging over them. We will call the former effect the 'marginal effect at the mean' (MEM) and only the latter the 'average marginal effect' (AME). It can easily be seen that the two coefficients are not identical:

$$\mathrm{MEM}_j = g(\mathbf{x}'\boldsymbol{\beta})\,\beta_j, \tag{8.29}$$

whereas

$$\mathrm{AME}_j = \frac{\sum_{i=1}^{N} g(\mathbf{x}'\boldsymbol{\beta})}{N}\beta_j. \tag{8.30}$$

It can be shown that the AME is not affected by scaling when the variables are normally distributed (e.g. Wooldridge, 2002, p. 470). The MEM, while easier to calculate, does not exhibit this property, and we therefore advise using the AME whenever possible. Monte Carlo simulations (see Best and Wolf, 2012) show that the AME indeed works very well in correcting for scaling, and only a minor bias remains when the variables are strongly skewed.

In addition to being relatively robust against scaling, AMEs have the advantage of allowing an easy and intuitive interpretation: they are simply an average effect on the probability. Hence, the probability of $y = 1$ on average changes by AME points when x_j increases by one unit. Of course, this is an average effect only, and it cannot express the model's non-linearity. Nonetheless, AME are superior to the (unrightfully) popular odds ratios in many respects (robustness, simplicity of interpretation, additivity).

Constant heterogenity

As the problem of scaling results from different variances of $\mathbf{x}'\boldsymbol{\beta}$ in different nested models, scaling in turn is no longer a problem if the variance of $\mathbf{x}'\boldsymbol{\beta}$ is equal, that is, if two nested models have the same 'explained variance'. This is the basic idea of a very simple yet effective method suggested by Karlson et al. (2012), and known as the KHB method. Let us assume a reduced model uses x as an explanatory variable, and additional variables z are entered into the model in a second step. Due to scaling, the coefficient of x is not comparable between the two models. Karlson et al. suggest estimating an OLS regression of z on x and using the residuals of this regression as additional explanatory variables in the first, reduced logit model. As x and the residuals are uncorrelated by definition, the partial effect of x remains unchanged by including the residuals. However, it requires the variance of $\mathbf{x}'\boldsymbol{\beta}$ to be equal to the variance in the full model. Scaling then is not an issue for a comparison of the coefficients. Karlson et al. (2012) recommend using their method in combination with AMEs to combine the advantages of both methods and gain comparability and the possibility of a simple interpretation at the same time.

The correction method builds on simple OLS regressions and therefore can easily be implemented in all statistical packages. In Stata the ado khb can be used to compute the coefficients in a convenient way (see Kohler et al., 2011). Monte Carlo simulations by the authors show that the KHB method works excellently in correcting for scaling in many settings.

Interaction effects

As logistic regression is a non-linear and non-additive probability model, the use and interpretation of interaction effects differ strongly from OLS regression. As discussed in the previous section, the effect of a given independent variable on $\Pr(y = 1)$ depends on the levels of all other variables. Such a dependence, however, is referred to usually as an interaction effect: the effect of a variable x_1 on y is moderated by a second variable x_2. In other words, logistic regression implicitly models model-inherent interaction effects on the probability (see Nagler, 1994), even if there is no explicit multiplicative term entered in the model. These model-inherent interaction effects mirror the dependence of effects on the proximity of $\widehat{y^*}$ to the threshold value τ (see the first section of this chapter), and do not point to an interaction in the linear model on the latent variable. It is therefore necessary to keep in mind the differences between model-inherent interaction effects which result from different base probabilities, and variable specific interaction effects that mirror moderated effects on the latent dependent variable. While the former are automatically modeled by logistic regression and require graphical inspection of the results, the latter need to be specified explicitly. Naturally, the same applies to squared terms and higher-order polynomials.

That said, variable-specific interaction effects and polynomials can be specified just as in OLS regression. Continuous variables are centered to reduce multicollinearity and a multiplicative term x_1x_2 is set up. This term is then included in a hierarchically well-defined model, that is, a model including all lower-order terms:

$$y^* = a + \beta_1 x_1 + \beta_2 x_2 + \beta_{12} x_1 x_2 + \varepsilon. \tag{8.31}$$

Simply placing x_2 outside brackets shows that the effect of x_2 now depends on x_1: it changes by β_{12} units when x_1 increases by one unit. β_{12} is the interaction effect. β_2 is a conditional effect and gives the slope conditional on $x_1 = 0$:

$$y^* = a + \beta_1 x_1 + (\beta_2 + \beta_{12} x_1) x_2 + \varepsilon. \tag{8.32}$$

By comparing models with and without a multiplicative term using suitable fit statistics, it is simple to assess whether the model has improved or not (see the subsection above on goodness of fit and model comparison) and whether the interaction is statistically significant (e.g. using a likelihood ratio test; see the subsection on statistical inference). However, it is important to keep in mind that the significance of the interaction as well as its interpretation are valid *with regard to the latent dependent variable* y^* *only*. Due to non-additivity and non-linearity of the logistic model of probabilities, the interaction can completely change with regard to the probabilities (see Ai and Norton, 2003, or Huang and Shields, 2000, for a more thorough discussion).

What can we do then? We suggest beginning with a theory as strong as possible and taking great care in specifying whether the interaction is due to the base probability level (model-inherent) or due to a moderation of the effect on the latent variable (variable-specific). We also strongly suggest using graphical tools when exploring interaction effects in logistic regression: profile plots and marginal effects plots can be of great help. The authors have found it helpful to plot the interaction in *multiple* settings, holding other variables constant at values leading to low, medium and high levels of probability. A 'low' level can be obtained by using high values for variables with a negative effect and low values for variables with a positive effect. Finally, it may be worthwhile to use the tools provided by Norton et al. (2004).

EXAMPLE ANALYSIS

In this section we demonstrate the application of logistic regression using the example of ethnocentric attitudes. We loosely build upon the paper by Greene (2009) and examine the effect of socio-demographic factors, political orientation and perceived threat on ethnocentric attitudes. We use round 1 (2002/2003) of the European Social Survey as our database but restrict the sample to Germany and Great Britain. The selection of two countries allows us to conduct cross-national comparison. Contrary to Greene, we use a dichotomous dependent variable which indicates whether a respondent has a preference to allow no immigrants with a different ethnic background into the country (a recoding of the originally continuous variable *imdfetn*). Ethnicity can be described as an ascribed criterion for migration policy, as compared to more meritocratic, acquired criteria such as skills. Hence, the preference for an anti-immigration policy based on ethnic background can be interpreted as a xenophobic attitude. The central independent variable of our analyses is 'perceived group threat', an attitude variable that captures possible economic and cultural consequences of migration.[6] The variable is scaled to range from 0 to 1, with high values pointing to a strongly perceived group threat.

We start by estimating a number of pooled logistic regression models (see Table 8.2). For all models we present unstandardized logit coefficients, the standard error and the average marginal effect. Model 1 includes socio-demographic variables and a dummy for Germany. We find that xenophobia decreases with years of education and household income, but increases with age. Xenophobia is more pronounced in large than in smaller settlements, and lower in Germany than in Great Britain. Most of these control variables are statistically insignificant, as can be seen from Table 8.2. Model 2 adds a variable for the general political position on a left–right axis, and model 3 includes an indicator for group threat. Not surprisingly, we find that persons with a left political position report less xenophobic attitudes, and that perceived group threat leads to higher xenophobia.

We would like to discuss some results of model 3 in greater detail, namely the effect of group threat and of the country dummy. The logit coefficient of the dummy for Germany points to a negative effect ($\beta = -0.752$) and is statistically significant with a standard error of 0.147. The coefficient, however, cannot be interpreted as an indicator for strength or effect size as it refers to the latent dependent variable only. Note that an odds ratio (0.47 in this case) would not be much more helpful as it cannot be interpreted in a meaningful way without knowing the base probability. The average marginal effect of -0.053 indicates that xenophobia is 5 percentage points lower in Germany than in Great Britain. The logit coefficient of group threat equals 7.891 and is statistically significant with a standard error of 0.522. Just like before, we can interpret the direction only and turn to the average marginal effect for a detailed interpretation. Here AME $= 0.561$ indicates that the probability of a preference for an anti-immigration attitude increases by 56 percentage points when the perceived group threat increases by one scale point. As the group threat variable only varies between 0 and 1, 56 percentage points is the effect group threat can have over its complete range. It may be more helpful in this case to calculate the effect of a standard deviation change in group threat: with a standard deviation of 0.16, the probability of xenophobia increases by 9 points for a one-standard-deviation change in group threat.

An alternative approach to the interpretation of logistic regression results are profile plots that plot the probability of $y = 1$ against values of x, keeping other variables constant. Figure 8.4 plots the predicted probabilities as group threat or country varies, holding the background at average.

Table 8.2 Logistic regression models

	Model 1		Model 2		Model 3	
	β (SE)	AME (SE)	β (SE)	AME (SE)	β (SE)	AME (SE)
Years of education	−0.142***	−0.012	−0.140***	−0.011	−0.081**	−0.006
	(0.022)	(0.002)	(0.022)	(0.002)	(0.028)	(0.002)
Relative HH income	−0.974***	−0.080	−1.020***	−0.084	−0.708**	−0.050
	(0.217)	(0.018)	(0.215)	(0.017)	(0.229)	(0.016)
Female	−0.061	−0.005	−0.028	−0.002	−0.009	−0.001
	(0.129)	(0.011)	(0.129)	(0.011)	(0.139)	(0.010)
Age	0.003	0.000	0.001	0.000	0.000	0.000
	(0.003)	(0.000)	(0.004)	(0.000)	(0.004)	(0.000)
Suburb	0.115	0.010	0.080	0.007	−0.218	−0.015
	(0.253)	(0.021)	(0.256)	(0.021)	(0.279)	(0.020)
Town	0.125	0.010	0.100	0.008	−0.200	−0.014
	(0.220)	(0.018)	(0.222)	(0.018)	(0.238)	(0.017)
Village	−0.073	−0.006	−0.115	−0.009	−0.345	−0.025
	(0.238)	(0.020)	(0.242)	(0.020)	(0.258)	(0.018)
Farm	0.090	0.007	0.057	0.005	−0.180	−0.013
	(0.460)	(0.038)	(0.465)	(0.038)	(0.488)	(0.035)
Germany	−0.619***	−0.051	−0.582***	−0.048	−0.752***	−0.053
	(0.129)	(0.011)	(0.132)	(0.011)	(0.147)	(0.010)
Left–right placement			0.112**	0.009	0.055	0.004
			(0.039)	(0.003)	(0.040)	(0.003)
Group threat					7.891***	0.561
					(0.522)	(0.035)
Intercept	0.613		0.146		−4.876***	
	(0.446)		(0.465)		(0.608)	
Nagelkerke R^2	0.088		0.094		0.273	
McFadden R^2	0.067		0.072		0.218	
AIC/N	0.579		0.577		0.488	
N	3331		3331		3331	

$^{\dagger}p \leq 0.1$; $^{*}p \leq 0.05$; $^{**}p \leq 0.01$; $^{***}p \leq 0.001$.
Reference categories: Male, City

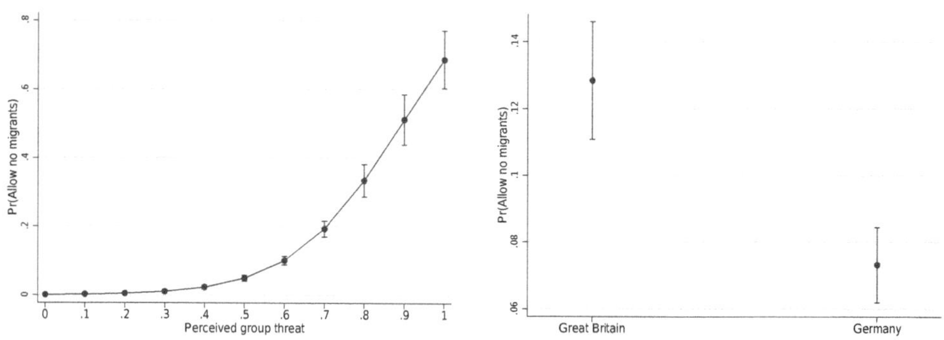

(a) by group threat

(b) by country

Figure 8.4 Predicted probabilities from model 3

Table 8.3 Logistic regression models with KHB correction

	Model 1a		Model 2a		Model 3	
	β (SE)	AME (SE)	β (SE)	AME (SE)	β (SE)	AME (SE)
Years of education	−0.184***	−0.013	−0.177***	−0.013	−0.081**	−0.006
	(0.029)	(0.002)	(0.028)	(0.002)	(0.028)	(0.002)
Relative HH income	−1.115***	−0.079	−1.194***	−0.085	−0.708**	−0.050
	(0.229)	(0.016)	(0.229)	(0.016)	(0.229)	(0.016)
Female	−0.045	−0.003	−0.009	−0.001	−0.009	−0.001
	(0.138)	(0.010)	(0.139)	(0.010)	(0.139)	(0.010)
Age	0.005	0.000	0.002	0.000	0.000	0.000
	(0.004)	(0.000)	(0.004)	(0.000)	(0.004)	(0.000)
Suburb	0.045	0.003	0.004	0.000	−0.218	−0.015
	(0.278)	(0.020)	(0.278)	(0.020)	(0.279)	(0.020)
Town	0.106	0.008	0.067	0.005	−0.200	−0.014
	(0.236)	(0.017)	(0.236)	(0.017)	(0.238)	(0.017)
Village	−0.039	−0.003	−0.091	−0.006	−0.345	−0.025
	(0.255)	(0.018)	(0.256)	(0.018)	(0.258)	(0.018)
Farm	−0.010	−0.001	−0.049	−0.003	−0.180	−0.013
	(0.487)	(0.035)	(0.488)	(0.035)	(0.488)	(0.035)
Germany	−0.788***	−0.056	−0.718 ***	−0.051	−0.752***	−0.053
	(0.145)	(0.010)	(0.146)	(0.010)	(0.147)	(0.010)
Left–right placement			0.159 ***	0.011	0.055	0.004
			(0.040)	(0.003)	(0.040)	(0.003)
Group threat					7.891***	0.561
					(0.522)	(0.035)
Intercept	0.785		0.081		−4.876***	
	(0.509)		(0.521)		(0.608)	
N	3331		3331		3331	

$^{\dagger}p \leq 0.1$; $^{*}p \leq 0.05$; $^{**}p \leq 0.01$; $^{***}p \leq 0.001$.
Reference categories: Male, City

It is easy to see from Figure 8.4a that the effect of group threat is non-linear and increases in strength as the group threat increases. There is almost no xenophobia as long as the perceived group threat is below 0.4, but after that point the probability starts to rise to a maximum of almost 0.7 when group threat reaches its maximum of one. Figure 8.4b shows the predicted probabilities for Great Britain and Germany along with their confidence intervals.

If a regression model has been built in a stepwise manner as in models 1–3, one would usually try to interpret the changes in coefficients as more variables are included in the models. In the case of logistic regression, this inter-model comparison is not trivial as coefficients may be biased by scaling effects due to neglected heterogeneity (see the previous section). For the comparison of coefficients between models, we advise the use of average marginal effects, the KHB correction or a combination of both. Table 8.3 therefore depicts the results from Table 8.2, corrected for bias due to scaling. Note that model 3 is identical in both tables.

Using this correction, we can compare the models in a meaningful way. For example, the negative effect of education changes only slightly when controlling for left–right self-placement, but drastically looses strength once the perceived group threat is included in the model: the absolute value of the logit coefficient decreases from −0.184 to −0.081, and the AME from −0.013 to −0.006, both corresponding to a reduction by about 55%. Note that the fit indices of the models are excluded from Table 8.3 as they are rendered uninformative by applying the KHB correction.

Table 8.4 Logistic regression models analyzing country differences

	Model 4		Model 5 (Germany)		Model 6 (GB)	
	β (SE)	AME (SE)	β (SE)	AME (SE)	β (SE)	AME (SE)
Years of education	–0.082**	–0.006	–0.041	–0.002	–0.120**	–0.010
	(0.028)	(0.002)	(0.039)	(0.002)	(0.040)	(0.004)
Relative HH income	–0.749**	–0.053	–0.979*	–0.058	–0.630*	–0.054
	(0.232)	(0.016)	(0.383)	(0.022)	(0.301)	(0.026)
Female	–0.022	–0.002	–0.087	–0.005	–0.032	–0.003
	(0.140)	(0.010)	(0.189)	(0.011)	(0.212)	(0.018)
Age	0.000	0.000	0.002	0.000	0.000	0.000
	(0.004)	(0.000)	(0.005)	(0.000)	(0.006)	(0.001)
Suburb	–0.214	–0.015	–0.338	–0.020	0.141	0.012
	(0.278)	(0.020)	(0.379)	(0.022)	(0.525)	(0.045)
Town	–0.173	–0.012	–0.125	–0.007	0.064	0.006
	(0.235)	(0.017)	(0.260)	(0.015)	(0.500)	(0.043)
Village	–0.314	–0.022	–0.572†	–0.034	0.212	0.018
	(0.255)	(0.018)	(0.300)	(0.018)	(0.530)	(0.046)
Farm	–0.161	–0.011	0.155	0.009	–0.126	–0.011
	(0.505)	(0.036)	(0.673)	(0.040)	(0.778)	(0.067)
Left–right placement	0.063	0.004	0.181***	0.011	–0.072	–0.006
	(0.041)	(0.003)	(0.054)	(0.003)	(0.061)	(0.005)
Group threat	9.190***	0.561	6.591***	0.387	9.150***	0.791
	(0.755)	(0.035)	(0.747)	(0.043)	(0.760)	(0.062)
Germany	0.812	–0.055				
	(0.672)	(0.011)				
Germany*Group threat	–2.442*					
	(1.030)					
Intercept	–5.701***		–5.696***		–4.976***	
	(0.720)		(0.835)		(0.963)	
Nagelkerke R^2	0.277		0.227		0.327	
McFadden R^2	0.221		0.186		0.253	
AIC/*N*	0.487		0.427		0.577	
N	3331		2000		1331	

†$p \leq 0.1$; *$p \leq 0.05$; **$p \leq 0.01$; ***$p \leq 0.001$.
Reference categories: Male, City, Great Britain

Finally we want to have a closer look at differences between Germany and Great Britain, with a special emphasis on the effect of group threat. We therefore estimate a model that includes an interaction effect between group threat and country, and two separate models for Germany and Great Britain (see Table 8.4). Model 4 includes the interaction effect. The coefficient of the interaction effect is negative and statistically significant. While the interpretation would be very straightforward in linear models, it is more complicated in logistic regression models. The interaction effect of -2.442 indicates the interaction with regard to the latent dependent variable only. As logistic regression is a non-linear probability model, this does not necessarily hold for the effect on probabilities. In extreme cases, the interaction may differ in strength, lose statistical significance, or even change its sign, depending on the values of covariates and the level of baseline probability. In order to interpret the interaction effect more thoroughly, we utilize graphical displays of predicted probabilities or marginal effects in Figure 8.5 (see Chapter 10 of this volume for a detailed discussion of the graphical representation of regression results). Figure 8.5a is a profile plot. It shows predicted probabilities of xenophobia as it varies

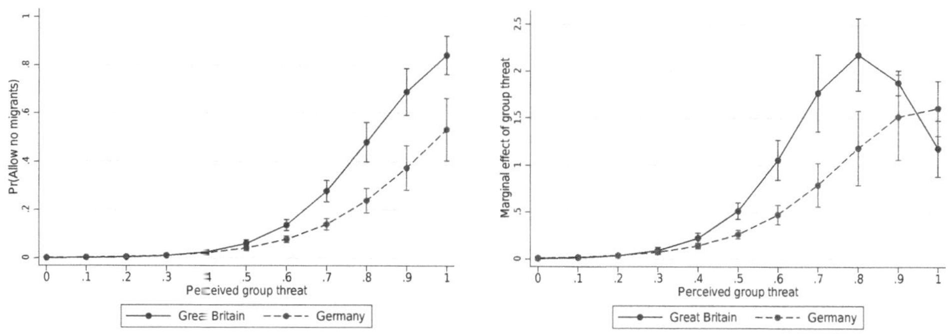

(a) predicted probabilities (b) marginal effects

Figure 8.5 Effect of perceived group threat by country

with perceived group threat, by country. It is evident that the curve for Great Britain is above the curve for Germany in the part of the value range of group threat above 0.4, and that the slope seems to be somewhat steeper.

A much more thorough interpretation can be made based on the marginal effects plot shown in Figure 8.5b. The plot allows a very detailed interpretation. It shows that the interaction indeed is negative for most values of group threat (averaged over the sample), that is, the marginal effect is smaller in Germany than in Great Britain. However, the strength of the interaction effect greatly varies over the range of group threat: it is very small for values below 0.3 to 0.4, then increases in strength, reaches its maximum at a group threat of about 0.8, and then decreases again. In the case of a very high perceived group threat, the interaction effect even changes sign: here the effect is stronger in Germany than in Great Britain.

A researcher interested in the effects of more than one variable may be tempted to estimate separate logit models – see models 5 and 6 in Table 8.4. When comparing the effects between separate models it is important to keep in mind that the logit coefficients (and odds ratios as well, for that matter) are scaled differently and hence cannot be compared directly. A valid comparison therefore should be based on probabilities rather than on the latent dependent variable. The easiest method of comparison between models compares average marginal effects. As AMEs are not strongly affected by scaling, a comparison between models is valid. We can therefore see that the effect of education is stronger (−0.010) in Great Britain than in Germany, where the effect is only 0.002. The AME of income is roughly equal in both countries, at 0.054 and 0.058. A more thorough interpretation would require estimating predicted probabilities from both models and plotting them in a profile plot, or including an interaction effect and using the methods described above.

CAVEATS AND FREQUENT ERRORS

In general, we regard logistic regression as a useful and relatively easy-to-use regression model for dichotomous dependent variables. Problems in usage and interpretation result most often from overestimating similarities with linear OLS regression.

Firstly, it is important always to keep in mind that logistic regression is a linear and additive model for the latent variable y^* only. Regarding probabilities, logit models describe *non-linear effects that cannot be expressed in a single coefficient.* Interpretations of logit coefficients are therefore much more prone to risk than OLS coefficients. A common error follows from assuming a linear relationship between x and $\Pr(y = 1)$ and interpreting the β coefficients as in OLS regression. At the same time, there is a risk of underutilizing the results when interpreting the direction of the effect only. We recommend using graphical tools for interpretation such as profile plots of the predicted probabilities. A good alternative – or rather an ideal addition – is calculating average marginal effects. AMEs give the average additive effect of x on $\Pr(y = 1)$. We absolutely do not recommend using odds ratios inasmuch as odds are related to probabilities in a non-linear way and a given odds ratio may translate into completely different relative risks depending on the base probability. Additionally, very few people have an intuitive understanding of odds (the authors do not), which leads to the hazard of interpreting the odds ratios explicitly as probability ratios or in a diffuse probabilistic way. In a paper published in the *American Journal of Sociology*, Cornwell and Laumann (2011, p. 195) interpret an odds ratio of 1.92 of the dichotomous variable 'partner betweenness' as follows: 'a man whose female partner has greater contact with some of his confidants than he does [this is 'partner betweenness', HB/CW] is about 92% more likely to have had trouble getting or maintaining an erection than a man who has greater access than his partner does to all of his confidants'. This interpretation is of course incorrect or misleading, depending on how one defines 'likely'. Such problems can be easily avoided by refraining from the use of odds ratios.

A second problem is due to the fact that logit coefficients are defined to scale only. Logit coefficients are defined to scale only and can be biased by neglected heterogeneity, that is, by the amount of unexplained variance, even if ε and x are uncorrelated. The actual size of the coefficient therefore is meaningless. Hence, an interpretation of β-coefficients or odds ratios as causal effects is impossible – even if the (conditional) independence assumption holds. For a causal interpretation of the regression results we recommend the careful use of AMEs (but see the remarks on causal inference based on cross sectional regression in Chapter 4 of this volume). The scaling problem also affects the comparison of coefficients between groups and nested models. We generally recommend using average marginal effects and for nested models addtionally the correction suggested by Karlson et al. (2012).

Finally, the use of interaction effects can be complicated. In principle, these are specified and interpreted just as in linear regression, but the interpretation grows more complicated when probabilities are the quantities of interest, instead of the latent y^*. Regarding probabilities, the effect of a variable x_1 already depends on the values of another variable x_2 even if there is no variable-specific interaction term in the model. Therefore, care should be taken when testing hypotheses concerning interactions in non-linear probability models. We recommend using graphical methods and additional robustness checks.

FURTHER READING

We believe Long (1997) still to be the best textbook on logistic regression. A more applied introduction using Stata is Long & Freese (2014). Wooldridge (2002) provides an excellent yet complex treatise; the same is true for Train (2009). We highly recommend both books for advanced users. For new users of logistic models Menard (1995) may be a good starting point.

NOTES

1 We acknowledge that attitudes are continuous rather than dichotomous in nature. However, we assume that, due to suboptimal measurement, we *observed* the binary state only.

2 Similarly, Gary King wrote on July 26, 2011 on the POLMETH list: 'I don't think the odds ratio makes any sense to report; it's used because it is a quantity that happens to be more convenient to estimate in some

specialized situations. If the outcome variable is very rare (e.g. almost all 1s and just a couple of 0s) then the odds ratio approximates a relative risk (e.g. Prob(war|democracy)/Prob(war|autocracy)), which does make sense, but in other situations this connection doesn't work and the odds ratio is merely confusing.'

3 When analyzing decisions, the latent variable could be interpreted as the subjectively expected utility of an action alternative.

4 The equation for estimating standard errors is given by Wooldridge (2002, p. 460).

5 Two models are nested when the parameters of one model are a *proper subset* of those of the second.

6 Generated from the variables *imwgdwn, imhecop, imtcjob, imbgeco, pplstrd, imueclt, and vrtrlg*. All variables load on a single factor in principal component analysis.

REFERENCES

Ai, C. and Norton, E. C. (2003). Interaction terms in logit and probit models. *Economics Letters*, 80, 123–129.
Akaike, H. (1973). Information theory and an extension of the maximum likelihood principle. In B. N. Petrov and B. F. Csaki (Eds), *Second International Symposium on Information Theory* (pp. 267–281). Budapest: Academiai Kiado.
Allison, P. D. (1999). Comparing logit and probit coefficients across groups. *Sociological Methods & Research*, 28(2), 186–208.
Best, H. (2009). Organic farming as a rational choice. Empirical investigations in environmental decision making. *Rationality and Society*, 21(2), 197–224.
Best, H. and Wolf, C. (2012). Modellvergleich und Ergebnisinterpretation in Logit- und Probit-Regressionen. *Kölner Zeitschrift für Soziologie und Sozialpsychologie*, 64(2), 377–395.
Cornwell, B. and Laumann, E. O. (2011). Network position and sexual dysfunction: Implications of partner betweenness for men. *American Journal of Sociology*, 117(1), 172–208.
Cox, D. R. and Snell, E. J. (1989). *The Analysis of Binary Data*. London: Chapman & Hall.
Cragg, J. G. and Uhler, R. (1970). The demand for automobiles. *Canadian Journal of Economics*, 3(3), 386–406.
Greene, E. (2009). Who can enter? a multilevel analysis on public support for immigration criteria across 20 european countries. *Group Processes & Intergroup Relations*, 12(1), 41–60.
Huang, C. and Shields, T. G. (2000). Interpretation of interaction effects in logit and probit analyses. *American Politics Research*, 28(1), 80–95.
Karlson, K. B., Holm, A., and Breen, R. (2012). Comparing regression coefficients between models using logit and probit: A new method. *Sociological Methodology*, 42(1).
Kohler, U., Karlson, K. B., and Holm, A. (2011). Comparing coefficients of nested nonlinear probability models. *The Stata Journal*, 11(3), 420–438.
Long, J. S. (1997). *Regression Models for Categorical and Limited Dependent Variables*. Thousand Oaks: Sage.
Long, J. S. and Freese, J. (2014). *Regression Models for Categorical Dependent Variables Using Stata*. Third edition College Station: Stata Press.
McFadden, D. (1973). Conditional logit analysis of qualitative choice behaviour. In P. Zarembka (Ed.), *Frontiers in Econometrics* (pp. 105–142). New York: Academic Press.
Menard, S. (1995). *Applied Logistic Regression*, volume 07-106 of *Quantitative Applications in the Social Sciences*. Thousand Oaks: Sage.
Nagler, J. (1994). Scobit: An alternative estimator to logit and probit. *American Journal of Political Science*, 38(1), 230–255.
Norton, E. C., Wang, H., and Ai, C. (2004). Computing interaction effects and standard errors in logit and probit models. *The Stata Journal*, 4(2), 103–116.
Raftery, A. E. (1995). Bayesian model selection in social research. *Sociological Methodolgy*, 25, 111–163.
Train, K. (2009). *Discrete choice methods with simulation*. Cambridge/New York: Cambridge University Press.
Winship, C. and Mare, R. D. (1984). Regression models with ordinal variables. *American Sociological Review*, 49, 512–525.
Wooldridge, J. M. (2002). *Econometric Analysis of Cross Section and Panel Data*. Cambridge: MIT Press.

9

Regression models for nominal and ordinal outcomes

J. Scott Long*

INTRODUCTION TO THE METHOD

Ordinal and nominal outcomes are common in the social sciences, with examples ranging from Likert items in surveys to assessments of physical health to how armed conflicts are resolved. Since the 1980s numerous regression models for nominal and ordinal outcomes have been developed. These models are essentially sets of binary regressions that are estimated simultaneously with constraints on the parameters. While advances in software have made estimation simple, the effective interpretation of these non-linear models is a vexingly difficult art that requires time, practice, and a firm grounding in the goals of your analysis and the characteristics of your model. Too often interpretation is limited to a table of coefficients with a brief discussion of signs and statistical significance. While the implications of a model are implicit in these parameters, post-estimation computations of probabilities and related quantities are essential for understanding the substantive impact of the regressors.

The goal in selecting a model is to find one that is parsimonious without distorting critical relationships. A too simple model risks bias, while an unnecessarily complex model is statistically inefficient. Models for nominal outcomes are sometimes avoided because of the number of parameters and the perceived difficulty in their interpretation. While nominal models might have more parameters, this complexity is transparent when probabilities are used for interpretation since software easily makes the computations. Ordinal models generally have fewer parameters, but this simplicity is achieved by imposing constraints that potentially distort the process being modeled. Understanding the substantive and theoretical context of your research, accompanied by an evaluation of the robustness of findings to alternative specifications, are fundamental to using regression models for nominal and ordinal outcomes.

What does ordinal or nominal mean?

Stevens (1946) provided the classic definitions of nominal and ordinal variables:

> *Nominal scales* assign numbers to categories as labels with no ordering implied by the numbers.
>
> *Ordinal scales* use numbers to indicate rank ordering on a single attribute.

Table 9.1 Models considered in the paper and whether they meet Anderson's criteria for being ordinal

Model		Is the model ordinal?
Multinomial logit	(MNLM)	No
Adjacent category logit	(ACLM)	Yes
Stereotype logit with 1 dimension	(SLM1)	Yes
Stereotype logit with 2+ dimensions	(SLM2, SLM3, ...)	No
Ordered logit	(OLM)	Yes
Generalized ordered logit	(GOLM)	No

Even though Stevens' taxonomy was hotly debated when it was proposed and has been critiqued since (Duncan, 1984; Velleman and Wilkinson, 1993), it is firmly established in the methods of many disciplines and is often used to classify models. Some variables commonly thought of as ordinal do not meet Stevens' criterion since they reflect multiple attributes rather than a single attribute as required by his definition. Consider political party affiliation which is used as an example in this paper. Affiliation was coded with the categories Strong Democrat (1=*SD*), Democrat (2=*D*), Independent (3=*I*), Republican (4=*R*), and Strong Republican (5=*SR*). On the attribute of left–right orientation, party is ranked from 1=*SD* to 5=*SR*. In terms of intensity of partisanship, the categories are ordered 1=*I*; 2=*R&D*; and 3=*SR&SD*. Anticipating results from the third section of this chapter, age might affect intensity of partisanship so that the probabilities of both *SD* and *SR* increase with age, while income affects left–right orientation but not intensity.

Ordinal regression models constrain the relationships between regressors and outcomes in a way that was elaborated by Anderson (1984). Suppose that the coefficient for x is positive. In an ordinal model, as x increases the probability of the lowest category decreases from 1 to 0 while the probability of the highest category increases from 0 to 1. The probability curves for other categories are bell-shaped with modes occurring at larger values of x for higher categories. This is illustrated in Figure 9.1. The definition of ordinal models proposed by Anderson depends on the relationship between regressors and outcomes, not the scale of measurement of the dependent variable. While nominal models can lead to predictions consistent with an ordinal model, they are not constrained to do so.

Table 9.1 lists the models reviewed in this paper and indicates which models are ordinal by Anderson's definition. The multinomial logit model (MNLM) is the most commonly used nominal regression model. The adjacent category logit model (ACLM) constrains the parameters in the MNLM so that the odds ratios for adjacent categories (e.g. 1 and 2, 2 and 3) are identical for a given regressor. The resulting model is ordinal. Anderson's (1984) stereotype logit model (SLM) modifies the MNLM to reduce the number of parameters. While the one-dimensional form of the model is ordinal, higher-dimensional versions are not. The most common ordinal regression model, often called *the* ordered logit model (OLM), is a set of logits on binary variables that divide the outcome into lower and higher categories (e.g. 1 versus higher categories; 1 and 2 versus higher categories). The coefficients for a given x_k are constrained to be equal in all equations. This makes the model ordinal, but can be unrealistic. In response to limitations of the OLM, the generalized ordered logit model (GOLM) allows the coefficients for x_k to differ across equations. The resulting model is no longer ordinal by Anderson's criteria. Models that fall between the OLM and the GOLM are considered briefly. While I focus on logit models, in most cases probit versions of these models are available and produce nearly identical predictions.

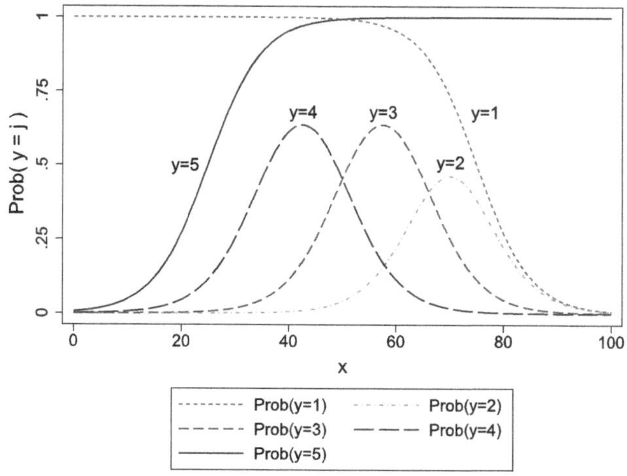

Figure 9.1 Illustration of Anderson's definition of an ordinal regression model

MATHEMATICAL FOUNDATIONS AND ADVANCED ASPECTS

Each model in Table 9.1 can be interpreted using predicted probabilities of the outcome categories and, for logit versions of the models, odds ratios. This section presents formulas for these quantities, with examples of their use given in the next section. While each model can be parameterized in several ways, I use parameterizations that emphasize the similarities among models. Outcome y has J categories with regressors $x_1, \ldots, x_K$. The intercept is β_0, with the linear combination of regressors and coefficients written as $\mathbf{x}'\boldsymbol{\beta} = \beta_1 x_1 + \cdots + \beta_K x_K$. Some models require an additional subscript such as $\beta_{k,m|n}$. To simplify the presentation, some models are initially shown with three outcomes and two regressors before a general form of the model is given.

Multinomial logit model

Multinomial logit is equivalent to the simultaneous estimation of a set of binary logits for all pairs of outcome categories. To see this, let y have the two categories D=Democrat and R=Republican. With two regressors the binary logit model is:

$$\ln \frac{\Pr(y = D \mid \mathbf{x})}{\Pr(y = R \mid \mathbf{x})} = \beta_{0,D|R} + \beta_{age,D|R}age + \beta_{income,D|R}income.$$

The notation for the βs makes explicit the categories being compared. If I add a third category, I=Independent, there are three binary comparisons:[1]

$$\begin{aligned}
\ln \frac{\Pr(y = D \mid \mathbf{x})}{\Pr(y = R \mid \mathbf{x})} &= \beta_{0,D|R} + \beta_{age,D|R}age + \beta_{income,D|R}income, \\
\ln \frac{\Pr(y = D \mid \mathbf{x})}{\Pr(y = I \mid \mathbf{x})} &= \beta_{0,D|I} + \beta_{age,D|I}age + \beta_{income,D|I}income, \\
\ln \frac{\Pr(y = R \mid \mathbf{x})}{\Pr(y = I \mid \mathbf{x})} &= \beta_{0,R|I} + \beta_{age,R|I}age + \beta_{income,R|I}income.
\end{aligned} \tag{9.1}$$

Begg and Gray (1984) show that estimates from the binary logits are consistent but inefficient estimates of the MNLM. Software for the MNLM obtains efficient estimates by simultaneously estimating all equations while imposing constraints that link the equations. These constraints can be seen in the mathematically necessary relationship,

$$\ln \frac{\Pr(y = D \mid \mathbf{x})}{\Pr(y = R \mid \mathbf{x})} = \ln \frac{\Pr(y = D \mid \mathbf{x})}{\Pr(y = I \mid \mathbf{x})} - \ln \frac{\Pr(y = R \mid \mathbf{x})}{\Pr(y = I \mid \mathbf{x})},$$

which implies that $\beta_{k,D|R} = \beta_{k,D|I} - \beta_{k,R|I}$. Accordingly, the coefficients from any two binary logits determine *exactly* the coefficients for the remaining logit. The smallest set of parameters that implies the parameters for all comparisons is called a *minimal set*. Often the minimal set consists of all comparisons relative to one of the categories which is called the *base category*. I assume the base category is J, but other values could be used.

Defining the odds of category j versus category J given $\mathbf{x}$ as $\Omega_{j|J}(\mathbf{x}) = \frac{\Pr(y=j|\mathbf{x})}{\Pr(y=J|\mathbf{x})}$ and $\mathbf{x}'\boldsymbol{\beta}_{j|J} = \beta_{1,j|J}x_1 + \cdots + \beta_{K,j|J}x_K$, the MNLM is

$$\ln \Omega_{j|J}(\mathbf{x}) = \ln \frac{\Pr(y = j \mid \mathbf{x})}{\Pr(y = J \mid \mathbf{x})} = \beta_{0,j|J} + \mathbf{x}'\boldsymbol{\beta}_{j|J}, \quad \text{for } j = 1, J.$$

Since $\Omega_{J|J}(\mathbf{x}) = 1$, it follows that $\beta_{0,J|J=0}$ and $\boldsymbol{\beta}_{J|J} = \mathbf{0}$. Taking the exponential,

$$\Omega_{j|J}(\mathbf{x}) = \exp\left(\beta_{0,j|J} + \mathbf{x}'\boldsymbol{\beta}_{j|J}\right),$$

with the odds ratio[2]

$$\mathrm{OR}_{k,j|J} = \frac{\Omega_{j|J}(\mathbf{x}, x_k + 1)}{\Omega_{j|J}(\mathbf{x}, x_k)} = \exp\left(\beta_{k,j|J}\right),$$

The odds for categories m and n are

$$\begin{aligned}\Omega_{m|n}(\mathbf{x}) &= \exp\left(\left[\beta_{0,m|J} + \mathbf{x}'\boldsymbol{\beta}_{m|J}\right] - \left[\beta_{0,n|J} + \mathbf{x}'\boldsymbol{\beta}_{n|J}\right]\right) \\ &= \exp\left(\beta_{0,m|n} + \mathbf{x}'\boldsymbol{\beta}_{m|n}\right),\end{aligned}$$

where $\beta_{0,m|n} = \beta_{0,m|J} - \beta_{0,n|J}$. The odds ratio for x_k is

$$\mathrm{OR}_{k,m|n} = \exp\left(\beta_{k,m|J} - \beta_{k,n|J}\right),$$

which can be interpreted as follows:

> For a unit increase in x_k, the odds of category m versus category n change by a factor of $\exp\left(\beta_{k,m|J} - \beta_{k,n|J}\right)$, holding other variables constant.

If OR is greater than one, I could say 'The odds are OR times larger'; if less than one, 'The odds are OR times smaller'.

The model can also be expressed in terms of probabilities. Solving the odds equations, the probability of category j is

$$\Pr(y = j \mid \mathbf{x}) = \frac{\exp\left(\beta_{0,j|J} + \mathbf{x}'\boldsymbol{\beta}_{j|J}\right)}{\sum_{q=1}^{J} \exp\left(\beta_{0,q|J} + \mathbf{x}'\boldsymbol{\beta}_{q|J}\right)} \quad \text{for } j = 1, \ldots, J. \tag{9.2}$$

Using estimates of the parameters, this equation is used to compute predicted probabilities and functions of these probabilities, such as marginal and discrete changes, which are used for interpretation as discussed below.

For a regressor to have no effect, the $J - 1$ coefficients associated with that variable must be simultaneously 0. In our example, the hypothesis that age has no effect is $H_{age} : \beta_{age,D|I} = \beta_{age,R|I} = 0$. If these two coefficients are 0, then $\beta_{age,D|R} = \beta_{age,D|I} - \beta_{age,R|I}$ must be also 0. But it is possible to reject H_{age} while not rejecting either $H_{D|I} : \beta_{age,D|I} = 0$ or $H_{R|I} : \beta_{age,R|I} = 0$. Suppose that age increases being a Democrat D relative to Independent I, but the effect is not large enough to be detected in the sample. Similarly, age decreases Republican R relative to Independent I but not significantly so. Since Democrat and Republican are further apart politically, the effect of age on D relative to R could be large enough to be significant within the sample. In general, the hypothesis that x_k has no effect is

$$H_{x_k} : \beta_{k,1|J} = \cdots = \beta_{k,J-1|J} = 0,$$

which can be tested with a Wald or likelihood ratio (LR) test with $J - 1$ degrees of freedom.

Independence of irrelevant alternatives

Independence of irrelevant alternatives (IIA) is the defining property of the MNLM that simplifies estimation and interpretation, but is potentially unrealistic. IIA implies that a person's choice between two alternatives (i.e. outcome categories) is unaffected by other alternatives. Suppose that a person is provided with a new alternative that is very similar to an existing alternative. Since these alternatives are similar, you would expect individuals to evenly divide their choice between the original and the new, similar alternative while the probability of dissimilar alternatives would not be affected. IIA requires that the probabilities of *all* alternatives, not just similar alternatives, decrease proportionately with the addition of the new alternative. This is behaviorally unrealistic.

Tests of IIA assess how estimates of coefficients change when the model is re-estimated with a restricted set of outcome categories (e.g. compare estimates using J categories to those obtained using $J - 1$ categories). If the test is significant, the assumption of IIA is rejected, indicating that the MNLM is inappropriate. The Hausman–McFadden test (see Hausman and McFadden, 1984) and the Small–Hsiao test (see Small and Hsiao, 1985) are the most common IIA tests. Using Monte Carlo experiments, Fry and Harris (1998, 1996) and Cheng and Long (2007) found these and other IIA tests have poor statistical properties in finite samples. They conclude that IIA tests are *not* useful for assessing violations of IIA. The best advice regarding IIA goes back to an early statement by McFadden (1973) that the MNLM should only be used when the alternatives 'can plausibly be assumed to be distinct and weighed independently in the eyes of each decision maker'. If you have alternatives that are very similar in how they are evaluated as choices, such as riding a red bus and riding a blue bus, combine the categories. Care in specifying the model to include distinct alternatives that are not substitutes is reasonable, albeit ambiguous, advice.

Multinomial probit model

For decades the multinomial probit model (MNPM) was proposed as a way to avoid the IIA assumption if computational problems could be solved. The MNPM is based on the normal distribution, which, unlike the logistic distribution used with the MNLM, allows alternatives to be dependent in the sense that a person can be more likely to select both alternative m and alternative n after controlling for regressors. This avoids the IIA assumption. Estimation using simulation is now practical, but several factors limit the model's usefulness. First, identification requires alternative-specific regressors. These are variables whose values depend on the outcome (i.e. alternative). For example, the choice of mode of transportation for commuting could depend on the time each alternative mode requires, where travel time varies by alternative. When alternative-specific variables are not available, the MNPM is not identified.[3] Second, even with alternative-specific regressors identification requires constraints on correlations among errors. Substantive motivation for these constraints is often unavailable. Finally, even when the model is *formally* identified, Keane (1992) found that identification is fragile, which requires additional restrictions to avoid the risk of unreliable estimates. My experience confirms Keane's (1992) statement: 'Given the lack of practical experience with [multinomial probit] models, however, there is a need to develop a "folklore" concerning the conditions under which the model performs well.' For full details on the MNPM and other models using alternative specific regressors, see Train (2009).

Summary of MNLM

The MNLM is a flexible model that imposes few restrictions on the relationships between regressors and outcomes. While it can lead to relationships between regressors and outcome probabilities that are consistent with Anderson's definition of an ordinal model, it is not constrained so that it must do so.

Adjacent category logit model

The adjacent categories logit model (Goodman, 1983; Clogg and Shihadeh, 1994) is an ordinal model that constrains the MNLM so that coefficients from adjacent ordinal categories are equal. For example, here is the MNLM for categories ordered 1, 2, and 3:

$$\ln\frac{\Pr(y=1\mid\mathbf{x})}{\Pr(y=2\mid\mathbf{x})}=\beta_{0,1|2}+\beta_{age,1|2}age+\beta_{income,1|2}income,$$
$$\ln\frac{\Pr(y=2\mid\mathbf{x})}{\Pr(y=3\mid\mathbf{x})}=\beta_{0,2|3}+\beta_{age,2|3}age+\beta_{income,2|3}income,$$
$$\ln\frac{\Pr(y=1\mid\mathbf{x})}{\Pr(y=3\mid\mathbf{x})}=\beta_{0,1|3}+\beta_{age,1|3}age+\beta_{income,1|3}income.$$

The ACLM assumes $\beta_{age,1|2}=\beta_{age,2|3}=\beta_{age}$ and $\beta_{income,1|2}=\beta_{income,2|3}=\beta_{income}$, leading to the model:

$$\ln\frac{\Pr(y=1\mid\mathbf{x})}{\Pr(y=2\mid\mathbf{x})}=\beta_{0,1|2}+\beta_{age}age+\beta_{income},income$$
$$\ln\frac{\Pr(y=2\mid\mathbf{x})}{\Pr(y=3\mid\mathbf{x})}=\beta_{0,2|3}+\beta_{age}age+\beta_{income}income.$$

Since

$$\ln\frac{\Pr(y=1\mid\mathbf{x})}{\Pr(y=3\mid\mathbf{x})}=\ln\frac{\Pr(y=1\mid\mathbf{x})}{\Pr(y=2\mid\mathbf{x})}+\ln\frac{\Pr(y=2\mid\mathbf{x})}{\Pr(y=3\mid\mathbf{x})},$$

it follows that

$$\begin{aligned}\ln\frac{\Pr(y=1\mid\mathbf{x})}{\Pr(y=3\mid\mathbf{x})}=&\left(\beta_{0,1|2}+\beta_{age}age+\beta_{income}income\right)\\&+\left(\beta_{0,2|3}+\beta_{age}age+\beta_{income}income\right)\\=&\left(\beta_{0,1|2}+\beta_{0,2|3}\right)+\left(2\beta_{age}\right)age+(2\beta_{income})income.\end{aligned}$$

Generally, the ACLM can be written as

$$\ln \frac{\Pr(y=j \mid \mathbf{x})}{\Pr(y=j+1 \mid \mathbf{x})} = \beta_{0,j|j+1} + \mathbf{x}'\boldsymbol{\beta}, \quad \text{for } j = 1, J-1,$$

where intercepts vary by j but the coefficients for x_k do not. Taking exponentials,

$$\Omega_{j|j+1}(\mathbf{x}) = \exp\left(\beta_{0,j|j+1} + \mathbf{x}'\boldsymbol{\beta}\right), \quad \text{for } j = 1, J-1,$$

with the odds ratio for adjacent categories equal to

$$\text{OR}_{k,j|j+1} = \frac{\Omega_{j|j+1}(\mathbf{x}, x_k + 1)}{\Omega_{j|j+1}(\mathbf{x}, x_k)} = \exp(\beta_k).$$

This can be interpreted as follows:

For a unit increases in x_k, the odds of adjacent categories change by a factor of $\exp(\beta_k)$, holding other variables constant.

Similarly to the MNLM, probabilities equal

$$\Pr(y=j \mid \mathbf{x}) = \frac{\exp\left(\beta_{0,j|j+1} + \mathbf{x}'\boldsymbol{\beta}\right)}{1 + \sum_{q=1}^{J-1}\left[\exp\left(\beta_{0,q|q+1} + \mathbf{x}'\boldsymbol{\beta}\right)\right]}, \quad \text{for } j = 1, J-1,$$

$$\Pr(y=J \mid \mathbf{x}) = 1 - \sum_{q=1}^{J-1} \Pr(y=q \mid \mathbf{x}).$$

The critical issue when using this model is whether it makes substantive sense that ORs for adjacent categories are equal. You can test this constraint with the LR test, comparing the ACLM to the MNLM. In my experience, the hypothesis is often rejected with a large chi-square. While I am not aware of software specifically for this model, estimation is possible with any program for the MNLM that allows constraints on the parameters.

Stereotype logit model

The stereotype logit model was proposed by Anderson (1984) in response to the restrictive assumption of parallel regressions in the ordered logit model (presented next) and to reduce the number of parameters in the MNLM. The MNLM with base J is

$$\ln \frac{\Pr(y=j \mid \mathbf{x})}{\Pr(y=J \mid \mathbf{x})} = \beta_{0,j|J} + \beta_{age,j|J}\,age + \beta_{income,j|J}\,income,$$

with $J-1$ parameters for each regressor, where the implied coefficient comparing m to n is $\beta_{k,m|n} = \beta_{k,m|J} - \beta_{k,n|J}$. To reduce the number of parameters, the SLM restricts the coefficients to vary by scale factors θ_j and ϕ_j for $j = 1, J$. The θs scale the intercepts as $\beta^*_{0,m|n} = (\theta_m - \theta_n)\beta_0$, while the ϕs scale the coefficients for x_k as $\beta^*_{k,m|n} = (\phi_m - \phi_n)\beta_k$. This leads to the one-dimensional SLM:

$$\begin{aligned}\ln \frac{\Pr(y=j \mid \mathbf{x})}{\Pr(y=J \mid \mathbf{x})} &= \beta^*_{0,j|J} + \beta^*_{age,j|J}\,age + \beta^*_{income,j|J}\,income \\ &= (\theta_j - \theta_J)\beta_0 + (\phi_j - \phi_J)\beta_{age}\,age + (\phi_j - \phi_J)\beta_{income}\,income.\end{aligned}$$

There is one coefficient for each regressor (e.g. β_{age}) with scale factors common to all regressors. Identification requires constraints on the scale factors such as $\phi_1 = 1$ and $\phi_J = \theta_J = 0$ (for details, see Long and Freese, 2014). With J categories and K regressors there are $2(J-2)+K+1$ parameters in the SLM1 compared to $(K+1)(J-1)$ in the corresponding MNLM. For example, with $J = 4$ outcomes and $K = 6$ regressors, the MNLM has 21 parameters, compared to 11 for the SLM1. To make the model ordinal, Anderson (1984) added the constraints $1 = \phi_1 > \phi_2 > \cdots > \phi_{J-1} > \phi_J = 0$. Since software for this model does not enforce these constraints, if you rearrange the order of the outcomes (e.g. renumber category 1 to 5 and category 5 to 1) the values of the ϕs switch. Substantively, the results are identical.

The general SLM1 with base J is

$$\begin{aligned}\ln\frac{\Pr(y=j \mid \mathbf{x})}{\Pr(y=J \mid \mathbf{x})} &= (\theta_j-\theta_J)\beta_0+(\phi_j-\phi_J)\mathbf{x}'\boldsymbol{\beta}\\ &= \theta_j\beta_0+\phi_j\mathbf{x}'\boldsymbol{\beta},\end{aligned}$$

where the last equality follows from the constraints $\theta_J = \phi_J = 0$. In terms of odds,

$$\Omega_{m|n}(\mathbf{x}) = \frac{\Pr(y=m)}{\Pr(y=n)} = \exp\left[(\theta_m-\theta_n)\beta_0+(\phi_m-\phi_n)\mathbf{x}'\boldsymbol{\beta}\right],$$

resulting in the odds ratio

$$\mathrm{OR}_{k,m|n} = \frac{\Omega_{m|n}(\mathbf{x},x_k+1)}{\Omega_{m|n}(\mathbf{x},x_k)} = \exp([\phi_m-\phi_n]\beta_k).$$

The odds ratios vary by the categories being compared, but since there are fewer parameters in the SLM1 the odds ratios do not vary as freely as in the MNLM. Similarly to the MNLM, the probabilities are computed as

$$\Pr(y=j \mid \mathbf{x}) = \frac{\exp(\theta_j\beta_0+\phi_j\mathbf{x}'\boldsymbol{\beta})}{\sum_{q=1}^{J}\exp(\theta_q\beta_0+\phi_q\mathbf{x}'\boldsymbol{\beta})}.$$

The two-dimensional model has two coefficients for each regressor:

$$\ln\frac{\Pr(y=j|\mathbf{x})}{\Pr(y=J|\mathbf{x})} = \theta_j\beta_0+\phi_j^{[1]}\mathbf{x}'\boldsymbol{\beta}^{[1]}+\phi_j^{[2]}\mathbf{x}'\boldsymbol{\beta}^{[2]},$$

where $\theta_J = \phi_2^{[1]} = \phi_J^{[1]} = \phi_1^{[2]} = \phi_J^{[2]} = 0$ and $\phi_1^{[1]} = \phi_2^{[2]} = 1$ for identification. Since the two odds ratios for x_k can operate in different directions (e.g. the OR for dimension 1 can be greater than 1 while the OR for dimension 2 is less than 1), the model is no longer ordinal.

The SLM can be extended by adding more dimensions until with $J-1$ dimensions it is identical to the MNLM. While the model has fewer parameters than the MNLM, full interpretation requires the evaluation of all comparisons. Since most of us cannot look at the scale factors and automatically compute the coefficients for the implied odds ratios, the smaller number of parameters does not simplify interpretation. Akaike information criterion (AIC) or Bayesian information criterion (BIC) statistics can be used to select among models with different numbers of dimensions.

Summary of ACLM and SLM

The ACLM and SLM are special cases of the MNLM. By imposing constraints they reduce the number of parameters while still allowing odds ratios to vary among all pairs of categories. I am unaware of applications of the ACLM or SLM in which the constraints were substantively motivated. Without such motivation the only advantages of these more restrictive models is to reduce the number of parameters. This does not, however, simplify the practical issues of interpretation that are described below.

The ordered logit model

The most commonly used model for ordinal outcomes was introduced by McKelvey and Zavonia (1975) as a probit model. McCullagh (1980) presented a logit version which is called the proportional odds model or the cumulative logit model. These models are also known as the parallel regression model and the grouped continuous model.

The model can be derived from a regression on an unobserved, continuous y^*:

$$y_i^* = \beta_0 + \mathbf{x}'\boldsymbol{\beta} + \varepsilon_i.$$

The ordered probit model assumes that ε is normal with mean 0 and variance 1, while the ordered logit model assumes that ε is logistic with mean 0 and variance $\pi^2/3$. Since the models provide nearly identical predictions, I only consider the OLM. The continuous y^* is divided into observed, ordinal categories using the thresholds $\tau_0, \ldots, \tau_J$:

$$y_i = j \text{ if } \tau_{j-1} \leq y_i^* < \tau_j \text{ for } j = 1, \ldots J,$$

where $\tau_0 = -\infty$ and $\tau_J = \infty$. For party affiliation, y^* is a continuous measure of left–right orientation with observed categories determined by the measurement model:

$$y_i = \begin{cases} 1 \Rightarrow SD & \text{if } \tau_0 = -\infty \leq y_i^* < \tau_1 \\ 2 \Rightarrow D & \text{if } \tau_1 \leq y_i^* < \tau_2 \\ 3 \Rightarrow I & \text{if } \tau_2 \leq y_i^* < \tau_3 \\ 4 \Rightarrow R & \text{if } \tau_3 \leq y_i^* < \tau_4 \\ 5 \Rightarrow SR & \text{if } \tau_4 \leq y_i^* < \tau_5 = \infty. \end{cases}$$

The simplest way to see the structure of the model is with *cumulative probabilities* of being less than or equal to category j:

$$\begin{aligned} \Pr(y \leq j \mid \mathbf{x}) &= \Pr(y^* < \tau_j \mid \mathbf{x}) \\ &= \Pr(\varepsilon < \tau_j - [\beta_0 + \mathbf{x}'\boldsymbol{\beta}] \mid \mathbf{x}), \quad \text{for } j = 1, J-1, \end{aligned}$$

where I substituted the equation for y^* and simplified. Since ε has a logistic distribution,

$$\Pr(y \leq j \mid \mathbf{x}) = \Lambda(\tau_j - \beta_0 - \mathbf{x}'\boldsymbol{\beta}), \quad \text{for } j = 1, J-1,$$

where Λ is the logistic cumulative density function (CDF). The probability of category j is the probability that $y \leq j$ minus the probability that $y \leq j - 1$:

$$\begin{aligned} \Pr(y = j \mid \mathbf{x}) &= \Pr(y \leq j \mid \mathbf{x}) - \Pr(y \leq j-1 \mid \mathbf{x}) \\ &= \Lambda(\tau_j - \beta_0 - \mathbf{x}'\boldsymbol{\beta}) - \Lambda(\tau_{j-1} - \beta_0 - \mathbf{x}'\boldsymbol{\beta}). \end{aligned}$$

The model is not identified, which can be seen by adding $\delta - \delta = 0$ within the CDF: $\Pr(y \leq j \mid \mathbf{x}) = \Lambda([\tau_j + \delta] - [\beta_0 + \delta] - \mathbf{x}'\boldsymbol{\beta})$. I can add any δ to τ_j while subtracting δ from β_0 without changing the probability. Identification is achieved by fixing the value of either the intercept or one of the thresholds. Assuming $\beta_0 = 0$, the identified model is

$$\Pr(y \leq j \mid \mathbf{x}) = \Lambda(\tau_j - \mathbf{x}'\boldsymbol{\beta}), \quad \text{for } j = 1, J-1. \tag{9.3}$$

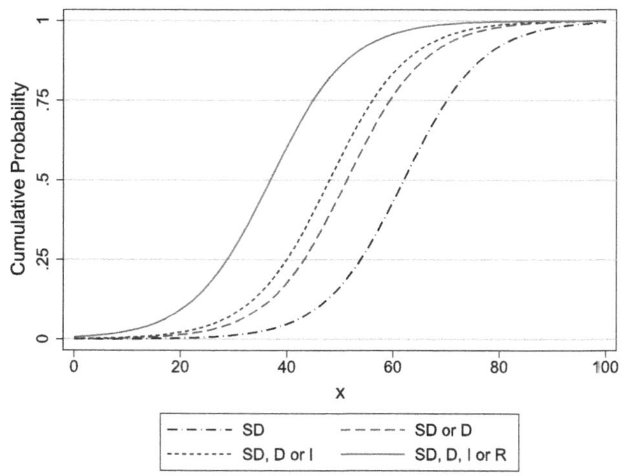

Figure 9.2 Parallel cumulative probability curves in the ordered regression model

For each j, equation (9.3) defines a binary logit on an outcome that divides the original categories between lower and higher values. The similarity to the binary logit model is easy to see by defining $\beta^*_{0,j} \equiv \tau_j$ and $\boldsymbol{\beta}^* \equiv -\boldsymbol{\beta}$ so that $\Pr(y \leq j \mid \mathbf{x}) = \Lambda\left(\beta^*_{0,j} + \mathbf{x}'\boldsymbol{\beta}^*\right)$. There is a binary logit for each of the $J - 1$ ways that the ordinal outcome can be dichotomized, with different intercepts for each binary model but identical slopes. The equality of slopes is known as the *parallel regression assumption*, which is shown by the parallel curves in Figure 9.2. This assumption is also called the parallel lines assumption, and, for the logit version of the model, the proportional odds assumption. Because of the identical slopes, adjacent categories can be combined and the estimates of the β_k will be consistent but inefficient (McCullagh, 1980). Combining categories might be necessary when estimation does not converge, since a category has a small number of cases.

The odds of being less than or equal to j compared to greater than j are

$$\Omega_{\leq j|>j}(x) = \frac{\Pr(y \leq j \mid \mathbf{x})}{1 - \Pr(y \leq j \mid \mathbf{x})} = \frac{\Lambda\left(\tau_j - \mathbf{x}'\boldsymbol{\beta}\right)}{1 - \Lambda\left(\tau_j - \mathbf{x}'\boldsymbol{\beta}\right)}, \quad \text{for } j = 1, J - 1.$$

Since $\Lambda\left(\tau_j - \mathbf{x}'\boldsymbol{\beta}\right) = \exp\left(\tau_j - \mathbf{x}'\boldsymbol{\beta}\right)/\left[1 + \exp\left(\tau_j - \mathbf{x}'\boldsymbol{\beta}\right)\right]$, this simplifies to

$$\Omega_{\leq j|>j}(\mathbf{x}) = \exp\left(\tau_j - \mathbf{x}'\boldsymbol{\beta}\right)), \quad \text{for } j = 1, J - 1.$$

The OR for x_k is

$$\text{OR}_{k,\leq j|>j} = \frac{\Omega_{\leq j|>j}(\mathbf{x}, x_k + 1)}{\Omega_{\leq j|>j}(\mathbf{x}, x_k)} = \exp(-\beta_k),$$

which can be interpreted as follows:

> For a unit increase in x_k, the odds of being in a category less than or equal to j change by a factor of $\exp(-\beta_k)$, holding other variables constant.

Since the odds ratio is the same for all j:

> For a unit increase in x_k, the odds of being in a lower category compared to a higher category change by a factor of $\exp(-\beta_k)$, holding other variables constant.

The odds ratio for a change of δ is $\exp(-\delta\beta_k)$. A standardized odds ratios is obtained by setting δ to the standard deviation of the regressor.

The parallel regression assumption leads to the elegant interpretation of the odds of higher and lower outcomes, but the assumption might be unrealistic. Score, LR, and Wald tests of the assumption are available. Essentially these tests compare the OLM estimates to those from binary logits on the dichotomized outcomes $y \leq j$ where the βs are not constrained to be equal. The model without constraints is called the generalized ordered logit model and is considered next. In my experience tests of parallel regressions are usually rejected. Allison (2012, p. 168) finds that the test is often significant when there are many regressors or the sample is large. A significant test suggests problems with the OLM, but does not imply that the more general GOLM is the appropriate model (Greene and Hensher, 2010, p. 188). It is prudent to compare the results of the OLM to those of the MNLM or the GOLM before accepting the conclusions from the OLM or deciding the model is inappropriate based on the parallel regression test.

Generalized ordered logit model

The generalized ordered logit model allows the β_k to vary by category so that $\mathbf{x}'\boldsymbol{\beta}_j = \beta_{1,j}x_1 + \cdots + \beta_{K,j}x_K$ and

$$\begin{aligned}
\ln \Omega_{\leq j|>j}(\mathbf{x}) &= \tau_j - \mathbf{x}'\boldsymbol{\beta}_j, \quad \text{for } j = 1, J-1, \\
\Omega_{\leq j|>j}(\mathbf{x}) &= \exp\left(\tau_j - \mathbf{x}'\boldsymbol{\beta}_j\right), \quad \text{for } j = 1, J-1, \\
\mathrm{OR}_{k,\leq j|>j} &= \exp\left(-\beta_{k,j}\right), \quad \text{for } j = 1, J.
\end{aligned}$$

The ORs can be interpreted as follows:

> For a unit increase in x_k, the odds of being less than or equal to j change by a factor of $\exp\left(-\beta_{k,j}\right)$, holding other variables constant.

Probabilities equal

$$\Pr(y = j \mid \mathbf{x}) = \frac{\exp\left(\tau_j - \mathbf{x}'\boldsymbol{\beta}_j\right)}{1 + \exp\left(\tau_j - \mathbf{x}'\boldsymbol{\beta}_j\right)} - \frac{\exp\left(\tau_{j-1} - \mathbf{x}'\boldsymbol{\beta}_{j-1}\right)}{1 + \exp\left(\tau_{j-1} - \mathbf{x}'\boldsymbol{\beta}_{j-1}\right)}, \quad \text{for } j = 1, J.$$

It is possible for predicted probabilities to be negative. McCullagh and Nelder (1989, p. 155) note that: 'If [negative probabilities] occur in a sufficiently remote region of the x-space, this flaw in the model need not be serious'. In the help file for gologit2, a Stata program for the GOLM, Williams (2006) reports that negative probabilities tend to occur when 'the model is overly complicated and/or there are very small N's for some categories of the dependent variable'. In these cases he suggests combining categories or simplifying the model.

Letting β_k vary by j avoids the parallel regression assumption, but leads to a model that is no longer ordinal and that has as many parameters as the MNLM. Since the MNLM is based on odds of individual categories (e.g. $\Omega_{1|2}(\mathbf{x})$) and the GOLM is based on odds for cumulative probabilities (e.g. $\Omega_{y\leq 1|y>1}(\mathbf{x})$), they do not produce identical predictions although they are often very similar.

There are several related models that reduce the number of parameters in the GOLM. The partial generalized ordered logit model lets the βs for some variables differ by j while others do not. Williams (2006) describes this model and stepwise procedures to select variables where coefficients are constrained to be equal (i.e. $\beta_{k,1} = \cdots \beta_{k,J-1} = \beta_k$). This allows the relationships between some regressors and the outcome to be ordinal while those for others are not. If stepwise procedures are used to select a model, it is important to report how your model was selected and to be careful that the model does not reflects peculiarities of the sample. Ideally, the sample should be randomly divided, using half of the data for selecting the model and the remaining half to confirm that the selected model is appropriate. Note that the GOLM is mathematically equivalent to the model proposed by Terza (1985) in which values of the thresholds vary across observations. While the substantive motivation for the heterogeneous threshold model differ from those of the GOLM, they are empirically indistinguishable. See Greene and Hensher (2010, pp. 209–14) for further details. Models with partial proportionality impose constraints that are similar to those in the SLM. This model was applied to educational outcomes by Hauser and Andrew (2006). See Fullerton (2009) for a review of this and related models.

Approaches to interpretation

This subsection reviews the ways in which probabilities and odds ratios can be used for interpretation. Examining how probabilities change as regressors change can effectively summarize the effects of regressors. Coefficients for the regressors, or transformations of these coefficients such as odds ratios, are limited since they do not indicate how much the probabilities change and do not always indicate even the direction of that change. Despite the advantages of probabilities, there is a catch. Since the models are non-linear, there might not be a simple way to summarize the results. In some cases a simple table or graph is effective, while in other cases a great deal of work is required to uncover and convey the findings. In practice it is often necessary to try multiple approaches in order to find the one that works best.

Probabilities

Let $\mathbf{x}^*$ contain specific values of the regressors. For example, $\mathbf{x}^*$ could hold observed values from the ith case ($\mathbf{x}^* = \mathbf{x}_i$) or hypothetical values such as the means ($\mathbf{x}^* = \overline{\mathbf{x}}$). The probability of outcome j evaluated at $\mathbf{x}^*$ is

$$\Pr(y = j \mid \mathbf{x}^*) = f(\mathbf{x}^*, \boldsymbol{\beta}^*), \quad \text{for } j = 1, J,$$

where $\boldsymbol{\beta}^*$ contains estimates of all parameters including slopes, intercepts, and scale coefficients. The function f depends on the model. Cumulative probabilities equal

$$\Pr(y \leq j \mid \mathbf{x}^*) = \sum_{k=1}^{j} \Pr(y = k \mid \mathbf{x}), \quad \text{for } j = 1, J.$$

Probabilities can be used in many ways to explain the relationships between regressors and outcomes. Most simply, predicted probabilities for individuals with characteristics of substantive importance, sometimes called ideal types, can be presented. Tables of predictions can show how changes in regressors affect the probabilities of outcomes. For example, I can examine the probabilities of party affiliation for men and compare them to those for women. Let $\mathbf{x}^*_{[-female]}$ contain specific values all regressors except *female*. The probability of category j for women at these values is

$$\Pr\left(y = j \mid \mathbf{x}^*_{[-female]}, female = 1\right),$$

and similarly for men,

$$\Pr\left(y = j \mid \overline{\mathbf{x}}_{[-female]}, female = 0\right).$$

Comparing the probabilities shows how men and women differ at specific values of the regressors.

For continuous regressors, graphs are useful. To show the effects of age, I can plot the probabilities of each affiliation as age changes while holding other variables at specific values. Let $\mathbf{x}^*_{[-age]}$ contain values for all regressors except age. Then

$$\Pr\left(y=j \mid \mathbf{x}^*_{[-age]}, \, age\right)$$

can be computed at values of age from 20 to 85. Plotting these probabilities against age shows how age is associated with party affiliation.

Since models for nominal and ordinal outcomes are non-linear, the researcher must decide where to hold other variables constant. While global means are often used, other values should be considered. For example, when comparing men and women, it might be more informative to compute probabilities for women holding other regressors at the means for women, and similarly use means for men when computing probabilities for men. This is illustrated below.

Changes in probabilities

The change in the probability of category j for a change in the x_k, holding other regressors constant, is a useful way to summarize the effect of a regressor. The critical idea is that only one regressor is changing while others are constant at specific values. I refer to these measures as *marginal effects*, although some authors use the term 'partial effects'. In contrast to linear models, the magnitude of a marginal effect depends on the amount of change in x_k, the value of x_k at the start of the change, and the values of other regressors as illustrated below.

There are two types of marginal effects that differ by the amount of change in x_k: discrete changes and marginal changes. A *discrete change* or *first difference* is the change in the probability for a discrete change in x_k, holding other regressors at specific values. For example, the change in the probability of *SR* as age increases from 30 to 40, holding other variables at their means, is a discrete change. Defining x_k^{Start} as the starting value for x_k and x_k^{End} as the ending value with $\mathbf{x}^*$ having specific values of other regressors, the discrete change is

$$\frac{\Delta \Pr(y=j \mid \mathbf{x}^*)}{\Delta x_k} = \Pr\left(y=j \mid \mathbf{x}^*, x_k = x_k^{End}\right) - \Pr\left(y=j \mid \mathbf{x}^*, x_k = x_k^{Start}\right).$$

Discrete changes can be computed with variables changing by any amount, such as from 0 to 1 for gender, 4 years for education, 15 points with IQ, or a standard deviation.

A *marginal change* is the partial derivative of the probability with respect to x_k:

$$\frac{\partial \Pr(y=j \mid \mathbf{x}^*)}{\partial x_k}.$$

It is the instantaneous rate of change in the probability for a change in x_k, holding other variables at $\mathbf{x}^*$. Roughly speaking, a marginal change is a discrete change as the amount of change in x_k approaches 0. While the estimated value of a discrete change of one and the estimate of the marginal change for x_k can be similar, this will not always be the case. Indeed, they can differ substantially in size. To the degree that the probability curve is linear near $\mathbf{x}^*$, the marginal change approximates how much the probability changes as x_k increases from x_k^* to $x_k^* + 1$. The more non-linear the curve in the region where x_k^* increases, the greater the difference between the marginal change and the discrete change.

There are several extensions that are important. First, the delta method can be used to compute standard errors for marginal effects, which allows confidence intervals and tests that the change is 0 (Xu and Long, 2005; Wooldridge, 2010, pp. 576–7). Second, marginal effects can be compared across groups. For example, the difference in the effects of x_k for men and women is a *second difference*,

$$\frac{\Delta^2 \Pr(y = j \mid \mathbf{x}^*)}{\Delta x_k, \mathit{female}} = \frac{\Delta \Pr(y = j \mid \mathbf{x}^*, \mathit{female} = 1)}{\Delta x_k} - \frac{\Delta \Pr(y = j \mid \mathbf{x}^*, \mathit{female} = 0)}{\Delta x_k}.$$

Third, while 'holding everything else constant' is fundamental to the idea of a marginal effect, an exception must be made for variables that are mathematically linked. For example, if $x_1 = \mathit{age}$ and $x_2 = \mathit{age\text{-}squared}$, you cannot change x_1 while holding x_2 constant. Instead, x_1 and x_2 must change together while holding all other variables constant. This is easy to illustrate with a discrete change in age from 20 to 30:

$$\frac{\Delta \Pr(y = j \mid \mathbf{x}^*)}{\Delta \mathit{age}(20 \to 30)} = \Pr\left(y = j \mid \mathbf{x}^*, x_1 = 30, x_2 = 30^2\right) - \Pr\left(y = j \mid \mathbf{x}^*, x_1 = 20, x_2 = 20^2\right).$$

Categorical regressors entered into a model as a set of indicators are also linked. Suppose that education has the categories: a respondent does not have a high school degree, high school is the highest degree, and a respondent has a college degree. Let $x_1 = 1$ if high school, else 0; and $x_2 = 1$ if college, else 0. If x_1 is 1, then x_2 must be 0 and vice versa. The effect of having college as the highest degree ($x_1 = 0, x_2 = 1$) compared to high school as the highest degree ($x_1 = 1, x_2 = 0$) is

$$\frac{\Delta \Pr(y = j \mid \mathbf{x}^*)}{\Delta x_1(0 \to 1), x_2(1 \to 0)} = \Pr(y = j \mid \mathbf{x}^*, x_1 = 0, x_2 = 1) - \Pr(y = j \mid \mathbf{x}^*, x_1 = 1, x_2 = 0).$$

This also has implications for where you hold variables constant. If $\mathbf{x}^*$ contains means and the regressors include $x_1 = \mathit{age}$ and $x_2 = \mathit{age\text{-}squared}$, you should use $x_1^* = \overline{\mathit{age}}$ and $x_2^* = (\overline{\mathit{age}})^2$, not $x_1^* = \overline{\mathit{age}}$ and $x_2^* = \overline{\mathit{age\text{-}squared}}$.

Summary measures of effects

Since the magnitude of a marginal effect depends on the amount of change and the values of all regressors, several decisions must be made about how to summarize the effect of a regressor. The first decision is the amount of change. For binary variables, the best choice is a change from 0 to 1. For continuous variables, fields such as economics prefer a marginal change. I prefer discrete changes since they indicate the actual amount of change in the probability for a specific change in x_k. For example, I prefer 'the probability of being *SR* increases by 0.05 for a four-year increase in education' to 'the rate of change in the probability of *SR* with respect to education is 0.02'.

Second, a decision must be made about where to hold other variables constant. Even the sign of the effect can differ at different values of the regressors. This decision is related to alternative methods for summarizing the effect of a regressor. There are three basic approaches. First, the effect can be computed with all variables held at their means, which is called the *marginal effect at the mean* (MEM). For a marginal change this is

$$\frac{\partial \Pr(y = j \mid \overline{\mathbf{x}})}{\partial x_k},$$

and for a discrete change

$$\frac{\Delta \Pr(y = j \mid \overline{\mathbf{x}})}{\Delta x_k}.$$

Second, the change can be computed when variables are fixed at values other than the mean, sometimes referred to as the *marginal effect at representative values* (MER). The MEM is a special case of the MER. Third, the *average marginal effect* (AME) is the mean of the marginal effect computed at the observed values of *all* observations. For a discrete change this equals

$$\text{mean}\frac{\Delta \Pr(y=j \mid \mathbf{x})}{\Delta x_k} = \frac{1}{N}\sum_{i=1}^{N}\frac{\Delta \Pr(y=j \mid \mathbf{x}_i)}{\Delta x_{ik}},$$

and for a marginal change

$$\text{mean}\frac{\partial \Pr(y=j \mid \mathbf{x})}{\partial x_k} = \frac{1}{N}\sum_{i=1}^{N}\frac{\partial \Pr(y=j \mid \mathbf{x}_i)}{\partial x_{ik}}.$$

Which measure should you use and why? While the average marginal effect and the marginal effect at the mean can be close in value, they are not asymptotically equivalent. Bartus (2005) and Verlinda (2006) show under what conditions they differ. Traditionally, the MEM was used most often, probably because it was simpler to compute (the AME requires N times more computation than the MEM) and was the default in most software. Newer software easily computes either measure, so this is no longer an issue. A common critique of the MEM is that the mean might not correspond to anyone in the sample or even in the population. For example, nobody can have the mean value of a binary regressor, such as being .47 female. While this is of course true, I find it reasonable to think of the marginal effect computed at the mean of a binary variable as a rough average of the effects for the two groups. Alternatively, effects can be computed at the modal values of the binary variables, but this ignores everyone who is in a less well-represented group. In general, the AME is a better approach. Since it averages the effects for every observation in the sample, it can be interpreted as the average size of the effect.

The AME is only a summary measure. Just as the mean of a regressor need not correspond to anyone in the sample, the average effect might not be close to the effect for anyone in the sample. For example, suppose that effects are positive for men and negative for women. The average effect does not indicate that men and women have effects in opposite directions. Indeed, the AME could be zero if the positive and negative effects cancel one another. Or suppose the average effect for an intervention is 0.20. If I am interested in the impact of the intervention when it is applied to everyone, the AME tells me what I need to know. If, however, I am interested in the effect of the intervention for a group with specific characteristics (e.g. high-risk youth), then the average is not what I want. The importance of looking at the distribution of effects is illustrated below. In general, no summary measure of effects is ideal for all situations. The best measure is the one that addresses the goals of your research.

Odds ratios or relative risk ratios

A common way to summarize the effects of regressors in logit models is the *odds ratio*, sometimes referred to as the *relative risk ratio*. The odds of p versus q is

$$\Omega_{p|q}(\mathbf{x}) = \frac{\Pr(p \mid \mathbf{x})}{\Pr(q \mid \mathbf{x})},$$

where p and q can be specific categories as in the MNLM or groups of categories such as $y \leq j$ and $y > j$ in the OLM. The odds ratio for x_k is the factor change in the odds as x_k increases by one, holding other variables constant, where β_k^* depends on the model used:

$$\text{OR}_{k,p|q} = \frac{\Omega_{p|q}(\mathbf{x}, x_k + 1)}{\Omega_{p|q}(\mathbf{x}, x_k)} = \exp\left(\beta_k^*\right).$$

The OR can be interpreted as follows:

> For a unit increase in x_k, the odds of p versus q change by a factor of OR_k, holding other variables constant.

For a change of δ, such as the standard deviation of x_k, $\text{OR}_k^\delta = \exp(\delta\beta_k^*)$ and can be interpreted as follows:

> For an increase of δ in x_k, the odds of m versus n change by a factor of OR_k^δ, holding other variables constant.

Unlike marginal effects, the odds ratio is the same at all values of the regressors. However, a specific factor change in the odds implies different changes in the probabilities of the outcome categories depending on the values of the regressors. For example, if the odds of R versus D are 100 to 1, doubling them to 200 to 1 implies a small change in probabilities. If the odds are 1 to 1, doubling the odds leads to a much larger change. Saying the odds double tells you little about the substantive process unless you know the value of the odds before they are doubled.

Summary

Models for nominal and ordinal outcomes are non-linear and there is no single method of interpretation that works in all applications. Odds ratios have a simple interpretation since their value does not depend on the levels of the regressors, but they do not indicate the magnitude of the change in the probabilities of the outcomes. Marginal effects on probabilities, whether discrete changes or marginal changes, have a direct interpretations, but the magnitude of the change depends on the values of the regressors. Full interpretation requires detailed post-estimation analyses to determine the most important findings and to find an elegant way to present them. Examples of these approaches to interpretation are provided in the next section.

EXAMPLE ANALYSIS: MODELING POLITICAL ATTITUDES

Data from the 1992 American National Election Study (ANES n.d.) are used to illustrate the interpretation of the models discussed above.[4] The dependent variable is party affiliation, which is coded as Strong Democrat (1=*SD*), Democrat (2=*D*), Independent (3=*I*), Republican (4=*R*), and Strong Republican (5=*SR*). As a reflection of left–right political orientation, the categories are ordered from 1=*SD* to 5=*SR*; as a reflection of intensity of partisanship, they are ordered 3=*I*, (2=*D*, 4=*R*), (1=*SD*, 5=*SR*). The distribution of categories is shown in Figure 9.3. Each of the models in Table 9.1 was estimated with the regressors described in Table 9.2 with the estimated parameters given in the Appendix. Too often the interpretation of models for nominal and ordinal outcomes is limited to brief comments about the sign and statistical significance of the coefficients which rarely provide an adequate understanding of the process being studied. More effective interpretation uses these parameters to compute probabilities, functions of probabilities, and odds ratios, as illustrated in this section. While I do not show each method for every model, I highlight the consequences of assuming ordinality by comparing the results from a nominal model to those from an ordinal model. While I usually compare the MNLM to the OLM, similar results are obtained by comparing any of the ordinal models to any of the nominal models.

Marginal effects

A useful place to begin exploring your model is with a summary of the effects of the regressors. I recommend using the average marginal effects as shown for the ordered logit model in Table 9.3. For continuous variables the table shows the average marginal change and the average discrete

Table 9.2 Descriptive statistics for regressors predicting party affiliations (N = 1382)

Variable	Mean	Standard deviation	Minimum	Maximum	Label
age	4.59	1.68	1.80	9.10	Age in decades.
income	3.75	2.78	0.15	13.13	Income in $10,000s.
black	0.14	—	0	1	1 if black, 0 if not.
female	0.49	—	0	1	1 if female, 0 if male.
hs#	0.58	—	0	1	High school is the highest degree.
college#	0.26	—	0	1	College is the highest degree.

Less than a high school degree is the excluded category.

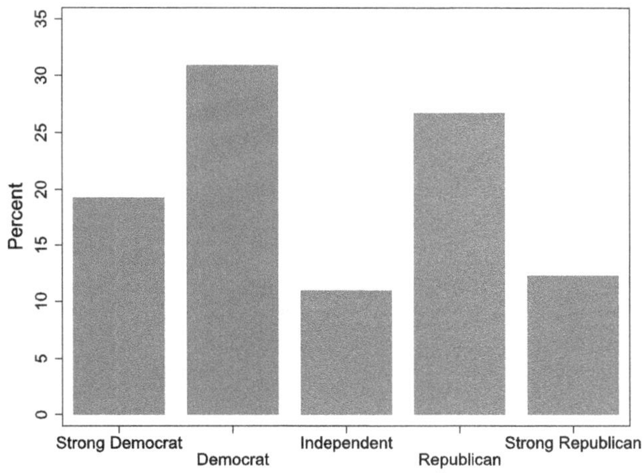

Figure 9.3 Distribution of party affiliation

change for a standard deviation change. Typically, I would only include one of these, but for didactic purposes I show both. For binary variables the table includes the average discrete change from 0 to 1. For education, it shows the average differences in the predicted probabilities for pairs of education levels.

Age increases affiliations on the left and decreases those on the right. The marginal changes are significant at the 0.05 or 0.10 level. For a standard deviation increase in age, the effects are all less than 0.02. The reason why these effects are so small is considered below. Income has larger effects, decreasing the probability of *SD* by 0.04 and increasing the probability of *SR* by 0.03. The effects are significant at the 0.01 level, except for the effect on *I*. Race has a large and significant effect on all outcomes. The average effect of being black is to increase the probability of *SD* by 0.28, while decreasing the probability of *R* by 0.18 and *SR* by 0.10. The pattern of effects for education shows that higher education significantly increases the probabilities of Republican affiliations while decreasing those for Democratic affiliations.

Table 9.4 compares the effects of income and age across models, with the nominal models listed first followed by the ordinal models. Discrete changes for a standard deviation increase in the regressors are shown to make it easier to assess the magnitude of the change in the probabilities. Income decreases the probabilities of Democratic categories and increase those for Republican categories. While the size and significance of these effects are similar in all models, the effects of age are distinctly different for nominal and ordinal models. For nominal

Table 9.3 Average marginal effects for the OLM

Regressor	Amount of change	Outcome				
		SD	*D*	*I*	*R*	*SR*
Age in decades	∂Pr/∂x	0.009**	0.005**	–0.000*	–0.007**	–0.007**
	ΔPr/Δx(*sd*)	0.015**	0.009**	–0.001*	–0.012**	–0.011**
Income in $10,000s	∂Pr/∂x	–0.013***	–0.008***	0.001**	0.011***	0.010***
	ΔPr/Δx(*sd*)	–0.037***	–0.023***	0.002**	0.030***	0.028***
Respondent is black?	ΔPr/Δ x(0→ 1)	0.275***	0.043***	–0.044***	–0.175***	–0.099***
Respondent is female?	ΔPr/Δx (0 → 1)	0.022	0.013	–0.001	–0.018	–0.016
HS degree vs no HS degree	ΔPr/Δ(*hs* → *college*)	–0.045*	–0.022**	0.005	0.037*	0.026**
College vs no HS degree	ΔPr/Δ(*hs* → *college*)	–0.090***	–0.056***	0.006*	0.077***	0.064***
College vs HS degree	ΔPr/Δ(*hs* → *college*)	–0.045***	–0.034***	0.001	0.040***	0.039***

Significance levels for two-tailed tests: ***0.01; **0.05;*0.10.

Table 9.4 Average discrete change for income and age in nominal and ordinal models

(a) Standard deviation change in income

Model	Outcome				
	SD	*D*	*I*	*R*	*SR*
MNLM	–0.042***	–0.019	–0.002	0.041***	0.022***
GOLM	–0.047***	–0.015	0.008	0.036***	0.018**
SLM2	–0.043***	–0.013*	–0.007	0.040***	0.023***
SLM1	–0.037***	–0.016***	–0.010***	0.038***	0.026***
ACLM	–0.035***	–0.027***	0.002***	0.032***	0.028***
OLM	–0.037***	–0.023***	0.002***	0.030***	0.028***

Significance levels for two-tailed tests of average marginal change: ***0.01; **0.05; *0.10.

(b) Standard deviation change in age

Model	Outcome				
	SD	*D*	*I*	*R*	*SR*
MNLN	0.051***	–0.030**	–0.025***	–0.027**	0.030***
GOLM	0.051***	–0.026**	–0.019***	–0.030***	0.024***
SLM2	0.051***	–0.032***	–0.022**	–0.028**	0.031***
SLM1	0.010	0.004	0.003	–0.010	–0.007
ACLM	0.010	0.008	–0.000	–0.009	–0.008
OLM	0.015**	0.009**	–0.001*	–0.012**	–0.011**

Significance levels for two-tailed tests of average marginal change: ***0.01; **0.05; *0.10.

models, age increases in the probabilities of the extreme categories *SD* and *SR*, with decreases in the probabilities of the middle categories. For ordinal models, age shows small increases in *SD* and small decreases in *SR*. Recall that ordinal models must show opposite effects for the extreme categories. Comparing the results from nominal and ordinal models suggests that age increases the degree of partisanship, but not left–right orientation.[5] This idea is explored later using graphs.

Table 9.5 compares average marginal effects to marginal effects at the mean using both discrete change and marginal change. In this example the two measures would lead to the same substantive conclusions. However, both measures are potentially limited since they are based

Table 9.5 Comparing measures of effect for age and income in the MNLM

Regressor	Amount of change	Measure of effect	Outcome				
			SD	*D*	*I*	*R*	*SR*
Age	ΔPr/Δx (*sd*)	AME	0.051***	−0.030**	−0.025***	−0.027**	0.030***
		MEM	0.054***	−0.030**	−0.026***	−0.026**	0.027***
	∂Pr/∂x	AME	0.031***	−0.018**	−0.015***	−0.016**	0.018***
		MEM	0.032***	−0.018**	−0.015***	−0.016**	0.016***
Income	ΔPr/Δx(*sd*)	AME	−0.042***	−0.019***	−0.002***	0.041***	0.022***
		MEM	−0.046***	−0.021	−0.003	0.048***	0.023***
	∂Pr/∂x	AME	−0.015***	−0.007	−0.001	0.015***	0.008***
		MEM	−0.017***	−0.008	−0.001	0.017***	0.008***

Significance levels for two-tailed tests: ***0.01; **0.05; *0.10.
AME: average marginal effect; MEM: marginal effect at mean.

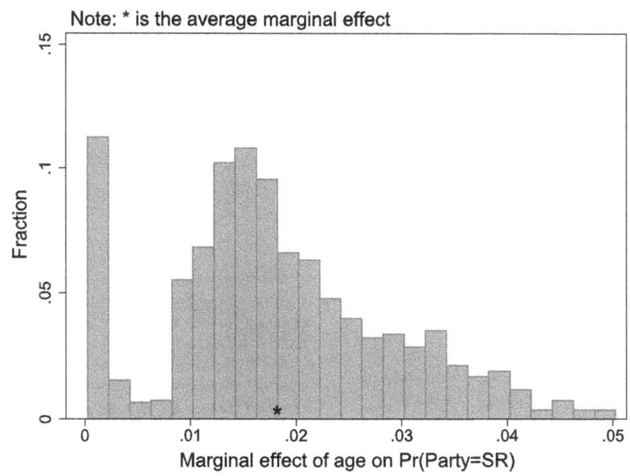

Figure 9.4 Distribution of marginal effects of age on the probability of *SR* for the MNLM

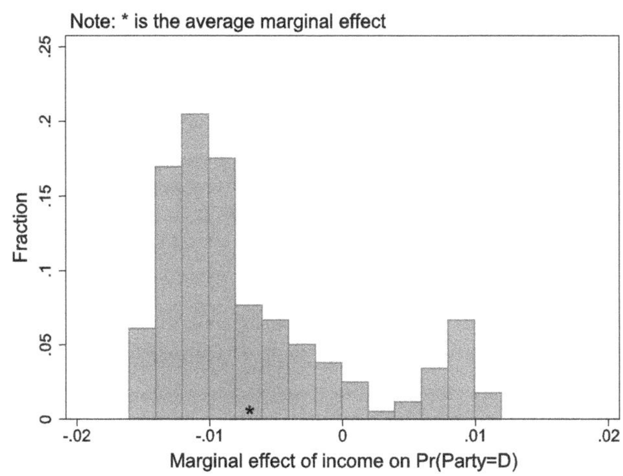

Figure 9.5 Distribution of marginal effects of income on the probability of *D* for the MNLM

on averages. If the average value of each regressor is a substantively interesting location in the data, the MEM is useful since it tells you the magnitude of the effect for someone with those characteristics. And, if you are interested in the average effect in the sample, the AME is appropriate. Regardless, for variables that are critical to the goals of the research it is important to examine the *distribution* of effects. To see why this is important, consider two of the effects from Table 9.5. The average marginal change for age on *SR* is 0.018 and for income on *D* is −0.007. Figure 9.4 shows the distribution of marginal changes for each observation with the AME indicated by the * near 0.02. The range of effects is from 0 to 0.05, with a spike near 0 followed by a gap till around 0.01. If age was a policy variable and the focus of the intervention was on individuals who had marginal effects near 0, the average would be misleading. Figure 9.5 shows that even the sign of the AME can be misleading. The distribution of effects is bimodal. with most of the sample having effects around −0.01 and a smaller group with positive effects near 0.01. Suppose the regressor was an intervention where the spikes correspond to two groups, say whites and blacks, and where the effect is negative for the larger group and positive for the smaller group. If substantive interest was in the effect of the intervention for the smaller group, the AME is misleading since it is dominated by the negative effects for the larger group.

Which measure of effect is most useful? First, in terms of the *amount* of change being assessed, I prefer discrete changes since the magnitude of the change in probabilities is clear. Second, I generally prefer the average marginal effects rather than the marginal effects at the mean. However, since the average summarizes individual effects that can differ widely in the sample, it is important to examine the distribution of effects for critical variables. Third, if specific groups are important to your research, it is valuable to compute marginal effects at values corresponding to these groups, perhaps at within-group means.

Plots of probabilities

Graphs can be effective for showing the effects of a variable over its entire range. These plots are constructed by computing predicted probabilities while changing one variable and holding other variables constant. Note, however, if your model includes linked variables, such as age and age-squared, you must change the values of both variables together (e.g. it does not make sense to change age while holding age-squared constant). Figures 9.6 and 9.7 plot the probabilities of party affiliations as income increases from $0 to $100,000 for the OLM and the MNLM. Consistent with the marginal effects shown above, income increases the probabilities of Republican affiliations while decreasing the probabilities of Democratic affiliations. The graphs for the OLM and MNLM are very similar. Figures 9.8 and 9.9 show corresponding graphs as age increases from 20 to 85, holding other variables at the mean. These graphs illustrate the constraints imposed by an ordinal regression model. For the OLM, as age increases from 20 to 85 the probability of *SD* increases from 0.15 to 0.21 while the probability of *SR* decreases from 0.12 to 0.08. For ordinal models, the probabilities of the highest and lowest categories must change in opposite directions. For the MNLM, as age increases the probability of *SD* increases from 0.10 to 0.33 and the probability of *SR* also increases, although more gradually, from 0.07 to 0.17. Since the MNLM is not ordinal, it does not force the changes in the extreme categories to be in opposite directions.

While these graphs are created holding other variables at their means, you could use other values. For example, if you were interested in gender differences, you could compute separate graphs for men and women using gender-specific means.

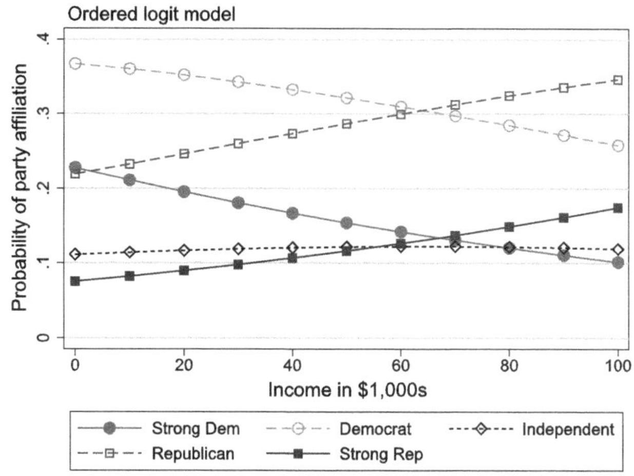

Figure 9.6 Predicted probabilities of party affiliation by income for the OLM, with other variables held at their means

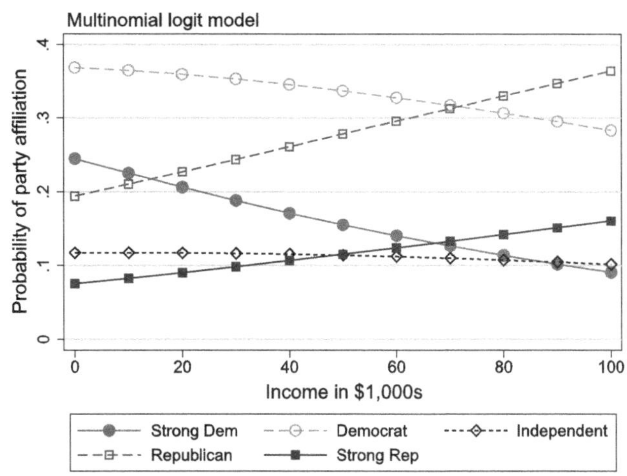

Figure 9.7 Predicted probabilities of party affiliation by income for the MNLM, with other variables held at their means

Tables of probabilities

Predicted probabilities can also be used in tables. For example, to show the effects of race and gender I computed probabilities at each combination of race and gender holding other variables at their means. The results for the MNLM and the OLM are shown in Table 9.6. For both models blacks are far more likely than whites to be *SD* or *D* and less likely to be *R* or *SR*, with much

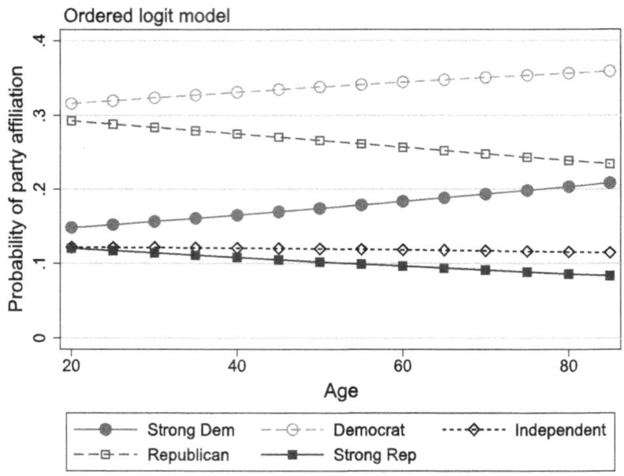

Figure 9.8 Predicted probabilities of party affiliation by age for the OLM, with other variables held at their means

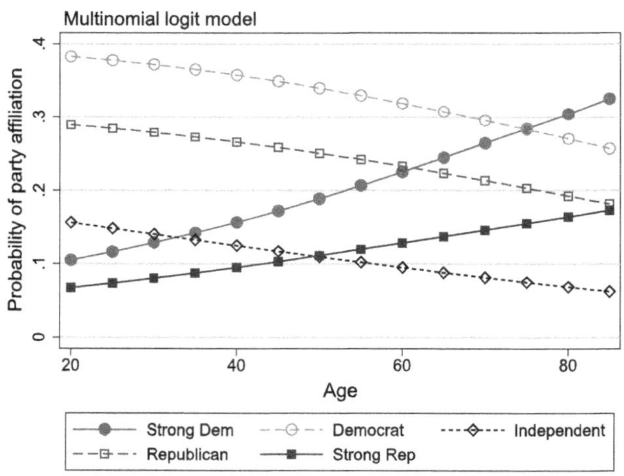

Figure 9.9 Predicted probabilities of party affiliation by age for the MNLM, with other variables held at their means

smaller gender differences. Gender-specific race differences can be computed with discrete changes. Let $\overline{\mathbf{x}}^*$ contain the means of all regressors except race and gender. For women,

$$\frac{\Delta \Pr(y=j \mid \mathbf{x}^*, female{=}1)}{\Delta\, black(0 \rightarrow 1)} = \Pr(y=j \mid \overline{\mathbf{x}}^*, female{=}1, black{=}1) - \Pr(y=j \mid \overline{\mathbf{x}}^*, female{=}1, black{=}0).$$

Discrete changes for men can be computed similarly. The results are shown in Table 9.7a for the MNLM. We find, for example, that the probability of *SD* is 0.273 larger for black women than

Table 9.6 Predicted probabilities of party affiliation by race and gender in the MNLM and OLM, with other variables held at their means

(a) MNLM

Group	Outcome				
	SD	*D*	*I*	*R*	*SR*
Black women	0.417	0.394	0.127	0.047	0.015
White women	0.138	0.354	0.092	0.304	0.111
Black men	0.440	0.328	0.162	0.049	0.021
White men	0.142	0.287	0.115	0.311	0.145

Other variables are held at their means.

(b) OLM

Group	Outcome				
	SD	*D*	*I*	*R*	*SR*
Black women	0.443	0.355	0.068	0.105	0.029
White women	0.154	0.321	0.122	0.287	0.116
Black men	0.405	0.367	0.075	0.119	0.034
White men	0.134	0.301	0.122	0.308	0.134

Other variables are held at their means.

Table 9.7 Discrete changes for race given gender in predicted probabilities of party affiliation in the MNLM and OLM, with other variables held at their means

(a) MNLM

Comparison	Outcome				
	SD	*D*	*I*	*R*	*SR*
Black women–white women	0.278***	0.040	0.035	−0.257***	−0.096***
Black men–white men	0.298***	0.041	0.047	−0.262***	−0.125***
Second difference	−0.0200	−0.001	−0.013	0.004	0.029*

Other variables are held at their means. Significance levels for two-tailed tests: ***0.01; **0.05; *0.10.

(b) OLM

Comparison	Outcome				
	SD	*D*	*I*	*R*	*SR*
Black women–white women	0.289***	0.034**	−0.054***	−0.182***	−0.087***
Black men–white men	0.270***	0.066***	−0.047***	−0.189***	−0.100***
Second difference	0.019	−0.031	−0.007	0.007	0.012

Other variables are held at their means. Significance levels for two-tailed tests: ***0.01; **0.05; *0.10.

for white women and 0.268 larger for black men than for white men. The changes are similar for men and women, and we can test if they are equal using a second difference,

$$\frac{\Delta \Pr(y=j \mid \mathbf{x}^*, female{=}1)}{\Delta\, black(0 \to 1)} - \frac{\Delta \Pr(y=j \mid \mathbf{x}^*, female{=}0)}{\Delta\, black(0 \to 1)}$$

None of the differences is significantly. Similar results for the OLM are shown in Table 9.7b.

Table 9.8 Probabilities of party affiliation by race and gender in the MNLM with other variables held at group means

Group	Outcome *SD*	*D*	*I*	*R*	*SR*
Black women	0.460	0.364	0.129	0.037	0.011
White women	0.145	0.354	0.093	0.298	0.109
Black men	0.461	0.310	0.168	0.044	0.017
White men	0.128	0.283	0.109	0.325	0.156

Comparison	Outcome *SD*	*D*	*I*	*R*	*SR*
Black women–white women	0.314***	0.009	0.036	−0.261**	−0.098***
Black men–white men	0.333***	0.028	0.060*	−0.281***	−0.139***
Second difference	−0.019	−0.019	−0.024**	0.020	0.041**

Other variables are held at group means. Significance levels for two-tailed tests: ***0.01; **0.05; *0.10.

Table 9.9 Discrete changes in probabilities of party affiliation for the young college graduates and older high school graduates in the MNLM

(a) Young college graduates

	Outcome *SD*	*D*	*I*	*R*	*SR*
Black women–white women	0.206***	0.139***	0.046*	−0.279***	−0.112***
Black men–white men	0.228***	0.135***	0.065**	−0.283***	−0.145***

Significance levels for two-tailed tests: ***0.01; **0.05; *0.10.

(b) Older high school graduates without college

	Outcome *SD*	*D*	*I*	*R*	*SR*
Black women–white women	0.324***	−0.016	0.015	−0.223***	−0.100***
Black men–white men	0.343***	−0.008	0.021	−0.227***	−0.129***

Significance levels for two-tailed tests: ***0.01; **0.05; *0.10.

To compute the predicted probabilities it was necessary to choose values for the other regressors. In Tables 9.8 and 9.9 variables were held at their means. The predictions compare white men, black women, white women, and black men who have the same values for age, income, and education. Since these groups are likely to differ on these variables, the predictions must be viewed as a 'what if' experiment: what would happen if these groups had the same characteristics? We could also compute predictions using within-group means, computing the probability of party affiliation for black men, holding other variables at the average for black men, and so on for other groups. This was done in Table 9.8. While general trends are the same, substantial differences are found for black men, which reflects that the means for this group differ substantially from the global means.

A similar technique that is often effective is to compute effects for targeted groups in which the levels of all variables are specified. To illustrate this, Table 9.9 considers two types of respondents.

Table 9.9a shows discrete changes for race by gender for 30-year-old college graduates with an income of $40,000. Table 9.9b has the results for 60-year-old high school graduates earning $25,000. While the general patterns are similar, the magnitudes of the effects differ substantially. For example, in Table 9.9a being black increases the probability of *SD* by 0.21 for women, while in Table 9.9b the effect is 0.32. If we examined the discrete change at other values of the control values, we would obtain different values.

Numerous variations on these methods can be made. Probabilities for ideal types representing characteristics of individuals of particular interest can be presented. Discrete changes can be computed with regressors changing by any amount of interest. For example, we could use second differences to compare the results in Table 9.9a to those in Table 9.9b.

Odds ratios

Since odds ratios are a standard method of interpretation in many fields, it is important to understand how they are used and their limitations. Table 9.10 contains odds ratios for adjacent categories of party affiliations. To save space, I do not include ORs for other comparisons such as *SD* versus *SR*. For the MNLM, increasing income decreases the odds of affiliations that are politically to the left relative to the adjacent category to the right. For example, a $10,000 increase in income decreases the odds of *SD* versus *D* by a factor of 0.93, holding other variables constant ($p \leq 0.10$ for a two-tailed test). The pattern of ORs shows that age affects intensity of partisanship but not left–right orientation. A 10-year increase in age increases the odds of *SD* versus *D* by a factor of 1.27 and similarly increases the odds of *SR* versus *R* by a factor of 1.25 (= 1/0.80). The different patterns of odds ratios for age and income suggest that our measure of party affiliation reflects both political orientation and intensity. This pattern could not be found using ordinal models. For example, the odds ratios for age in the SLM1 (see Table 9.11) are near 1 and non-significant, while the odds ratios for income show a pattern similar to those for the MNLM.

The ordinal ACLM constrains the ORs for adjacent categories to be equal. For a $10,000 increase in income the odds of being in a party to the left compared to the adjacent party to the right decrease by a factor of 0.96. While the results for income are similar to those for the MNLM, the ORs for age differ substantially. The two-dimensional stereotype model (SLM2) has two coefficients for each regressor in addition to the scaling coefficients. The combined effects from dimension 1 and 2 for income are close to those for the MNLM. The two dimensions have ORs in different directions for three pairs of categories. For example, dimension 1 decreases the odds of *SD* versus *D* by a factor of 0.84, while dimension 2 increases the odds by a factor of 1.10. Dimension 2 for age increases the odds of more extreme partisanship, while dimension one has a weaker effect on left–right orientation. Overall, the conclusions from SLM2 and MNLM are similar.

The odds ratios in Table 9.11 are for the cumulative odds of being in parties to the right versus to the left. With the OLM, the odds ratios are identical for each of the four ways in which left and right are split, reflecting the proportional odds assumption of the model. A $10,000 increase in income increases the odds of being more to the right by a factor of 1.10, holding other variables constant ($p \leq 0.05$). A 10-year increase in age decreases the odds of being more right-oriented by a factor of 0.94 ($p \leq 0.05$). A Brant test of the parallel regression assumption is significant ($X^2_{18} = 89.84, p \leq 0.01$) suggesting that the ORs might vary by where the outcome is divided between right and left. The GOLM, which allows this, shows that ORs for income gradually decrease as the dividing point moves from *SD* to *R*, but these differences are not significant ($X^2_3 = 2.09, p = 0.55$). The ORs for age are significantly different ($X^2_3 = 36.1, p \leq 0.01$).

Table 9.10 Odds ratios for adjacent categories of political affiliation for income and age

(a) Odds ratios for a $10,000 increase in income

				SLM2		
Odds of	*MNLM*	*ACLM*	*SLM1*	Dim 1+2	Dim 1	Dim 2
SD vs D	0.93*	0.96***	0.94***	0.92***	0.84***	1.10***
D vs I	0.99	0.96***	1.02	1.01	0.99	1.02
I vs R	0.93*	0.96***	0.91***	0.91***	0.94	0.97
R vs SR	0.99	0.96***	0.97	0.98	1.08**	0.91**

Significance levels for two-tailed tests: ***0.01; **0.05;*0.10

(b) Odds ratios for a ten-year increase in age

				SLM2		
Odds of	*MNLM*	*ACLM*	*SLM1*	Dim 1+2	Dim 1	Dim 2
SD vs D	1.27***	1.02	1.03	1.27***	1.03	1.24***
D vs I	1.08	1.02	0.99	1.06	1.00	1.06
I vs R	0.93	1.02	1.05	0.95	1.01	0.95
R vs SR	0.80***	1.02	1.01	0.80***	0.99	0.80***

Significance levels for two-tailed tests: ***0.01; **0.05; *0.10

Table 9.11 Odds ratios for cumulative outcomes for income and age

	Income		Age	
Odds of	*OLM*	*GOLM*	*OLM*	*GOLM*
(SR+R+I+D) vs SD	1.10**	1.13***	0.94**	0.80***
(SR+R+I) vs(D+SD)	1.10**	1.10***	0.94**	0.94*
(SR+R) *vs* (I+D+SD)	1.10**	1.10***	0.94**	0.98
SR vs (R+I+D+SD)	1.10**	1.07***	0.94**	1.15***

Significance levels for two-tailed tests: ***0.01; **0.05; *0.10

Age increases the odds of more partisan party affiliation, whether on the right or the left. While the results of the GOLM and MNLM are consistent, I find it easier to see how age affects the extreme categories relative to less extreme categories in the MNLM since categories are not combined.

Odds ratios provide a single number that summarizes how a change of a regressor affects the relative probabilities of two outcomes. This is rarely sufficient since it does not indicate the magnitude or even necessarily the direction of the change in the probabilities of the outcomes. For example, if x_k increases the odds of m relative to n, the probabilities of both categories could increase, both could decrease, or m could increase while n decreases. If the odds ratio is 2 for 30-year-old white male high school graduates earning $20,000, the odds ratio is also 2 for 60-year-old black women with college degrees earning $50,000. However, the meaning of the OR in terms of changes in probabilities differs for these individuals. Still, ORs are useful for uncovering how variables affect the probability of one outcome compared to another, holding other variables constant. For example, odds ratios in the MNLM can be used to test whether a regressor differentiates between two categories (e.g. whether race affects the odds of being *SR* compared to *R*). Graphical methods that exploit this information are found in Long (1997, pp. 170–5) and Long and Freese (2014).

CAVEATS AND FREQUENT ERRORS

Assess ordinality with sensitivity analysis Ordinal models impose constraints on how regressors are related to the outcomes. Having an ordinal dependent variable does not imply that these constraints are appropriate for the process being modeled. Accordingly, it is important to compare the predictions from an ordinal model to those from a nominal model. If nominal models provide predictions that differ noticeably from those from the ordinal model, you should evaluate whether an ordinal model is appropriate for your application.

Using caution with stepwise procedures Stepwise procedures are often used to find simpler models. If the parallel regression assumption is violated, stepwise methods might be used to find a partial generalized ordered logit model that relaxes the assumption for some but not all variables. Or, to avoid the complexity of the MNLM, tests can be run to determine whether categories can be combined. Or you might use the AIC or BIC to find the SLM with the smallest number of dimensions. If such procedures are used, I recommend randomly dividing your sample into two subsamples. Use the first subsample as the exploration sample where stepwise procedures are used to select your candidate model. After you have selected a model, verify its fit with the second, verification subsample. If the results differ markedly, your model could be reflecting peculiarities of the sample rather than the underlying process.

Lack of interpretation Do not limit interpretation to a table of coefficients with a brief discussion of the signs and significance. Presenting odds ratios without information about the magnitude of the effects on probabilities is incomplete. Use predicted probabilities and related measures to show the magnitude of the effects. Some authors question the feasibility of using predictions to interpret complex models such as the MNLM. For example, Greene and Hensher (2010, p. 188) write: 'The multinomial logit model for unordered choices produces coefficients, but it would be arduous at best to translate them into something meaningful to describe the behavior of an ordered random variable, such as the outcome of an attitude survey.' The previous section shows that if probabilities are used, it is no more difficult to interpret MNLM with many more parameters than simpler models such as the OLM. The hard part is deciding which information is most important for the substantive purposes of the analysis and finding an elegant way to present it.

Look at all of the parameters implied by your model Standard software for the MNLM shows estimates for the minimal set of parameters. This set might not include comparisons that are of greatest interest. To estimate parameters for all comparisons you can recode the values for your outcome and re-estimate the model (see Long and Freese, 2014, for details). If a significance test that a regressor has no effect is not rejected, it is possible that coefficients for specific comparisons are significant. If those comparisons are substantively important, you should test the individual coefficients. If you use the SLM, you should compute the odds ratios from the estimated parameters.

Independence of irrelevant alternatives There is no commonly available nominal model that avoids the IIA property. Tests for IIA do not have good properties and often produce contradictory results, with some tests rejecting the null hypothesis while others accept the hypothesis. The OLM model avoids IIA but has other limitations. Mixed logit models (see Train, 2009) do not impose the IIA assumption, but require intensive calculation to estimate and involve more complicated data structures. At the time of writing, software is not readily available.

Test of parallel regressions Tests of the parallel regression assumption in the ordered regression model often reject the hypothesis. Some evidence suggests that tests are sensitive to issues unrelated to the parallel regression assumption (Greene and Hensher, 2010, pp. 182–8). If the null hypothesis is rejected, compare the predictions from the ordered regression model to those from the GOLM or the MNLM to determine if there are substantively meaningful differences in the predictions of the two models. If not, it is reasonable to use the ordered regression model.

Non-linearity of the model and on the right-hand side Models for nominal and ordinal outcomes are inherently non-linear, which leads to complications in interpretation as discussed above. Even with the non-linearity of the models, it is important to consider non-linearities on the right-hand side as well. This includes the use of quadratic terms such as age, age-squared, and age-cubed and the inclusion of interactions.

CONCLUSIONS

This chapter reviews the most common regression models for nominal and ordinal outcomes. Most of the applications in the social sciences use either the multinomial logit model or the ordered logit or probit models. With advances in software to estimate models such as the generalized ordered logit model and the stereotype model, their use is increasing. While ordinal models can simplify interpretation and the added information from ordinality allows more efficient estimates, it is critical to assess whether the restrictions implicit in ordinal models are appropriate for your substantive application. Before selecting an ordinal model, compare the results from that model to those from a nominal model.

I have not considered heterogeneous choice models, also known as location–scale models (see Williams, 2009, for a review). While these models are theoretically promising, Keele and Park (2006) and Williams (2009) find that they are highly sensitive to specification of the variance function. If the specification of the variance equation is uncertain, it may be better to use a standard logit or probit model. Another model of potential interest is the continuation ratio model that is appropriate when the outcome reflects stages that individuals pass through in sequence, such as the ranks of assistant professor, to associate professor, to full professor, to named professorship. Extensions of these models and issues of interpretation are discussed by Mare (2006).

FURTHER READING

- Agresti (2010) provides a detailed discussion of regression models for ordinal variables as well as models for the analysis of contingency tables with ordinal variables.
- Long (1997) provides a more technical discussion of regression models for ordinal and nominal outcomes as well as binary and count variables. Methods of interpretation using predictions are discussed in detail.
- Long and Freese (2014) focus on the estimation and interpretation of regression models using Stata, including most of the models discussed in this chapter.
- O'Connell (2006) focuses on logit models for ordinal outcomes with examples of programs in SAS, SPSS and Stata. Many examples are given.
- Train (2009) includes a detailed discussion of models for discrete choice, including new models that can be estimated by simulation. Models with alternative-specific regressors are considered.

APPENDIX

Table 9.12 Estimates from the multinomial logit model and the adjacent category logit model ($N = 1382$)

	MNLM				ACLM			
Regressor	*SD*	*D*	*I*	*R*	*SD*	*D*	*I*	*R*
age	0.0282	−0.208***	−0.288***	−0.217***	0.080	0.060	0.040	0.020
income	−0.175***	−0.102***	−0.090*	−0.013	−0.173***	−0.130***	−0.0864***	−0.043***
black	3.075***	2.080***	2.291***	0.106	3.002***	2.252***	1.501***	0.751***
female	0.237	0.478**	0.048	0.245	0.314*	0.236*	0.157*	0.079*
hs#	−0.555	−0.210	−0.602	−0.183	−0.502*	−0.377*	−0.251*	−0.125*
college#	−1.375***	−0.746**	−1.758***	−0.696*	−1.049***	−0.786***	−0.524***	−0.262***
constant	1.182**	2.332***	2.226***	2.093***	0.831**	1.366***	0.261	1.000***
Log-likelihood	−1960.911				−2009.038			
AIC	3977.821				4038.076			
BIC	4124.297				4090.389			

Coefficients are relative to base category *SR*. #: Not completing high school is the excluded category.
Significance levels for two-tailed test: ***0.01, **0.05, *0.10.

Table 9.13 Estimates from the stereotype logit models with one and two dimensions ($N = 1382$)

Regressor	SLM1	SLM2	
		Dimension 1	Dimension 2
age	−0.0778	−0.0256	0.217***
income	0.174***	0.177***	0.096**
black	−3.105***	−3.525***	−2.425***
female	−0.170	−0.196	−0.275
hs#	0.615**	0.525*	0.333
college#	1.277***	1.189***	1.018***
ϕ_1	1^C	1^C	0°
ϕ_2	0.627***	0°	1^C
ϕ_3	0.732***	−0.081	1.257***
ϕ_4	0.164*	−0.436**	0.974***
ϕ_5	0^B	0^B	0^B
θ_1	0.980**	1.142**	
θ_2	1.459***	2.588***	
θ_3	0.451	1.811***	
θ_4	0.963***	2.017***	
θ_5	0^B	0^B	
Log-likelihood	−1995.095	−1968.932	
AIC	4016.19	3977.865	
BIC	4084.196	4082.49	

#Not completing high school is the excluded category.
Significance levels for two-tailed test: ***0.01, **0.05,*0.10.
C = constrained; B = base outcome; O = omitted

Table 9.14 Estimates from the ordered logit and generalized ordered logit models (N = 1382)

Regressor	OLM	GOLM			
		SD vs *D,I,R,SR*	*SD,D* vs *I,R,SR*	*SD,D,I* vs *R,SR*	*SD,D,I,R* vs *SR*
age	−0.0636**	−0.225***	−0.066*	−0.017	0.140***
income	0.096***	0.125***	0.098***	0.093***	0.064**
black	−1.476***	−1.463***	−1.388***	−2.412***	−2.001***
female	−0.157	−0.044	−0.277**	−0.098	−0.261
hs#	0.294*	0.233	0.106	0.305*	0.338
college#	0.642***	0.455*	0.204	0.733***	0.971***
constant	—	2.170***	0.121	−0.858***	−3.152***
cut_1	−1.457@				
cut_2	0.147@				
cut_3	0.638@				
cut_4	2.275@				
Log-likelihood	−2010.198	−1962.094			
AIC	4040.395	3980.189			
BIC	4092.708	4126.665			

#Not completing high school is the excluded category. Significance levels for two-tailed test: ***=0.01, **0.05, *0.10. @test is not appropriate.

NOTES

* I thank Henning Best, Andy Fullerton, Kristian Karlson, Tom VanHeuvelen, Rich Williams, and Mike Vasseur for their comments.

1 The redundant comparisons *R* versus *D*, *I* versus *D*, and *I* versus *R* are not shown.

2 The ratio of probabilities is sometimes called the 'relative risk' in which case the odds ratios is referred to as the 'relative risk ratio'.

3 There is a version of the MNPM that maintains the IIA assumption and does not require alternative specific regressors to be identified. This model produces very similar results to the MNLM but requires substantially more computation to estimate.

4 Models were estimated using Stata 13.1 (StataCorp, 2013). The do-files and data can be obtained by entering the Stata commands: (1) net from http://www.indiana.edu/~jslsoc/stata/ (2) net get cdanomord.

5 Since this is cross-sectional data, this could be a cohort effect rather than an effect of aging.

REFERENCES

Agresti, A. (2010). *The Analysis of Ordinal Cateorigal Data*. New York: Wiley.

Allison, P. D. (2012). *Logistic Regression Using the SAS System: Theory and Application*. Cary, NC: SAS Institute Inc.

Anderson, J. A. (1984). Regression and ordered categorical variables (with discussion). *Journal of the Royal Statistical Society, Series B*, 46, 1–30.

Bartus, T. (2005). Estimation of marginal effects using margeff. *Stata Journal*, 5(3), 309.

Begg, C. B. and Gray, R. (1984). Calculation of polychotomous logistic regression parameters using individualized regressions. *Biometrika*, 71, 11–18.

Cheng, S. and Long, J. S. (2007). Testing for IIA in the multinomial logit model. *Sociological Methods & Research*, 35(4), 583–600.

Clogg, C. C. and Shihadeh, E. S. (1994). *Statistical Models for Ordinal Variables*. Thousand Oaks, CA: Sage.

Duncan, O. D. (1984). *Notes on Social Measurement: Historical and Critical*. New York: Russell Sage Foundation.

Fry, T. R. L. F. and Harris, M. N. (1996). A Monte Carlo study of tests for the independence of irrelevant alternatives property. *Transportation Research Part B: Methodological*, 30(1), 19–30.

Fry, T. R. L. F. and Harris, M. N. (1998). Testing for independence of irrelevant alternatives: some empirical results. *Sociological Methods & Research*, 26(3), 401–423.
Fullerton, A. S. (2009). A conceptual framework for ordered logistic regression models. *Sociological Methods & Research*, 38, 306–347.
Goodman, L. A. (1983). The analysis of dependence in cross-classifications having ordered categories, using log-liner models for frequencies and log-linear models for odds. *Biometrics*, 39, 149–160.
Greene, W. H. and Hensher, D. A. (2010). *Modeling Ordered Choices: A Primer*. New York: Cambridge University Press.
Hauser, R. M. and Andrew, M. (2006). Another look at the stratification of educational transitions: the logistic response model with partial proportionality constraints. *Sociological Methodology*, 36(1), 1–26.
Hausman, J. and McFadden, D. (1984). Specification tests for the multinomial logit model. *Econometrica*, 52(5), 1219–1240.
Keele, L. and Park, D. K. (2006). Difficult choices: An evaluation of heterogeneous choice models. Paper presented at the 2004 Meeting of the American Political Science Association.
Keane, M. P. (1992). A note on identification in the multinomial probit model. *Journal of Business and Economic Statistics*, 10, 193–200.
Long, J. S. (1997). *Regression Models for Categorical and Limited Dependent Variables*. Thousand Oaks, CA: Sage.
Long, J. S. and Freese, J. (2014). *Regression Models for Categorical Dependent Variables Using Stata*. Third Edition College Station, TX: Stata Press.
Mare, R. D. (2006). Statistical models of educational stratification: Hauser and Andrew's models for school transitions. *Sociological Methodology*, 36(1), 27–37.
McCullagh, P. (1980). Regression models for ordinal data (with discussion). *Journal of the Royal Statistical Society, Series B*, 42(2), 109–142.
McCullagh, P. and Nelder, J. A. (1989). *Generalized Linear Models*. New York: Chapman & Hall.
McFadden, D. (1973). Conditional logit analysis of qualitative choice behaviour. In P. Zarembka (Ed.), *Frontiers in Econometrics* (pp. 105–142). New York: Academic Press.
McKelvey, R. D. and Zavonia, W. (1975). A statistical model for the analysis of ordinal level dependent variables. *Journal of Mathematical Sociology*, 4, 103–120.
O'Connell, A. A. (2006). *Logistic Regression Models for Ordinal Response Variables*. Thousand Oaks, CA: Sage.
Small, K. A. and Hsiao, C. (1985). Multinomial logit specification tests. *International Economic Review*, 26, 619–627.
StataCorp (2013). *Stata: Release 13. Statistical Software*. College Station, TX: StataCorp LP.
Stevens, S. S. (1946). On the theory of scales of measurement. *Science*, 7, 677–680.
Terza, J. V. (1985). Ordinal probit: a generalization. *Communications in Statistics: Theory and Methods*, 14(1), 1–11.
Train, K. (2009). *Discrete Choice Methods with Simulation*. Cambridge/New York: Cambridge University Press.
Velleman, P. F. and Wilkinson, L. (1993). Nominal, ordinal, interval and ratio typologies are misleading. *American Statistician*, 47(1), 65–72.
Verlinda, J. A. (2006). A comparison of two common approaches for estimating marginal effects in binary choice models. *Applied Economics Letters*, 13(2), 77–80.
Williams, R. (2006). Generalized ordered logit/partial proportional odds models for ordinal dependent variables. *Stata Journal*, 6(1), 58–82.
Williams, R. (2009). Using heterogeneous choice models to compare logit and probit coefficients across groups. *Sociological Methods & Research*, 37(4), 531–559.
Wooldridge, J. (2010). *The Econometrics of Cross-Section and Panel Data*. Cambridge, MA: MIT Press.
Xu, J. and Long, J. S. (2005). Confidence intervals for predicted outcomes in regression models for categorical outcomes. *Stata Journal*, 5(4), 537–559.

10

Graphical display of regression results

Gerrit Bauer

INTRODUCTION TO THE METHOD

The world is inherently multivariate. Statistical models in all scientific disciplines are therefore usually similarly multivariate, but whenever researchers want to display their data or their results on a sheet of paper, or even on a video screen, they have to reduce the multidimensionality to a much smaller number of dimensions. This reduction in information is the key issue whenever we attempt to visualize multivariate information such as regression results. Of course, scientists simultaneously want to highlight the most relevant aspects of variation found in the data under consideration. Designing regression graphs thus requires careful thought regarding the selection, grouping and representation of information. Edward R. Tufte, a pioneer in the field of data visualization, describes the fundamental problem scientists face whenever they try to display multivariate information:

> The world is complex, dynamic, multidimensional; the paper is static, flat. How are we to represent the rich visual world of experience and measurement on mere flatland? (Tufte, 2008, p. 9)

Tufte further asserts that a number of general principles, each with specific visual consequences, aid in identifying design excellence. Researchers who recognize and implement those guidelines will create better displays. This chapter therefore starts with the principles guiding data visualization in general and discusses how disregarding those rules may violate 'graphical integrity'. Those guidelines refer to all kinds of graphs and figures in scientific publications (such as frequency charts, histograms or bivariate scatter plots). The chapter then discusses challenges arising particularly in visualizing multivariate regression results. It goes on to provide examples of how graphical tools in the Stata software can be used to visualize complex regression models. Two (intentionally) over-complicated models with numerous interaction effects and transformed data will be highly difficult to interpret. Displays should in such situations lead to straightforward interpretations, whereas tables of regression coefficients will likely lead to misinterpretations. Visualization is therefore an important and powerful tool in facilitating an appropriate evaluation of regression estimates. Using European Social Survey (ESS) data, the example illustrations deal with attitudes towards homosexuality (ordinary least squares (OLS) model) and church attendance (logit model), and ask how those variables are influenced by religiosity, age, country of residence, employment status, gender, education and political views (left–right placement). Note that all graphs in the examples section aim to display regression results. There are numerous charts that serve regression diagnostic purposes, that is to say, those displays are designed to detect the violation of underlying assumptions (such as linearity, homoscedasticity, no autocorrelation) and are used as supplements to or sometimes even replacements of formalized statistical tests. Chapter 5 in this volume discusses regression diagnostic figures in considerable detail. Finally, the present chapter continues with a short compendium of caveats and frequent errors and provides recommendations for further reading.

General principles of information display

All statistical graphs share a common aim: to depict (observed or estimated) data using combinations of dots, lines, numbers, symbols, words, shadings, colours, and often a coordinate system. Although the history of science provides examples of ancient, creative and now-unusual types of information displays (Tufte, 2008), most types of graphs in today's publications were developed about 200 years ago: William Playfair (1759–1823) made use of bar and pie charts to depict information usually presented in distribution tables. Playfair also developed early forms of time-series graphs, illustrating the time-dependency of outcomes and the processual nature of change (for a detailed description of historical development in information design, see Wainer, 2000, 2005). Regression graphs rely on these forms of information representation (with the exception of pie charts, which are rare in regression graphics). Bar charts may depict regression coefficients or the distribution of a predicted outcome. Modifications of time-series graphs may show how change over time (or any other continuous variable) affects an outcome variable. Regression displays are therefore not a special type of chart; virtually any chart can be used to display regression-relevant information: not only observed values will be visualized, but also predicted values, error terms (the differences between predicted and observed values) and regression estimates (coefficients). In cases where regression models are based on a random sample, charts can also contain the confidence intervals surrounding coefficients or predictions.

All graphs that one considers for presentation or publication — be they regression graphs or distribution charts — should describe statistical findings in a manner understandable to readers with different levels of quantitative interest and expertise. The following guidelines and the central principles of information display discussed by authors such as Tufte (2001) or Miller (2004, 2005) thus highlight the important goal that

(1) the beholder should understand a graph presented on its own.

The reader should be able to comprehend what is displayed without reading the text and without regarding regression coefficients. In most cases, however, (regression) graphs cannot completely substitute coefficients in table form, as the latter documents the precise estimation procedure. Nonetheless, the example analysis below will illustrate that coefficients alone can be misleading and that graphical displays assist in quickly delivering the most relevant information. In order to design inherently understandable information displays, those graphs necessarily contain contextual information: an appropriate title leads to conclusions with regard to the underlying research question, and axis titles and labels of variables, conspicuous observations, or composed groups and categories inform viewers about the type of data being analysed.

All of the following principles ensure that information is displayed in a clear and precise manner, and that figures are descriptive and memorable. Such displays are characterized by what Tufte (2001) refers to as 'graphical integrity': information graphs should inform about reality and thus draw a simplified but unbiased picture of the world. Graphs violating these principles are imposing examples of how to lie with statistics. The second general principle thus refers to the

(2) appropriate labelling of data values and axes and the selection of axes' ranges.

In order to keep graphs simple and readable, authors should use data labels sparingly (numeric information adjacent to points, bars or lines in chart). The main purpose of a chart is not to indicate specific values but to depict general patterns – and those are visible without labelling specific values (Miller, 2005, pp. 123ff.). While such data labels are not used very frequently, it is common practice to denote the values of the dependent variable on the vertical dimension (y) and those of the explanatory variable (predictor) on the horizontal dimension (x) and to label the axes properly. Each graph should outline what information is depicted on which axis, and how (i.e. in what units) information was measured. The axes' values may not necessarily cover the entire range from the minimum to the maximum because small but meaningful effects/differences might then not be visible. Of course, an inappropriately selected range will deceive readers who will likely over- or underestimate effect sizes. It is usually good to display typical rather than extreme values. If a distribution is skewed and if extreme values are included in a graph, one might consider depicting a variable's mean value (e.g. by including a reference line). If multiple graphs appear next to each other and show the same dependent or independent variables, the range of the axes should be identical. As a matter of course, the space between depicted values on the axes (approximately five to ten values for each axis appears reasonable) must be proportional to the intervals of the values (without stretching or compressing parts of the axes). This directly leads to the third principle of information design, requiring

(3) proportionality of effect sizes and their graphical representations.

In some charts, area or volume representations are used instead of linear scales. Such graphs often depart from the proportionality guideline and tend to exaggerate differences. Area representations (e.g. rectangles, circles) and volume representations (e.g. cuboids, cones or spheres) fool readers, simply because an increase in the (linear) size leads to a square of the increase for areas and to a cube of the increase for volumes (Tufte, 2001). Thus, choosing such a display type likely results in a violation of the proportionality postulation. Charts with a second dimension (like squares instead of bars) or third dimension (like cuboids or spheres) often violate the fourth fundamental principle of information display, that

(4) the dimensionality of a graph must be in accordance with the dimensionality of the information presented.

Multiple dimensions found in the data can of course be presented with multidimensional graphs, but if we consider, for a simple example, a frequency table about the distribution of a variable with five distinct outcomes (e.g. a variable capturing restrictive attitudes towards lesbian, gay, bisexual, and transgender (LGBT) living arrangements), there is no reason why this distributional information should be depicted with a multidimensional graph: one can easily show the five values on one dimension and the corresponding (relative) frequencies on the second dimension. A third dimension, such as in cuboids or spheres, adds no information – only stylistic variation. Anyone examining the chart has to consider which stylistic variation depicts data variation and which does not. We can therefore formulate a final guideline, stating that scientific charts containing statistical information should

(5) express variation of data, not variation of stylistic affection (no artwork).

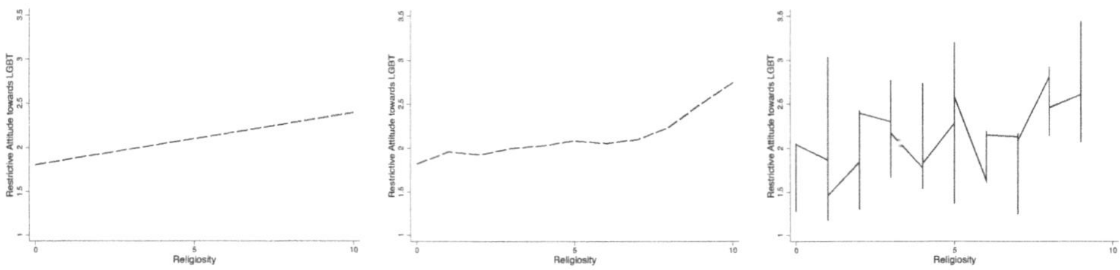

(a) bivariate OLS regression (b) mean regression (c) multivariate OLS regression

Figure 10.1 Visualization of predicted values (profile plots) in OLS regression models

This principle refers to more than the number of dimensions shown in a graph: for example, charts often rely on more colours than necessary. If colour is used instead of black and white print, it should have only one purpose — to express data variation. Colourful backgrounds, shadings and colour gradients often express stylistic variation only. If one does employ colours to express data variation, one should remember that this variation should also be visible if printed or photocopied in greyscale (dotted or dashed lines, shaded bars, etc.).

ADVANCED ASPECTS

We have thus far discussed five central principles for visualizing quantitative information, derived from the extensive work of Tufte (2001, 2008) and the more practically oriented guidebooks by Miller (2004, 2005). Turning to specific challenges arising whenever scientists decide to visualize multivariate regression models, we have to consider techniques of dimensional compression, that is, strategies for escaping 'flatland' (Tufte, 2008). We must also decide what sort of information involved in regression analysis to depict, and what information to omit.

Multivariate results and dimensional compression

As long as regression models exploit variation from one dependent and one independent variable only, displaying such regression results does not demand much effort: a regression line connects the predicted values of y conditional upon x. Figure 10.1 depicts such bivariate regression models. The dependent variable measures restrictive attitudes towards LGBT: 'Homosexual women and men should be free to live their lives as they wish.' It is measured on a five-point scale, ranging from 1 (totally agree) to 5 (totally disagree). The independent variable captures self-reported religiosity, ranging from 1 (not religious at all) to 10 (very religious). Figure 10.1a shows a linear (OLS) regression model, Figure 10.1b a simple mean regression: for this model, I have calculated the mean value of the dependent variable conditional upon the values of the explanatory variable. The graph simply connects those calculated values, sorted by ascending religiosity. But what would the picture look like if the prediction of y depended on a multivariate regression model? Figure 10.1c again illustrates predictions of an OLS regression, but now controls for a number of additional variables (age, left–right placement, education, etc.). Here, the predicted values cannot be connected in a straight line: covariates that are included in the estimation produce additional variation, and this variation produces the zigzag depicted in Figure 10.1c.

Given one particular value of x (religiosity), the outcome varies due to controlled covariates. Even extremely religious people will not necessarily have strong attitudes against homosexuality, since their attitude (y) also depends on other characteristics, such as country, political orientation, age, gender, and education received. How can we solve the graphical 'problem' caused by multivariate regression models? Figure 10.1c is certainly not an appropriate display of the calculated regression model. Some authors present rotating point clouds on computer screens and screenshots of these graphs to visualize multidimensionality (e.g. Cook and Weisberg, 1994). Such displays do allow viewers to assess three-dimensional data variation, but even this technique does not solve the zigzag curves in Figure 10.1c – the model is not 'three-variate' but indeed multivariate. Depicting such models requires strategies of dimensional compression and careful consideration about which variations to show or not. Researchers wishing to display multivariate regression results will thus have to make choices and compromises:

> Nearly every escape from flatland demands extensive compromise, trading off one virtue against another; the literature consists of partial, arbitrary, and particularistic solutions; and neither clever idiosyncratic nor conventionally adopted designs solve the inherent general difficulties of dimensional compression. Even our language, like our paper, often lacks immediate capacity to communicate a sense of dimensional complexity. (Tufte, 2008, p. 15)

In principle, the solution is rather simple: variables included in the model but not depicted in a graph must be fixed at certain values whenever y is predicted. In the example above, we could fix age, education, gender, church attendance, country, employment, and age at specific values and run predictions on restrictive attitudes towards LGBT. We would yield a linear regression line with one value of y for each value of x, and the slope of the line would be equivalent to the regression coefficient of religiosity in the multivariate OLS model. Fixing all covariates at one specific value solves the problem of dimensional compression in a radical manner. Is there any solution that requires less extensive compromise in dimensional compression?

Layering, grouping and separating information

As discussed above, graphs not only consist of a coordinate system to express variation of data, but also may employ stylistic elements to depict variance. The most common method of adding a third dimension to a line chart is to draw two or more lines within the chart. Multiple lines within one chart (or bars, etc.) can depict variation introduced by a third variable. In the example analysis, I could have estimated separate predictions of restrictive attitudes against homosexuality depending on both religiosity and gender. Fixing all other variables at specific values, two regression lines, one for women and one for men, would display the calculated values of y depending on x (religiosity). In linear regression models without interaction effects, this procedure would yield two parallel lines (only the intercepts depend on the values of the fixed variables). In non-linear models, and in linear models involving interaction effects, the slopes of the curves or lines, respectively, would likely differ (such charts are thus often referred to as 'conditional effect plots'). In the example just discussed, it is clear at what values one should fix the variable that opens up the third dimension, say 0 for male respondents and 1 for female respondents. If the third variable involved is not categorical but continuous, one has to choose a number of plausible values (usually two to four). I recommend using one value that is close to the variable's mean (or median), one value that is significantly below the mean and one significantly above the mean. However, the lower and the higher value should still be realistic. If a researcher wants to display more than three dimensions, it is often recommended to draw subgraphs and not to depict too much information within one line chart (see examples below, especially Figures 10.3d, 10.4a and 10.7).

Deciding what to display: Data, coefficients, predicted values

The type of regression graph most frequently employed depicts, as just illustrated, predicted values of variables or predicted probabilities of characteristics/events. This type of information display is called a 'predicted value plot', 'predicted probability plot', 'profile plot', and sometimes a 'conditional effect plot' (the latter is quite confusing; effects in such plots are not necessarily conditional upon other variables). This information can then be displayed in line charts, but also in any other type of information display. In addition to predictions, some social scientists depict observations, usually in scatter plots and often in combination with regression lines or smoothers. Showing real data provides additional information about the distribution of the two variables. In multivariate analyses, a scatter-plot matrix can provide this information (see, for example, Cook and Weisberg, 1994, pp. 47ff.).

Statistical packages also estimate confidence intervals (surrounding predicted values) and error terms. The former can be used to test the hypothesis that values calculated based on a random sample differ significantly from some specified null value. Confidence intervals in regression plots require careful interpretation and should only be displayed if such an interpretation is provided (see Payton et al., 2003, for a discussion about the meaning of overlapping confidence intervals). In regression diagnostic plots, the errors (i.e. the differences between observed data and predicted data) are often displayed in order to check for the violation of assumptions (normality of errors, constant variance of errors, size of errors invariance with respect to x, etc.). However, displaying errors is generally unhelpful in presenting regression results.

Finally, researchers might want to display regression coefficients (such as OLS coefficients, logits, odds ratios, or marginal effects in non-linear models). Those numbers can easily be included in dot plots or bar charts. Again, confidence intervals may be displayed in such figures, showing whether an effect differs significantly from 0. If coefficients have not been standardized, they should not be compared. Note that confidence intervals surrounding multiple regression coefficients do not provide tests for the equivalence of effect sizes but usually test whether an effect is significantly below or above 0.

EXAMPLE ANALYSIS

The example analysis presented in this chapter consists of two parts: I first show graphical displays for linear regression models and then discuss how to visualize binary (here logit) regression models. Both model specifications presented in this section are by design extremely complex. Because of multiple interaction effects and non-linear relationships their interpretation will likely lead to mistakes. One frequent cause of misinterpretations is that the reference categories to which the main effects refer are difficult to identify in models with multiple interactions. As is the case for most other chapters in this volume, the data used here is from the European Social Survey (ESS). Rounds 1 to 4 have been accumulated and data from 16 countries participating in the ESS is exploited (Austria (AT), Belgium (BE), Bulgaria (BG), the Czech Republic (CZ), Germany (DE), Estonia (EE), Finland (FI), France (FR), the United Kingdom (GB), Greece (GR), Ireland (IE), Israel (IL), the Netherlands (NL), Norway (NO), Russia (RU), and Sweden (SE)). The complex sampling features of the ESS were ignored for purposes of illustration in this chapter (for analysis with such weights, see Chapter 11 in this volume). The following variables are included: restrictive attitudes towards LGBT (five-point scale); self-reported religiosity; placement on a left–right political scale (both variables ranging from 1 to 10); gender (female=1, male=0); employment status (a set of dummy variables: paid work, retired, housework, other); a dummy variable capturing frequent church attendance (frequent=1, meaning in church at least once a month); education (in years); respondent's age; and a dummy variable for each country. All (by assumption) continuous independent variables (religiosity, left–right placement, education, and age) have been centred (i.e. their mean is 0). The aim of the following example analyses is the visual interpretation of complex models with a variety of different interaction effects. Although a theoretical model should usually guide the regression model specification, this is not the case in the following pages. The model specification was intentionally guided by the aim of creating complexity. The Stata code used here is available on the companion website of this book.

A complex linear regression model

In theory one might, for whatever reason, postulate that respondents' attitudes towards homosexuality depend on religiosity, church attendance, political orientation, employment, age, and education. One might further assume country differences (level effects) in attitudes towards homosexuality. We expect the effect of gender to be conditional upon employment and on church attendance (thus, the model contains interaction effects between gender and church attendance, gender and employment, church attendance and employment, and gender and church attendance and employment). Further, the effect of religiosity is expected to be non-linear and thus captured by a linear and squared specification. The model is even more complicated because this non-linear effect is conditional upon gender. Finally, we let the effect of the left–right placement vary among countries. This model specification is presented in Table 10.1 (the table continues in two more parts on the following pages).

Although the regression model is linear, interpreting the effects in Table 10.1 is difficult. For instance, the female-versus-male effect does not indicate that women in general have −0.31 units less restrictive attitude towards homosexuality, because gender is included in several interaction effects. The effect of −0.31 refers to women who are employed, do not attend church regularly (0=less than once a month, 1=at least once a month) and have mean values on the religiosity measurement (because religiosity is centred). Likewise, the depicted effects of church attendance and employment status are effects estimated for cases in which interacted variables take the reference value. How can statistical software packages allow for a straightforward interpretation of results? Simply illustrating the regression coefficients from Table 10.1 is not likely to lead to appropriate interpretations and will not make it easier to assess the relevant information. Figure 10.2a shows the average marginal effects of education, age, left–right placement, and religiosity: a one-unit (i.e. year) increase in education decreases restrictive attitudes towards LGBT by −0.04. A one-year increase in age leads – on average – to an increase in restrictive attitudes by the value of 0.005. A one-unit move to the right on the political left–right scale increases restrictive attitudes by 0.04. Such an increase in religiosity also affects the dependent variable by about 0.04. Average marginal effects are calculated by assessing the effect of a certain variable for each observation, given the empirically observed variation with regard to other covariates. If interaction effects are involved or if non-linear models are specified, the effect sizes differ between observations and are thus averaged to a single coefficient (see Chapter 8 in this volume for mathematical details).

An example illustrates the difference between regression coefficients and average marginal effects. The effect of left–right placement on the restrictive attitudes towards LGBT is about 0.11 in the regression table, but the marginal effect in Figure 10.2a is only 0.04. The reason for this difference is that the effect in Table 10.1 refers to the reference category of the variable interacting with the left–right scale (here: country). For respondents from Austria (the reference

Table 10.1 Denial of homosexuality, part 1

	Coefficient (OLS)	CI (95%)
Sex of respondent		
Male	Reference	
Female	–0.3086	[–0.3321,–0.2852]
Employment status		
Paid work	Reference	
Retired	0.1284	[0.0841,0.1728]
Housework	–0.0839	[–0.1840,0.0162]
Other	0.0078	[–0.0202,0.0358]
Sex–employment interactions		
Female × paid work	Reference	
Female × retired	–0.0398	[–0.1011,0.0214]
Female × housework	0.1772	[0.0720,0.2824]
Female × other	–0.0238	[–0.0627,0.0152]
Church attendance		
Less than once a month	Reference	
At least once a month	0.2861	[0.2491,0.3230]
Female–church interaction	0.0286	[–0.0224,0.0795]
Employment–church interactions		
Paid work × Church	Reference	
Retired × Church	–0.0494	[–0.1429,0.0441]
Housework × Church	–0.0355	[–0.2590,0.1880]
Other × Church	0.1316	[–0.0615,0.2018]
Female–church–employment interactions		
Female × church × paid work	Reference	
Female × church × retired	0.0480	[–0.0799,0.1760]
Female × church × housework	0.0695	[–0.1618,0.3008]
Female × church × other	–0.0433	[–0.1405,0.0539]
Religiosity		
linear	0.0389	[0.0349,0.0430]
squared	0.0029	[0.0016,0.0042]
Female–religiosity interaction		
Female × religiosity	–0.0024	[–0.0078,0.0031]
Female × religiosity × religiosity	0.0030	[0.0013,0.0047]
Age of respondent (centred)	0.0055	[0.0049,0.0061]
Education (in years, centred)	–0.0372	[–0.0394,–0.0350]
Age–education interaction	–0.0002	[–0.0003,–0.0000]

Table continues below

category for country), the effect is indeed 0.11, but it is much more moderate and sometimes even negative for respondents residing in other European countries. On average, the effect is 0.04. Note that all confidence intervals shown in Figure 10.2a can be used for testing the hypothesis that an effect is significantly below or above 0 (I therefore decided to show a reference line at this value). The intervals should not be used for comparisons between coefficients. Likewise, because variables have not been standardized, the effect sizes depicted in the chart cannot be compared (except for those variables measured on scales with identical range).

Country (level) differences, compared to Austria, are shown in Figure 10.2b. Like Figure 10.2a, this depicts average marginal effects, now displayed in the form of a bar chart. It again includes confidence intervals. The effects of country dummies are, due to the interaction effects included in the model, conditional upon the placement on the left–right scale. Since the latter variable is centred on its mean, the average marginal effects of the country dummies are, unlike the effects

Table 10.1 (cont) Denial of homosexuality, part 2

Country	Coefficient (OLS)	CI (95%)
AT	Ref.	
BE	−0.1614	[−0.2005,−0.1224]
BG	0.4473	[0.3927,0.5018]
CZ	0.3833	[0.3400,0.4266]
DE	0.1036	[0.0669,0.1402]
EE	0.9595	[0.9134,1.0056]
FI	0.1681	[0.1295,0.2067]
FR	−0.1946	[−0.2340,−0.1552]
GB	0.0719	[0.0336,0.1101]
GR	0.2739	[0.2319,0.3160]
IE	−0.1862	[−0.2262,−0.1463]
IL	0.3995	[0.3553,0.4438]
NL	−0.3290	[−0.3672,−0.2908]
NO	0.0572	[0.0187,0.0957]
RU	1.3268	[1.2750,1.3786]
SE	−0.0734	[−0.1119,−0.0349]

Table continues below

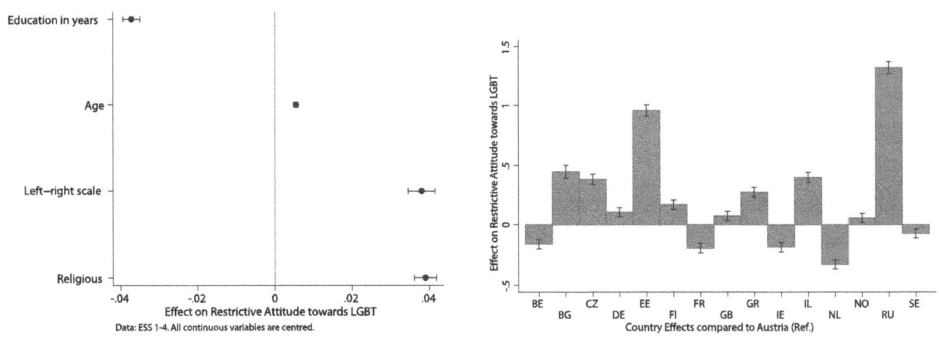

(a) dot plot (b) bar chart

Figure 10.2 Displays of selected regression coefficients in OLS regression models

of left–right scale in Figure 10.2a, in perfect accordance with the regression coefficients listed in the table.

Figure 10.2b shows that attitudes towards homosexuality differ between countries; all effects are, compared to Austria, statistically significant at the 5% level. Respondents residing in eastern European countries report on average higher values on the scale than respondents living in central Europe. People living in the Benelux countries and in France show the lowest values on average.

The next two graphs display not regression coefficients (or the corresponding average marginal effects) but predicted values of y conditional upon a number of categorical variables and interactions between those predictors. In Stata, the statistical package used for the example analysis, such plots are called 'profile plots'. Figure 10.3a shows two dimensions: employment status (x) and attitude towards LGBT (y). The four predicted values (which Stata calls 'margins') are the averages of the predictions over the estimation sample, holding the employment status to each of the four levels. The other covariates remain at the observed empirical values when the software calculates the predicted values of y.

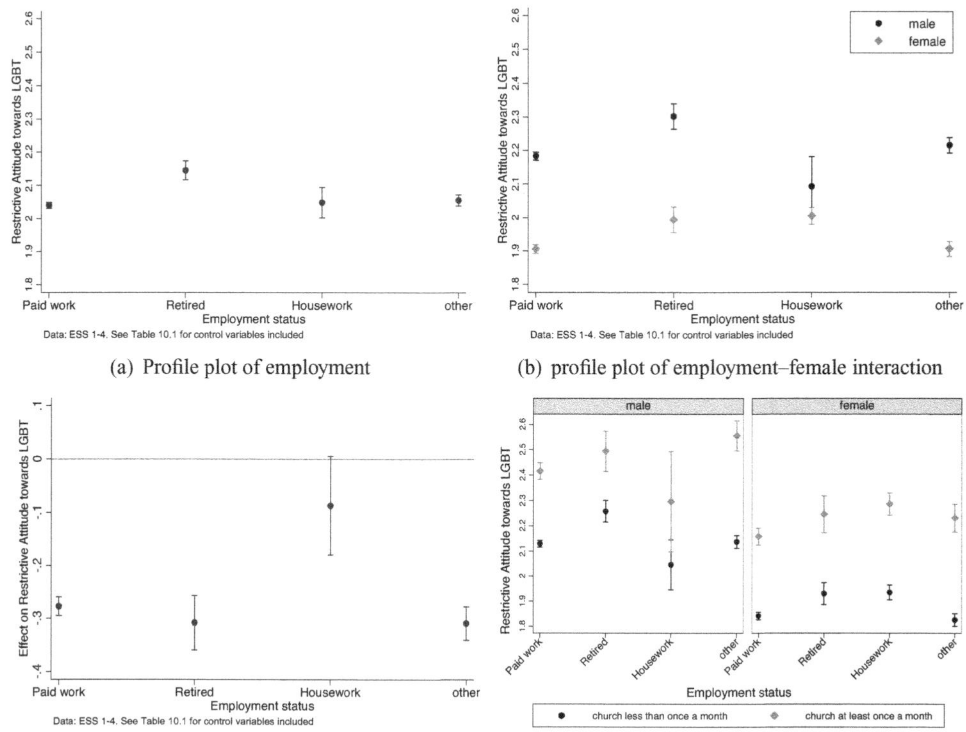

(a) Profile plot of employment

(b) profile plot of employment–female interaction

(c) conditional effect plot of effect of gender conditional on employment

(d) profile plot of employment–female–church interaction

Figure 10.3 Visualization of predicted values and effect sizes in OLS regression models

Figure 10.3b adds a third dimension: gender. Since such a plot now requires different symbols for depicting predictions for males and females, I had to add a legend to the chart. Because of the employment–gender interaction effect included in the estimation, the effect of employment status differs by gender, and vice versa (the effect of gender differs by employment status). Sometimes, but not in Figures 10.3a and 10.3b, lines connecting the predicted values are used to aid in visualizing differences between effects. Note, however, that the pattern could look significantly different if the employment status (not an ordinal scale) had appeared in another order (see Figure 10.7 for an example with lines connecting predicted values). We have just noticed that the effect of gender is conditional upon employment. How can we visualize the conditional effect of gender? Figure 10.3c shows such a conditional effect plot. All one has to do is to calculate the differences between predicted values for males and females separately for each value of the employment status. For example, in Figure 10.3b, the model has predicted a value of 1.91 for men in paid work and of 2.18 for employed women, resulting in an effect size of −0.27. The latter is shown in Figure 10.3c. If women are not employed but housewives, the effect of gender is weaker (about −0.09). Note that the confidence intervals in Figure 10.3c refer to the hypothesis that effects differ from 0. Hence, when employment status is paid work, retired and other, the effect of gender, female versus male, is significantly negative and hence different from 0. Gender does not have a significant effect if respondents' employment status is doing housework.

Table 10.1 (cont) Denial of homosexuality, part 3

	Coefficient (OLS)	CI (95%)
Placement on left–right scale (LRS)	0.1141	[0.0997,0.1286]
LRS–country interactions		
LRS × AT	Ref.	
LRS × BE	–0.0902	[–0.1098,–0.0705]
LRS × BG	–0.1511	[–0.1742,–0.1279]
LRS × CZ	–0.1654	[–0.1848,–0.1459]
LRS × DE	–0.0431	[–0.0617,–0.0244]
LRS × EE	–0.1417	[–0.1654,–0.1181]
LRS × FI	–0.0330	[–0.0521,–0.0139]
LRS × FR	–0.0638	[–0.0822,–0.0454]
LRS × GB	–0.0698	[–0.0904,–0.0492]
LRS × GR	–0.0647	[–0.0847,–0.0447]
LRS × IE	–0.1000	[–0.1211,–0.0788]
LRS × IL	–0.1057	[–0.1240,–0.0873]
LRS × NL	–0.0724	[–0.0915,–0.0534]
LRS × NO	–0.0533	[–0.0723,–0.0342]
LRS × RU	–0.1750	[–0.2012,–0.1489]
LRS × SE	–0.0819	[–0.1001,–0.0636]
Constant	2.0109	[1.9785,2.0432]
Observations	77484	

Finally, in the last chart dealing with the three categorical variables, I want to add a fourth dimension. The effect of employment varies with gender and also with church attendance. In order to avoid drawing too many lines with different symbols in one graph, Figure 10.3d is divided into subgraphs, a graph with predicted values for males on the left and one with predicted values for females on the right. Of course, I could have also made two subgraphs by church attendance or four subgraphs divided by employment status. Usually, theoretical predictions as to which comparisons are most important in such situations prove useful in identifying the best way to separate, combine, and arrange information.

Having examined effects of categorical variables, we now turn to the next two figures, which illustrate effects of a continuous variable (left–right scale), and how this effect is conditional upon country. The third part of Table 10.1 shows the relevant regression coefficients of the interaction effects.

The main effect of the respondents' placement on the left–right scale (0.11) refers to the reference category, Austria. The effect is weaker in all other countries because all interaction effects are negative. Figures 10.4a and 10.4b both illustrate how the effect of political beliefs on restrictive attitudes towards homosexuality is conditional upon the country. Figure 10.4a depicts country-specific regression lines. It not only shows how the slopes of the regression lines vary, but also displays the country-specific intercepts. The chart thus incorporates country-level effects that we have already graphically addressed in Figure 10.2b. However, Figure 10.4a provides no information about the statistical significance of country-specific effects. The regression coefficients and their confidence intervals in the third part of Table 10.1 indicate how effects differ with regard to Austria. The effect of left–right placement is significantly weaker in Belgium than in Austria, but does the slope in Belgium remain significant, that is, different from 0?

The next bar chart, Figure 10.4b, displays the average marginal effect of the left–right scale, conditional upon country. The confidence intervals shown here allow the following interpretation: left–right placement does not significantly affect the denial of homosexuality in Ireland

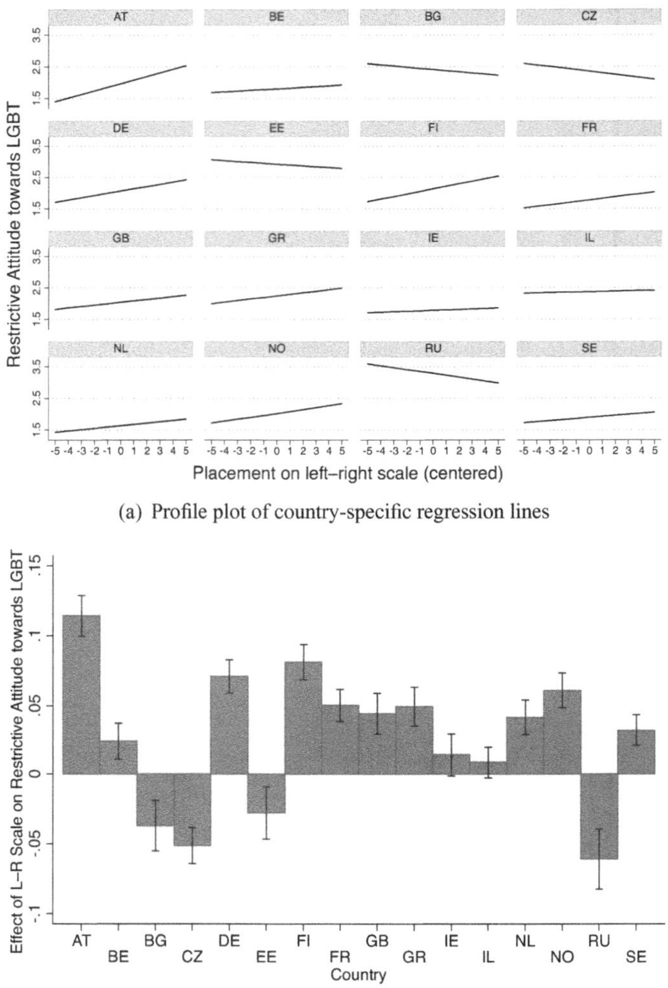

(a) Profile plot of country-specific regression lines

(b) Conditional effect plot of effect of left–right placement by country

Figure 10.4 Regression lines and corresponding coefficients in OLS regression models

and Israel. The effect is significantly negative in Bulgaria, the Czech Republic, Estonia and the Russian Federation. In all other countries, reporting a right placement positively relates to restrictive attitudes towards LGBT, whereas respondents placed on the left have less dismissive attitudes towards homosexual women and men.

We have thus far visualized interaction effects (via plotting the marginal effects or predicted values) of up to three categorical variables, and conditional effects between one dichotomous and one continuous variable. The regression model estimated above also contains interaction effects between two continuous predictors: age and education. Both variables have been centred on their mean value. Centring the variables avoids multicollinearity and allows one to interpret the effect of interacting variables at the other variable's mean. The effect of age (Table 10.1, part 1) is 0.0055 (per year) for observations with mean education. Conversely, the effect of the educational level is −0.0372 (also per year) if age is fixed at its mean value. The regression estimates further show a small negative effect for the age–education interaction. As outlined in the next subsection, plotting interaction effects between two continuous variables on a two-dimensional surface requires that we fix one of the interacting variables at a small number of plausible values (here three). Figure 10.5a depicts age on the x-axis, and education takes the mean value (0, because the variable is centred) and the mean plus/minus 10 years (rather extreme but still plausible values). The dashed line in the middle of Figure 10.5a displays the slope of the regression coefficient in Table 10.1, whereas the solid line shows predictions calculated at a schooling level of 10 years above average and the dotted line at an educational level of 10 years below average. The three lines look almost parallel, but because of the negative interaction effect, the slope is just slightly steeper at low levels of education. Figure 10.5b depicts how the effect of age is conditional upon education. The figure shows effects on the y-axis and not predicted values. It thus depicts the derivatives of the regression lines in Figure 10.5a. At the mean level of education (dashed line), the slope of the regression line is 0.0055, which is exactly the value listed in Table 10.1. If the educational level is 10 years below average, the age effect is stronger, the coefficient being about 0.007 (this is the slope of the dotted regression line). If, on the other hand, the educational level is 10 years above the average, the effect is somewhat weaker (about 0.004, the slope of the solid line). Figure 10.5b shows the effect sizes of age not only at education levels of -10, 0 and 10, but also for the whole continuum of the continuous education variable. At all values of education in the interval -10 to 10, the effect of age on restrictive attitudes is positive and statistically significant, that is, the displayed confidence interval does not include 0.

Of course, the interaction effect can also be understood as the effect of education being conditional upon age (instead of the age effect being conditional upon education). The same interaction effect is thus displayed in Figure 10.5c, but now plotted against education on the x-axis, with age fixed at its mean and at values of 25 years above and below the mean. We see that the three lines are again almost parallel, but the slope of the education effect is somewhat steeper for younger than for rather older respondents (conditional effect plot, Figure 10.5d).

Finally, the linear regression model included religiosity as a predictor of attitudes towards homosexuality, and the effect of this predictor was expected to have a non-linear form. The model therefore contained religiosity in a linear as well as in a squared specification. Without visualization, the (combined) effect of both variables, religious and religious2, is hardly conceivable. Figure 10.6a makes this effect easy to understand: religiosity affects the denial of homosexuality positively, but the strength of the effect depends upon religiosity. Although this interpretation may sound a bit unusual, the squared specification of religiosity is indeed an interaction effect similar to that between age and education as discussed above. Here, religiosity interacts with religiosity. Therefore, Figure 10.6b again shows the derivative of the regression curve: an increase in one unit of religiosity leads to a greater change in attitude towards homosexuality if respondents are rather more religious (e.g. an increase from 4 to 5 on the x-axis), and leads to less, but still significantly positive change if respondents are rather less religious (e.g. an increase from -3 to -2). A change in religiosity from -5 to -4 would not lead to a significant change in restrictive attitudes towards LGBT, as the confidence interval includes the null value.

The OLS regression model specifies that the non-linear effect of religiosity should be conditional upon gender. Interpreting these effects without a graphical display and only showing regression coefficients in a table is a pure waste of paper! Figure 10.6c illustrates the predicted values of y, conditional upon religiosity and gender. We can again see level differences by gender, but also the functional form of the religiosity effect differs between women and men. The corresponding derivatives in Figure 10.6d make this difference clear: the effect of religiosity on

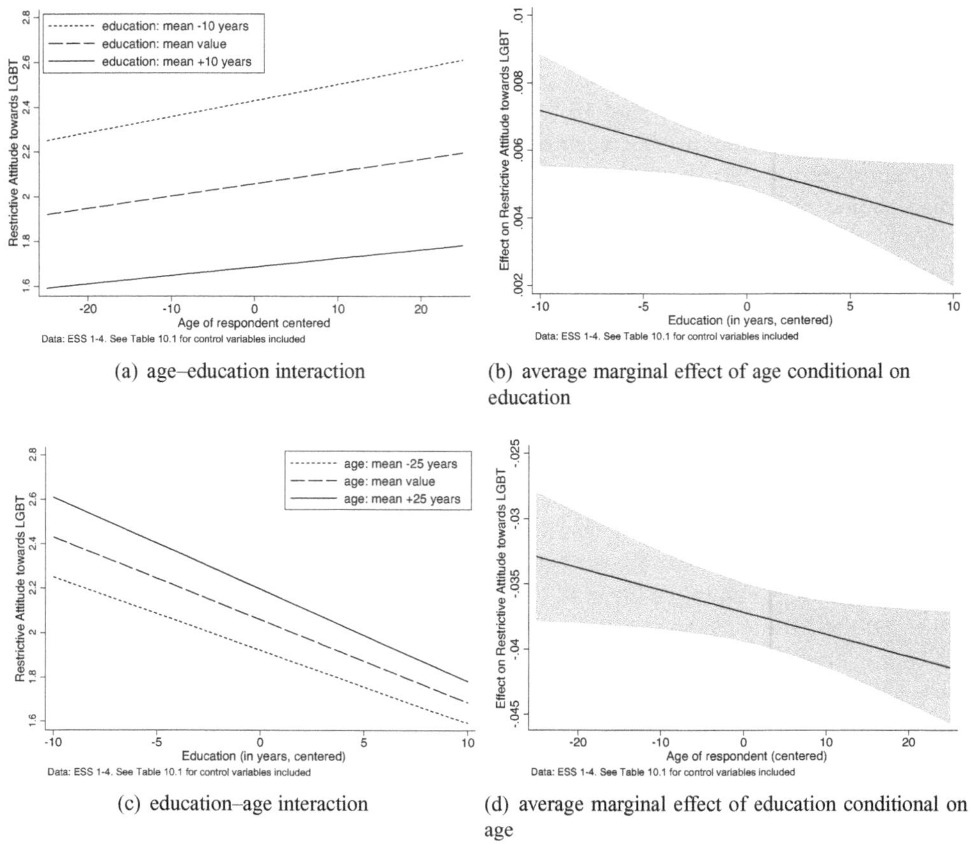

(a) age–education interaction

(b) average marginal effect of age conditional on education

(c) education–age interaction

(d) average marginal effect of education conditional on age

Figure 10.5 Interaction effects in OLS regression models

y is less strong for women than for men if they are not particularly religious. The confidence intervals even show that religiosity does not significantly affect the dependent variable below values of -4 if the respondent is male. If, on the other hand, the self-reported religiosity is above the average, any further increase in religiosity has a stronger effect on restrictive attitudes among women than among men.

Logistic regression models

The example analysis has already demonstrated that even the interpretation of linear regression models profits from visualization if interaction effects are specified. We will now turn to models with binary dependent variables and calculate a logistic regression. In such models, effects of various predictors are not additive; even in cases when interaction effects are not explicitly modelled, effects on probabilities are conditional upon the values of the other covariates. In principle, however, we can display regression results of logit models in very similar ways to those I have just shown.

Table 10.2 documents the estimation of two different model specifications: how do religiosity, age, employment status, political orientation, and the country of residence affect the probability of frequent church (or temple or mosque) attendance? Model 1 does not contain any explicit interaction effects; in model 2, an interaction effect between age and the placement on the

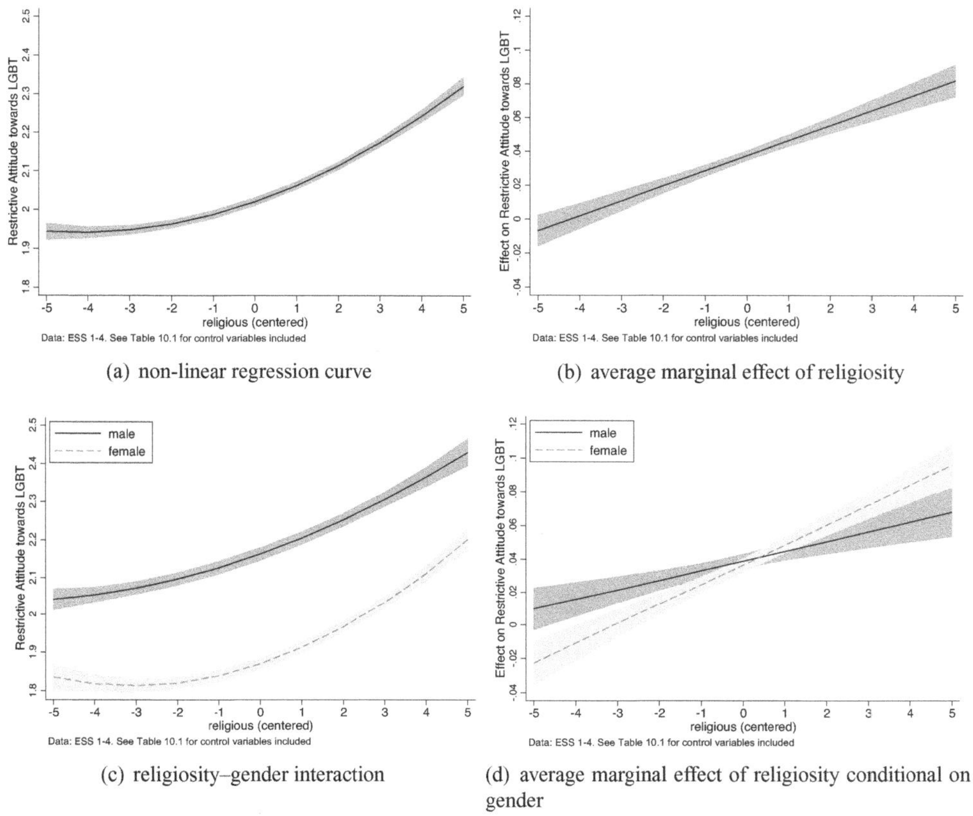

(a) non-linear regression curve

(b) average marginal effect of religiosity

(c) religiosity–gender interaction

(d) average marginal effect of religiosity conditional on gender

Figure 10.6 Profile plots and conditional effects in logistic regression models

left–right scale, and interaction effects between the four dummy variables measuring the employment status and 16 country dummies are included. The country–employment interactions are, for the sake of brevity, not shown in Table 10.2.

In model 1, we find that probability of regular church attendance increases with religiosity. The logit coefficient is 0.59; hence, the effect is positive. We could also interpret the odds ratios: any increase of religiosity by one unit would lead to an increase in the odds of frequent church visits by a factor of $\exp(0.59) = 1.8$. With regard to employment status, housekeepers and retired people differ significantly from respondents in paid work. Being a housewife or househusband increases the odds of frequent church attendance by $\exp(0.36) = 1.43$. Logits and odds, however, can be misleading. We thus refrain from any further interpretation and predict probabilities, as those allow a straightforward interpretation.

Figure 10.7 shows predicted probabilities on the y-axis (instead of predicted values in the case of OLS models). The values have been calculated for each country separately. The predictions in Figure 10.7a are based on model 1 in Table 10.2 (and thus calculated without explicit interaction effects between employment and country); the values in Figure 10.7b stem from model 2 and thus are based on explicit interaction effects. Interpreting Figure 10.7a, one can identify a number of remarkable country differences: irrespective of employment status, the share of respondents attending church at least once a month is highest in Ireland (over 50%) and rather

Table 10.2 Church attendance at least once a month

	Model 1		Model 2	
	Coefficient (logit)	CI (95%)	Coefficient (logit)	CI (95%)
Religiosity	0.5913	[0.5798,0.6029]	0.5924	[0.5808,0.6040]
Employment status				
Paid work	Ref.		Ref.	
Retired	0.1767	[0.0884,0.2651]	0.2196	[−0.0225,0.4617]
Housework	0.3594	[0.2895,0.4294]	0.4245	[0.1801,0.6690]
Other	0.0010	[0.0626,0.0645]	−0.2334	[−0.4674,0.0006]
Country				
AT	Ref.		Ref.	
BE	−1.2085	[−1.3279,−1.0892]	−1.2495	[−1.4053,−1.0937]
BG	−0.3986	[−0.5652,−0.2320]	−0.2459	[−0.4546,−0.0372]
CZ	−0.4341	[−0.5831,−0.2851]	−0.4677	[−0.6661,−0.2693]
DE	−0.4583	[−0.5621,−0.3544]	−0.3563	[−0.4884,−0.2242]
EE	−1.2457	[−1.4183,−1.0730]	−1.2369	[−1.4508,−1.0230]
FI	−1.7317	[−1.8513,−1.6121]	−1.9044	[−2.0590,−1.7497]
FR	−0.9078	[−1.0354,−0.7803]	−0.9565	[−1.1247,−0.7883]
GB	−0.3785	[−0.4893,−0.2677]	−0.3409	[−0.4802,−0.2016]
GR	−0.2260	[−0.3316,−0.1203]	−0.4592	[−0.5958,−0.3225]
IE	1.6652	[1.5628,1.7677]	1.7092	[1.5803,1.8381]
IL	−0.0981	[−0.2205,0.0243]	−0.0344	[−0.1926,0.1238]
NL	−0.7933	[−0.9004,−0.6861]	−0.7798	[−0.9171,−0.6425]
NO	−0.9582	[−1.0816,−0.8348]	−1.0203	[−1.1716,−0.8689]
RU	−0.6658	[−0.8267,−0.5049]	−0.6448	[−0.8439,−0.4457]
SE	−1.0766	[−1.2081,−0.9452]	−1.1087	[−1.2669,−0.9505]
Ctry–Employ interaction	not included in the model		included, but not shown	
Left–right scale (LRS)	0.0585	[0.0472,0.0698]	0.0555	[0.0440,0.0669]
Age of respondent	0.0099	[0.0080,0.0118]	0.0095	[0.0075,0.0114]
Age–LRS interaction	not included in the model		0.0007	[0.0001,0.0016]
Constant	−1.8987	[−1.9799,−1.8175]	−1.8845	[−1.9828,−1.7863]
Observations	77484		77484	

low in the three Scandinavian countries (under 20%). Without explicitly modelling interaction effects, only minor differences appear between persons in different employment positions. In model 2, Table 10.2, explicit interaction effects are included in the estimation. Those conditional effects are clearly visible in Figure 10.7b: in some countries, such as Greece, employment status has, compared to the other countries, rather strong effects. In Greece, being a housewife or a househusband positively (but in some other countries negatively) relates to probabilities of frequently attending religious ceremonies. The dashed lines connecting the four values of the employment status variable help in identifying country differences. Note, however, that employment status is not an ordinal scale and patterns would look quite different if the statuses were to appear in a different order.

Interpreting the coefficients of interaction effects in logit and probit models can be misleading. The final example (Figure 10.8) illustrates this. How does the respondents' placement on the left–right scale affect the probability of frequent church, temple or moscue attendance? And how does this effect relate to age? Figure 10.8a shows the predicted probabilities (y) against left–right placement (x), separately for three values of age (mean and plus/minus 25 years). It shows three (almost) parallel lines: the probability of church attendance rises with political opinions towards the right side of the scale. The average marginal effect of the left–right placement is

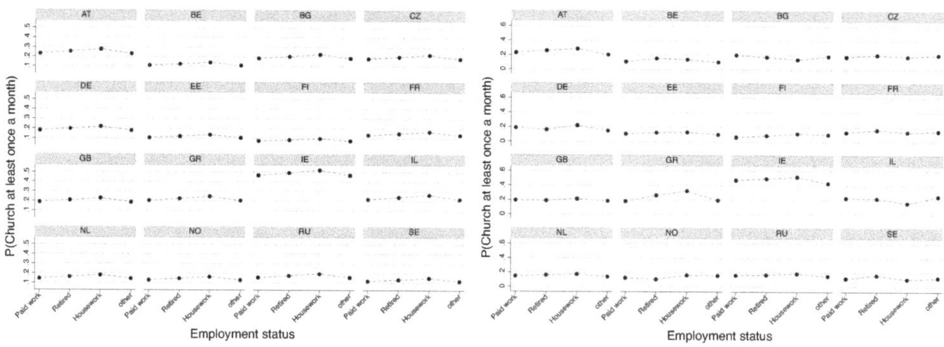

(a) profile plot without interaction term (b) profile plot with interaction term

Figure 10.7 Profile plots and interaction effects in logistic regression models

0.006, that is, a one-unit move to the right on the 10-point left–right scale leads, on average, to an increase in the probability of frequent church attendance by 0.6%. Also, the effect of age is clearly visible in Figure 10.8a: older respondents have higher probabilities of frequent participation in religious services than younger women and men. The three lines appear parallel, but the slopes are not completely identical – even though no interaction effect between age and left–right placement was included in the regression model (model 1 in Table 10.2). Thus, the effect of left–right placement is conditional upon age. Figure 10.8b shows this interdependency in greater detail and depicts the conditional average marginal effects. Even though the marginal effects are now conditional upon age, they are still averaged: the effect of the left–right scale would also be conditional upon all other covariates that are included in model 1 in Table 10.2, and those variables have, when predicting y, been fixed to the empirically observed value for each observation. Figure 10.8b now demonstrates that the average marginal effect of the left–right scale on church attendance varies between 0.005 and 0.006, conditional upon the value of age (in an interval of 25 years above and below the arithmetic mean).

In the second logistic regression model in Table 10.2, an interaction effect between age and left–right placement leads to greater variation in the average marginal effect of the left–right scale. Figure 10.8c depicts this, and, compared to Figure 10.8a, the slopes of the predictions now clearly differ among the three constructed age groups. In Figure 10.8d, the average marginal effect varies between a 0.003 (0.3%) and approximately a 0.008 (0.8%) increase per unit on the left–right scale, conditional upon age. Interaction effects in logistic regression models thus do not make effects conditional upon other variables, as is the case in linear regression models. Effects are conditional upon other covariates even without including interaction effects (Figures 10.8a and 10.8b). It is difficult to determine how the interaction effect modifies the slope by considering only the coefficients. Visual displays are therefore extremely helpful in interpreting non-linear regression models.

Note that we could also display regression coefficients for logit models, in the same way as I have suggested for linear regressions in Figures 10.2a and 10.3b. In such graphs, even in simple models without interaction effects, one would probably want to depict average marginal effects instead of logit coefficients and instead of odds ratios (see Chapter 8 in this volume for a discussion of how to interpret different coefficients in logistic regression models and the work of Ai and Norton (2003) for interaction effects and their standard errors in logit regressions).

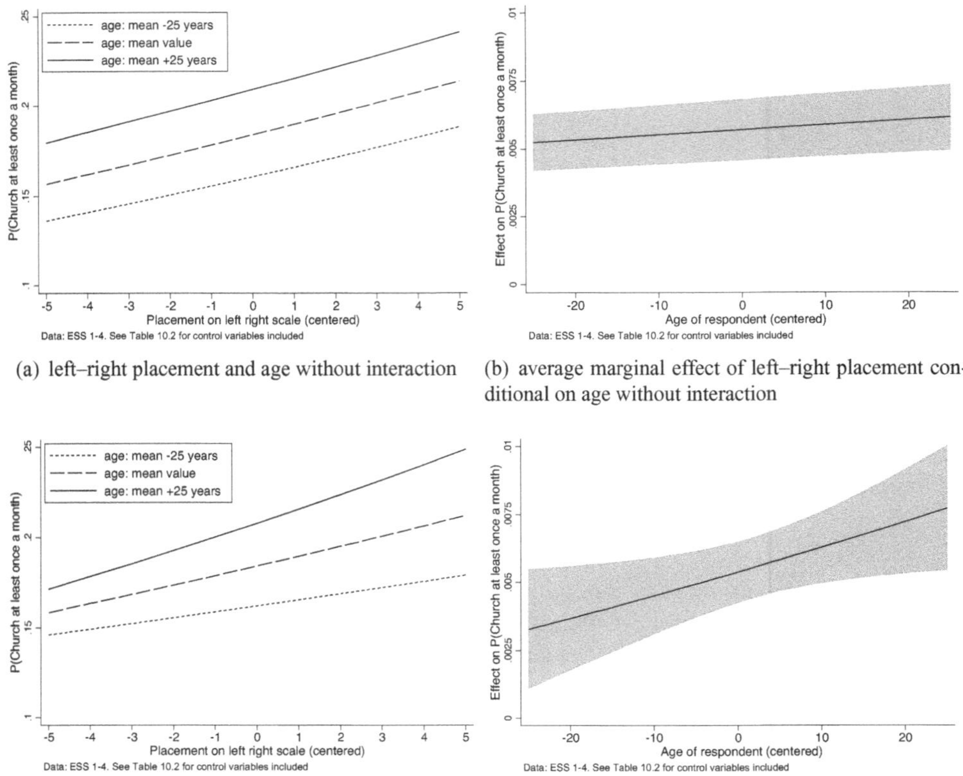

(a) left–right placement and age without interaction

(b) average marginal effect of left–right placement conditional on age without interaction

(c) left–right placement and age with interaction

(d) average marginal effect of left–right placement conditional on age with interaction

Figure 10.8 Visualization of interaction effects in logistic regression models

Other non-linear regression models

Even though other non-linear regression models, such as ordered logistic or multinomial regressions, are rather difficult to interpret, their visualization is quite simple and the procedure does not, in principle, differ from visualizing a logit or a probit model. The only difference in ordered logistic and multinomial models is a higher number of predictions: in both types of model, there is more than one outcome for which the model can make predictions. Graphical displays thus need to depict the probabilities of $y(1), \ldots, y(n)$, depending on the number of categories of y (see Bauer, 2010, for some examples). Additionally, we could visualize an increasing number of average marginal effects. The average marginal effect of x will differ with regard to the categories of y. For example, the effect of a family's high social status (x) on children's type of school (y) will certainly differ by the outcome category of y. A high social status might have a strongly negative correlation to low schooling level, less negatively for medium education, and might have a strong positive effect on successfully completing high levels of education, such as receiving a university degree. For more examples of how to visualize complex regression models with categorical dependent variables, see Chapter 9 in this volume.

CAVEATS AND FREQUENT ERRORS

The main aim of the visual display of regression results is clearly to avoid typical errors in the interpretation of regression coefficients. Such errors occur frequently in cases where non-linear relationships and interaction effects are involved in regression analysis. Social scientists can avoid misinterpretations (by authors and by readers) by visualizing the results as I have shown in the examples above. The typical shortcomings in displaying regression results have been discussed in the first section of this chapter. Following the guidelines outlined in the opening section of this chapter will produce charts of high 'graphical integrity', i.e. those displays will meet scientific standards. Nonetheless, designing expressive figures often requires active intervention when using standard statistical software. For example, most software packages automatically adapt the range of a scale (especially y) to the slope of a regression coefficient, which makes small differences appear large and meaningful. Besides adhering to the five general principles, the prediction of values from the regression model may cause some (technical) difficulties. All software packages can predict values or probabilities from regression models. The main challenge is to keep covariates constant at specific values, since this is the only strategy available for the reduction of dimensional complexity. Simply predicting values for each observation and drawing a figure for a specified subgroup (e.g. for respondents at age 25 with religiosity values of 5, living in Germany, etc.) will often not produce the desired chart because the number of observations selected is likely to be too small. Statistical software offers solutions for specifying such predictions. The solutions differ between the available products and are especially poor in the frequently used software package SPSS.

FURTHER READING

Readers interested in principles of information design in general should consider the work of Tufte (2001, 2008), which offers brilliant examples selected from a variety of scientific fields. With regard to the graphical display of multivariate data, the guidebooks by Miller (2004, 2005) provide extensive checklists for designing quantitative information displays, and also an extensive range of good examples. With regard to practical implementation with standard statistical software, I recommend Stata users take a look at the guidebook on interpreting and visualizing regression models by Mitchell (2012) and the script and do-files on cross-sectional regression by Brüderl (2012). The book and the script rely on the use of Stata's new 'margins' and 'marginsplot' commands that have been used for Figures 10.2–10.8. In older versions of Stata, these two commands were unavailable. Alternative solutions for fitting complex graphs (and some more display types not discussed here, such as odds ratio plots) are discussed in the book by Long and Freese (2006), which focuses on models with categorical dependent variables. Readers using the software package R will find an introduction to regression graphs in the textbook by Fox and Weisberg (2010). SPSS users should consult the introduction to visual information display in the textbook by Field (2009). With regard to regression with graphics, SPSS users can refer to the textbook by Hamilton (1991). The website `http://www.ats.ucla.edu/stat/spss/examples/rwg` provides SPSS syntax for replicating Hamilton's examples.

REFERENCES

Ai, C. and Norton, E. C. (2003). Interaction terms in logit and probit models. *Economics Letters*, 80(1), 123–129.

Bauer, G. (2010). Graphische Darstellung regressionsanalytischer Ergebnisse. In C. Wolf and H. Best (Eds), *Handbuch der sozialwissenschaftlichen Datenanalyse* (pp. 905–927). Wiesbaden: VS Verlag für Sozialwissenschaften.

Brüderl, J. (2012). *Angewandte Regressionsanalyse mit Stata (Script and Stata Do-Files on Cross-Sectional Regression)*. Munich: University of Munich (LMU).

Cook, D. R. and Weisberg, S. (1994). *An Introduction to Regression Graphics*. New York: John Wiley & Sons.

Field, A. (2009). *Discovering Statistics using SPSS*. Los Angeles: Sage.

Fox, J. and Weisberg, S. (2010). *An R Companion to Applied Regression*. Los Angeles: Sage.

Hamilton, L. C. (1991). *Regression with Graphics. A Second Course in Applied Statistics*. Belmont, CA: Duxbury Press.

Long, J. S. and Freese, J. (2006). *Regression Models for Categorical Dependent Variables Using Stata*. College Station, TX: Stata Press.

Miller, J. E. (2004). *The Chicago Guide to Writing about Numbers*. Chicago: University of Chicago Press.

Miller, J. E. (2005). *The Chicago Guide to Writing about Multivariate Analysis*. Chicago: University of Chicago Press.

Mitchell, M. N. (2012). *Interpreting and Visualizing Regression Models Using Stata*. College Station, TX: Stata Press.

Payton, M. E., Greenstone, M. H. and Schenker, N. (2003). Overlapping confidence intervals or standard error intervals: What do they mean in terms of statistical significance? *Journal of Insect Science*, 3(34), 1–6.

Tufte, E. R. (2001). *The Visual Display of Quantitative Information*. Cheshire, CT: Graphics Press.

Tufte, E. R. (2008). *Envisioning Information*. Cheshire, CT: Graphics Press.

Wainer, H. (2000). *Visual Revelations: Graphical Tales of Fate and Deception from Napoleon Bonaparte to Ross Perot*. Hillsdale, NJ: Lawrence Erlbaum Associates.

Wainer, H. (2005). *Graphic Discovery. A Trout in the Milk and Other Visual Adventures*. Princeton, NJ: Princeton University Press.

11

Regression with complex samples

Steven G. Heeringa, Brady T. West and Patricia A. Berglund

INTRODUCTION

A short history of regression analysis and inference for complex sample survey data

The science of survey sampling, survey data collection methodology and the analysis of survey data is less than a century old. The basic theory for 'design-based' inference for descriptive population parameters such as means, proportions and totals was laid down in a landmark paper by Jerzy Neyman (1934). Following the publication of Neyman's paper, there was a major proliferation of new work on sample design, estimation of population parameters and variance estimation techniques required to develop confidence intervals for sample-based inference, or what in more recent times has been labeled *design-based inference* (Deming, 1950; Hansen et al., 1953; Sukatme, 1954; Yates, 1960; Kish, 1965; Cochran, 1977). The World War II years established the probability sample survey as a tool for describing population characteristics, beliefs and attitudes. As early as the late 1940s, social scientists led by sociologist Paul Lazarsfeld of Columbia University began to move beyond using survey data to estimate population means, proportions and totals to using these data to explore relationships among the variables measured (see Kendall and Lazarsfeld, 1950; Klein and Morgan, 1951). Kish and Frankel (1974) were two of the first to empirically study and discuss the impact of complex sample designs on inferences related to regression coefficients. Binder (1983) focused on the sampling distributions of estimators for regression parameters, including parameters in generalized linear models in finite populations, and defined robust variance estimators for pseudo-maximum likelihood estimates of regression model parameters. Skinner et al. (1989) summarized estimators of the variances for regression coefficients that allowed for complex designs (including linearization estimators). Fuller (2002) provided a modern summary of regression estimation methods for complex sample survey data. The application of the new regression modeling techniques to complex sample survey data was not without controversy. Most of the controversy centered on the use of survey weights (DuMouchel and Duncan, 1983; see also Pfeffermann, 2011) in the estimation of regression model parameters. The question of weighting in regression analysis and some of the historical background are addressed in a later section of this chapter.

Complex sample design effects

In the finite world of population sampling, the closest analogue to the assumption that the data are independent and identically distributed (iid) draws from a distributional model is simple random sampling (SRS), in which each element in the target population is assumed to have an equal and independent chance (greater than zero) of being selected for observation. A subtle theoretical difference that we will for the most part ignore is that our survey samples are actually drawn from a finite population (e.g. adults in the Russian Federation) that is assumed to be one possible realization of a finite population from a 'superpopulation model' (Heeringa et al., 2010, Chapter 3). Survey data collections such as the European Social Survey (ESS) are typically not based on simple random sampling. Instead, the probability sample designs for large survey programs often feature stratification of the target population, multi-stage cluster sampling, and disproportionate sampling in the sample selection. Survey organizations use these 'complex' design features to optimize the variance–cost ratio of the final design or meet precision targets for subpopulations of the survey population. The term 'complex' arises from the more complex features of these sample designs relative to SRS, including variation in inclusion probabilities, sample stratification of samples, and sampling of clusters of elements (which introduces homogeneity within sample clusters).

Relative to SRS, the need to apply weights to complex sample survey data changes the approach to estimation of population statistics or model parameters. Also relative to SRS designs, stratification, cluster sampling and weighting all influence the size of standard errors for survey estimates. At any chosen sample size, the effect of sample stratification is generally a reduction in standard error relative to SRS. Clustering of sample elements and designs that require weighting for unbiased estimation generally tend to yield estimates with larger standard errors than an SRS sample of equal size.

Survey designers can employ *cluster sampling* (which we use to refer to both single-stage and multi-stage cluster samples) for several reasons, but a primary reason is that geographic clustering of elements for household surveys reduces interviewing costs by amortizing travel and related expenditures over a group of observations. By definition, multi-stage sample designs such as those employed in the ESS incorporate cluster sampling at one or more stages of the sample selection. While cluster sampling can reduce survey costs or simplify the logistics of the actual survey data collection, clustered selection of elements affects approaches to variance estimation and developing inferences from the sample data (see the section on methods for inference below). Special approaches are required to estimate the correct standard errors. The 'iid' or SRS variance estimation formulae and approaches incorporated in the standard programs of most statistical software packages no longer apply, because they are based on assumptions of independence for variable values collected from the sample observations, and such values from within the same cluster generally tend to be correlated (e.g. students within a classroom, or households within a neighborhood).

The general increase in variance of sample estimates due to either single-stage or multi-stage clustered sampling is caused by correlations (non-independence) of observations within sample clusters. Many characteristics measured on sample elements within naturally occurring clusters, such as children in a school classroom or adults living in the same neighborhood, are correlated. Socio-economic status, access to health care, political attitudes and even environmental factors such as the weather are all examples of characteristics that individuals in sample clusters may share to a greater or lesser degree. When such group similarity is present, the amount of 'statistical information' contained in a cluster sample of n persons is less than in an independently selected simple random sample of the same size. Hence, cluster sampling increases the standard errors of estimates relative to an SRS of equivalent size, because there is not as much unique information available to compute variance estimates; the use of standard SRS formulae to estimate variances will therefore tend to understate estimates of variances. A statistic that is frequently used to quantify the amount of homogeneity that exists within sample clusters is the *intraclass correlation* ρ (Kish, 1965).

Strata are non-overlapping groupings of population elements or clusters of elements that are formed by the sample designer prior to the selection of the probability sample. Stratification can be used to sample elements or clusters of elements. In survey practice, stratified sampling serves several purposes, enabling increased precision for overall estimates of population parameters (provided that the strata are as homogeneous as possible; Cochran, 1977) or disproportionate allocation of the sample to strata that define subpopulations of interest.

Because stratified sampling selects independent samples from each of the $h = 1, \ldots, H$ explicit strata, any sampling variance attributable to differences among strata is eliminated from the sampling variance of the estimate. Consequently, stratification in complex samples works to reduce sampling variances relative to non-stratified samples of the equivalent size.

Under probability sampling and design-based inference, weighting of the survey data, for which sample inclusion probabilities for individual observations vary, is required to 'map' the sample back to an unbiased representation of the survey population. Generally, the final analysis weights in survey data sets are the product of the sample selection weight (w_{sel}), a non-response adjustment factor (w_{nr}) and a post-stratification adjustment factor (w_{ps}):

$$w_{\text{final},i} = w_{\text{sel},i} \times w_{\text{nr},i} \times w_{\text{ps},i}. \qquad (11.1)$$

Under the theory of design-based inference for probability samples, weighted estimation using the 'inverse probability' weight factors, w_{sel}, will yield unbiased (or nearly unbiased) estimates of population statistics. For example:

$$\widehat{B} = \frac{\sum_{i=1}^{n} w_{\text{sel},i} y_i x_i}{\sum_{i=1}^{n} w_{\text{sel},i} x_i^2} \quad \text{is an unbiased estimate of } B, \qquad (11.2)$$

where B is the finite population value of the population simple linear regression slope.

Throughout this chapter, weighted estimation of population parameters will follow a similar approach, even for procedures as complex as the pseudo-maximum likelihood (PML) estimation of the coefficients in a multivariate logistic regression model.

Weighting by w_{sel} will only yield unbiased estimates of population parameters if all n elements in the original sample are observed. Unfortunately, due to *survey non-response*, observations are only collected for r cases of the original probability sample of n elements (where $r \leq n$). Therefore, survey data producers must develop statistical models of the conditional probability that a sample element will be an observed case. In general terms, the non-response adjustment factor, w_{nr}, in the analysis weight is the reciprocal of the estimated conditional probability that the sample case responds. The objective in applying non-response factors in survey weights is to attenuate bias due to differential non-response across sample elements. A price that may be paid for the bias reduction through non-response weighting takes the form of increases in standard errors for the weighted estimates (due to increased variation in the adjusted weights).

Most survey data producers also introduce the post-stratification factor w_{ps} into the final weight or perform other calibration of the weights to external population data. As its label implies, the post-stratification factor is an attempt to apply stratification corrections to the observed sample after the survey data have been collected. The use of post-stratification weight factors can lead to reduced standard errors (variance) for sample estimates or attenuate any sampling biases that may have entered the original sample selection due to sample frame non-coverage or omissions that occurred in implementing the sample plan.

Using the methods and software presented in this chapter, the survey data analyst will be able to compute confidence intervals and test statistics that incorporate the estimates of standard errors corrected for the complex sample design. Nevertheless, general knowledge of the existence of these *design effects* (Kish, 1965) and their influence on standard errors permits the analyst to understand why the sampling plan for their data has produced efficiency losses relative to simple random sampling and to identify features such as extreme clustering or weighting influences that might affect the stability of the inferences that they will draw from the analysis of the data.

The reader is encouraged to see Heeringa et al. (2010, Chapter 2) for a more in-depth review of these 'design effects' and the influence that they have on the precision of sample estimates. Readers may also be interested in recent work by Lohr (2014), which specifically addresses design effects for estimated regression parameters in clustered samples.

Sampling error calculation models and replicate weights

Correct estimation of standard errors for estimates computed from complex sample survey data requires the specification of a *sampling error calculation model*. Along with the final survey weight described above, the data producer must generally supply a *sampling error stratum* and a *sampling error cluster* variable for each observation. In the ESS (Round 4) Russian Federation data set used for the example exercises below, the survey weight variable is labeled DWEIGHT, the primary stage strata are coded in the variable STRATIFY and the primary stage cluster units (PSUs) are coded in the variable PSU (see the example section for more details).

These *sampling error variables* identify the primary stage strata and 'clusters' that the survey respondents belong to, approximating the original sample design as closely as possible while at the same time conforming to the analytical requirements of several methods for estimating variances from complex sample survey data. Sampling error stratum and cluster codes are required for the Taylor series linearization method for variance estimation (discussed in the following section). These same sampling error variables can be specified in the analysis software setup to create the 'replicate' samples and weights required for replicated variance estimation using the balanced repeated replication or jackknife repeated replication methods. If replicate weights are provided with the data set, the analyst specifies only the variable names for the replicate weights and the full sample weight, and separate coding of primary stage strata and clusters is not required.

STATISTICAL FOUNDATIONS

We focus in this section on the following topics, all in the context of fitting regression models to survey data collected from samples with complex designs: model specification; model building; estimation; model evaluation; and inference. We describe how all of these aspects of regression modeling change (if at all) when analyzing complex sample survey data. For more background on each of these topics, we refer readers to Chapters 2, 4, 5, 8, and 9 of this volume.

Model specification and model building

In general, the process of specifying a regression model for a finite population of interest and then building that model (or reducing the model for purposes of parsimony) does not change when analyzing complex sample survey data. What changes are the mathematical techniques used to estimate the regression parameters in the model, the methods for estimating the variances of estimated parameters and constructing confidence intervals for the target parameters, and the statistical tests used to test parameters for significance. We recommend the following general sequence of steps for specifying and then building a regression model:

1. Determine an appropriate distribution for the dependent variable of interest (e.g. normal, log-normal, binomial, Poisson, negative binomial), along with an appropriate link function that enables straightforward interpretation of the estimated coefficients and yields appropriate predicted values. For example, an analyst working with binary survey data might assume that the dependent variable follows a Bernoulli distribution governed by a parameter p, and use a logit link to ensure that predicted values remain between 0 and 1 (see Chapter 8 of this volume).
2. Determine the measurement types for the predictor variables of scientific interest (continuous, categorical, etc.). Create binary indicators for levels of the categorical predictors, and examine initial scatter plots to determine approximate functional relationships of the continuous predictors with the dependent variable.
3. Given that the general objective of fitting regression models to survey data is to estimate a theoretical model that defines relationships between variables in some larger (and conceptual) finite population, include fixed regression coefficients for all predictor variables posited to have relationships with the outcome variable in the finite population, and make sure that functional forms of the relationships between continuous predictors and the dependent variable are correctly specified based on Step 2. Consider including interactions between predictors, depending on the scientific question of interest.
4. Fit the specific model using a statistical software procedure that is capable of correctly implementing the estimation and inference techniques for complex sample survey data discussed later in this section (e.g. `svy: regress` in Stata, or `svyglm()` in R). Compare model estimates incorporating complex sampling features and also ignoring them (for more detail on this step and why this should be done, see the section below on caveats).
5. Model evaluation (discussed later in this section) can be used to assess whether or not the predicted values and residuals based on a given model appear to reflect the specified statistical distribution, but software tools for model evaluation that recognize features of complex samples are fairly limited at the time of this writing (as will be discussed).
6. Examine tests of significance and confidence intervals for the regression parameters of interest, and remove non-significant parameters from the model if variable selection is the goal. For other practical references on model building and variable selection in regression modeling (with complex samples or otherwise), we refer readers to Chapters 4 and 5 of this volume, Harrell (2001), or Hosmer and Lemeshow (2000).

Estimation

Analysts fitting regression models to complex sample survey data need to decide whether or not to incorporate the final survey weights that are often provided in the data sets into their estimation routines. We discuss this choice in more detail later; here, we describe the methods that are used to compute estimates of the parameters in a specified regression model both with and without the use of weights. We divide regression models into two classes: linear regression models for normally distributed continuous outcomes (e.g. systolic blood pressure in mmHg), and generalized linear models for non-normal outcomes (e.g. whether or not a person votes for a certain candidate in an election).

Linear regression models

When fitting linear regression models to continuous outcomes and *ignoring* the final survey weights, ordinary least squares (OLS) is typically used for estimation of the regression parameters. This technique is covered in detail in Chapters 2, 3 and 4 of this volume; we briefly describe OLS estimation here for comparison with estimation methods incorporating the weights. The OLS method focuses on estimating the unknown vector of regression parameters $\boldsymbol{\beta}$ in a specified linear regression model (say, $y_i = \mathbf{x}_i\boldsymbol{\beta} + \epsilon_i$) by finding values of the parameters that minimize the residual sum of squares (or sum of squared errors, SSE) based on the model:

$$SSE = \sum_{i=1}^{n}(y_i - \mathbf{x}_i\boldsymbol{\beta})^2 . \tag{11.3}$$

The estimate of $\boldsymbol{\beta}$ is obtained analytically based on the sample of size n using the following algebraic result, where $\mathbf{X}$ is the $n \times p$ matrix of values on the vector of p predictor variables x_i (where the design matrix of p predictor variables generally includes a vector of 1s for models with intercepts), and $\mathbf{y}$ is the vector of values on the dependent variable:

$$\widehat{\boldsymbol{\beta}} = (\mathbf{X}^T\mathbf{X})^{-1}\,\mathbf{X}^T\mathbf{y} \tag{11.4}$$

As discussed in Chapter 2 of this volume, the OLS estimator of $\boldsymbol{\beta}$ has several important properties. This estimator of $\boldsymbol{\beta}$ is unbiased and has the lowest variance among all other unbiased estimators that are also linear functions of the values of the dependent variable, making it the *best linear unbiased estimator* (BLUE). In addition, assuming normally distributed errors, the least squares estimates are equal to estimates derived based on maximum likelihood estimation.

When using the final survey weights to fit regression models to complex sample survey data collected from a finite population, we choose estimates of the finite population regression parameters $\mathbf{B}$ that minimize the following objective function for a population of size N:

$$f(\mathbf{B}) = \sum_{i=1}^{N}(y_i - \mathbf{x}_i\mathbf{B})^2 \tag{11.5}$$

We can think of this objective function $f(\mathbf{B})$ as a finite population 'residual' sum of squares, SSE_{pop}; this objective function also defines a population total for the squared residual values. An unbiased sample estimate of this total incorporating the final survey weights can be written as follows:

$$\widehat{WSSE}_{\text{pop}} = \sum_{h}^{H}\sum_{\alpha}^{a_h}\sum_{i=1}^{n_{h\alpha}} w_{h\alpha i}(y_{h\alpha i} - \mathbf{x}_{h\alpha i}\mathbf{B})^2 . \tag{11.6}$$

In this expression, h is a stratum index, α is a cluster (or primary sampling unit) index, and i is an index for elements within the αth cluster. When survey weights are provided in a complex sample survey data set, an analyst can use *weighted least squares* (WLS) to compute the necessary weighted estimates, $\widehat{\mathbf{B}}$, of the parameters in a specified linear regression model for a finite population (more details on making this decision are provided later) that minimize the unbiased sample estimate above. The WLS estimates are obtained analytically using the closed-form estimator

$$\widehat{\mathbf{B}} = (\mathbf{X}^T\mathbf{W}\mathbf{X})^{-1}\,\mathbf{X}^T\mathbf{W}\mathbf{y}. \tag{11.7}$$

In this estimator, $\mathbf{W}$ is an $n \times n$ diagonal matrix, where the final survey weights for each of the cases in the sample, $w_{h\alpha i}$, are on the diagonal and all values off of the diagonal are zero.

Generalized linear regression models

Analysts fitting linear regression models to complex sample survey data can employ the convenient set of closed-form estimators defined in equations (11.4) and (11.7). Analysts fitting generalized linear regression models to non-normal outcomes need to rely on iterative pseudo-maximum likelihood estimation techniques, as closed-form estimators are generally no longer available for this broader class of models. We use logistic regression models in this section to illustrate the adaptation of maximum likelihood estimation techniques to complex sample survey data.

In general, maximum likelihood estimation techniques for simple random samples of size n with iid data seek estimates of regression parameters in generalized linear models that maximize the likelihood function

$$L(\boldsymbol{\beta}|\mathbf{y},\mathbf{X}) = \prod_{i=1}^{n} f(y_i|\boldsymbol{\beta},\mathbf{x}_i). \tag{11.8}$$

In 11.8, $f(y_i|\boldsymbol{\beta},\mathbf{x}_i)$ is a density function for a non-normal dependent variable y defined by a given generalized linear model. In the case of logistic regression (see Chapter 8 of this volume), the likelihood function for a simple random sample of n observations on a binary variable y with possible values 0 and 1 is based on the density function for the binomial distribution,

$$L(\boldsymbol{\beta}|\mathbf{y},\mathbf{X}) = \prod_{i=1}^{n} \pi(\mathbf{x}_i)^{y_i}\,[1 - \pi(\mathbf{x}_i)]^{1-y_i}, \tag{11.9}$$

where $\pi(\mathbf{x}_i)$ is the probability that $y_i = 1$. This probability is linked to the logistic regression model coefficients through the logistic cumulative distribution function (CDF):

$$\pi(x_i) = \frac{\exp(x_i\boldsymbol{\beta})}{1 + \exp(x_i\boldsymbol{\beta})}. \tag{11.10}$$

Iterative mathematical procedures (e.g. the Newton–Raphson algorithm) are applied by statistical software to find estimates of the regression parameters that maximize the likelihood function above (i.e. maximize the probability of the data that were collected).

When the survey data have been collected under a complex sample design, straightforward application of maximum likelihood estimation procedures is no longer possible for several reasons. First, the probabilities of selection (and responding) for the $i = 1, \ldots, n$ sample observations are generally no longer equal. Sampling weights are thus required to estimate the finite population values of the logistic regression model parameters. Second, the stratification and clustering of survey respondents violate the assumption of independence of observations that is crucial to the standard maximum likelihood estimation approach for estimating the sampling variances of the model parameters and choosing a reference distribution for the likelihood ratio test statistic.

Two general approaches have been developed to estimate the logistic regression model parameters and standard errors for complex sample survey data. Grizzle et al. (1969) first formulated an approach based on weighted least squares estimation. The WLS estimation method was originally programmed for logistic regression in the GENCAT software package (Landis et al., 1976) and still remains available as an option in programs such as SAS PROC CATMOD. Later, Binder (1981, 1983) presented a second general framework for fitting logistic regression models and other generalized linear models to complex sample survey data. Binder proposed *pseudo-maximum likelihood estimation* (PMLE) as a technique for estimating the model parameters. The PMLE approach to parameter estimation was combined with a linearized estimator of the variance–covariance matrix for the parameter estimates, taking complex sample design features into account (see the subsection on methods for inference below). Further details and evaluation of the PMLE approach are presented in Roberts et al. (1987), Morel (1989) and Skinner et al. (1989). The PMLE approach is now the standard method for generalized linear regression modeling in all of the major software systems that support analysis of complex sample survey data (e.g. `svy: logit` in Stata and `svyglm()` in R).

Following the PMLE approach for logistic regression models, estimates of the finite population regression parameters **B** are obtained by finding the values of the parameters that maximize the following unbiased estimate of the population likelihood, which is a weighted function of the observed sample data and the $\pi(\mathbf{x}_i)$ values (note the use of the final survey weights, w_i):

$$PL(\mathbf{B}|\mathbf{y},\mathbf{X}) = \prod_{i=1}^{n} \left\{ \pi(\mathbf{x}_i)^{y_i} \left[1 - \pi(\mathbf{x}_i)\right]^{1-y_i} \right\}^{w_i}, \qquad (11.11)$$

with $\pi(\mathbf{x}_i) = \exp(\mathbf{x}_i\mathbf{B})/(1 + \exp(\mathbf{x}_i\mathbf{B}))$.

Like the standard maximum likelihood estimation procedure, this weighted pseudo-likelihood function can be maximized using the iterative Newton–Raphson method or related algorithms. Application of the PMLE procedure for other generalized linear models simply requires a definition of the density function for a given non-normal dependent variable, given the predictors and regression parameters in a specified regression model; see Heeringa et al. (2010, Chapter 9) for additional examples.

Model evaluation

In this subsection, we provide a summary of the techniques that have been proposed in the literature to date for evaluating regression models that have been fitted to complex sample survey data. Unfortunately, many of these techniques are relatively new additions to the literature and have yet to be implemented in popular statistical software; whenever possible, we provide references to available computing procedures where these techniques are implemented.

R^2 and pseudo-R^2 statistics for assessing model fit

A standard measure of the 'fit' of a linear regression model to the survey data is the *coefficient of multiple determination* (R^2), which is interpreted as the proportion of variance in the dependent variable explained by the independent variables:

$$R^2 = 1 - \frac{SSE}{SST}. \qquad (11.12)$$

Here, *SST* refers to the total sum of squares, or the sum of squared differences between the values on the dependent variable and the mean of the dependent variable, and *SSE* is given in 11.3 (after plugging in estimates of the regression parameters). The use of R^2 as a measure of explained variance carries forward to regression modeling of complex sample survey data, although the statistic that is output by the analysis software is actually a *weighted* version, where each squared difference contributing to the sums is weighted by the corresponding final survey weight:

$$R^2_{\text{weighted}} = 1 - \frac{WSSE}{WSST}. \qquad (11.13)$$

Although in theory it could be argued that this weighted R^2 statistic estimates the proportion of population variance explained by the population regression of y on x, in practice it is safe to simply view it as the fraction of explained variance in y attributable to the regression on x. The weighted R^2 statistic is easily computed using procedures in Stata (`svy: regress`) and R (`svyglm()` and `svyvar()`).[1]

When fitting generalized linear models (such as logistic regression models) to simple random samples using maximum likelihood methods, one can compute an approximate version of the R^2 statistic known as the *pseudo R^2 statistic*, which in its simplest form[2] represents the proportionate decrease in the log-likelihood for a model including the predictors of interest relative to an intercept-only model excluding any predictors. This statistic is also referred to as a *negative log-likelihood* or *entropy* statistic, and should not be confused with the R^2 statistic from linear regression, which represents the proportion of variance in a given continuous outcome explained by a set of predictors. When fitting generalized linear models, these approximate measures of fit should only be used to compare competing models to see which does a better job of maximizing the likelihood of the observed data.

Few alternatives have been proposed in the literature for generalized linear models fitted to complex sample survey data, largely because of the use of PMLE methods. These alternatives include the negative log-likelihood measure above, only based on the pseudo log-likelihood functions for a given pair of models (the 'full' model and the 'reference' model with no predictors); the *likelihood ratio (Cox–Snell) pseudo R^2 statistic*, defined as

$$1 - \exp\left\{\frac{2}{\widehat{N}}\left(l(0) - l(\widehat{\mathbf{B}})\right)\right\}, \tag{11.14}$$

where $\widehat{N}$ is the sum of the final survey weights for the sample (or an estimate of the population size), $l(0)$ is the pseudo log-likelihood for an intercept-only model, and $l(\widehat{\mathbf{B}})$ is the value of the pseudo log-likelihood evaluated at the weighted PML estimates of the regression parameters; and the *likelihood ratio pseudo R^2 statistic* (Estrella, 1998),

$$1 - \left[\frac{l(\widehat{\mathbf{B}})}{l(0)}\right]^{-\frac{2}{\widehat{N}}l(0)}. \tag{11.15}$$

These alternatives have not been widely programmed in the software procedures capable of fitting generalized linear models to complex sample survey data to date (the WesVar software does provide these statistics; see `http://www.westat.com/Westat/expertise/information_systems/WesVar/index.cfm`). Even if they do become more widely available, we recommend their use only to choose between competing models (where higher values indicate models with better fits), and they should not be interpreted as percentages of variance explained in a given non-normal dependent variable.

Testing goodness of fit

Archer and Lemeshow (2006) and Archer et al. (2007) have extended the standard Hosmer and Lemeshow (2000) goodness-of-fit test for logistic regression models for application to complex sample survey data. The Archer–Lemeshow procedure is a modification of the standard Hosmer–Lemeshow test for goodness of fit that takes the sampling weights and the stratification and clustering features of the complex sample design into account when assessing the residuals $y_i - \widehat{\pi}(\mathbf{x}_i)$ based on the fitted logistic regression model. The papers above should be consulted for more details, but this procedure is currently implemented in the Stata software, where it can be executed by calling the command `estat gof` after fitting a logistic regression or probit regression model to a binary dependent variable using `svy: logit` or `svy: probit`. At the time of this writing, this procedure has not yet been implemented in the `survey` package of the R software.

Additional goodness-of-fit tests for other generalized linear models that incorporate complex sampling features have not been developed to date, and we expect this to be an active area of research in the near future.

Regression diagnostics

Analysts fitting linear regression models to complex sample survey data sets should certainly examine the standard set of diagnostic analyses based on the residuals from a fitted model (e.g. fitted residual plots, quantile–quantile plots, added variable plots, etc.; see Chapter 5 of this volume or Chapter 7 of Heeringa et al., 2010). These informal diagnostic analyses can be used to identify problems with the structure of the model in the larger conceptual finite population, violations of distributional assumptions for the dependent variable, or outliers that might have large leverage and/or influence on the fit of a model. Stata and R both make preparation of these standard plots quite easy after residuals and predicted values based on a fitted model have been saved as new variables.

Survey statisticians have only recently begun to examine how the computation and interpretation of other common diagnostic tools for linear regression models should be adapted to accommodate complex sampling features, and a small literature is developing in this area. Li and Valliant (2009) were the first to describe how survey weights should be incorporated into the computation of hat matrices and leverage statistics, and Li and Valliant (2011a) also describe the identification of influential observations using the forward search method in linear regression models fitted to complex samples. Li and Valliant (2011b) also discuss how the DFBETAs, DFFITs, and Cook's *D* statistics for checking the influence of individual sample cases on fitted models should be computed to incorporate survey weights.

Even more recently, survey statisticians have examined how the computation of common diagnostics for multicollinearity should be adapted to incorporate complex sampling features. Liao and Valliant (2012b) describe the computation of variance inflation factors for parameter estimates in models fitted to complex samples, and Liao and Valliant (2012a) also describe the computation of design-adjusted condition indices for global examinations of problems with multicollinearity in linear regression models.

While these recent publications demonstrate that common diagnostic tools can be adapted to incorporate complex sampling features, these updated tools have yet to make their way into any of the available statistical software for analysis of complex sample survey data. In addition, related diagnostic tools for generalized linear models (e.g. Poisson regression models) have yet to be developed. The evaluation of regression models fitted to complex samples is a ripe area for research by survey statisticians, and rapid developments are likely in the next few years. In the meantime, we aim to keep analysts appraised of software developments in the area of regression diagnostics on the website for the book *Applied Survey Data Analysis* (`http://www.isr.umich.edu/src/smp/asda`).

Methods for inference

At first glance, the tools for inference in regression analysis of complex sample survey data are familiar. However, due to the complex features of the sample data, methods for estimation of the variances and covariances of the estimated regression parameters, the construction of confidence intervals for population parameters and the nature of hypothesis test statistics differ from the standard regression setting.

Variance estimation

Estimates of the variance and covariances of estimated regression coefficients can be obtained using one of two classes of robust methods: the Taylor series linearization (delta) method, or replication methods.

As its name implies, the Taylor series linearization (TSL) method transforms the non-linear weighted estimates of the regression coefficients into a linear function. The variance of the 'linearized estimator' can then be calculated as a linear combination of variances and covariances of (possibly weighted) sample totals, for which closed-form formulae are available (Heeringa et al., 2010). For linear regression models and generalized linear models such as logistic regression models, Binder (1983) derived a sandwich estimator of the variance–covariance matrix for pseudo-maximum likelihood estimates of the regression parameters. Binder's linearization approach is now the default method for estimating the variances and covariances of estimated regression parameters in many popular software procedures for fitting regression models to complex samples, including those in Stata, R, SAS, SPSS and SUDAAN.

An alternative non-parametric approach to estimation of variance for complex sample survey data is through the application of the balanced repeated replication (BRR) or jackknife repeated replication (JRR) methods. These replicated approaches to variance estimation are based on a four-step algorithm: (1) creation of $r = 1, \ldots, R$ replicate subsamples of the full survey sample; (2) creating 'replicate weights' by adjusting the full survey weights to reflect the subsampling used to create each replicate; (3) separate estimation of the regression model for the full sample and each replicate sample; (4) application of simple replicated variance estimation formulae to compute the variances and covariances of the estimated regression parameters. As a simple example, the BRR expression for estimating the variance of a single regression parameter estimate is:

$$\mathrm{var}_{\mathrm{BRR}}(\widehat{B}) = \frac{1}{R}\sum_{r=1}^{R}(\widehat{B}_r - \widehat{B})^2 \qquad (11.16)$$

where $\widehat{B}_r$ is the regression parameter estimate for replicate $r = 1, \ldots, R$; and $\widehat{B}$ is the regression parameter estimate based on the full sample.

The primary distinction between the BRR and JRR methods lies in the rules used to create the R replicates (for more details, see Wolter, 2007; Heeringa et al., 2010). In practice, variance estimates computed using the linearization and BRR or JRR methods will be very similar, and the choice among these variance options should not generally affect the inferences drawn from a wide variety of regression analyses. The bootstrap is another popular replication-based variance estimation approach that can be used to estimate variances in complex samples, and applications of this technique have received more research focus in recent years; see Kolenikov (2010) for more details.

Confidence intervals

When constructing the confidence interval (or the pivotal hypothesis test statistic) for a single regression parameter estimated from complex sample survey data, two aspects of the inferential process change: the standard error of the estimated regression parameter, $se(\widehat{B})$, is *estimated* using a non-parametric technique such as TSL, BRR or JRR; and 2) the degrees of freedom for the Student t reference distribution must be adjusted to reflect the reduced degrees of freedom for the complex sample estimate of $se(\widehat{B})$. The 'design degrees of freedom' are approximated by $df = \sum_h a_h - H$, or the number of primary stage clusters minus the number of primary stage strata.[3] For example, in the ESS Russian Federation data set, the approximation to the design degrees of freedom is $134 - 10 = 124$ (see below). The correct estimate of the standard error and degrees of freedom for the Student t reference distribution can be used to develop a design-based $100(1 - \alpha)\%$ confidence interval for the regression parameter of interest, given by

$$\widehat{B} \pm t_{1-\alpha/2,df} \times se(\widehat{B}). \tag{11.17}$$

Hypothesis tests

The t-statistic for a two-sided test of the null hypothesis that a single regression parameter is equal to zero, $H_0 : B_j = 0$, is given by

$$t = \frac{\widehat{B}_j}{se(\widehat{B}_j)}.$$

The 'p-values' generally printed by software procedures alongside the test statistics are the probability that $t_{df} \geq t$, where df is defined above. As noted earlier, multi-parameter tests must be adapted to the complex design features of the sample. *Wald test statistics* (Judge et al., 1985) replace the overall and partial F-tests for the OLS linear regression model and the equivalent likelihood ratio tests that are employed in hypothesis testing of generalized linear regression models and their parameters. The multi-parameter Wald test statistics can be calculated as follows:

$$\text{overall: modified } X^2_{W,\text{overall}} = \frac{\widehat{\mathbf{B}}^T \widehat{\sum}(\widehat{\mathbf{B}})^{-1}\,\widehat{\mathbf{B}}}{p} = F_{W,\text{overall}}, \tag{11.18}$$

$$\text{partial: modified } X^2_{W,\text{partial}} = \frac{\widehat{\mathbf{B}}_2^T \widehat{\sum}(\widehat{\mathbf{B}}_2)^{-1}\,\widehat{\mathbf{B}}_2}{p_2} = F_{W,\text{partial}}, \tag{11.19}$$

where $\widehat{\mathbf{B}}, \widehat{\mathbf{B}}_2$ are vectors of p and p_2 estimated regression parameters; and $\widehat{\sum}(\widehat{\mathbf{B}}), \widehat{\sum}(\widehat{\mathbf{B}}_2)$ are the estimated variance–covariance matrices.

Under the null hypothesis $H_0 : B = 0$, the overall modified Wald test statistic, $F_{W,\text{overall}}$, follows an F distribution with numerator degrees of freedom equal to p and denominator degrees of freedom equal to the design degrees of freedom (df). Likewise, to test $H_0 : B_2 = 0$, or the null hypothesis that the p_2 parameters are all equal to 0 in the nested model, the modified Wald partial test statistic is referred to the critical value of the F distribution with p_2 and df degrees of freedom.

Wald tests can also be used to test more general hypotheses regarding linear combinations of regression model parameters. Consider the null hypothesis $H_0 : \mathbf{CB} = \mathbf{0}$, where $\mathbf{C}$ is a matrix that defines specific linear combinations of the regression parameters in the vector $\mathbf{B}$. In this case, a version of the Wald test statistic that follows a chi-square distribution with degrees of freedom equal to the rank of the matrix $\mathbf{C}$ under the specified null hypothesis can be written as follows:

$$X^2_W = [\mathbf{C}\widehat{\mathbf{B}}]' \left[\widehat{\mathbf{C}\sum(\widehat{\mathbf{B}})\mathbf{C}'}\right]^{-1} [\mathbf{C}\widehat{\mathbf{B}}]. \tag{11.20}$$

These more general hypothesis tests are available in a variety of software packages that can fit regression models to complex sample survey data sets, including Stata and R. Dividing these more general chi-square test statistics by the number of linear contrasts being tested will result in test statistics that follow F distributions. We next consider examples of how to specify these tests for multiple parameters.

EXAMPLE ANALYSES

In this section, we present examples illustrating linear and logistic regression analyses of survey data collected from a sample with a complex design. Specifically, we consider data collected in the Russian Federation as a part of the fourth round (2008) of the European Social Survey. We use available software procedures in Stata and R to fit the models, perform design-adjusted hypothesis tests, and examine model diagnostics.

Overview of data sets and survey design variables

The ESS sample design for the Russian Federation is broadly described as a stratified four-stage probability cluster sample (see the ESS Round 4 Documentation Report on the Russian Federation, `http://ess.nsd.uib.no/ess`, for more details). Each record in the ESS Russian Federation data file includes a STRATIFY variable containing codes for 10 geographic zone strata, with a total of 134 sampled primary stage sampling units (PSUs) or settlements (cities, towns, villages) nested within the 10 stratification zones. Codes for these 134 clusters are contained in the PSU variable. Since the ESS Russian Federation sample design includes multiple (more than two) PSUs per design strata, either the TSL, JRR or bootstrap variance estimation methods can be used in the analysis of these complex sample survey data. The BRR method can only be applied when the design coding specifies exactly two PSUs per stratum. In our examples, the default TSL variance estimation method will be used.

Each data record also contains a survey weight value stored in DWEIGHT. The current ESS convention is that DWEIGHT variables only include the sample selection weight factor, W_{sel}, for each sample case and do not incorporate additional adjustments for either non-response or post-stratification to country population controls. ESS public use data files include a second population size weight (PWEIGHT). The population size weight is simply a linear scaling of the DWEIGHT values that is intended to be used when analysts wish to combine multiple ESS country samples in a single analysis. In such pooled analysis of multi-country data, PWEIGHT weights each country's influence in proportion to its adult population age 15+. Except when the statistics of interest are population totals, the results from weighted analysis of survey data are invariant to any such linear scaling of the analysis weight. Therefore, the example analyses conducted using the ESS DWEIGHT and PWEIGHT values yield identical results. The example syntax that is available on the book website for each example will use DWEIGHT.

The tabular information below provides a description of the STRATIFY and PSU variables through use of the `svyset` and `svydes` commands in Stata. For 'setting' the survey variables prior to use with Stata `svy` commands, a numeric version of the STRATIFY variable was created and named NSTRATIFY. This numeric variable is subsequently used in the `svyset` command as the strata variable, the PSU variable is identified as the cluster variable, and the DWEIGHT variable is specified as the `pweight` or probability weight. These Stata commands set the survey variables and final survey weight in a global manner for the upcoming analyses; that is, this only needs to be done once, and all subsequent `svy` commands used to fit the models will refer to these design variables.

The Stata `svydes` output below provides a summary of the characteristics of the survey design variables for the 2008 ESS Russian Federation sample of $n = 2512$ observations including the number of strata (10) and number of PSUs nested within the strata (134). Overall, there is

Stratum	#Units included	#Units omitted	#Obs with complete data	#Obs with missing data	#Obs per included Unit min	mean	max
1	11	0	240	0	9	21.8	87
2	22	0	551	0	12	25.0	171
3	8	0	141	0	8	17.6	29
4	8	0	135	0	14	16.9	20
5	17	0	287	0	9	16.9	22
6	19	0	327	0	11	17.2	24
7	20	0	345	0	9	17.3	26
8	14	0	253	0	12	18.1	38
9	8	0	123	0	10	15.4	18
10	7	0	110	0	5	15.7	22
10	134	0	2512	0	5	18.7	171
			2512				

an average of 18.7 observations per PSU and 13.4 PSUs per stratum. A similar process of defining the survey variables and weights to be used with the `survey` package in R is presented in the syntax examples on the book website.

Overview of analysis examples

The following examples illustrate linear and logistic regression analyses of complex sample survey data. The reader can find annotated Stata and R `survey` package command syntax used to produce these results on the book website.

For the first linear regression example, we consider three modeling approaches: a 'naive' approach using OLS, with the weights and complex sample design variables completely ignored; an approach that again ignores the weights but uses the two complex sample design variables for variance estimation; and an approach using the weights for WLS estimation of the model parameters and also using the complex sample design variables to estimate variances. The three linear regression models provide a comparison of how incorporating none, some, or all of the complex sample design/weighting features affects overall model performance and prediction of satisfaction with life (STFLIFE). The comparison of results from fitting the second and third models also permits the calculation of the inefficiency measure proposed by Korn and Graubard (1999, p. 175), described in the next section of this chapter.

The STFLIFE dependent variable is measured on a 0–10 scale with 0 = extremely dissatisfied and 10 = extremely satisfied and with codes of 77, 88 or 99 used for missing data. Although this variable is not strictly continuous, initial descriptive analyses reveal a symmetric distribution, and the variable serves as a suitable continuous type of dependent variable with an assumed latent continuous distribution. The predictor variables included in each linear regression model are marital status, age, highest education level, self-rated health, and gender. Each predictor variable is treated as categorical except age, which represents continuous age in years.

The third model of the set uses both the complex sample design variables to estimate variances of the estimated parameters and the final survey weight to implement WLS estimation of the parameters, and correctly reflects all features of the complex sample design. It is therefore used for each additional analysis. With this model, we test scientifically interesting and viable interactions of gender with each other predictor, evaluate further inclusion of interactions in the final model using complex sample adjusted hypothesis tests, and either drop or include any significant interaction terms in the final model. Once model building is complete, we estimate the final model, based on the above model building process, and produce a simple set of linear regression model diagnostics, including a plot of residuals versus predicted values, a histogram of residuals and a normal quantile–quantile plot. All of the model building and evaluation analyses are performed with the `svy: regress` command in Stata and the `svyglm` function in R, along with the post-estimation commands `test` (in Stata) and `regTermTest` (in R).

The logistic regression example uses a constructed binary outcome variable named HAMPERED, predicted by categorical predictors measuring age, marital status, gender, and education. The outcome variable is based on the raw variable HLTHHMP and is coded 1 = hampered a lot or to some extent and 0 = not hampered, with codes of 7, 8, or 9 used for missing data. The HLTHHMP variable gathers information about how much physical illness/disability/mental illness/infirmity hampers the respondent's ability to perform daily activities. The logistic regression examples begin with a model including main effects and the interaction of gender with each other predictor in the model. We then use complex sample adjusted significance tests to assess whether any of the gender interactions are significantly different from zero in the overall model. After evaluation of the significance of the gender interactions and the resultant inclusion or exclusion of the interactions, a final logistic regression model is fitted using the `svy: logistic` command (in Stata) or the `svyglm` function with the `family=quasibinomial` option (in R). We also once again use the `test` (Stata) or `regTermTest` (R) commands, post-estimation, for design-adjusted Wald or F-tests of whether selected parameters are equal to zero in a given model of interest.

Linear regression example

Table 11.1 presents the results generated by fitting these three models in Stata and R (see the book website for the complete annotated command syntax). We see consistent evidence of individuals with lower health and males having lower life satisfaction, regardless of the modeling approach used. Estimates of the variances based on TSL (incorporating the complex sampling features) do have a tendency to increase relative to the variance estimates based on OLS, which would be expected (due to intra-cluster homogeneity in the features being measured). Finally, using the weights to estimate the parameters results in some changes in inference, especially for age (where we have evidence of a significant negative relationship based on WLS estimation), gender (where there is a larger gap between males and females), and education. Given that the changes in estimates of the standard errors are not substantial when using the weights, we prefer the weighted model, given that we have unbiased estimates of the parameters in the specified model.

We now consider the inclusion of interactions of gender with the other predictors in the final weighted model, and we use design-adjusted Wald tests to assess the importance of these interactions. Table 11.2 presents the results of these design-adjusted multi-parameter tests. Overall, we do not find significant evidence in favor of adding any of these interactions to the model from Table 11.1.

Finally, we consider some simple model diagnostics to assess the fit of the final model. As discussed in the previous section, model diagnostics for complex samples are an active area of current research, and more advanced procedures for examining model diagnostics will likely be available in Stata and R in the near future. In both Stata and R, we first refit the final model. In Stata, we then generate new variables (`resid1` and `yhat1`) containing the model-based residuals and predicted values, and examine plots of these saved variables. In R, a simple `plot()` function can be used in conjunction with the model fit object (`designwgt`), post-estimation, to obtain simple diagnostics plots.

Figure 11.1 shows the resulting plots, generated by Stata. We see little concern with an assumption of normality for the residuals, and the fitted-residual plot has a 'slanting' pattern

Table 11.1 Estimates of the regression parameters in selected models for satisfaction with life in the 2008 ESS, Russian Federation

	Without weight or complex sample design variables				Without weight and with complex sample design variables				With weight and complex sample design variables				
	Coeff	SE	t	$P > t$	Coeff	Lin. SE	t	$P>t$	Coeff	Lin. SE	t	$P > t$	DEFF
Marital status													
Currently married (ref)	–	–	–	–	–	–	–	–	–	–	–	–	
Previously married	−0.311	0.119	−2.620	0.009	−0.311	0.131	−2.380	0.019	−0.322	0.162	−1.980	0.049	1.599
Never married	−0.042	0.146	−0.290	0.776	−0.042	0.161	−0.260	0.796	−0.053	0.194	−0.270	0.784	1.723
	F(2,2436) = 3.45, p = 0.0320				*F(2,123)=2.83, p=0.0627*				*F(2,123)=1.98, p=0.1423*				
Age in years	−0.006	0.004	−1.520	0.129	−0.006	0.005	−1.280	0.204	−0.011	0.005	−2.050	0.042	1.679
	F(1,2436)=2.31, p=0.1286				*F(1,124)=1.63, p=0.2039*				*F(1,124)=4.22, p=0.0421*				
Gender													
Male (ref)	–	–	–	–	–	–	–	–	–	–	–	–	
Female	0.356	0.100	3.540	<0.001	0.356	0.099	3.590	<0.001	0.447	0.105	4.260	< 0.001	1.135
	F(1,2436)=12.53, p=0.0004				*F(1,124)=12.90, p=0.0005*				*F(1,124)=18.13, p < 0.0001*				
Education													
Less than secondary	−0.257	0.158	−1.620	0.105	−0.257	0.185	−1.390	0.167	−0.052	0.209	−0.250	0.803	1.580
Secondary education	−0.504	0.132	−3.810	<0.001	−0.504	0.141	−3.570	0.001	−0.454	0.156	−2.900	0.004	1.472
Some college or technical training	−0.249	0.127	−1.960	0.050	−0.249	0.114	−2.180	0.031	−0.257	0.135	−1.910	0.059	1.157
Bachelor degree or higher (ref)	–	–	–	–	–	–	–	–	–	–	–	–	
	F(3,2436)=4.84, p=0.0023				*F(3,122)=4.42, p=0.0055*				*F(3,122)=3.20, p=0.0260*				
Self-rated health													
Very good (ref)	–	–	–	–	–	–	–	–	–	–	–	–	
Good	−1.425	0.262	−5.440	< 0.001	−1.425	0.321	−4.440	< 0.001	−1.484	0.332	−4.470	< 0.001	2.235
Fair	−2.312	0.264	−8.770	< 0.001	−2.312	0.333	−6.940	< 0.001	−2.329	0.342	−6.810	< 0.001	2.230
Bad	−2.872	0.292	−9.820	< 0.001	−2.872	0.361	−7.950	< 0.001	−2.811	0.386	−7.290	< 0.001	2.010
Very bad	−3.875	0.384	−10.090	< 0.001	−3.875	0.479	−8.080	< 0.001	−3.844	0.525	−7.330	< 0.001	1.462
	F(4,2436)=39.07, p < 0.0001				*F(4,121)=25.30, p < 0.0001*				*F(4,121)=20.03, p < 0.0001*				
Intercept	7.89	0.30	26.10	< 0.001	7.89	0.357	22.10	<0.001	8.05	0.367	21.96	< 0.001	1.74
	Notes: $n = 2448$, $R^2 = 0.115$, $F(11,2436) = 29.83$, $p < 0.0001$				Notes: $n = 2448$, $R^2 = 0.119$, $F(11,114) = 27.32$, $p < 0.0001$				Notes: n=2448, $R^2 = 0.121$, $F(11,114) = 25.3$, $p < 0.0001$				

– indicates reference category.
Source: Analysis based on ESS Round 4 Russian Federation data.

Table 11.2 Design-adjusted *F*-tests of interactions of gender and other predictors in final weighted model

Interaction terms tested	*F* statistic	*p*-value
Marital status–Gender	$F(2,123)=0.41$	0.660
Age–Gender	$F(1,124)=3.29$	0.070
Education–Gender	$F(3,122)=2.20$	0.092
Health status–Gender	$F(4,121)=1.72$	0.149

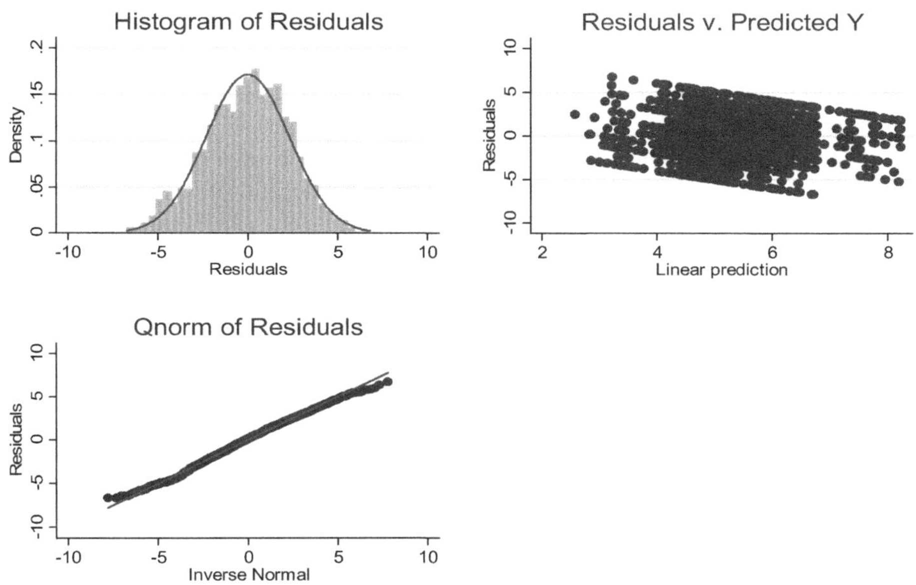

Figure 11.1 Linear regression diagnostic plots

due to the ordinal dependent variable being bounded by 1 and 10. Nevertheless, we do not see any evidence of significant outliers or problems with non-constant residual variance.

Logistic regression example

In this example, we use pseudo-maximum likelihood estimation in Stata (`svy: logistic`) and R (`svyglm` with `family = quasibinomial`) to compute weighted estimates of the parameters in the aforementioned logistic regression model for the probability of being hampered in terms of daily activities, with variance estimates for the estimated parameters computed using TSL to reflect the stratification and clustering. We first fit a model including all of the two-way interactions of gender with each other predictor (marital status, age, and education), and tested the significance of these interactions using `test` and `regTermTest`, as illustrated above. None of these interactions were significant, so we refit the model excluding the interactions, used the `estat gof` post-estimation command in Stata to perform the design-adjusted Archer–Lemeshow goodness-of-fit test for this reduced model, and then performed multi-parameter Wald tests for all of the parameters defined by each categorical predictor.

As a comparison, we fitted the final main effects logistic model in Stata (`logistic`) and R (`glm` with `family = quasibinomial`) ignoring the ESS weights or complex sample design variables to compute unweighted estimates of the parameters without design-based variance estimation. In addition, we used the post-estimation command `estat gof` to perform a Hosmer–Lemeshow goodness-of-fit test for the reduced model and then performed multi-parameter chi-square tests for each set of categorical predictors.

Table 11.3 presents the estimated parameters in this final model, first for the naive logistic regression analysis performed without weights or complex sample design variables, and second using the complex sample design corrections described above along with the design-adjusted Wald tests for the categorical predictors and the result of the Archer–Lemeshow goodness-of-fit test.

Both goodness-of-fit tests indicate that the estimated models produce predicted probabilities that tend to be consistent with the observed data, suggesting a good fit (recall that the null hypothesis for this test is that the model has a close fit to the observed data; we fail to reject this null hypothesis). Examining the complex sample design-adjusted parameter estimates in the absence of the two-way interactions, we see that higher education consistently reduces the odds of being hampered, with those having a bachelor's degree or higher having 52% lower odds of being hampered relative to those with less than secondary education (controlling for the other predictors). People who were previously married have significantly higher odds of being hampered (relative to those who are currently married), possibly due to a lack of support from a spouse. Females also have significantly higher odds of being hampered, as do older individuals. These overall conclusions do not change when considering estimates based on the naive model that ignores the survey weights and the complex sample design. As expected, the standard errors from the naive model are smaller compared to the corrected, design-adjusted TSL standard errors. This set of socio-demographics therefore has relationships with the probability of being hampered in the Russian population that one might expect in theory.

CAVEATS AND FREQUENT ERRORS

Given modern advances in the theory, methods and software introduced above, regression analysis of complex sample survey data is a topic that could fill a complete volume. It is hard to do it justice in a single chapter. The purpose of this section is to tie up a few loose ends – important caveats and also information that bears on several important issues in contemporary regression analysis of survey data.

Subpopulation analysis

Researchers are often interested in fitting a regression model to only selected cases from the full population sample. For example, a logistic regression model for a diagnosis of prostate cancer would be fitted only to the male subpopulation. In complex sample surveys, the distribution of the subpopulation sample size is a random variable – varying both in its size and its distribution across the design strata and clusters. The correct approach to analyzing subpopulations of complex samples is to perform an *unconditional subpopulation analysis*, in which inferences do not condition on the observed distribution of subpopulation cases to particular strata and clusters of the full sample design. That is, conducting a *conditional subpopulation analysis* by simply using the data management capabilities of statistical software to delete or filter out those cases that do not fall into the subpopulation of interest (e.g. females) can produce biased inferences. Point estimates of population parameters will be identical under both subclass analysis

Table 11.3 Logistic regression models predicting the probability of being hampered

	Without weights or complex sample design variables						With weights and complex sample design variables					
	OR	SE	a	*P* > *z*	95% CI		OR	Lin. SE	*t*	*P* > *t*	95% CI	
Age groups												
15–29 (ref)	–	–	–	–	–	–	–	–	–	–	–	–
30–44	2.136	0.417	3.89	<0.001	1.456	3.132	2.587	0.660	3.730	<0.001	1.561	4.285
45–59	4.741	0.916	8.05	<0.001	3.246	6.923	5.927	1.455	7.250	<0.001	3.647	9.635
60+	15.461	3.108	13.62	<0.001	10.426	22.927	17.798	4.740	10.810	<0.001	10.507	30.150
	ChiSq(3)=261.85, Prob > ChiSq = < 0.0001								*F(3,122)=49.45, p < .0001*			
Gender												
Male (ref)	–	–	–	–	–	–	–	–	–	–	–	–
Female	1.344	0.141	2.82	0.005	1.095	1.651	1.359	0.180	2.320	0.022	1.045	1.766
	ChiSq(1)=7.97, Prob > ChiSq = 0.005								*F(1,124)=5.36, p = 0.0223*			
Marital status												
Currently married (ref)	–	–	–	–	–	–	–	–	–	–	–	–
Previously married	1.616	0.183	4.23	<0.001	1.294	2.019	1.469	0.177	3.190	0.002	1.157	1.864
Never married	1.353	0.241	1.69	0.09	0.954	1.919	1.482	0.400	1.460	0.148	0.868	2.529
	ChiSq(2)=18.65 Prob > ChiSq = < 0.0001								*F(2,123)=6.31 p = 0.0025*			
Education												
< Secondary (ref)	–	–	–	–	–	–	–	–	–	–	–	–
Secondary	0.602	0.095	−3.21	0.001	0.442	0.821	0.570	0.100	−3.200	0.002	0.403	0.807
Some college or technical	0.508	0.079	−4.38	<0.001	0.375	0.688	0.535	0.103	−3.250	0.002	0.365	0.783
Bachelor degree or higher	0.522	0.084	−4.02	<0.001	0.380	0.716	0.482	0.098	−3.600	<0.001	0.322	0.719
	ChiSq(3)=21.63, Prob > ChiSq = < 0.0001								*F(3,122)=4.92, p = 0.0029*			
Intercept	0.153	0.035	−8.26	<0.001	0.098	0.239	0.128	0.038	−6.950	<0.001	0.072	0.231
Goodness-of-fit test	*Pearson ChiSq(84)=90.16, Prob > ChiSq = 0.3032*								*F(9,116)=0.69, p = 0.7207*			

– indicates reference category.

approaches. However, because a conditional analysis does not incorporate the variance in the subpopulation sample size, it tends to result in *underestimates* of standard errors, which lead analysts to overstate the precision of estimated survey statistics for subclasses. This requires some additional preparation on the part of the data analyst.

Software systems differ in their syntax for specifying a correct unconditional subpopulation analysis. In Stata, the analyst first creates an indicator of subpopulation membership:

$$I_{S,i} = \begin{cases} 1 & \text{if case } i \text{ is a member of the subpopulation of interest,} \\ 0 & \text{if case } i \text{ is not a member.} \end{cases} \tag{11.21}$$

The analyst then includes the `subpop(indicator)` option in a command statement (e.g. `svy, subpop(indicator): regress`) to invoke the correct unconditional subpopulation analysis. R users can use the `svyby()` function after the indicator variable has been generated to implement the correct analyses. Readers are encouraged to see West et al. (2008) or Heeringa et al. (2010, Chapter 4) for more information on this topic.

Should weights be used when fitting regression models to complex sample survey data?

The use of final survey weights to estimate descriptive parameters (e.g. means and proportions) in finite populations is generally accepted by survey statisticians without controversy (see Korn and Graubard, 1999, Chapter 4). However, the best approaches for analysts interested in regression modeling are subject to a great deal of debate and controversy (see Gelman, 2007, for a summary of the general issues).

Korn and Graubard (1999, Chapter 4) emphasize the importance of model specification in answering this question. Using the simple case of estimating a linear regression model that specifies a linear relationship between two variables that clearly have a quadratic (non-linear) relationship, these authors show that weighted and unweighted estimates will be quite different when a model has not been correctly specified. In this case, a weighted estimate of a regression parameter (based on WLS or PMLE methods) at least has the advantage of estimating a population parameter, but this is a nearly unbiased (and design-consistent) estimate of a population parameter *for a poorly specified population model.* Of course, analysts will typically have no idea whether or not they have specified the 'correct' model for a given population, but empirical evidence suggests that differences between weighted and unweighted estimates will be much smaller when models are 'correctly' (or nearly correctly) specified. Subject matter knowledge is therefore extremely important when specifying models for finite populations.

What happens if a model has actually been correctly specified and one still uses the weights for WLS or PML estimation? In this situation, weights can make estimates of the regression parameters less efficient (i.e. increase their variance unnecessarily). Korn and Graubard (1999, p. 175) recommend a simple inefficiency calculation to gauge the loss in efficiency from using weights to estimate regression parameters, defined as 1 minus the ratio of the estimated variance of an unweighted estimate to the estimated variance of a weighted estimate (where both variance estimates use linearization or replication methods to recognize complex sampling features such as stratification and clustering). Computing this inefficiency measure for our example linear regression analyses above in the previous section, the largest loss of efficiency due to the use of WLS estimation is for the parameter representing the mean difference between those who are previously married and those currently married. When ignoring the weights and using TSL, the estimated variance of this parameter estimate is $0.131^2 = 0.017$, and when using WLS estimation along with TSL (now using weighted sample totals), the estimated variance of this parameter estimate is $0.162^2 = 0.026$. The resulting inefficiency calculation is $1 - 0.017/0.026 = 0.346$, which suggests that ignoring the weights would decrease the variance of this parameter estimate by nearly 35% (standard errors increasing by roughly 12%). Given the small change in this parameter estimate when using the weights, we may prefer the unweighted analysis. However, we also have to take into account what happens to all of the other parameter estimates in the model when using the weights, and there are much larger changes for other predictors. That said, we would likely be OK with this loss of efficiency.

Another key issue to consider in this debate is whether or not the sampling is *informative* (i.e. probabilities of selection are related to the variables in the population model). For example, if cases with larger values on the dependent variable y have a higher probability of being selected, ignoring the weights in estimation will lead to biased estimates of regression parameters *even if the population model is correctly specified*. The documentation for public-use data sets will often provide detailed descriptions of the sampling that was performed and whether or not cases with particular values on proxies of key variables in regression models were over-sampled. If this is the case, the use of weights in estimation of regression models is generally recommended. In fact, unless losses in efficiency are excessive (see Korn and Graubard, 1999, Section 4.6), the use of weights in estimation is generally not problematic, and results in nearly unbiased design-based variance estimates (given large population sizes) when making finite-population inferences, even when population models are misspecified (see Binder, 2011, or Preffermann, 2011, for more on this issue).

So how can an analyst determine whether weighted and unweighted estimates are significantly different? Fuller (1984) describes a method that can be used with standard survey software to test whether unweighted estimates of linear regression parameters are significantly biased. We have prepared a detailed example implementing this methodology using standard statistical software on the website for *Applied Survey Data Analysis* (see the Supplementary Materials section for the document 'Example of Fuller (1984) Method.pdf'). This method only applies to linear regression models estimated using OLS and WLS; to our knowledge, related methods for GLMs have not yet been developed. Comparisons of weighted and unweighted estimates in GLMs are therefore left to the discretion of the researcher, and should be performed taking differences in variance estimates (e.g. Korn and Graubard's inefficiency measure) into account.

Accounting for weighting and complex sample designs in multilevel models

An entirely different 'model-based' approach can be taken to fitting regression models to complex samples: one that ignores the notion of a finite population and assumes that the survey data arise from an infinite data generation process governed by a probability model, where estimation of the parameters that define that model is the focus of the analysis. These so-called 'model-based' approaches to fitting regression models to complex sample survey data (see Kott, 1991) have generally come to rely on multilevel (or hierarchical linear) models (see Chapter 7 of this volume). The complex sampling features essentially become predictors in these models, entering as either fixed effects (for strata that are fixed by design across hypothetical repeated samples) or random effects (for randomly sampled clusters) and help to define the assumed probability distribution for the dependent variable of interest. In addition, depending on how informative the sampling is, the analyst needs to decide how to handle the survey weights: use the weights to estimate the parameters of the probability model, or include the weights as covariates to 'control' for the effects of features used to define the weights (see Korn and Graubard, 1999, Section 4.5). This decision is not clearly guided by any theoretical results, and is currently a source of controversy among statisticians (Gelman, 2007).

The theory and methods for incorporating sampling weights into the estimation of the fixed effect and covariance parameters defining a multilevel model were initially described by Pfefferman et al. (1998), and later expanded on by Asparouhov (2006), Rabe-Hesketh and Skrondal (2006) and Carle (2009). When fitting multilevel models to cross-sectional survey data, survey respondents generally constitute level 1 of the data hierarchy (where the dependent variable is also measured), and the randomly sampled first-stage PSUs (or ultimate clusters) define level 2 of the data hierarchy (as respondents are nested within the clusters). A key feature of weighted estimation approaches is the need for: *conditional* weights at level 1 of the data hierarchy, which indicate inverses of the probability of selection *conditional* on a given PSU being sampled; and PSU-level weights at level 2, representing inverses of the probabilities of selection for the PSUs. In addition, the level 1 weights that are specific to each cluster need to be *scaled* or *normalized* across all PSUs, to reduce the varying magnitudes of these weights across PSUs; Rabe-Hesketh and Skrondal (2006) describe alternative methods for performing weight scaling. The final survey weights provided in a survey data set typically represent inverses of the products of the probabilities of selection at *all* stages of a complex sample design; computation of the conditional weights at level 1 requires dividing the final weights by the PSU-level weights at level 2 to determine the inverse of the conditional level 1 probabilities required for estimation. Both of these weights are required for maximum likelihood estimation of the multilevel model parameters to ensure unbiased estimation of the fixed effect and covariance parameters.

At present, these approaches for weighted estimation of multilevel models (and the alternative methods for scaling the level 1 weights) are not widely implemented across statistical software packages; see West and Galecki (2011) for a review. The `mixed` procedure in Stata (Version 13+) presently allows analysts fitting multilevel models to continuous outcomes to incorporate weights at both levels, using `mixed ... [pw = wt1] || cluster: ..., pweight(wt2) pwscale(scale_method)`, where `wt1` is the conditional respondent-level weight, `wt2` is the PSU- (or cluster-) level weight, and `scale_method` is one of three possible methods for scaling the weights proposed in the literature (a general recommendation is to try all three approaches to see how sensitive inferences are to the weight scaling). Additional software implementing these approaches for both linear and generalized linear regression models includes HLM, MLwiN, Mplus, and the `gllamm` command in Stata (`www.gllamm.org`). Another practical issue, aside from software availability, is the reluctance of survey data producers to include the PSU-level sampling weights in public-use data files (to prevent identification of the sampled PSUs). In the absence of this information, analysts are forced to assume equal-probability sampling of PSUs, setting the second-level weights to 1; this has obvious implications for inference, but alternatives are not presently available.

An alternative model-based approach involves simply including some functional form of the final survey weight as a predictor in a multilevel model (Korn and Graubard, 1999, Section 4.5). Conceptually, including fixed effects of the final survey weights in this manner will ‘explain’ any variance in a given outcome variable that is arising due to informative sampling, allowing analysts to identify other predictors that have significant relationships with a given outcome when holding the weights fixed. Proponents of these model-based approaches argue that this is the best way to account for any information present in the sample design features about the dependent variables of interest; the complex sampling features essentially become an important part of the specification of a ‘correct’ probability model for the observed data. In addition, this approach is simple to implement in existing software and free of many of the aforementioned issues surrounding the use of survey weights in the estimation of multilevel models.

Finally, we do not discuss multilevel models for complex sample *panel* survey data in this chapter. See Heeringa et al. (2010, Chapter 12) for more background on this topic.

FURTHER READING

The presentation in this chapter has only touched the high points in the range of methods and issues that apply in regression analysis of complex sample survey data. A more extensive, yet intermediate-level, coverage of these topics is provided in Chapters 7–12 of the authors' own text (Heeringa et al., 2010). Excellent volumes that provide an applied treatment of complex sample survey design and analysis are Korn and Graubard (1999) and Lohr (1999). Readers who are interested in a more rigorous, mathematical treatment of the analysis of complex sample survey data are referred to the many excellent papers cited in the References or to the classic edited volumes by Skinner et al. (1989) and Chambers and Skinner (2003).

NOTES

1 In R, after fitting a linear regression model using `svyglm()`, the `summary()` function (when applied to the model fit object) will indicate the residual dispersion parameter for the Gaussian family; this is the WSSE. The `svyvar()` function can then be applied to estimate the total variance of the dependent variable; this is the WSST. The resulting weighted R^2 statistic can then be computed.

2 This definition of the pseudo R^2 statistic can be attributed to McFadden (1974). Many competing alternatives have been proposed in the literature for simple random samples; see Estrella (1998).

3 Recent work by Valliant and Rust (2010) has called this simple approximation into question. Valliant and Rust propose an alternative estimator of the degrees of freedom that is shown to produce confidence intervals with better coverage in certain situations.

REFERENCES

Archer, K. J. and Lemeshow, S. (2006). Goodness-of-fit test for a logistic regression model estimated using survey sample data. *Stata Journal*, 6(1), 97–105.

Archer, K. J., Lemeshow, S. and Hosmer, D. W. (2007). Goodness-of-fit tests for logistic regression models when data are collected using a complex sample design. *Computational Statistics and Data Analysis*, 51, 4450–4464.

Asparouhov, T. (2006). General multilevel modeling with sampling weights. *Communications in Statistics: Theory and Methods*, 35, 439–460.

Binder, D. A. (1981). On the variances of asympototically normal estimators from complex surveys. *Survey Methodology*, 7, 157–170.

Binder, D. A. (1983). On the variances of asymptotically normal estimators from complex surveys. *International Statistical Review*, 51, 279–292.

Binder, D. A. (2011). Estimating model parameters from a complex survey under a model-design randomization framework. *Pakistan Journal of Statistics, Festschrift for Ken Brewer*, 27, 371–390.

Carle, A. C. (2009). Fitting multilevel models in complex survey data with design weights: Recommendations. *BMC Medical Research Methodology*, 9: 49.

Chambers, R. L. and Skinner, C. J. (eds) (2003). *Analysis of Survey Data*. Chichester: John Wiley & Sons.

Cochran, W. G. (1977). *Sampling Techniques*. New York: John Wiley & Sons.

Deming, W. E. (1950). *Some Theory of Sampling*. New York: John Wiley & Sons.

DuMouchel, W. H. and Duncan, G. S. (1983). Using sample survey weights in multiple regression analyses of stratified samples. *Journal of the American Statistical Association*, 78, 535–543.

Estrella, A. (1998). A new measure of fit for equations with dichotomous dependent variables. *Journal of Business and Economic Statistics*, 16(2), 198–205.

Fuller, W. (1984). Least squares and related analyses for complex survey designs. *Survey Methodology*, 10, 97–118.

Fuller, W. A. (2002). Regression estimation for survey samples (with discussion). *Survey Methodology*, 28(1), 5–23.

Gelman, A. (2007). Struggles with survey weighting and regression modeling. *Statistical Science*, 22(2), 153–164.

Grizzle, J., Starmer, F. and Koch, G. (1969). Analysis of categorical data by linear models. *Biometrics*, 25, 489–504.

Hansen, M. H., Hurwitz, W. N. and Madow, W. G. (1953). *Sample Survey Methods and Theory*, volumes I and II. New York: John Wiley & Sons.

Harrell, F. E. (2001). *Regression Modelling Strategies*. New York: Springer.

Heeringa, S. G., West, B. T. and Berglund, P. A. (2010). *Applied Survey Data Analysis*. Boca Raton, FL: Chapman & Hall/CRC.

Hosmer, D. W. and Lemeshow, S. (2000). *Applied Logistic Regression*. New York: John Wiley & Sons.

Judge, G. G., Griffiths, W. E., Hill, R. C. and Lee, T.-C. (1985). *The Theory and Practice of Econometrics*. New York: John Wiley & Sons.

Kendall, P. L. and Lazarsfeld, P. F. (1950). Problems of survey analysis. In R. K. Merton and P. F. Lazarsfeld (eds), *Continuities in Social Research: Studies in the Scope and Method of 'The American Soldier'*. Chicago: Free Press.

Kish, L. (1965). *Survey Sampling*. New York: John Wiley & Sons.

Kish, L. and Frankel, M. R. (1974). Inference from complex samples. *Journal of the Royal Statistical Society, Series B*, 36, 1–37.

Klein, L. R. and Morgan, J. N. (1951). Results of alternative statistical treatment of sample survey data. *Journal of the American Statistical Association*, 46, 442–460.

Kolenikov, S. (2010). Resampling variance estimation for complex survey data. *Stata Journal*, 10(2), 165–199.

Korn, E. L. and Graubard, B. I. (1999). *Analysis of Health Surveys*. New York: John Wiley & Sons.

Kott, P. S. (1991). A model-based look at linear regression with survey data. *American Statistician*, 45, 107–112.

Landis, R. J., Stanish, W. M., Freeman, J. L. and Koch, G. G. (1976). A computer program for the generalized chi-square analysis of categorical data using weighted least squares (gencat). *Computer Programs in Biomedicine*, 6, 196–231.

Li, J. and Valliant, R. (2009). Survey weighted hat matrix and leverages. *Survey Methodology*, 35, 15–24.

Li, J. and Valliant, R. (2011a). Detecting groups of influential observations in linear regression using survey data: Adapting the forward search method. *Pakistan Journal of Statistics, Festschrift for Ken Brewer*, 27, 507–528.

Li, J. and Valliant, R. (2011b). Linear regression diagnostics for unclustered survey data. *Journal of Official Statistics*, 27, 99–119.

Liao, D. and Valliant, R. (2012a). Condition indexes and variance decompositions for diagnosing collinearity in linear model analysis of survey data. *Survey Methodology*, 38, 189–202.

Liao, D. and Valliant, R. (2012b). Variance inflation factors in the analysis of complex survey data. *Survey Methodology*, 38, 53–62.

Lohr, S. L. (1999). *Sampling: Design and Analysis*. Pacific Grove, CA: Duxbury Press.

Lohr, S.L. (2014). Design effects for a regression slope in a cluster sample. *Journal of Survey Statistics and Methodology*, 2, 97–125.

McFadden, D. (1974). Conditional logit analysis of qualitative choice behavior. In P. Zarembka (Ed.), *Frontiers in Econometrics* (pp. 105–142). New York: Academic Press.

Morel, G. (1989). Logistic regression under complex survey designs. *Survey Methodology*, 15, 202–223.

Neyman, J. (1934). On the two different aspects of the representative method: The method of stratified sampling and the method of purposive selection. *Journal of the Royal Statistical Society*, 97, 558–606.

Pfefferman, D., Skinner, C. J., Holmes, D. J., Goldstein, H. and Rasbash, J. (1998). Weighting for unequal selection probabilities in multilevel models. *Journal of the Royal Statistical Society, Series B*, 60(1), 23–40.

Pfeffermann, D. (2011). Modelling of complex survey data: Why model? Why is it a problem? How can we approach it? *Survey Methodology*, 37(2), 115–136.

Rabe-Hesketh, S. and Skrondal, A. (2006). Multilevel modelling of complex survey data. *Journal of the Royal Statistical Society, Series A*, 169, 805–827.

Roberts, G., Rao, J. N. K. and Kumar, S. (1987). Logistic regression analysis of sample survey data. *Biometrika*, 74, 1–12.

Skinner, C. J., Holt, D. and Smith, T. M. F. (1989). *Analysis of Complex Surveys*. New York: John Wiley & Sons.

Sukatme, P. V. (1954). *Sampling Theory of Surveys, With Applications*. Ames: Iowa State College Press.

Valliant, R. and Rust, K. F. (2010). Degrees of freedom approximations and rules-of-thumb. *Journal of Official Statistics*, 26(4), 585–602.

West, B. T., Berglund, P. A. and Heeringa, S. G. (2008). A closer examination of subpopulation analysis of complex-sample survey data. *Stata Journal*, 8(4), 520–531.

West, B. T. and Galecki, A. T. (2011). An overview of current software procedures for fitting linear mixed models. *American Statistician*, 65(4), 274–282.

Wolter, K. M. (2007). *Introduction to Variance Estimation*. New York: Springer Verlag.

Yates, F. (1960). *Sampling Methods for Censuses and Surveys*. London: Griffin.

PART III

Causal Inference and Analysis of Longitudinal Data

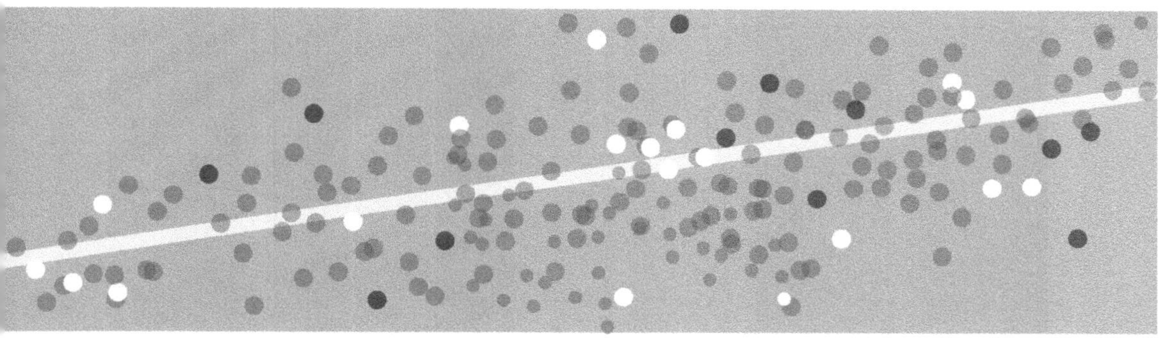

12

Matching estimators for treatment effects

Markus Gangl

INTRODUCTION

Matching estimators have gained in popularity as flexible tools for estimating treatment effects in observational studies as insights from biometrics, epidemiology and statistics have increasingly spread into the social sciences. Fundamental to all matching estimators is the construction of a control group that is as similar as possible to the treatment group of interest with respect to observed covariates. If observed covariates are sufficient to eliminate the impact of potential confounders of treatment, matching estimators consistently identify and empirically estimate the causal effect of treatment on outcomes. Compared to regression analysis, matching estimators rest on minimal mathematical foundations that are easily accessible to the applied researcher and that result in readily interpretable parameter estimates. Practical implementation of matching estimators is aided by the increasing availability of canned routines in standard statistical software packages.

As estimators of treatment effects under maintained exogeneity, matching estimators share many features with conventional regression methods. However, matching methods differ from regression in so far as they avoid the specification of a fully parametric model for outcomes, but estimate treatment effects non-parametrically from the comparison of outcome distributions across matched samples. Accordingly, instead of focusing on many or all potential determinants of outcomes, it is the precise definition of treatment counterfactuals and the specification of the assignment model predicting treatment status that assume center stage with matching estimators. Ensuing concerns about the theoretical validity of the assignment model and the construction of appropriately matched samples directly relate to core principles of research design for supporting credible causal inference with observational data (for reviews, see Morgan and Winship, 2007; Gangl, 2010).

In fact, matching estimators may be seen as a natural implementation of the *effects-of-causes* approach to causal analysis (Holland, 1986), where the sole focus of the analysis is on the convincing isolation of a specific and well-defined causal effect of interest. Regression modeling may evidently be utilized for the same purpose, yet matching estimators have a conceptual clarity about them that is bound to assist applied researchers in appreciating key issues in causal inference, as well as in communicating empirical results to academic and non-academic audiences. At the same time, straightforwardness of application and interpretation should not delude social scientists into conceiving of matching estimators as yet another hoped-for panacea for causal inference. As is discussed in more detail below, matching estimators do indeed form a versatile class of non- and semi-parametric techniques for comparing outcome distributions across comparison groups comprised of observationally similar units. However, the validity of causal inferences derived from any matching estimator critically hinges on the validity of the underlying assignment model. Absent randomized experimentation, assessing the latter inevitably requires subject-matter knowledge and hence to some extent transcends the strictly statistical considerations at the heart of the present chapter.

Fundamental assumptions

Although matching estimators come with fewer statistical assumptions than standard regression models, they remain bound to the inferential challenges associated with the identification of causal effects from observational data. Fundamentally, causal statements imply statements about counterfactual states of the world that would materialize if some condition D were to be changed. It is logically impossible, however, to directly observe the causal effect of D on outcomes Y in empirical research since any particular unit of observation i may only be observed in one particular treatment condition $D_{it} = d$ at any single point in time t. It is precisely the attempt to tackle this *fundamental problem of causal inference* (Holland, 1986) that distinguishes descriptive from causal inference, as well as pure statistics from subject-matter empirical analysis. Intuitively, and abstracting from important subtleties and qualifications discussed elsewhere (e.g. Morgan and Winship, 2007; Gangl, 2010), the necessary condition for any successful identification of a causal effect of interest is to be able to conduct a comparison of outcomes Y across *behaviorally* equivalent groups of observations in an expected outcome sense that differ in terms of (degree of) actual exposure to the treatment condition D of interest. This condition is met in successfully randomized experiments, where exposure to treatment D is both actively manipulated by the researcher and distributed randomly, that is, independently of any potential confounder Z, in the sample. Causal inference in observational studies is significantly more involved since treatment exposure is observed ex post instead of being actively manipulated, and since covariate controls are an inherently imperfect substitute for randomization. To sustain a causal interpretation of some estimate in an observational study, researchers need to be willing to maintain that observable covariates permit sufficiently extensive control for potential confounders Z that are antecedent correlates of both treatment status D and outcomes Y. Available covariate data, in other words, need to be sufficiently rich to capture real-world allocation to treatment conditions D to such an extent that residual variation in treatment status may plausibly be considered (as if) exogenously assigned conditional on the vector of observable covariates Z.

This identifying assumption of (conditional) exogeneity of treatment assignment (also known as selection on observables or conditional independence of treatment and outcomes, and referred to henceforth as the conditional independence assumption, CIA) is not germane to matching estimators, but is similarly invoked in causal interpretations of standard regression parameters. Maintaining the CIA in any observational study is equivalent to the theoretical statement that Figure 12.1 accurately describes the structure of observations in the study at hand. If and only if Figure 12.1 holds, observable covariates Z represent a sufficiently rich vector of (temporally or logically) antecedent correlates or causes of treatment status D. Then, conditional on Z, error terms u and e are uncorrelated or, equivalently, expected outcomes Y are equal across the comparison groups in the analysis given the absence of actual treatment D. If so, consideration

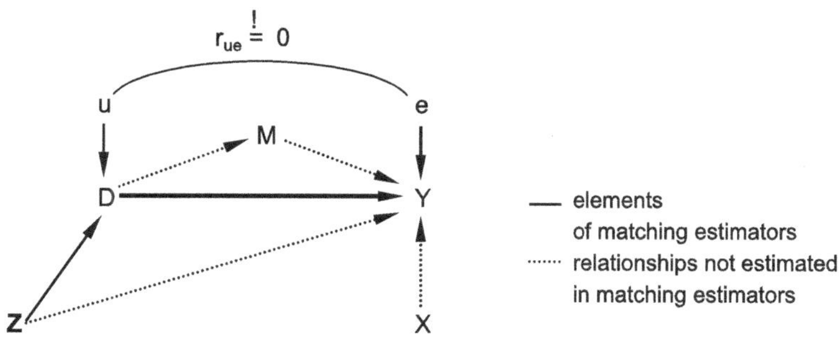

Figure 12.1 Identification of causal effects under conditional exogeneity of treatment

of additional predictors X of outcomes Y that are (conditionally) unrelated to D does not aid the identification of the causal effect of interest. In addition, mediating factors such as M should remain outside the estimated model since the causal effect of interest corresponds to the total effect of D on Y, whereas the direct effect of D on Y in a model including M would only correspond to the residual treatment effect not mediated (i.e. unexplained) through M. Under these quite restrictive circumstances, the causal effect of D on Y may be estimated from observational data.

Treatment effects as estimands of interest

As estimators of treatment effects under exogeneity, these fundamental considerations apply to matching and regression analysis alike. Compared to regression, however, matching estimators render identification issues exceptionally transparent by virtue of the way the empirical analysis is conducted. As one aspect, matching estimators are appealing due to their unique focus on parameter estimates that are directly related to key counterfactual quantities of interest, and which carry correspondingly straightforward interpretations in consequence.

Fundamentally, a causal or treatment effect describes the change in outcomes Y given a change in exposure to treatment condition D. Treatment effects can be thought of in principle as applying at the level of individual units of observation; in practice, however, social scientists will always be estimating average treatment effects for (subgroups of observations in) the sample data at hand. Furthermore, modern theory conceives of treatment effects as potentially heterogeneous in the population, which provides another rationale for focusing on average treatment effects in the empirical analysis. Two particularly important parameters are the *average treatment effect* (ATE) in the sample (or the target population) and the *average treatment effect on the treated* (ATT), that is, the average impact in the sample of units actually exposed to treatment (dose) $D = d$. Though more circumscribed than ATE, the ATT parameter might be of considerable substantive interest in many applications (e.g. in program evaluation or inequality decomposition), and is also empirically identified under slightly less restrictive conditions than those depicted in Figure 12.1 (see the section on mathematical foundations below). Going beyond estimates of average effects, it may also be of interest to examine the *distribution* of treatment effects in the population via *quantile treatment effects* (QTE and QTT) defined, for example, at the median or the lower and upper quartiles of the distribution of treatment effects. And it may be of interest to examine treatment effects separately by population subgroup, which is equivalent to estimating *conditional average treatment effects* (CATE and CATT) or respective quantile parameters.

While these parameters describe key quantities of interest that are independent of the specific estimation method, the choice of matching estimators does have an immediate consequence for what is actually being estimated in the concrete analysis. Specifically, the non-parametric character of matching estimators implies that any causal parameter is only estimable across the *common support* in the sample data, that is, within the range of (the joint distribution of) covariate data over which there is overlap across the comparison groups of the analysis. In the absence of an explicit parametric model for outcomes, non-parametric estimators are unable to extrapolate counterfactual outcomes into those areas of the covariate space that lack observations from one comparison group. In consequence, matching estimators require sample data that actually provides observations on (sufficiently) similar members from both (or at least two) comparison groups in order to produce any estimate of empirical treatment effects. A comparison of matching and regression estimates thus often provides a useful sensitivity analysis on the extent to which causal inference may be considered primarily data-driven or critically reliant on assumptions about the functional form of the regression model.

Typical steps in using matching estimators for causal inference

Matching estimators for treatment effects comprise three prototypical stages of analysis. The first stage concerns the determination of relevant controls that are considered antecedent causes or correlates of treatment status, and hence confound the observed relationship between treatment and outcomes unless properly adjusted for. With propensity score matching, this includes estimation of a separate *assignment model* that predicts treatment status D from antecedent covariates Z, whereas alternative exact matching estimators operate at the level of covariate data directly. Based on the assignment model, the second stage of the analysis consists of utilizing an appropriate matching algorithm to balance the distribution of covariates across comparison groups. Finally, given the CIA and sufficient homogeneity of treatment and control group observations, the causal effect of D on Y is estimated non-parametrically as the simple weighted difference in outcome distributions across the matched samples.

These three stages have evident links with the main elements of the counterfactual model of causal inference, and much of the appeal of matching estimators stems from the fact that their very setup makes respective concerns unusually transparent. Matching estimators practically force the analyst to be explicit about key aspects of research design, which permits easier communication of empirical results but also provides the ground for scientific scrutiny of and rigor about causal inference in observational studies. In fact, the benefits of matching have long been evident to social scientists, and informal descriptions of matching estimators feature prominently in the research design sections of many introductory methods textbooks. Practical application of matching estimators has long been hampered, however, due to the sparse-data problem associated with forming exact matches across multidimensional covariate spaces. The large number of covariates in typical social science applications, combined with the typical mix of qualitative and quantitative measurements, quite simply requires very large data sets to render the construction of control groups via (even reasonably) exact matching an empirically feasible estimation strategy. In a foundational paper, Rosenbaum and Rubin (1983) were able to decisively simplify the construction of comparison groups in multivariate matching estimators by showing that consistent estimation of the treatment effect of D on Y is ensured by balancing a suitable linear combination of antecedent covariates Z across comparison groups. The distance measure known as the *propensity score* has been the cornerstone of applied matching ever since as it reduces the task of control group construction from a multidimensional matching problem in full covariate space to one of matching observational units along a one-dimensional metric.

MATHEMATICAL FOUNDATIONS AND KEY ASPECTS OF MATCHING ESTIMATORS

Identification of treatment effects under exogeneity

The notion that causal effects represent the differences in potential outcomes under alternative conditions D is fundamental to the modern counterfactual framework of causal inference. In the canonical case of a binary treatment D, the unit causal effect Δ_i of treatment D on outcome Y is defined as the difference

$$\Delta_i \equiv Y_{1i} - Y_{0i} \tag{12.1}$$

between outcomes Y_{1i} and Y_{0i} that would be observed if unit i were exposed to alternative conditions $D \in \{0, 1\}$. Implicit in this definition is the fundamental existence assumption that unit causal effects represent a structurally invariant feature of (social) reality, which is known as the *stable unit treatment value assumption* (SUTVA). The SUTVA is far reaching in so far as it rules out general equilibrium effects, but also any impact of, for example, social interactions between treatment and control groups, or of the probability and (social) distribution of treatment conditions, on the relationship between treatment and outcomes; the SUTVA, in other words, rules out that anything about the (members of the) treatment group affects expected outcomes among non-treated units. When the SUTVA cannot be maintained, unit causal effects are non-existent and a causal interpretation does not apply to either matching or regression estimates. In that case, matching may at best identify local treatment effects that occur within a specific social setting or, if applied at the systemic level, equilibrium effects within the interaction environment, whether that may usefully be defined at the family, neighborhood, community or some wider social level.

Assuming that the SUTVA holds, it is possible to define the average treatment effect

$$ATE \equiv E[\Delta_i] = E[Y_{1i} - Y_{0i}] = E[Y_{1i}] - E[Y_{0i}], \tag{12.2}$$

the average treatment effect on the treated

$$ATT \equiv E[\Delta_i|D_i = 1] = E[Y_{1i} - Y_{0i}|D_i = 1] = E[Y_{1i}|D_i = 1] - E[Y_{01}|D_i = 1], \tag{12.3}$$

and the average treatment effect on the untreated

$$ATU \equiv E[\Delta_i|D_i = 0] = E[Y_{1i} - Y_{0i}|D_i = 0] = E[Y_{1i}|D_i = 0] - E[Y_{0i}|D_i = 0] \tag{12.4}$$

as key quantities of interest. These parameters describe the average unit treatment effect in the population, in the subpopulation of units actually exposed to treatment (i.e. $D_i = 1$), and in the subpopulation of units not exposed to treatment (i.e. $D_i = 0$). Defining $\pi = E[D]$ as the proportion of the population exposed to treatment, the average treatment effect is the weighted sum

$$ATE = \pi ATT + (1 - \pi) ATU \tag{12.5}$$

of the two subpopulation average treatment effects. If the population can be partitioned into different strata S (e.g. various socio-demographic groups), these quantities can naturally be defined at the strata level, resulting in the conditional average treatment effect

$$CATE \equiv E[\Delta_i|S = s] = E[Y_{1i} - Y_{0i}|S = s] = E[Y_{1i}|S = s] - E[Y_{0i}|S = s], \tag{12.6}$$

and the corresponding parameters for the subpopulations of treated and untreated units. In this case, the population-level average treatment effect is given as the weighted sum

$$ATE \equiv \sum_{s \in S} \Pr(S = s) E[\Delta_i | S = s] \tag{12.7}$$

over the strata-specific conditional average treatment effects, and equivalent expressions apply to the average treatment effect on the treated and the untreated.

Since the empirical world reveals only the one observable outcome

$$Y_i^* \equiv D_i Y_{1i} + (1 - D_i) Y_{0i} \tag{12.8}$$

for each unit of observation, namely the outcome Y_i^* that is realized under the condition D_i unit i is empirically exposed to, causal effects cannot be observed empirically, but need to be estimated. As an illustration of this fundamental problem of causal inference, it is possible to decompose the average treatment effect into the expression

$$\begin{aligned} E[\Delta_i] &= \pi E\left[Y_1^* - Y_0 | D_i = 1\right] + (1 - \pi) E\left[Y_1 - Y_0^* | D_i = 0\right] \\ &= \pi E\left[Y_1^* | D_i = 1\right] + (1 - \pi) E[Y_1 | D_i = 0] \\ &\quad - \left\{\pi E[Y_0 | D_i = 1] + (1 - \pi) E\left[Y_0^* | D_i = 0\right]\right\}, \end{aligned} \tag{12.9}$$

which derives the average treatment effect as a weighted sum of observable and unobservable (i.e. counterfactual) terms. To arrive at an empirical estimate of the average treatment effect, the unobservable quantities $E[Y_1|D_i = 0]$ and $E[Y_0|D_i = 1]$ have to be replaced with empirically observable and credible substitutes.

Now if one assumes that potential outcomes Y_1 and Y_0 are generally determined by observed (Z) as well as unobserved (U) factors according to

$$\begin{aligned} Y_{0i} &= \mu_0(Z_i) + U_{0i}, \\ Y_{1i} &= \mu_1(Z_i) + U_{1i}, \end{aligned} \tag{12.10}$$

the ATE parameter is identified whenever the two conditions

$$\begin{aligned} &\text{A-1}: E\left[Y_1^* | Z, D_i = 1\right] = E[Y_1 | Z, D_i = 0] \\ &\quad \Leftrightarrow E[Z|D_i = 1] = E[Z|D_i = 0] \cap E[U_1|D_i = 1] = E[U_1|D_i = 0], \\ &\text{A-2}: E[Y_0 | Z, D_i = 1] = E\left[Y_0^* | Z, D_i = 0\right] \\ &\quad \Leftrightarrow E[Z|D_i = 1] = E[Z|D_i = 0] \cap E[U_0|D_i = 1] = E[U_0|D_i = 0] \end{aligned} \tag{12.11}$$

can be maintained.[1] In other words, the ATE is identified if it may be plausibly postulated that expected potential outcomes $E[Y]$ are independent of empirical treatment status D, conditional on observed covariates Z, among both those units empirically not exposed to treatment (A-1) and those units empirically exposed to treatment (A-2). A-1 and A-2 jointly correspond to the assumption of *conditional mean independence*, according to which the ATE is identified if and only if the expected impact of unobservables U on outcomes Y is exactly equal across comparison groups. This is the assumption of exogenous assignment to treatment status given observable covariates Z.[2]

Matching as sample reweighting

Matching estimators achieve the required conditioning by appropriately reweighting the treatment and control samples. Irrespective of the specific matching algorithm, the matching estimator of the ATT parameter can be expressed as

$$\begin{aligned} ATT_{\mathrm{M}} &= \frac{1}{N_{D_1}} \sum_{i \in D_1 \cap S} w_i \left[Y_{1i} - \sum_{j \in D_0 \cap S} W_{i,j} Y_{0j} \right] \\ &= \frac{1}{N_{D_1}} \sum_{i \in D_1 \cap S} w_i Y_{1i} - \frac{1}{N_{D_1}} \sum_{i \in D_1 \cap S} \left[w_i \sum_{j \in D_0 \cap S} W_{i,j} Y_{0j} \right], \end{aligned} \tag{12.12}$$

that is, as the sample average of the differences between observed outcomes Y_{1i} in the treatment sample ($D_i = 1$) and the appropriately weighted observed outcomes Y_{0j} in the control sample ($D_i = 0$), potentially using design or other survey weights w_i (cf. Heckman et al., 1998). By the same logic, the corresponding matching estimator of the ATU parameter is

$$ATU_{\mathrm{M}} = \frac{1}{N_{D_0}} \sum_{j \in D_0 \cap S} w_j \left[\sum_{i \in D_1 \cap S} W_{i,j} Y_{1i} - Y_{0j} \right] \tag{12.13}$$

and the matching estimator of the ATE simply is the weighted sum

$$ATE_{\mathrm{M}} = \pi ATT_{\mathrm{M}} + (1 - \pi) ATU_{\mathrm{M}} \tag{12.14}$$

of (12.12) and (12.13). To serve as an appropriate estimate of the distribution of counterfactual outcomes, the weights $W_{i,j}$ need to ensure that the distribution of relevant confounders Z is balanced across the matched samples, so that the required assumption of equality of expected outcomes net of treatment may plausibly be sustained in each case. Weights $W_{i,j}$ in practice express the similarity of (or inverse distance between) individual members of the comparison groups in the analysis.

Once appropriate weights have been determined, the causal parameter of interest may be estimated non-parametrically as the average difference between observed and reweighted counterfactual outcomes, or from the simple difference in reweighted group means in the particular case.[3] Importantly, this non-parametric estimator is defined over the common support S only, that is, over that part of the covariate space for which the comparison groups of the analysis overlap. Short of any parametric model for outcomes, matching estimators provide no way to extrapolate outcome data beyond the covariate space represented in the actual data. Depending on the covariate distribution in the sample data, matching estimators may hence result in treatment effect estimates that apply to a very specific subpopulation that may be far from representing the target population. Assessing common support is hence of evident importance with respect to the external validity of results in any concrete application, and lack of common support one of the main reasons for differences between matching and regression estimators of treatment effects. To put it another way, a divergence of regression and matching estimates that use comparable specifications may often serve as a useful indication of the extent to which estimates are reliant on regression extrapolation into off-support covariate space, that is, functional form assumptions.

Constructing the counterfactual: Propensity score versus exact matching

Within the above framework, alternative matching estimators differ in the specific algorithm for matching observations from the treatment and control sample, or in the implicit weight function $W_{i,j}$ of the estimator. In that respect, a basic difference occurs between exact matching algorithms that construct weights from observational equivalence in the covariate space Z, and propensity score based algorithms that construct weights using (estimated) propensity scores as a distance metric. In addition, important variants such as classical covariate (Mahalanobis) matching or the more recent entropy balancing matching algorithm take an intermediate position in so far as observational similarity between units is determined using alternative distance functions directly in the covariate space Z. Since key considerations in algorithm choice can be usefully illustrated by contrasting exact and propensity score matching estimators, I focus on these polar strategies for the present discussion. Table 12.1 summarizes the properties and implicit weighting functions for selected core algorithms.

Among these, exact matching is the historical and didactical epitome of all matching algorithms. Exact matching results in pairwise matches of observationally identical units from the treatment and control sample, defined in the space of observed covariates Z. Expressed in terms of the weighting function for ATT_{M}, exact matching loops over all treatment group observations and assigns a weight of one to that control sample observation that is observationally identical to a treatment sample observation, and zero to all others; in case of multiple exact matches, positive weights are the inverse of the number of exact matches. Weights are summed if matching is done with replacement, that is, if any particular observation from the control sample is permitted to serve as a matched control case multiple times. While perfectly intuitive as an algorithm, the practical problems with exact matching are both evident and severe. In typical social science applications, matching needs to proceed across potentially large sets of confounders, including any mix of categorical and continuous covariates. Hence, finding exact matches, even for just subsets of the treatment sample, is likely to require unrealistically large samples in applied research, making the exact matching estimator largely infeasible in practice.

As a practically feasible alternative, algorithms that use the propensity score have greatly contributed to the increased popularity of matching estimators. Following the fundamental insight of Rosenbaum and Rubin (1983) that covariate balance may be achieved by matching on a suitable linear combination of observed covariates, the propensity score

$$P(Z) = \Pr(D = 1|Z_i) \tag{12.15}$$

is defined as the conditional probability of an individual receiving treatment (or being exposed to treatment condition) D given observed covariates Z_i. In practice, the propensity score is unknown except in rare cases, and has to be estimated from an assignment model for $P(Z)$, usually by way of a parametric probability model such as the logit model

$$\widehat{P}(Z) = \frac{\exp(Z_i\beta)}{1 + \exp(Z_i\beta)} \tag{12.16}$$

predicting treatment status.[4] Within the assignment model, the role of observed covariates Z is purely predictive, with estimated regression coefficients β providing importance weights for individual covariates in determining observational similarity or distance. Usually, specification of the assignment model is an iterative process that seeks to balance model parsimony and goodness of fit, and that may be aided by a wide range of regression diagnostics. Importantly, however, the covariate vector Z cannot include perfect predictors of treatment status, since matching estimators cannot extrapolate to off-support areas of the covariate space and hence

Table 12.1 Implicit weight functions in alternative matching algorithms

	Description	Weight function $W_{i,j}$ (with ATT_M)
	Exact matching algorithms	
Exact matching	Pair matching between treatment group observation i and control group observations that are observationally identical in Z	$W_{i,j} = \begin{cases} 1/N_j & \text{if } Z_i = Z_j \\ & \text{otherwise} \end{cases}$
Coarsened exact matching	Stratification of the sample into k cells defined over appropriately discretized covariate vector Z	$W_{i,j} = \begin{cases} 1/N_k & \text{if } j \in k \\ 0 & \text{otherwise} \end{cases}$
	Propensity score matching algorithms	
Stratification (interval matching)	Stratification of the sample into k intervals defined over P	$W_{i,j} = \begin{cases} 1/N_k & \text{if } j \in k \\ 0 & \text{otherwise} \end{cases}$
Nearest-neighbor matching	Pair matching between treatment group observation i and m most similar members of the control group sample over P	$W_{i,j} = \begin{cases} 1/m & \text{if } j \in \arg\ \min_m\{\lvert P_i - P_j\rvert\} \\ 0 & \text{otherwise} \end{cases}$
Caliper matching	Pair matching between treatment group observation i and m most similar members of the control group sample within caliper $c(P)$ around P_i	$W_{i,j} = \begin{cases} 1/m_i & \text{if } j \in \{\lvert P_i - P_j\rvert \leq c\} \cap \arg\ \min_m\{\lvert P_i - P_j\rvert\} \\ 0 & \text{otherwise} \end{cases}$
Radius matching	Matching of treatment group observation i and all members of the control group sample within range $r(P)$ around P_i	$W_{i,j} = \begin{cases} 1/P_r & \text{if } j \in \{\lvert P_i - P_j\rvert \leq r\} \\ 0 & \text{otherwise} \end{cases}$
Kernel matching	Counterfactual estimate as the distance-weighted average of control sample observations using kernel function $K(\cdot)$ over P and within bandwidth h	$W_{i,j} = \frac{K[(P_j - P_i)/h]}{\sum_{k \in C} K[(P_k - P_i)/h]}$

break down for lack of overlap in covariate distributions in case of (near) perfect sample separation. If covariates are available, they are likely to be in accordance with the requirements of regression discontinuity designs and instrumental variable estimation instead.

With predicted propensity scores at hand, various matching algorithms as well as any combination thereof may be utilized in the construction of conditioning weights. Nearest-neighbor matching provides the equivalent to exact matching in the propensity score metric in so far as matches are formed between treatment and control observations that are most similar with respect to $P(Z)$. Potentially, matching may be limited to just one nearest neighbor or to any fixed number of k members of the control group that are most similar to observation i in the treatment sample. Caliper matching extends nearest-neighbor matching by setting a maximum dissimilarity (or minimum similarity) c for matching, and radius matching involves accepting all available control group observations within maximum dissimilarity radius r around i, even if this results in an unbalanced number of matched controls per treatment observation. More sophisticated algorithms such as kernel matching (or closely related variants such as local linear matching) use specific distance functions to assign weights to control group observations within bandwidth h around i that decline with the absolute distance of control group observations to i. At its simplest, matching may be performed by stratifying the propensity score distribution and then weight controls by the inverse number of control group observations within the strata of i. Net of any matching algorithm, it is also possible to construct the inverse probability estimator weight

$$W_{i,j} = \frac{1}{N_{D=0}} \times \frac{P(Z)}{1 - P(Z)} \tag{12.17}$$

directly from the estimated propensity score when estimating the average treatment effect on the treated (Hirano et al., 2003).

These alternative algorithms are equivalent asymptotically, yet empirical researchers are often left with difficult choices in practice since behavior of the algorithms varies in finite (especially small to medium) samples, so that algorithm choice should depend on specific features of the sample and the problem at hand. Fundamentally, there is a trade-off between bias and efficiency (or variance) of the resulting estimator, but also a related trade-off between bias and scope of the estimator to consider in applied research. Algorithms such as exact matching or nearest-neighbor matching with replacement and within small calipers tend to minimize bias, that is, the imbalance of covariate distributions between the comparison groups of the analysis, at the price of potentially considerable losses in efficiency (since only a subset of the available sample information is utilized, e.g. when pairwise matching algorithms systematically discard information from suitable but second-best control group observations) and scope of the estimator (since common support within the subsample of very good matches may be a small subset of the covariate space only). Nearest-neighbor matching with multiple controls, but especially radius, kernel and related matching algorithms that tend to utilize the sample data more comprehensively, generally achieve a lower variance of the resulting estimator, although potentially at some loss of covariate balance. Improper trade-offs may be avoidable with reasonably large sample sizes and favorable ratios of the number of control and treatment group observations that permit the use of multiple matches within closely circumscribed calipers or bandwidth. Often, combining different principles of matching also helps to minimize trade-offs in practice, for example by utilizing stratified propensity score based estimators that perform propensity score matching within strata defined by exact matches on key covariates of the analysis. Also, since the use of the propensity score (or another distance metric) partly compensates for problems of data sparseness by smoothing weights across adjacent regions of the covariate space, propensity score based estimators tend to achieve a broader scope of the resulting estimates (i.e. produce overlap across a broader range of covariate constellations, or a larger fraction of the sample) relative to exact matching.[5]

These general recommendations notwithstanding, the development of alternative matching algorithms continues to be a very dynamic field of research. Noting the potential for bias due to covariate imbalance and the dependence on correct model specification in the assignment model of propensity score estimators, recent contributions have sought to develop algorithms that avoid estimating the propensity score altogether or that seek to achieve optimal balance in the covariate space Z directly. The coarsened exact matching (CEM) estimator of Iacus et al. (2011, 2012) deserves special mention for combining practical feasibility and the appeal of exact matching algorithms. The CEM algorithm improves on classical exact matching by requiring exact matches within appropriately stratified (i.e. categorically coarsened) distributions of covariates Z only, thus yielding a much more practically feasible estimator. At the same time, CEM retains the simplicity and efficiency of classical exact matching in adjusting for entire covariate constellations, which typically results in superior algorithm performance with respect to balancing higher moments of covariate distributions (variance and skew) as well as multivariate dependence patterns (interactions) across Z relative to propensity score based estimators using standard (e.g. main effects) parametric specifications of the assignment model.

Assessing sample balance and support

Among the three desirable properties of matching estimators, minimizing bias certainly takes precedence in studies aiming for causal inference. As matching estimators require the balancing of (expected) counterfactual outcomes net of treatment across the comparison groups for valid causal inference, the empirical degree of covariate balance in the sample becomes an important benchmark in the iterative process of choosing an adequate matching algorithm in the concrete application. To assess the quality of matching in this respect, Rosenbaum and Rubin's (1985) *standardized bias*,

$$SB(Z) = \frac{\overline{Z}_{D_1} - \overline{Z}_{D_0}}{\sqrt{\frac{1}{2}\left[V_{D_1}(Z) + V_{D_0}(Z)\right]}}, \tag{12.18}$$

the difference in covariate means normalized by the square root of the averaged variances, has become a widely accepted measure to express univariate group differences in covariate distributions, the extent of remaining covariate imbalance post-matching and the extent of bias reduction relative to the raw sample data. Often, the rule of thumb is given that remaining bias of the order of 3–5% should be acceptable in practice (e.g. Rubin, 2006), yet recent Monte Carlo and benchmark study evidence suggests that much higher levels of balance, certainly on key covariates, and potentially also balance on higher moments and multivariate dependencies in covariate distributions may be advisable in order to ensure valid causal inference. Generalizations of the standardized bias measure as well as other multivariate metrics are available (notably the L1 metric of King et al., 2011), but have yet to see more widespread use in practice. In contrast, the fairly widespread use of significance testing to assess covariate balance – whether through two-sample *t* tests or goodness-of-fit tests of the assignment model – is ill-founded since covariate balance is a sample rather than a population characteristic in the context of matching estimators (see Imai et al., 2008).[6]

In fact, an emphasis on bias reduction as the primary goal of matching estimators is also behind the resurgence of interest in exact matching and related algorithms that avoid specification of an explicit assignment model. Under the reasoning that standard parametric regression specifications are unlikely to reflect all essential detail of the empirical differences in covariate distributions, fully non-parametric estimators such as CEM are conceptually very attractive since they avoid model dependence. Alternatively, several recent optimizing algorithms such as optimal matching (Rosenbaum, 2002), genetic matching (Diamond and Sekhon, 2013), and entropy balanced matching (Hainmüller, 2012) all implement machine learning tools to minimize alternative distance metrics, and thus are likely to represent considerable improvements over the received wisdom of iteratively finding satisfactory assignment model through manual trial and updating of increasingly flexible specifications. That said, it is also important to emphasize that balance checking is not equivalent to a formal validity test of the CIA. Theoretically, full covariate balance is not even a necessary condition for the CIA to hold, at least in its mean-independence form required to estimate average treatment effects. What is required for causal inference instead is balance of expected counterfactual outcomes, and this may at the same time involve additional unobserved covariates and only a subset of observed covariates, so that the CIA ultimately cannot be assessed on the basis of statistical evidence alone.[7]

These qualifications notwithstanding, what has been underappreciated in the matching literature so far is that methods designed for assessing covariate imbalance between matched samples may also be very usefully employed to characterize the scope of the matching estimator, that is, the discrepancy between the available samples over and off common support. Since matching estimators differ in the degree of smoothing over areas of data sparseness, the extent to which the target quantity becomes redefined by estimating treatment effects over common support only is evidently critical with respect to the external validity of the resulting estimate. Traditionally, histograms or densities of the estimated propensity score have been utilized, but the empirical analysis below will illustrate how balance checking techniques may be adopted for this purpose.

Statistical inference for matching estimators

In addition to the point estimate of some treatment effect in the sample at hand, researchers will usually be interested in determining associated standard errors or confidence intervals for population inference. Unfortunately, large-sample theory for matching estimators is in its infancy, and straightforward analytical results exist for relatively few – and typically quite simple – matching algorithms (Imbens, 2004). As a result, approximation methods, notably bootstrapping techniques, dominate in applied research. When bootstrap estimates are being constructed, it is important to realize that bootstrap replications need to comprise all stages of the matching estimator – that is, estimation of the assignment model, construction of the matched samples and computation of treatment effects of interest – since otherwise sample variation in estimated propensity scores, common support, and, with nearest neighbor without replacement, the sort order of sample observations becomes omitted from the computations.

The use of bootstrap methods in the context of matching has, moreover, come under some criticism in the econometric literature. Nearest-neighbor algorithms with a fixed small number of control observations have long been known to fall short in terms of efficiency, and Abadie and Imbens (2008) have recently demonstrated that this also implies severe inefficiency of the bootstrap estimator in this case. On the other hand, since bootstrap failure – or more correctly, the conservative nature of resulting standard error estimates – is closely linked to the efficiency loss surrounding the overly restrictive use of available control group observations, alternative algorithms such as radius, kernel or local linear matching are unlikely to be seriously affected, especially if samples with favorable ratios of control to treatment group observations are available. Also, it is important to emphasize that non-algorithmic estimators such as the inverse probability weighted estimator are unaffected by this particular issue.

On the other hand, there has been progress in terms of analytical results and practically feasible variance estimators for matching algorithms. Noting the reweighting representation of matching (and related) estimators, Abadie and Imbens (2006) in particular derive the conditional variance of a treatment effect estimator as

$$\mathrm{Var}(\Delta|D,Z) = \sum_i W_{i,j}^2 \sigma_{D_i}^2(Z), \tag{12.19}$$

and propose a feasible non-parametric estimator for the variance term $\sigma_D^2(Z)$. As an appealing general estimator, it seems likely that the Abadie–Imbens variance estimator may become the future standard in the field. At present, however, alternative software implementations of matching estimators utilize different variance approximations and will hence often produce inconsistent estimates of the standard error of some treatment effect estimate. If possible, it may thus be advisable to base statistical inference on the results of alternative routines in applied research, especially of course when considering borderline cases. Also, whatever the specific approximation or variance estimator, standard errors and confidence intervals around a treatment effect estimate established through matching should generally be expected to be considerably inflated relative to a comparable parametric regression specification.

Extensions and advanced aspects

Although the present exposition, like much applied work, has been cast in terms of the impact of a binary treatment D on a quantitative outcome Y, matching estimators in fact represent a versatile class of non-parametric estimators that is suitable for addressing a broad range of empirical questions. Once it is noted that $E(Y|D,X) = \Pr(Y = 1|D,X)$ in the case of binary outcomes, matching estimators accommodate the estimation of treatment effects on categorical outcomes in straightforward ways within the framework presented here, and are readily adapted to ordinal outcomes by either discretization or focusing on appropriate quantile effects. Also, matching estimators can readily be extended to accommodate polychotomous, ordinal or appropriately discretized quantitative treatment indicators by examining multiple binary contrasts using the methods described above. Alternatively, it is also possible to match on the index function of an appropriate statistical model for ordinal data or to rely on the coarsened exact matching estimator in order to avoid estimating a whole set of propensity score equations to describe the non-random assignment to various treatment statuses or conditions of exposure. Evidently, the empirical analysis can nevertheless become unwieldy with many-valued treatment indicators, when it may be advisable to focus on selected contrasts of particular theoretical interest.

Also, matching estimators are easily adapted to accommodate special features of the sample data at hand. In particular, as with traditional regression estimators, longitudinal and hierarchical data may insulate the empirical analysis against key inferential threats, principally by the availability of measures of biographical or peer-group or contextual covariate information. Besides, the availability of longitudinal data naturally permits researchers to ensure maintenance of the proper time order between measurement of treatment, controls and outcomes, including the use of time-varying controls to accommodate differences in the onset of treatment exposure (a.k.a. dynamic treatment selection; see Brand and Xie, 2007). In addition, longitudinal data allows the analyst to estimate richer sets of treatment effects, notably point-in-time treatment effects, for example by elapsed time since onset of treatment (exposure), but also treatment effects defined according to duration of or by period of exposure. Similarly, the availability of hierarchical data enables researchers to define treatment effects by respondents' peer-group status and contextual features.

Most importantly, however, these richer data structures improve on the analyst's ability to account for the impact of unobserved confounders of outcomes, and thereby potentially greatly increase the viability of the CIA inherent in the design of the study and the availability of observed covariates. In particular, *difference-in-differences* (DiD) matching estimators along the lines of

$$ATT_{\text{DiD}} = \frac{1}{N_{D_1}} \sum_{i \in D_1 \cap S} w_i \left[(Y_{1i,t} - Y_{1i,t-1}) - \sum_{i \in D_0 \cap S} W_{i,j} (Y_{0j,t} - Y_{0j,t-1}) \right] \tag{12.20}$$

(cf. Heckman et al., 1998) have regularly been employed in applied research to account for unobserved heterogeneity between individual units and, ultimately, comparison groups in the analysis. It is also possible to define the closely related fixed-effects (within) matching estimator

$$ATT_{\mathrm{FE}} = \frac{1}{N_{D_1}} \sum_{i \in D_1 \cap S} w_i \left[(Y_{1i,c} - \overline{Y}_{1c}) - \sum_{i \in D_0 \cap S} W_{i,j} (Y_{0j,c} - \overline{Y}_{0c}) \right] \qquad (12.21)$$

that accommodates both longitudinal and hierarchical data structures (Gangl, 2012), yet so far this estimator has not seen applications in practice. Irrespective of the specific data structure at hand, several techniques for conducting sensitivity analyses are available that assess the robustness of causal inferences relative to the presence of an unobserved confounder of specified features (cf. Rosenbaum, 2002). Naturally, it is particularly advisable to conduct such sensitivity analyses with cross-sectional designs or whenever theoretically relevant controls are unobserved.

ILLUSTRATION: MATCHING ESTIMATES OF RETURNS TO EDUCATION IN GERMANY

To illustrate the practical application of matching estimators, I describe an analysis of earnings returns to higher education in Germany. The analysis uses the cross-sectional 2008 (wave Y) sample of the German Socio-Economic Panel (GSOEP; cf. Wagner et al., 2007) combined with information on social background from the GSOEP biography module. The dependent variable of the analysis will be the log of respondents' annual gross earnings, including self-employed income. Respondents' level of education will be considered the treatment variable, and respondents' age, gender, immigrant status, region (East or West Germany), and social background will serve as potential confounders that need to be adjusted for. Social background will be measured via parental highest level of education, parental highest (International Socio-Economic Index, ISEI; Ganzeboom and Treiman, 1996) socio-economic status during the respondent's adolescence, and whether the respondent's mother was employed during the respondent's adolescence. Evidently, covariate selection is for illustrative purposes only and omits potentially important factors such as family income, number of siblings or the quality of parent–child relationships that one may want to consider in a full analysis.

Before embarking on the actual statistical analysis, the use of any matching estimator characteristically necessitates a clear definition of the estimand of interest. In the concrete case, level of education corresponds to the case of a multi-valued 'treatment' condition, that is, corresponds to a situation where multiple specific contrasts may be of legitimate analytical interest so that a precise definition of the counterfactual of interest is required. In the following illustration, I will focus on one particular contrast, namely on the returns to full university education (five-year diploma and master's degrees) relative to applied professional degrees (four-year degrees) available from Germany's professional colleges (*Fachhochschulen*). Naturally, many other interesting contrasts could be evaluated – whether the returns of university degrees relative to vocational training in the German apprenticeship system or the returns of an apprenticeship relative to leaving the education and training system with a school certificate only – following from the principle of appropriately discretizing multi-valued treatment conditions. Besides evaluating returns to education at the top end of the educational hierarchy, I will also limit attention to estimating returns to university education in the sense of the average treatment effect on the treated. In other words, the following analysis will be interested in the economic value of university degrees *for those respondents who actually did complete one*. By focusing on the ATT, the analysis will hence attempt to answer the question of whether completion of a full academic degree was economically justified in the population of university graduates, that is, among those who decided to pursue university education relative to the option of pursuing a shorter applied professional degree only. Naturally, the analysis could be extended to asking, for example, how much, if anything, respondents who obtained applied professional degrees could have gained on average by completing a full university degree, or how much, if anything, respondents with an *Abitur* (grammar school) certificate who pursued vocational training instead of an academic education could have gained by choosing the latter option. The latter questions would be two particular ways of defining an average treatment effect on the untreated of interest.

Covariate imbalance and assignment model

Within these confines, I retain a sample of 2076 GSOEP respondents aged 25–64 with valid earnings and covariate data for analysis who either completed an applied professional or a full university degree. According to the GSOEP data, about one half of grammar school graduates (respondents holding an *Abitur* certificate) with some advanced degree had completed full university degrees, one quarter had completed an applied professional degree, and one quarter had completed a vocational training degree. Since the following is focused on the contrast between the two major academic pathways, however, only the first two groups of respondents are retained in the analysis, which gives $N = 1457$ observations in the treatment sample and $N = 619$ observations in the control group.

Having defined the comparison groups of the analysis, it is useful to assess covariate imbalance in the two samples and, if propensity score matching is to be performed, to specify the assignment model of the estimator. Naturally, I will simply presume in the following that the available covariates were sufficient to ensure the validity of the CIA required for a causal interpretation of matching estimates; similarly, standard causal inference presupposes the SUTVA to hold, that is, that university graduates' earnings are unaffected by the presence of graduates with applied college degrees in the labor market and vice versa; needless to say, any strict reading of the SUTVA violates standard economic price theory, where wages reflect the relative scarcity of worker skills, so a more realistic interpretation is to assume that it will at best be possible to identify local (or partial equilibrium) treatment effects in the sense of returns to (marginal investment in) education under the current macroeconomic equilibrium.

That said, Table 12.2 indeed demonstrates that the two comparison groups differ in terms of covariate distributions. Relative to graduates with applied degrees, respondents with full university degrees tend to be slightly older, are more likely to be female, first-generation immigrants and from East Germany. Also, university degree holders tend to come from families with higher levels of parental education, higher socio-economic status, and from families where mothers were more likely to have worked during respondents' adolescence. On all these indicators, the Rosenbaum–Rubin standardized bias measure suggests covariate imbalance of the order of $SB = 10\%–35\%$. Compared to other problems, the range of bias estimates indicates that only moderately difficult adjustments are required, yet the unfavorable ratio of almost 3 : 1 between treatment and control group observations may be expected to generate problems for any nonparametric estimator due to data sparseness and a relative lack of control group observations.[8] Moving beyond univariate distributions, one could also examine (aspects of) the multivariate covariate distribution to note, for example, imbalances with respect to higher levels of parental education or a mild overrepresentation of East Germans among university graduates in the female sample specifically. Given the relatively large sample, group differences on all covariates that show covariate imbalance of the order of $SB = 10\%$ are also statistically significant on conventional two-sample t tests. Interestingly, there is no appreciable group difference with

Table 12.2 Sample differences in covariate distributions, standardized bias and assignment models

	University degree ($D = 1$) (1)	Applied professional degree (FH) ($D = 0$) (2)	Stand. bias (1) – (2) (3)	Assignment model 1: main effects specification (4)	Assignment model 2: two-way interactions (5)
Univariate distributions					
Female	0.486	0.433	0.106	0.226*	–1.393
			(0.027)	(0.101)	(2.037)
First generation	0.076	0.032	0.192	1.102**	0.917**
migrant			(0.000)	(0.258)	(0.335)
Second generation	0.033	0.036	–0.014	0.253	0.243
migrant			(0.764)	(0.277)	(0.367)
East Germany	0.263	0.178	0.206	0.479**	–0.202
			(0.000)	(0.128)	(2.617)
Age	45.49	43.47	0.198	0.031**	–0.074
(years)	(10.35)	(10.12)	(0.000)	(0.005)	(0.063)
Parental level of	13.72	12.61	0.343	0.078**	0.094**
education (years)	(3.44)	(3.01)	(0.000)	(0.020)	(0.029)
Parental occupational	53.87	48.29	0.322	0.013**	0.017**
status (ISEI)	(17.55)	(17.07)	(0.000)	(0.004)	(0.005)
Mother employed in	0.485	0.404	0.163	0.199	–0.007
child's adolescence			(0.001)	(0.103)	(0.147)
Multivariate distributions (selected aspects only)					
Female ×	6.67	5.55	0.161	–	0.007
parental education	(7.27)	(6.68)	(0.001)		(0.042)
Female ×	0.139	0.115	0.072	–	–0.783**
East Germany			(0.140)		(0.273)
LR-Test 2-way				N/A	4.61
gender interaction					(df=7)
LR-Test 2-way region				N/A	33.62**
(East/West) interaction					(df=5)
LR-Test vs model 1				N/A	48.68**
(main effects spec.)					(df=14)
Pseudo-R^2				0.055	0.074
N	1457	619		2076	2076

Notes: Columns (1)–(2), standard deviations of continuous covariates in parentheses; columns (3)–(5), statistical significance levels in parentheses (* $p < 0.05$, ** $p < 0.01$). Columns (4)–(5), logit regression coefficients; model 2 also includes second-order polynomial terms for age.

Source: German Socio-Economic Panel, wave Y (2008), unweighted data

respect to second-generation immigrants to Germany, who are but a small minority in either sample.

The assignment model, and in consequence the estimated propensity score derived from it, is a tool to map covariate imbalance on all these dimensions onto a one-dimensional distance metric. For the present analysis I will be working with two versions of the assignment model. Model 1 is a plain main effects logit model, whereas model 2 follows the received wisdom in the literature to incorporate additional non-linear and interaction terms to improve goodness of fit. More specifically, model 2 includes a second-order polynomial for age and also the two-way interactions between East Germany and all other covariates save migration status, and between gender and all other covariates. Increasing the logit model's goodness of fit from a pseudo-R^2 of 5.4% to 7.4%, model 2 indeed performs better than model 1 on this and other standard measures. Importantly, however, the specification of the more elaborate model is entirely ad hoc here, but should follow from a specification search that systematically explores patterns of non-linearity and interactions in the empirical data in any real application. Finally, it should be emphasized that the goodness of fit of the assignment model is merely one device to assess the relative performance of alternative specifications of the assignment model. Specifically, since the methodological purpose of the assignment model is to partition variation in treatment status into its endogenous and exogenous components, the absolute level of any goodness-of-fit statistic will not be informative about the quality of the research design. Thus, there can be no generally applicable critical threshold for a useful assignment model since the relative importance of exogenous variation in treatment conditions will depend on the substantive application.

Assessing balance and support

To illustrate the properties of alternative matching estimators, I employ several algorithms to construct the counterfactual outcome distribution and compare their relative performance. Specifically, I provide results for simple exact matching on the available covariates, two variants of coarsened exact matching, and four propensity score based algorithms, each using both assignment models from the previous section. Among the coarsened exact matching estimators, I distinguish between a finely balanced algorithm that uses Sturge's rule,

$$c_{st} = \frac{\max Z_i - \min Z_i}{\ln n + 1}, \tag{12.22}$$

to determine bin width c for all quantitative covariates, and a coarsened algorithm that matches on five equidistant age and parental ISEI groups, and three groups defined in terms of parental education (less than 12 years of education, 12 to less than 16 years of education, and 16 or more years of education). Matching will be exact on categorical covariates in either case. Among propensity score based estimators, I will compare the behavior of a simple nearest-neighbor algorithm to nearest-neighbor caliper matching with caliper $c = 0.001$, radius matching using $r = 0.001$, and kernel matching using bandwidth $h = 0.001$. All propensity score based algorithms are run twice, once using the main effects specification (model 1) and once using the more elaborate gender and region interaction effects specification (model 2) of the assignment model. All propensity score estimators match on the estimated propensity score within common support, and all matching algorithms are run with replacement.[9] Also, it should be noted that the chosen caliper of $c = 0.001$ is equivalent to less than 1% of the standard deviation of the propensity score distribution, and as such amounts to a much stricter similarity requirement than default recommendations often found in the literature to, for example, set c equal to one quarter of the standard deviation. Naturally, the ideal estimator combines unbiasedness, full scope (representativeness) and efficiency.

As a first step in assessing the empirical behavior of the different algorithms, Table 12.3 provides key results from balancing tests for the specifications. In Table 12.3, I follow conventional practice of presenting results for standardized bias and two-sample t tests but note once more that algorithm performance is best judged by the *change* in either quantity relative to the raw data results or the *absolute level* of remaining standardized bias. Also, I focus on a few selected covariate constellations for illustrative purposes here, but note that the distribution of remaining standardized bias across covariate (constellations) is an excellent measure of global imbalance reduction achieved by any matching algorithm; Rubin's rule of thumb suggests that remaining

Table 12.3 Balancing tests by matching algorithm and assignment model specification

	$N_D \in D = 1$	Standardized bias (*p*-value)				
		Female	Female × East Germany	Parental ISEI	Parental education	Female × parental education
Raw data	1457	0.106*	0.072	0.322**	0.343**	0.161**
		(0.027)	(0.140)	(0.000)	(0.000)	(0.000)
Exact matching	90	0.000	0.000	0.000	0.000	0.000
Coarsened exact matching						
Fine binning	431	0.000	0.000	0.002	0.001	0.001
				(0.983)	(0.987)	(0.995)
Coarse binning	1111	0.000	0.000	0.029	0.002	0.000
				(0.579)	(0.976)	(0.992)
Propensity score matching, main effects specification						
Nearest neighbor, k=1	1456	–0.001	–0.045	–0.093**	–0.043	–0.033
		(0.970)	(0.249)	(0.014)	(0.269)	(0.396)
Nearest neighbor, k = 1, c = 0.001	1275	–0.009	0.000	–0.086*	–0.041	–0.037
		(0.812)	(1.000)	(0.031)	(0.313)	(0.370)
Radius matching, *r* = 0.001	1275	–0.013	–0.027	–0.049	–0.029	–0.040
		(0.747)	(0.517)	(0.220)	(0.482)	(0.323)
Kernel matching, *h* = 0.001	1275	–0.009	–0.022	–0.059	–0.039	–0.035
		(0.820)	(0.597)	(0.142)	(0.346)	(0.396)
Propensity score matching, two-way interaction specification						
Nearest neighbor, k = 1	1446	–0.093*	–0.029	–0.058	–0.123**	–0.127**
		(0.013)	(0.459)	(0.127)	(0.002)	(0.001)
Nearest neighbor, k = 1, c = 0.001	1210	–0.090*	–0.030	–0.030	–0.037	–0.102*
		(0.028)	(0.458)	(0.463)	(0.379)	(0.015)
Radius matching, r = 0.001	1210	–0.066	–0.026	–0.042	–0.030	–0.082
		(0.104)	(0.520)	(0.306)	(0.481)	(0.052)
Kernel matching, h = 0.001	1210	–0.068	–0.027	–0.036	–0.036	–0.084*
		(0.096)	(0.503)	(0.386)	(0.393)	(0.047)

Notes: Statistical significance levels for two-sample *t*-tests in parentheses (* $p < 0.05$, ** $p < 0.01$).
Source: German Socio-Economic Panel, wave Y (2008), unweighted data

imbalance is satisfactorily minimized if bias is of the order of 3–5% on any dimension considered. I also omit balancing tests for the propensity score in Table 12.3 since all propensity score based algorithms achieve perfect balance in this case.

In more substantive terms, the trade-off between covariate balance and scope of the estimator is readily apparent from the results of Table 12.3. Exact matching results in just that, yet at the price of being able to find matches for a mere 90 treatment cases, that is, covering the counterfactual outcome distribution for about 5% of the sample of treated cases only. Coarsened exact matching performs much better in comparison. The finely binned algorithm results in near perfect balance in the matched sample for at least 431 treatment group observations (i.e. about 30% of the full sample), and if more coarsening is permitted, the second CEM algorithm provides excellent balance – with relatively mild imbalance on parental socio-economic status the sole exception among the five covariate constellations considered – for three quarters of the treatment group observations. Relative to CEM and exact matching, the propensity score based algorithms all exhibit higher levels of covariate imbalance, yet achieve a broader representativeness of the matched samples. Plain nearest-neighbor matching evidently matches all treatment observations within common support, yet even within a strict caliper of $c = 0.001$, the propensity score based algorithms succeed in matching controls to around 85% of the treatment group observations.

Also, while the propensity score based algorithms generally reduce covariate imbalance to levels consistent with Rubin's rule, there are some noteworthy features and exceptions. For one thing, all propensity score algorithms tend to overcompensate for imbalance in the concrete analysis since measures of remaining bias are consistently negative. Furthermore, as with the CEM algorithm, parental ISEI turns out to be a covariate for which it seems relatively difficult to achieve adequate balance with a main effects specification of the assignment model. Relative to nearest-neighbor matching with its well-known susceptibility to random error in matching, radius and kernel matching algorithms significantly improve the situation, and are sufficient to bring standardized bias down to 5–6%. What is much more discomforting, however, is that the elaborate specification of model 2 does not clearly improve covariate balance in the concrete example, despite superior goodness of fit on all standard measures. While imbalance with respect to parental socio-economic status is improved, covariate balance with respect to gender, parental level of education in the female sample and, for simple nearest-neighbor matching, overall parental level of education has clearly deteriorated relative to the simpler assignment model specification. This result might be due to particular features of the samples and the problem at hand, yet it suggests that standard advice to include higher-order interactions as a default for adequate assignment model specification should be taken with a grain of salt. Since higher-order interactions still require linearity, the flexibility of the regression specification might improve only very modestly, and it might have been more worthwhile to systematically explore non-linearities in first- and higher-order relationships instead. Given similarity in the size of the matched sample, the superior balancing performance of the coarsely binned CEM algorithm certainly suggests that this is exactly the case in the present application.

As a flip side to the assessment of covariate balance, it is also important to consider the scope of the resulting estimate, not the least because this implicitly (re)defines the actual quantity being estimated. Clearly, the stark differences in the size of the matched samples already suggest that the various algorithms differ strongly with respect to sample representativeness. Another way to examine the issue is to inspect the kernel density estimate (or, alternatively, the histogram) of the propensity score distribution in the original data and in the matched samples. Figure 12.2 provides the data, using estimated propensity scores from the main effects assignment model (model 1) as a summary measure to evaluate the behavior of the exact matching algorithms. Evidently, the propensity score based algorithms are far more successful in this respect: nearest-neighbor matching results in exactly the original treatment sample distribution (minus a few off-common support cases) and is therefore not shown separately in Figure 12.2, and radius (or any other of the caliper-based algorithms) closely aligns with the original sample up until about $P(Z) = 0.85$ and higher up in the upper tail of the distribution, that is, among groups of respondents who are empirically (nearly) exclusively observed to complete full university degrees. In comparison, the distribution for the finely binned CEM algorithm has clearly moved to the left, being centered in the area around $0.5 \leq P(Z) \leq 0.85$, the core area of overlap between the samples where suitable controls are relatively abundant. The sample distribution resulting from the exact matching algorithm evidently borders on the bizarre, showing clear bimodal features near the modes of the two raw data distributions even with the smoothing implied by the kernel density estimation.

Since the propensity score summarizes the multivariate relationships between covariates and treatment status, it is very difficult to use the propensity score distribution (as in Figure 12.2) to characterize the target population that results from any of the matching algorithms. Instead, it is usually more instructive to apply the logic of the balancing test to the issue, and Table 12.4 compares covariate support between the original (full) treatment observation sample and the matched samples using both the standardized bias measure and the conventional t test for group

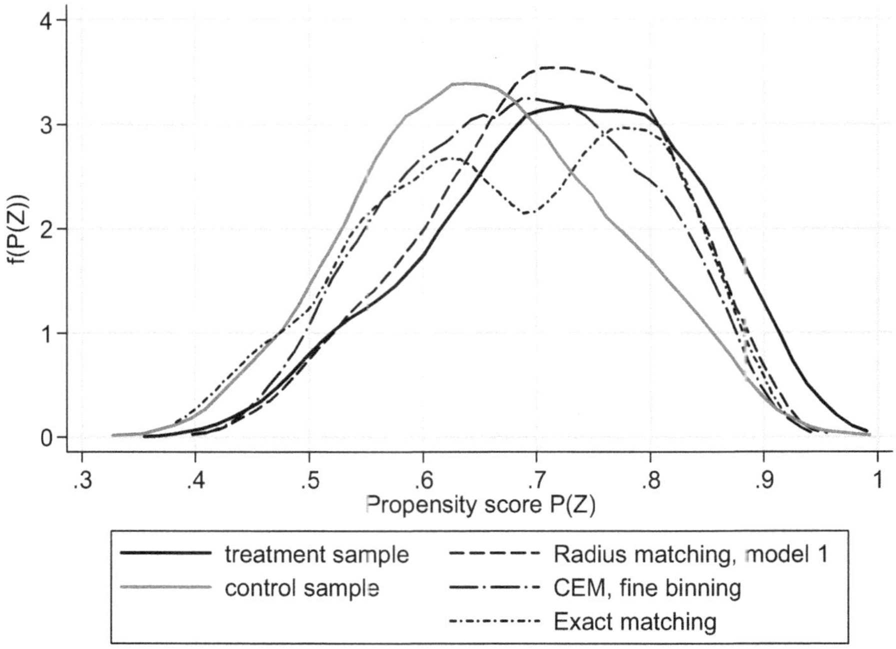

Figure 12.2 Kernel density estimates of the propensity score distribution in different samples

Notes: Exact and CEM matched samples evaluated using estimated propensity scores from assignment model 1.
Sources: German Socio-Economic Panel, wave Y (2008), unweighted data

mean differences. Unsurprisingly, the results are the exact reverse of Table 12.2, with propensity score based algorithms that are able to smooth (i.e. match) across areas of data sparseness clearly outperforming exact matching algorithms that refrain from doing so. Nearest-neighbor matching results in fully representative matched samples by definition, yet differences in sample characteristics surface once (fairly stringent) calipers or coarsened exact matching is imposed. Within calipers and coarsely binned CEM strata, matched samples are less likely to include observations for high-ISEI and high-education parental backgrounds since there are few if any respondents with these characteristics who have chosen to complete applied professional degrees instead of full university degrees – which makes the counterfactual non-identified with any non-parametric estimator. Clearly, the situation much deteriorates with finely binned CEM, let alone the plain exact matching algorithm. The finely binned CEM algorithm results in a matched sample that is disproportionately lacking female respondents, and especially so for East Germany and for women from high-parental education backgrounds. Naturally, biases of estimator scope are largest with exact matching, yet the results of Table 12.4 may at least help to illustrate the warning that absolute significance levels may be misleading indicators of covariate imbalance. The exact matching sample exhibits major imbalance relative to the full sample on all dimensions but parental ISEI, yet only the underrepresentation of women from East Germany is so severe as to result in a statistically significant t test in the small sample.

Table 12.4 Common support tests by matching algorithm and assignment model specification

	$N_D \in D=1$	Standardized bias (*p-value*)				
		Female	Female × East Germany	Parental ISEI	Parental education	Female × parental education
Raw data	1457	0.486	0.139	53.87	13.72	6.67
Exact matching	90	−0.196	−0.382**	0.033	0.097	−0.113
		(0.072)	(0.000)	(0.758)	(0.384)	(0.311)
Coarsened exact matching						
Fine binning	431	−0.228**	−0.193**	0.011	0.025	−0.183**
		(0.000)	(0.000)	(0.835)	(0.651)	(0.001)
Coarse binning	1111	0.013	0.005	−0.058	−0.101*	−0.011
		(0.747)	(0.898)	(0.142)	(0.011)	(0.791)
Propensity score matching, main effects specification						
Nearest neighbor,	1456	0.001	0.000	−0.001	−0.001	0.001
k = 1		(0.986)	(0.994)	(0.987)	(0.982)	(0.986)
Nearest neighbor,	1275	−0.020	−0.020	−0.054	−0.080*	−0.038
k = 1, c = 0.001		(0.607)	(0.600)	(0.162)	(0.038)	(0.319)
Propensity score matching, two-way interaction specification						
Nearest neighbor,	1446	0.007	0.003	−0.003	−0.003	0.007
k = 1		(0.842)	(0.934)	(0.932)	(0.933)	(0.851)
Nearest neighbor,	1210	−0.005	−0.061	−0.036	−0.046	−0.018
k = 1, c = 0.001		(0.899)	(0.115)	(0.357)	(0.234)	(0.640)

Notes: Statistical significance levels for two-sample *t*-tests in parentheses (*$p < 0.05$, **$p < 0.01$). Common support for radius and kernel matching algorithms is equivalent to those for nearest-neighbor caliper matching using identical calipers.
Source: German Socio-Economic Panel, wave Y (2008), unweighted data

Parameter estimation

As in the current illustration, it would seem likely that suitably specified caliper, radius or kernel propensity score algorithms or appropriately coarsened CEM estimators may represent appealing compromises between the twin goals of ensuring covariate balance and sample representativeness in most practical applications of matching estimators in social science research. Table 12.5, which finally provides the treatment effect estimates of original interest to the analysis, nevertheless continues to contain estimates from all algorithms discussed here, not least as a demonstration that much of the above recommendation also carries over to the case of estimator efficiency. As a rule, propensity score estimators tend to have lower variance than exact matching algorithms since the size of the resulting matched samples will be larger. Among the propensity score based algorithms, radius, kernel and related algorithms make more comprehensive use of the available sample data, and hence tend to exhibit lower variance.

Leaving aside extensions such as doubly robust estimation at this point, the ATT treatment effect estimate is simply the difference in average earnings between the treatment sample and the reweighted counterfactual sample of control group observations. Interestingly, and despite some quite strong differences in algorithm behavior observed above, all matching estimates of the ATT (but one) converge on the range of an earnings return to full university degrees of some 16–20%, which also closely corresponds to the estimates from a comparable linear regression model.[10] The matching estimates thus consistently suggest that observable earnings differences in the raw data are a misleading estimate of the causal impact of university education on earnings due to sizeable suppressor effects which are, upon closer inspection, mostly due to the gender imbalance in graduation patterns. In all likelihood, the fact that the exact matching estimator is such a clear outlier relative to all other estimates is best interpreted as being related to the

Table 12.5 ATT parameter estimates by matching algorithm

		Conditional treatment effects			
	ATT University degree	CATT men	CATT women	CATT non-acad. origins	CATT academic origins
Raw data	0.096*	0.118*	0.156*	0.037	0.312**
	(0.048)	(0.055)	(0.072)	(0.053)	(0.104)
Exact matching	0.048	0.106	–0.042	–0.001	0.107
	(0.200)	(0.265)	(0.306)	(0.220)	(0.353)
Coarsened exact matching					
Fine binning	0.166	0.282**	–0.027	0.079	0.287
	(0.089)	(0.104)	(0.162)	(0.094)	(0.177)
Coarse binning	0.171**	0.180*	0.161	0.155*	0.202
	(0.061)	(0.081)	(0.094)	(0.063)	(0.137)
Propensity score matching, main effects specification					
Nearest neighbor,	0.194*	0.492**	–0.121	0.062	0.397**
k = 1	(0.082)	(0.088)	(0.113)	(0.074)	(0.155)
Nearest neighbor,	0.137*	0.437**	–0.193*	0.058	0.281**
r = 1, c = 0.001	(0.069)	(0.085)	(0.094)	(0.072)	(0.126)
Radius matching,	0.170**	0.487**	–0.179*	0.104	0.290**
r = 0.001	(0.065)	(0.074)	(0.088)	(0.070)	(0.111)
Kernel matching,	0.169**	0.487**	–0.181*	0.099	0.296**
h = 0.001	(0.066)	(0.079)	(0.085)	(0.069)	(0.108)
Propensity score matching, two-way interaction specification					
Nearest neighbor,	0.178*	0.475**	–0.132	0.104	0.292*
k = 1	(0.083)	(0.096)	(0.103)	(0.081)	(0.130)
Nearest neighbor,	0.163*	0.430**	–0.118	0.101	0.267*
k = 1, c = 0.001	(0.071)	(0.083)	(0.098)	(0.081)	(0.121)
Radius matching,	0.195**	0.492**	–0.116	0.154	0.265*
r = 0.001	(0.072)	(0.084)	(0.097)	(0.080)	(0.120)
Kernel matching,	0.181*	0.470**	–0.121	0.134	0.262*
h = 0.001	(0.072)	(0.086)	(0.098)	(0.077)	(0.120)

Notes: Bootstrap standard errors in parentheses, $N = 250$ replications; statistical significance levels for t-tests indicated at $^{*}p < 0.05$, $^{**}p < 0.01$.

Source: German Socio-Economic Panel, wave Y (2008), unweighted data

circumscribed nature of the resulting target population. Naturally, all of the above interpretation presupposes that the available covariates have been sufficient to maintain the CIA, that is, to balance expected outcomes net of treatment across the comparison groups, so as to enable valid causal inference.

With that qualification, it is also instructive to note that the various matching algorithms diverge considerably as far as more specific conditional ATT estimates are concerned. In particular, inference about gender differences in returns to university education, but also about differences by parental educational background considerably depend on the estimator. The fine binning version of the CEM algorithm and all propensity score based algorithms agree that earnings returns to full university degrees are modest for graduates from non-academic social backgrounds, but quite considerable – typically as much as about twice or three times as large – among graduates from academic backgrounds; with the coarsely binned CEM algorithm, the respective estimates are quite closely aligned, however. Similar, if not more striking, differences surface with respect to gender. Here, estimates from the finely binned CEM and all propensity score based algorithms agree that university education has significant earnings returns among male graduates only, yet zero returns among female graduates. Equally consistently, gender differences in the estimates are much more pronounced with propensity score matching, and the coarsely binned CEM variant once more deviates completely by signaling no gender difference in ATT parameters at all. While full resolution of these differences is beyond the scope of the current chapter, it is evident that such divergence of results indicates that counterfactual estimates clearly depend on how that information is constructed, that is, which control observations are being relied upon and what target population the estimate is referring to, since extrapolation to cases of less than perfect matching is a requirement in any practically relevant research in the social sciences. Here, different researchers may legitimately take different stances about which estimation strategy – say, kernel matching versus the coarsened CEM algorithm in the present example – might be preferable on statistical or substantive grounds. The fact that concerns for a close relationship between statistical analysis and (implicit) substantive theory are practically forced on both the analyst and her audience should be considered a major virtue of matching estimators.

PITFALLS IN APPLIED RESEARCH

As with any other statistical technique, matching estimators are neither a silver bullet nor immune to misinterpretation and malpractice. With the primary goal of valid causal inference in mind, it seems useful to sharply distinguish between statistical and broader inferential pitfalls in applied research using matching methods. On the statistical side, it would seem that matching estimators in many respects are designed to take the mathematical machinery out of causal inference, which is an element likely to be attractive to the applied social scientist. As non-parametric estimators, matching techniques minimize the role of assumptions about functional forms, distributions of error terms or treatment effect homogeneity that are often very hard to motivate on substantive grounds and hence constitute a major source of misapplication or misinterpretation of regression models in applied research. Nevertheless, as is evident in the illustration above, matching estimators clearly face trade-offs between the goals of covariate balance, estimator scope and efficiency that are similar to related issues in regression modeling. Proper specification of the assignment model for propensity score matching is an art in itself and typically requires exploratory specification searches guided by both model diagnostics and sufficient attendance to salient properties of the empirical covariate distribution, besides any subject-matter input on what actually constitutes the set of appropriate covariates to identify a treatment effect of interest. In matching on the covariate distribution directly, recent alternatives such as the coarsened exact matching algorithm may often provide a useful alternative that avoids the explicit specification of an assignment model (and the potential for mistakes that comes with this), yet as the empirical example has illustrated, great care might be needed to ensure that the analysis results in a treatment effect estimate that is sufficiently close to representing the intended target population. In any case, the appropriate choice of an algorithm to construct the counterfactual observation weights is evidently the critical step in any matching estimator, and new developments in this dynamic field of research – such as entropy balancing and optimal matching algorithms – may be expected to further assist applied researchers with the statistical considerations involved.

In addition to any statistical considerations, the use of matching estimators for causal inference implies substantial inferential pitfalls related to inattentive analysis and overconfidence about causal assertions. Neither of these is necessarily germane to matching, and in fact one might argue that matching estimators again require analysts to attend to and confront core issues in causal inference – the choice of covariates, the balancing of samples, and the clear definition of the estimand of interest – as a natural byproduct of the technique. Relatedly, Rubin (2006) considers the separation of the design step – the specification of the assignment model and the balancing of covariates across comparison groups – from the actual estimation of the treatment effect of interest as one of the key methodological advantages to matching since the setup of any typical matching estimator reduces the chances of pure curve fitting in applied research. On a more general level, however, causal inference based on matching estimates is subject to the usual qualifications and assumptions of causal inference using observational data. Matching estimators are easily applied for overconfident causal assertions if taken at face value and without concern about the validity of the underlying – explicit or implicit – assignment model that justifies the exogeneity assumption. As with related techniques, that assessment goes beyond purely statistical considerations but requires subject-matter input, and different analysts may in fact differ in their assessment as to the conditions under which exogeneity of treatment assignment may plausibly be asserted. It is a definitive virtue of matching estimators, however, that analyst choices and assumptions become exceptionally transparent – and hence subject to scientific criticism – in the process of the empirical analysis.

FURTHER READING

Morgan and Winship (2007) provide an excellent introduction to the counterfactual model of causal inference, including a comprehensive overview of the fundamentals of matching estimators and their relationship with regression models. Rosenbaum (2002), Rubin (2006), Imbens (2004) and Heckman et al. (1998) provide major reviews of the statistical and econometric approaches to causal inference using matching estimators; Rosenbaum and Rubin (1985, 1983) developed the fundamentals of propensity score matching in two papers in the 1980s. Morgan and Harding (2006), Smith and Todd (2005), Dehejia and Wahba (2002), Caliendo and Kopeinig (2008) and Iacus et al. (2012) discuss various aspects in the practical application of matching estimators. Canned routines to implement matching estimators are increasingly becoming available for standard statistical packages, including Stata (`psmatch2`, `pscore`, `nnmatch`, `cem`) and R (`matchit`, `cem`).

NOTES

1 Via the non-parametric function $\mu(\cdot)$, equation (12.10) is intended to describe a completely general relationship between observable covariates and potential outcomes. The impact of observed as well as unobserved factors may depend on treatment status D in principle (i.e. in general, $\mu_0(\cdot) \neq \mu_1(\cdot)$, $U_{0i} \neq U_{1i}$ and $E(U_0) = E(U_1) = 0$).

2 Applied to the case of the ATT parameter, the equivalent expression to equation (12.9) demonstrates that the average treatment effect on the treated is identified under slightly less demanding conditions. Specifically, there is only one unobservable quantity, namely $E[Y_0|D_i = 1]$, that needs to be estimated, so being able to maintain assumption A-2 is sufficient to identify the parameter. In the case of quantile or other distributional treatment effects, strict ignorability of assignment (i.e. independence strictly at the individual level instead of in a conditional expectations sense) is required for identification.

3 It is also possible to estimate the ATT or a related parameter using regression analysis on the matched sample (Rubin and Thomas, 2000). Such *doubly robust* estimators seek to minimize bias relative to standard matching and regression analysis since specification bias in each standalone estimator may partly cancel out in combination.

4 Alternatively, probit or linear probability models are used. There is broad consensus that the choice of probability model is of minor importance in matching estimators unless a considerable fraction of the treatment sample is in the tails of the propensity score distribution. In fact, though most applications match on the propensity score, it is also possible to utilize either the predicted index or the predicted odds of treatment for the purpose, especially for additional differentiation when, again, a considerable fraction of the sample is in the tails of the distribution.

5 There is also an efficiency aspect to using estimated propensity scores. Since matching on estimated propensity scores involves conditioning on the systematic association between observed covariates and treatment status only, the resulting estimator is purged from unsystematic measurement error in the assignment model (see Rosenbaum, 1987).
6 One particularly irritating consequence of using significance tests to assess covariate balance is that 'sufficient' balance (i.e. a non-significant result on the chosen test) may be achieved simply by reducing the sample size. At given sample size, *differences* in significance levels of balancing tests across alternative specifications of the assignment model remain a useful indicator of their relative adequacy, however.
7 Expressed differently, a theoretical argument is required to decide whether divergence between propensity score based and exact estimators in terms of covariate balance is an indication of an inadequately specified assignment model (resulting in bias with propensity score matching) or of permissible data smoothing across empirically irrelevant predictors (for which random matching given the propensity score is the efficient response).
8 In comparison, many reference data sets used in the methodological literature comprise much more extreme counterfactuals. For example, relative to the Current Population Survey, the LaLonde experimental benchmark data requires standardized bias adjustment of the order of $SB = 2.5$ in the case of some categorical covariates (e.g. since the wide majority of training program participants in that study were African Americans), and up to $SB = 49$ in case of pre-treatment earnings since many training program participants had experienced limited earnings and extensive unemployment histories prior to program participation (Dehejia and Wahba, 2002). On the other hand, the treatment–control ratio in Deheija and Wahba's study was of the order of 1:80.
9 Due to the well-behaved distribution of the estimated propensity score and the considerable overlap in score distributions between the comparison groups (see Figure 12.2), common support is simply defined in what follows by $[\max\{\min(P(Z)|D=0), \min(P(Z)\mid D=1)\}, \min\{\max(P(Z)|D=0), \max(P(Z)|D=1)\}]$. With stronger group separation or multimodal distributions in the empirical data, it would be more advisable to use a trimming rule to exclude sparse areas of the propensity score distribution from the analysis (Heckman et al., 1998).
10 Using the full GSOEP sample, the linear regression marginal effect estimate for full university degrees is $b = 0.172$ (s.e. = 0.043) in the main effects specification, and $b = 0.203$ (*s.e.* = 0.043) with additional gender and region interactions.

REFERENCES

Abadie, A. and Imbens, G. W. (2006). Large sample properties of matching estimators for average treatment effects. *Econometrica*, 74(1), 235–267.

Abadie, A. and Imbens, G. W. (2008). On the failure of the bootstrap for matching estimators. *Econometrica*, 76(6), 1537–1558.

Brand, J. E. and Xie, Y. (2007). Identification and estimation of causal effects with time-varying treatments and time-varying outcomes. *Sociological Methodology*, 37, 393–434.

Caliendo, M. and Kopeinig, S. (2008). Some practical guidance for the implementation of propensity score matching. *Journal of Economic Surveys*, 22(1), 31–72.

Dehejia, R. H. and Wahba, S. (2002). Propensity score-matching methods for nonexperimental causal studies. *Review of Economics and Statistics*, 84(1), 151–161.

Diamond, A. and Sekhon, J. S. (2013). Genetic matching for estimating causal effects: A general multivariate matching method for achieving balance in observational studies. *Review of Economics and Statistics*, 95, 932–945.

Gangl, M. (2010). Causal inference in sociological research. *Annual Review of Sociology*, 36, 21–47.

Gangl, M. (2012). Fixed effects matching: nonparametric causal inference with longitudinal and hierarchical data. Unpublished.

Ganzeboom, H. B. G. and Treiman, D. J. (1996). Comparable measures of occupational status for the 1988 International Standard Classification of Occupations. *Social Science Research*, 25(3), 201–239.

Hainmüller, J. (2012). Entropy balancing for causal effects: a multivariate reweighting method to produce balanced samples in observational studies. *Political Analysis*, 20(1), 25–46.

Heckman, J. J., Ichimura, H. and Todd, P. E. (1998). Matching as an economic evaluation estimator. *Review of Economic Studies*, 65, 261–294.

Hirano, K., Imbens, G. W. and Ridder, G. (2003). Efficient estimation of average treatment effects using the estimated propensity score. *Econometrica*, 71(4), 1161–1189.

Holland, P. W. (1986). Statistics and causal inference. *Journal of the American Statistical Association*, 81(396), 945–960.

Iacus, S. M., King, G. and Porro, G. (2011). Multivariate matching methods that are monotonic imbalance bounding. *Journal of the American Statistical Association*, 106(493), 345–361.

Iacus, S. M., King, G. and Porro, G. (2012). Causal inference without balance checking: Coarsened exact matching. *Political Analysis*, 20(1), 1–24.

Imai, K., King, G. and Stuart, E. A. (2008). Misunderstandings between experimentalists and observationalists about causal inference. *Journal of the Royal Statistical Society, Series A*, 171(2), 481–502.

Imbens, G. W. (2004). Nonparametric estimation of average treatment effects under exogeneity: a review. *Review of Economics and Statistics*, 86(1), 4–29.

King, G., Nielsen, R., Coberley, C., Pope, J. E. and Wells, A. (2011). Comparative effectiveness of matching methods for causal inference. Unpublished.

Morgan, S. L. and Harding, D. J. (2006). Matching estimators of causal effects. Prospects and pitfalls in theory and practice. *Sociological Methods & Research*, 35(1), 3–60.

Morgan, S. L. and Winship, C. (2007). *Counterfactuals and Causal Inference*. New York: Cambridge University Press.

Rosenbaum, P. R. (1987). Model-based direct adjustment. *Journal of the American Statistical Association*, 82(398), 387–394.

Rosenbaum, P. R. (2002). *Observational Studies*. New York: Springer.

Rosenbaum, P. R. and Rubin, D. B. (1983). The central role of the propensity score in observational studies for causal effects. *Biometrika*, 70, 41–55.

Rosenbaum, P. R. and Rubin, D. B. (1985). Constructing a control group using multivariate matched sampling methods that incorporate the propensity score. *American Statistician*, 39, 33–38.

Rubin, D. B. (2006). *Matched Sampling for Causal Effects*. Cambridge: Cambridge University Press.

Rubin, D. B. and Thomas, N. (2000). Combining propensity score matching with additional adjustments for prognostic covariates. *Journal of the American Statistical Association*, 95(450), 573–585.

Smith, J. A. and Todd, P. E. (2005). Does matching overcome LaLonde's critique of nonexperimental estimators? *Journal of Econometrics*, 125, 305–353.

Wagner, G. G., Frick, J. R. and Schupp, J. (2007). The German Socio-Economic Panel Study (SOEP) – scope, evolution and enhancements. *Schmollers Jahrbuch*, 127(1), 139–169.

13

Instrumental variables regression

Christopher Muller, Christopher Winship and Stephen L. Morgan

INTRODUCTION

Because of the presence of unobserved variables, social scientists are often unable to use standard methods such as stratification, matching, or regression to estimate the causal effect of a treatment D on an outcome Y. Instrumental variables (IV) regression provides one potential solution to this problem. IV regression involves finding an exogenous variable, the instrument Z, that affects Y only through the causal variable D. The researcher identifies the causal effect of D on Y by estimating the portion of variation in the outcome Y associated with the treatment D that is attributable to the exogenous variable Z.

In this chapter we introduce IV regression. Our discussion is based in part on the presentation in chapters 3 and 7 of Morgan and Winship (2007).[1] Our primary goal is to provide the reader with a good conceptual understanding of IV regression.[2] In order to provide conceptual clarity, we rely in part on directed acyclic graphs (DAGs) to represent the causal relationships among variables (see Pearl, 2000, 2009). Our secondary goal is to demonstrate the broad applicability of IV regression. To this end, we present a number of examples from economic history, several of which use natural experiments (Diamond and Robinson, 2010; Rosenzweig and Wolpin, 2000), rather than the standard examples from labor economics research.

We begin with a brief introduction to DAGs. We then discuss binary and continuous instruments. Throughout, we provide both algebraic and graphical introductions to IV regression. We then discuss whether IV assumptions can be tested with data, ways to adjudicate between ordinary least squares (OLS) and IV estimation, how IV assumptions can be relaxed, and two-stage least squares estimation. We close with a didactic example using European Social Survey (ESS) data on Italy and a brief consideration of some important limitations of IV regression.

INTRODUCTION TO THE METHOD

Introduction to directed acyclic graphs

In his 2000 book, *Causality: Models, Reasoning, and Inference*, Judea Pearl lays out a powerful and extensive graphical theory of causality. Here, we present only the elements that are relevant to understanding IV estimators. Pearl has shown that graphs provide a powerful way of thinking about causal systems of variables and the identification strategies that can be used to estimate the effects within them.

As with standard path models, the basic goal of drawing a causal system as a DAG is to represent explicitly all causes of the outcome of interest. Each symbol in a causal graph represents a random variable. To the reader acquainted with traditional linear path models, much of this material will be familiar. There are, however, important and subtle differences between traditional path models and DAGs.

In this chapter, symbols that are placed in brackets are unobserved random variables. All other symbols represent observed random variables. Causes are represented by directed edges $\rightarrow$ (i.e. single-headed arrows), such that an edge from one symbol to another signifies that the variable at the origin of the directed edge causes the variable at the terminus.[3] Missing arrows encode the identifying assumptions of an analysis. The absence of an arrow between two variables represents the strong assumption of the absence of a causal relation.

DAGs are more general than standard paths models in that the functional relationship between variables need not be linear.[4] However, unlike standard path models, a DAG does not permit the representation of simultaneous causation. Only directed edges are permissible: direct causation can run in only one direction, as in $X \rightarrow Y$. Furthermore, a DAG is defined to be an acyclic graph. Accordingly, no directed paths emanating from a causal variable can return to that variable.

In some circumstances, it is useful to use a curved and dashed bidirected edge as a shorthand device to indicate that two variables are mutually dependent on one or more (typically unobserved) common causes. When this representation is used, the resulting graph is no longer formally a DAG. Because the bidirected edge is simply a semantic substitution, however, the graph can usually be treated as if it were a DAG. Such shorthand can be helpful in suppressing a complex set of background causal relationships that are irrelevant to the empirical analysis at hand. Nonetheless, these bidirected edges should not be interpreted in any other way than as we have just stated. They are not indicators of mere correlations between the variables that they connect, nor do they signify the possibility that the two variables have a direct effect on each other.

Figure 13.1 uses DAGs to present the three basic patterns of causal relationships that can occur for any three variables that are related to each other: a chain of mediation, a fork of mutual dependence, and an inverted fork of mutual causation. The first two types of relationship are conventional. For the graph in Figure 13.1a, A affects B through A's causal effect on C and C's causal effect on B. This type of a causal chain renders the variables A and B unconditionally associated. For the graph in Figure 13.1b, A and B are both caused by C. Here, A and B are also unconditionally associated because they mutually depend on C.

For the third graph in Figure 13.1c, A and B are again connected by a pathway through C. But now A and B are both causes of C. Pearl labels C a 'collider' variable. Formally, a variable is a collider along a particular path if it has two arrows running into it. Figuratively, in the third graph the causal effects of A and B 'collide' with each other at C. Collider variables are common in social science applications: any endogenous variable that has two or more causes is a collider along some path.

A path containing a collider variable does not generate an unconditional association between the variables that cause the collider variable. For the mutual causation graph in Figure 13.1c, the pathway between A and B through C does not generate an unconditional association between A and B. Pearl's language is quite helpful here. The path $A \rightarrow C \leftarrow B$ does not generate an association between A and B because the collider variable C 'blocks' the possible causal effects of A and B on each other.

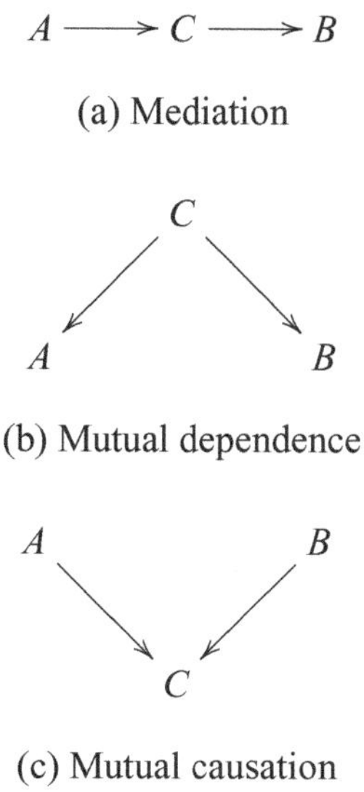

Figure 13.1 Basic patterns of causal relationships for three variables

The importance of considering collider variables is a key insight of Pearl's framework, and it is closely related to the familiar concerns of selecting on the dependent variable and conditioning on an endogenous variable.[5] Even though collider variables do not generate unconditional associations between the variables that determine them, the incautious handling of colliders can create conditional dependence that can sabotage a causal analysis. Understanding collider variables also helps clarify when conditioning strategies can and cannot be used to identify causal effects (Elwert and Winship, 2014).

One of the most common modeling strategies for estimating causal effects in quantitative research is to analyze a putative causal relationship within groups defined by one or more variables. Whether referred to as subgroup analysis, subclassification, stratification, or tabular decomposition, the usual motivation is to analyze the data after conditioning on membership in groups identified by values of a variable that is thought to be related to both the causal variable and the outcome variable.

From a graphical perspective, the result of such a modeling strategy is to generate simplified subgraphs for each subgroup or stratum of the data that correspond to each value of the conditioning variable. This procedure is analogous to disconnecting the conditioning variable from all other variables it points to in the original graph and rewriting the graph as many times as there are values for the conditioning variable. In the mutual dependence graph in Figure 13.1b, conditioning on C results in separate graphs for each value of C; in each of these subgraphs, A and B are disconnected. The reasoning here should be obvious: if analysis is carried out for a group in which all individuals have a particular value for the variable C, then the variable C is constant within the group and cannot be associated with A or B. Thus, from a graphical perspective, conditioning means transforming one graph into a simpler set of component graphs where fewer causes are represented.

As a technique for estimating a causal effect, conditioning is a very powerful and very general strategy. But one very important qualification must be noted: conditioning on a collider variable induces an association between its causes. Rather than eliminating it, conditioning on a collider creates association in the new subgraph where it previously did not exist.

The intuition behind this consequence of conditioning on a collider can be demonstrated with a simple example of selecting on the dependent variable. Consider a hypothetical university where in order to get tenure one must either work hard or do innovative work, but not necessarily both. Hard work and innovative work are separately sufficient to get tenure, but neither is necessary. Assume that among faculty at this hypothetical university there is no relationship between working hard and doing innovative work. Now consider only the faculty who are tenured. If a faculty member does not work hard, we know that he must do innovative work; similarly, if he does not do innovative work, we know that he must work hard. In short, among tenured faculty, working hard and doing innovative work are negatively associated.

Conditioning on a variable therefore has very different effects depending on whether the variable is a collider or a non-collider variable. In the first case, conditioning on a variable potentially eliminates association. In the second case, association is induced. A simple way of thinking about this is that if one conditions on a non-collider variable, this 'blocks' any association that is transmitted across the paths to which it is connected. Paths containing collider variables do not transmit association because a collider 'blocks' the path. However, if one conditions on a collider, this unblocks the paths of which they are a part, enabling these paths to transmit association.

Causal effect estimation with a binary instrumental variable

Consider the causal regression setup

$$Y = \alpha + \delta D + \epsilon, \tag{13.1}$$

where Y is the outcome variable, D is a binary treatment variable, α is an intercept, and ϵ is a summary random variable that stands for all other causes of Y. When equation (13.1) is considered to be a structural or causal model, the parameter δ represents the causal effect of D on Y. If δ does not have a subscript (i) either explicitly or implicitly, then the model indicates that the causal effect of D on Y is constant across individuals. As we discuss later in this chapter, this is a strong assumption that, if false, has important implications for how an IV estimate should be interpreted.

Now consider a binary variable, Z, such that the probability that D is equal to 1 rather than 0 is a function of Z. Figure 13.2 depicts two possible ways the variable Z could be related to both D and Y. Note first that, for both causal diagrams, the presence of the set of paths denoted by $D \leftarrow\text{-}\text{-}\text{-}\text{-}\rightarrow \epsilon \rightarrow Y$ prevents a least squares regression of Y on D (or any other conditioning strategy) from generating a consistent estimate of the effect of D on Y. For the graph in Figure 13.2a, Z has an association with Y only through D. For the graph in Figure 13.2b, however, Z has an association with Y through D and also through common causes that determine both Z and ϵ.[6]

Consider how Z can be used to estimate the causal effect of D on Y in the first but not the second causal diagram. We know that for the set of causal relationships in both parts of Figure 13.2, the probability that D is equal to 1 rather than 0 is a function of the value of Z. But D still varies when conditioning on the value of Z because there are common causes that determine both D and ϵ. In addition, in Figure 13.2b, Z directly affects Y. In this case, Z should be included in equation (13.1) as a cause of Y. If we continue to maintain the assumption that the effect of D on Y is a constant structural effect δ, then it is not necessary to relate all of the variation in D to all of the variation in Y in order to obtain a consistent estimate of the causal effect. The

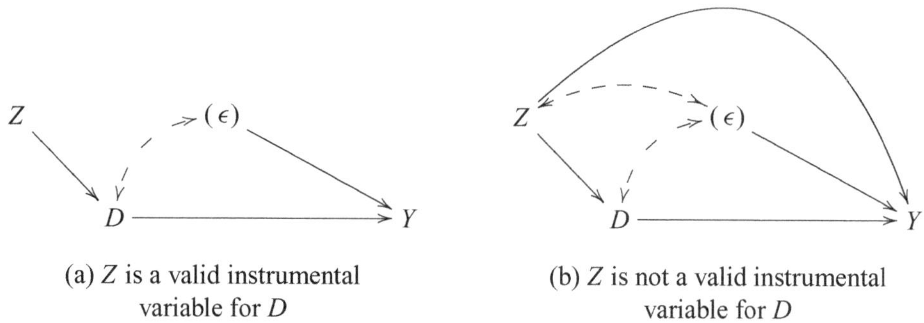

Figure 13.2 Causal diagrams for *Z* as a potential IV

covariation in D and Y that is generated by the common causes of D and ϵ can be ignored if a way of isolating the variation in D and Y that is causal can be found.

Can the variation in D and Y that is generated by variation in Z be used to consistently estimate the causal effect of D on Y in this way? The answer to this question depends crucially on whether or not Z has an association with Y independent of its association with Y through D. For the graph in Figure 13.2a, this strategy will succeed because Z is associated with Y only through D. In the language of econometrics, the instrument satisfies the 'exclusion restriction' with respect to equation (13.1). For the graph in Figure 13.2b, this strategy will fail because Z and Y share common causes, as represented by the bidirected edge $Z \leftarrow\!-\!-\!-\!\rightarrow \epsilon$ and because Z has its own direct causal effect on Y. As a result, the variation that Z appears to generate in both D and Y is also generated by unobserved common causes of Z and by Z's direct effect on Y.

To see this result a bit more formally, take the population-level expectation of equation (13.1), $E[Y] = E[\alpha + \delta D + \epsilon] = \alpha + \delta E[D] + E[\epsilon]$, and rewrite it as a difference equation in Z:

$$\begin{aligned} &E[Y|Z=1] - E[Y|Z=0] \\ &\qquad = \delta(E[D|Z=1] - E[D|Z=0]) + (E[\epsilon|Z=1] - E[\epsilon|Z=0]). \end{aligned} \tag{13.2}$$

Equation (13.2) is now focused narrowly on the variation in Y, D, and ϵ that exists across levels of Z.[7] Now take equation (13.2) and divide both sides by $E[D|Z=1] - E[D|Z=0]$. This yields

$$\begin{aligned} &\frac{E[Y|Z=1] - E[Y|Z=0]}{E[D|Z=1] - E[D|Z=0]} \\ &\qquad = \frac{\delta(E[D|Z=1] - E[D|Z=0]) + (E[\epsilon|Z=1] - E[\epsilon|Z=0])}{E[D|Z=1] - E[D|Z=0]}. \end{aligned} \tag{13.3}$$

If the data are generated by the set of causal relationships depicted in Figure 13.2a, then Z has no association with ϵ, and $E[\epsilon|Z=1] - E[\epsilon|Z=0]$ in equation (13.3) is equal to 0. Consequently, the right-hand side of equation (13.3) simplifies to δ:

$$\frac{E[Y|Z=1] - E[Y|Z=0]}{E[D|Z=1] - E[D|Z=0]} = \delta. \tag{13.4}$$

Under these conditions, the ratio of the population-level association between Y and Z and between D and Z is equal to the causal effect of D on Y. This result suggests that, if Z is in fact associated with D but not associated with ϵ (or with Y, except through D), then the following sample-based estimator will equal δ in an infinite sample:

$$\hat{\delta}_{\text{IV, Wald}} = \frac{E_N[y_i|z_i = 1] - E_N[y_i|z_i = 0]}{E_N[d_i|z_i = 1] - E_N[d_i|z_i = 0]}. \tag{13.5}$$

As suggested by its subscript, this is the IV estimator, which is known as the Wald estimator when the instrument is binary. Although the Wald estimator is consistent for δ in this scenario, the assumption that δ is an invariant structural effect is crucial for this result.[8] As already noted, this is a strong assumption. We explain when and how this assumption can be relaxed later in this chapter.[9]

More insight can be gained about the Wald estimator by separately considering the definitions of the numerator and denominator in equation (13.4). The denominator is equal to what is known as the standard or naïve estimator of the causal effect of Z on D (Morgan and Winship, 2007, p. 44). That is, it is the mean outcome for D when $Z = 1$ minus the mean outcome for D when $Z = 0$. The numerator is equal to the standard or naïve estimator for the causal effect of Z on Y. That is, it is the mean outcome for Y when $Z = 1$ minus the mean outcome for Y when $Z = 0$. Equation (13.4) shows that the ratio of these two estimates provides an estimate of δ, the causal effect of D on Y.

For completeness, return to consideration of Figure 13.2b, in which Z has a non-zero association with ϵ. In order to simplify the algebra, assume that Z does not directly affect Y, but that it has a non-zero association with ϵ. In this situation, $E[\epsilon|Z = 1] - E[\epsilon|Z = 0]$ in equations (13.2) and (13.3) cannot equal 0, and thus (13.3) does not reduce further to (13.4). Instead, it reduces only to

$$\frac{E[Y|Z = 1] - E[Y|Z = 0]}{E[D|Z = 1] - E[D|Z = 0]} = \delta + \frac{E[\epsilon|Z = 1] - E[\epsilon|Z = 0]}{E[D|Z = 1] - E[D|Z = 0]}. \tag{13.6}$$

Here, the ratio of the population-level association between Y and Z and between D and Z does not equal the causal effect of D on Y but rather the causal effect of D on Y plus the last term on the right-hand side of equation (13.6). The Wald estimator in equation (13.5) is not consistent or asymptotically unbiased for δ in this situation. Instead, it converges to the right-hand side of equation (13.6), which is equal to δ plus a bias term that is a function of the net association between Z and ϵ.

A binary instrumental variable example from economic history

Banerjee and Iyer (2005) study the long-term effect of historical institutions on economic development. In particular, they are interested in the legacy of colonial land tenure systems established in British India. The authors propose that historically landlord-based districts will economically underperform historically non-landlord-based districts today because landlords' historical control over revenue collection generated a durable class-based antagonism that frustrates collective action.[10] They estimate the effect of historical non-landlord control over revenue collection on development today using the timing of British control as an instrumental variable.

Using historical evidence, Banerjee and Iyer (2005) make a strong case that there are no unobserved common causes of districts' historical use of landlord-based tenure systems and low levels of economic development today. In fact, they show that the districts with the lowest levels of historical development were those least likely to adopt a landlord system of revenue collection. Thus, if anything, OLS would be expected to underestimate the historical effect of a district's colonial land tenure system. The authors therefore estimate the effect of land tenure using both OLS and IV.

The authors argue that because of the influence of individual colonial administrators, districts where the British took over land revenue collection between 1820 and 1856 were much more likely to have non-landlord systems. Administrator Sir Thomas Munro, for instance, believed that the non-landlord *raiyatwari* system prevented landlords from absorbing surplus agricultural yields. Imposing the *raiyatwari* system, he claimed, would increase agricultural productivity by improving cultivators' incentives. After the Madras Board of Revenue, which favored landlords, prevailed over Munro, he 'traveled to London ... and managed to convince the Court of Directors of the East India Company of the merits of the individual-based *raiyatwari* system; they then ordered the Madras Board of Revenue to implement this policy all over the province after 1820' (Banerjee and Iyer, 2005, p. 1195). Since the assignment of non-landlord land-tenure systems to Madras districts was based largely on the ideologies of individual administrators such as Munro until there was a policy reversal in 1857, Banerjee and Iyer (2005, p. 1191) conclude that 'areas where the land revenue collection was taken over by the British between 1820 and 1856 (but not before or after) are much more likely to have a non-landlord system, for reasons that have nothing to do with factors that directly influence agricultural investment and yields' (see also Wilson, 2011). This statement entails removing the bidirected edge $Z \leftarrow\text{----}\rightarrow \epsilon$ and the directed edge $Z \rightarrow Y$ from Figure 13.2b.

Banerjee and Iyer (2005) construct an instrument, Z, scoring one if a district was colonized between 1820 and 1856 and zero otherwise, and use this instrument to predict their treatment, D, the proportion of a district under non-landlord tenure.[11] Key to their IV identification strategy is the assumption, argued using historical evidence, that there is no bidirected edge connecting Z and their outcome Y, which indexes agricultural production, nor any direct effect of Z on Y. Munro and his allies, in other words, did not select districts for the *raiyatwari* system based on their inherent capacity for high agricultural yields.[12]

Table 13.1 reproduces the OLS and IV estimates for eight different dependent variables measuring agricultural investments and productivity. It shows that, consistent with the authors' hypothesis, the more districts historically relied on non-landlord systems of colonial land tenure, the more developed they are today. The coefficients of the IV estimates are larger than those of the OLS estimates, although not statistically significantly so. The authors argue that the OLS estimates are biased downward because the districts most likely to adopt non-landlord systems were initially the least agriculturally productive.

As we discuss later in this chapter, when district-level effect heterogeneity is present, the particular causal effect the IV estimates reported in Table 13.1 identify is not the average causal effect for all districts. Instead, these estimates identify the average causal effect for districts whose system of land tenure was dictated by the date of British control. Before discussing this interpretation of IV estimates in more detail, we turn to a more complete accounting of the traditional IV literature.

Traditional instrumental variable estimators

As detailed by Goldberger (1972), Bowden and Turkington (1984), and Heckman (2000), IV estimators were developed first in the 1920s by biologists and economists analyzing equilibrium price determination in market exchange (see Working, 1925, 1927; Wright, 1921, 1925). After subsequent development in the 1930s and 1940s (e.g. Haavelmo, 1942; Reiersol, 1941; Schultz, 1938; Wright, 1934), IV estimators were brought into widespread use in economics by researchers associated with the Cowles commission (see Hood and Koopmans, 1953; Koopmans and Reiersol, 1950). The structural equation tradition in sociology shares similar origins to that of the IV literature (see Duncan, 1975). The most familiar deployment of IV estimation in the

Table 13.1 Regression of agricultural investments and productivity on non-landlord proportion

Dependent variable$^{\alpha}$	OLS	IV
Agricultural investments		
Proportion of gross cropped area irrigated	0.065*	0.216
	(0.034)	(0.137)
Fertilizer use (kg/ha)	10.708***	26.198**
	(3.345)	(13.244)
Proportion of rice area under HYV	0.079*	0.411**
	(0.044)	(0.163)
Proportion of wheat area under HYV	0.092**	0.584***
	(0.046)	(0.163)
Proportion of other cereals area under HYV	0.057*	0.526***
	(0.031)	(0.129)
Agricultural productivity		
log (yield of 15 major crops)	0.157**	0.409
	(0.071)	(0.261)
log (rice yield)	0.171**	0.554*
	(0.081)	(0.285)
log (wheat yield)	0.229***	0.706***
	(0.067)	(0.214)
Number of districts	166	166

$^{\alpha}$ Each cell represents the coefficient from a regression of the dependent variable on the non-landlord proportion. Standard errors, corrected for district-level clustering, are in parentheses. HYV = high-yielding varieties. All models include year fixed effects, geographic controls, and controls for the date of British land revenue control. $^{*}p < 0.05$; $^{**}p < 0.01$; $^{***}p < 0.001$.
Source: Banerjee and Iyer (2005).

extant sociological research is as the order condition for identification of a system of structural equations (see Bollen, 1989, 1995, 1996a,b, 2001; Fox, 1984).

Consider the same basic ideas presented earlier for the Wald estimator in equation (13.1):

$$Y = \alpha + \delta D + \epsilon. \tag{13.7}$$

The OLS estimator of the regression coefficient on D is

$$\hat{\delta}_{\text{OLS, bivariate}} \equiv \frac{\text{Cov}_N(y_i, d_i)}{\text{Var}_N(d_i)}, \tag{13.8}$$

where $\text{Cov}_N(\cdot)$ and $\text{Var}_N(\cdot)$ denote unbiased, sample-based estimates from a sample of size N of the population-level covariance and variance.

Again suppose that a correlation between D and ϵ renders the least squares estimator biased and inconsistent for δ in equation (13.7). If least squares cannot be used to effectively estimate δ, an alternative IV estimator can be attempted, with an instrument Z, as in

$$\hat{\delta}_{\text{IV}} \equiv \frac{\text{Cov}_N(y_i, z_i)}{\text{Cov}_N(d_i, z_i)} \tag{13.9}$$

where Z can now take on more than two values. If the instrument Z is correlated with D but uncorrelated with ϵ, then the IV estimator in equation (13.9) is consistent for δ in equation (13.7).[13]

One way to see why IV estimators yield consistent estimates is to again consider the population-level relationships between Y, D, and Z, as in equations (13.1)–(13.4). Manipulating equation (13.1) as before, one can write the covariance between the outcome Y and the instrument Z as

$$\text{Cov}(Y, Z) = \delta \text{Cov}(D, Z) + \text{Cov}(\epsilon, Z), \tag{13.10}$$

again assuming that δ is a constant structural effect. Dividing by $\text{Cov}(D, Z)$ then yields

$$\frac{\text{Cov}(Y, Z)}{\text{Cov}(D, Z)} = \frac{\delta \text{Cov}(D, Z) + \text{Cov}(\epsilon, Z)}{\text{Cov}(D, Z)}, \tag{13.11}$$

which is directly analogous to equation (13.3). When $\text{Cov}(\epsilon, Z)$ is equal to 0 in the population, then the right-hand side of equation (13.11) simplifies to δ. This suggests that

$$\frac{\text{Cov}(y_i, z_i)}{\text{Cov}(d_i, z_i)} \xrightarrow{p} \delta \tag{13.12}$$

if $\text{Cov}(\epsilon, Z) = 0$ in the population and if $\text{Cov}(D, Z) \neq 0$. This would be the case for the causal diagram in Figure 13.2a. But here the claim is more general and holds for cases in which Z is many-valued (and, in fact, for cases in which D is many-valued as well, assuming that the linear specification in equation (13.7) is appropriate).

In economic history, IVs are considered all-purpose remedies for least squares estimators that yield biased and inconsistent estimates, whether because of obvious omitted variables or subtle patterns of self-selection. In the following section we describe three examples of traditional IV estimation from economic history.

Traditional instrumental variables examples from economic history

Nunn (2008) studies the long-term effect of the slave trades on economic development in Africa, using African countries' overland and sailing distance from slaves' ultimate destination as instrumental variables. Drawing on historical documents and shipping records, he constructs estimates of the number of slaves exported from the territories of current African countries. Normalizing these estimates using countries' current land area, he finds a robust negative relationship between historical slave exports and per capita GDP in 2000.

One potential concern with using an OLS estimate to summarize the relationship between historical slave exports and current economic performance is that there might be an unobserved common cause of both the treatment and the outcome. Perhaps Europeans, for instance, sought out regions with historically low levels of development for involvement in the slave trades. Like Banerjee and Iyer (2005), Nunn (2008) uses historical evidence to show that selection, if anything, ran in the opposite direction of the treatment effect.[14]

As a second strategy for addressing his concern about omitted variables, Nunn (2008) uses instrumental variables. He measures the overland and sailing distance of each country from slaves' ultimate destination. The validity of this instrument rests on the assumption that slaves were taken, for example, from the west coast of Africa because this region was close to the West Indies. If instead the West Indies were selected as a slave destination because of their proximity to the west coast of Africa, the instrument, Z, would be an outcome rather than a cause of the causal variable D, rendering it invalid. This would be akin to removing the edge $Z \rightarrow D$ from Figure 13.2 and replacing it with the edge $Z \leftarrow$ D. Nunn (2008) provides considerable historical evidence that regions of slave demand were selected because their climates were suitable for sugar and tobacco cultivation or because they contained gold and silver mines,

Table 13.2 Regression of log real per capita GDP in 2000 on slave exports

Causal variable[α]	OLS	IV	OLS	IV
ln(exports/area)	−0.076***	−0.286*	−0.108***	−0.248***
	(0.029)	(0.153)	(0.037)	(0.071)
First-stage *F*-statistic		1.73		4.01
Number of observations	52	52	42	42

[α]Standard errors are in parentheses. All models include colonizer fixed effects and geographic controls. Columns three and four report estimates from the same model using a smaller sample excluding island and North African countries. $^{*}p < 0.05$; $^{**}p < 0.01$; $^{***}p < 0.001$.
Source: Nunn (2008).

Table 13.3 Regression of log real per capita GDP in 1995 on current institutions

Causal variable[α]	OLS	IV
Average protection against expropriation risk	0.40	1.10
	(0.06)	(0.46)
Number of observations	64	64

[α]Standard errors are in parentheses. 'Average protection against expropriation risk 1985–1995' is the measure of current institutions. All models include indicator variables for Asia, Africa, and other continents as well as controls for latitude.
Source: Acemoglu et al. (2001).

not because they were geographically proximate to the regions that supplied slaves. Since an African country's distance from the historical location of slave demand should have no effect on its current development other than through its effect on the country's historical involvement in the slave trades, Nunn (2008) argues that the exclusion restriction is satisfied.

Table 13.2 reproduces the paper's main OLS and IV estimates. As in Banerjee and Iyer (2005), both the IV coefficients and standard errors are larger than the corresponding OLS coefficients and standard errors, but here the difference in the estimates is statistically significant.[15] The downward bias in the OLS estimates in Nunn (2008) should be expected given historical evidence that the most prosperous societies selected into the slave trades.

Papers estimating the effect of historical conditions such as slavery or colonial institutions on development often take their cue from an influential paper by Acemoglu et al. (2001). The authors estimate the effect of colonial institutions on the economic development of countries colonized by Europeans, using an index of settler mortality as an IV. The problems with using OLS estimation in this analysis should be familiar by now: 'It is quite likely that rich countries can choose or afford better institutions', or that 'economies that are different for a variety of reasons will differ both in their institutions and their income per capita' (Acemoglu et al., 2001, pp. 1369ff.). For Acemoglu et al. (2001), achieving causal identification therefore necessitates circumventing the bidirected edge, $D \leftarrow\text{----}\rightarrow \epsilon \rightarrow Y$, connecting institutions, D, to development, Y, in Figure 13.2.

The authors propose that countries colonized by Europeans received one of roughly two types of institutions. The first type, *extractive* states, provided little support for property rights and few checks and balances against expropriation by elites or the government. The second type, *settler* states, following the European example, championed property rights and placed substantial constraints on executive power. Settler states promoted stronger institutions that, according to the argument, generate durable economic prosperity.

Whether a country received an extractive or a settler state, the authors argue, depended in part upon whether Europeans ultimately could settle there. Some countries' disease environment prohibited extensive European settlement. Assuming that there is a strong relationship between historical and current institutions, an index of European settler mortality can therefore be used as an instrument for current institutions. The edge $Z \rightarrow Y$ can be removed because the indigenous population had developed immunity to the diseases, such as malaria and yellow fever, that most commonly took the lives of European settlers. Table 13.3 reproduces the results of the most parameterized models reported in Acemoglu et al. (2001). As in the previous two examples, the IV estimates are larger and less precisely estimated than the OLS estimates, but they retain statistical significance.[16]

A final example of instrumental variables regression in economic history is provided by Tabellini (2010). Rather than investigate the effects of historical institutions, Tabellini estimates the effect of culture on economic performance, using two measures of historical institutional strength as IVs. Whereas the identification strategies of Banerjee and Iyer (2005), Nunn (2008), and Acemoglu et al. (2001) could be straightforwardly summarized using Figure 13.2, Tabellini's analysis is slightly more complex. We therefore depict the author's assumptions in the DAG drawn in Figure 13.3.

Like Banerjee and Iyer (2005), Tabellini (2010) studies regions rather than countries. The author constructs a regional measure of culture by extracting the principal component from individual respondents' answers to survey questions about their levels of trust, regard for values such as obedience and respect, and belief that they have control over their lives. Like the effect of institutions on development, the effect of culture on development might be confounded by an unobserved common cause of both variables. One possibility is that historical economic prosperity generated both more trust – as well as other relevant cultural attributes – *and* a higher level of current development. To block the possible backdoor path running from culture to output, Tabellini (2010) conditions on a measure of urbanization in 1850.[17] Figure 13.3 shows that conditioning on urbanization blocks the path Culture $\leftarrow$ Urbanization $\rightarrow$ Output.

The concern that a separate unobserved common cause of culture and output might confound the analysis motivates Tabellini's (2010) second identification strategy. Since the author is concerned that there are unobserved common causes of culture and current development other than historical development, we might also be concerned that there are unobserved common causes, U, of culture and historical development. Common causes, W, of historical and current development, moreover, might confound the relationship between urbanization and output. Figure 13.4 adds these unobserved causes to the DAG drawn in Figure 13.3. This DAG shows that in the presence of these unobserved causes, conditioning on Urbanization in 1850 will unblock the path Culture $\dashleftarrow$ U $\dashrightarrow$ Urbanization $\dashleftarrow$ W $\dashrightarrow$ Output on which it is a collider.

Aware of this limitation, Tabellini (2010) adopts an instrumental variables strategy. He constructs two instruments for culture: one measuring a region's historical literacy rate and the other measuring the quality of its historical institutions. The author acknowledges that historical literacy rates, in violation of the exclusion restriction, might have a direct effect on output. Reasoning that a region's historical literacy rate will only affect output through its effect on culture or current education, he conditions on current school enrollment to block the path Culture $\leftarrow$ Literacy $\rightarrow$ Enrollment $\rightarrow$ Output. In addition, all of the paper's regressions include country fixed effects. This rules out the possibility that differences in national historical institutions drive the results. Banerjee and Iyer (2005), however, show that historical institutions at the regional level can have an effect on output independent of what is captured by the measures of culture

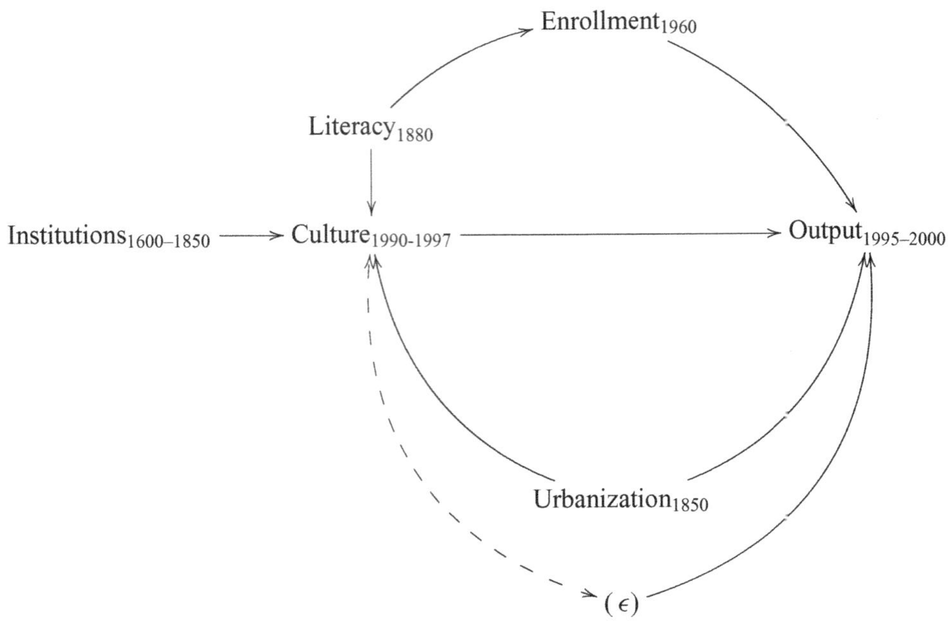

Figure 13.3 The assumptions of Tabellini (2010) in graphical form

used in Tabellini (2010). Were there a direct effect of historical regional institutions on output, the backdoor path Institutions --→ Output, as depicted in Figure 13.4, would remain unblocked, in violation of the exclusion restriction.

Our objective in this discussion is not to question the results reported in Tabellini (2010).[18] The plausibility of the identifying assumptions depends on the reader's substantive knowledge of the case. Rather, we aim to show how the simple graphical language of DAGs can render these assumptions transparent, better enabling readers to assess their validity. This is especially important for analyses using instrumental variables, since a commonly used test of the IV assumption yields misleading answers.[19] We turn to this issue in the next section.

ADVANCED ISSUES

Can data be used to test the IV identifying assumption?

The basic assumption underlying the three studies just summarized is that Z has no causal effect on Y except indirectly through D. It may seem intuitive that this assumption can be tested in the following way. If Z has no direct effect on Y, we may suppose that there will be no association between Z and Y conditional on D. Thus, if we regress Y on Z and D, we should find no association between Z and Y if the IV assumption is correct.

In Figure 13.2a, an unblocked path connects D and Y, but Z is a valid IV that identifies the causal effect of D on Y. The instrument is valid because it causes D and because it is unconditionally unassociated with the common cause of D and Y. To test this assumption, some authors estimate the relationship between Z and Y, using D as a control variable. The rationale for this test is that conditioning on D will block the indirect relationship between Z and Y through D. Accordingly, if the only association between Z and Y is indirect through D, there

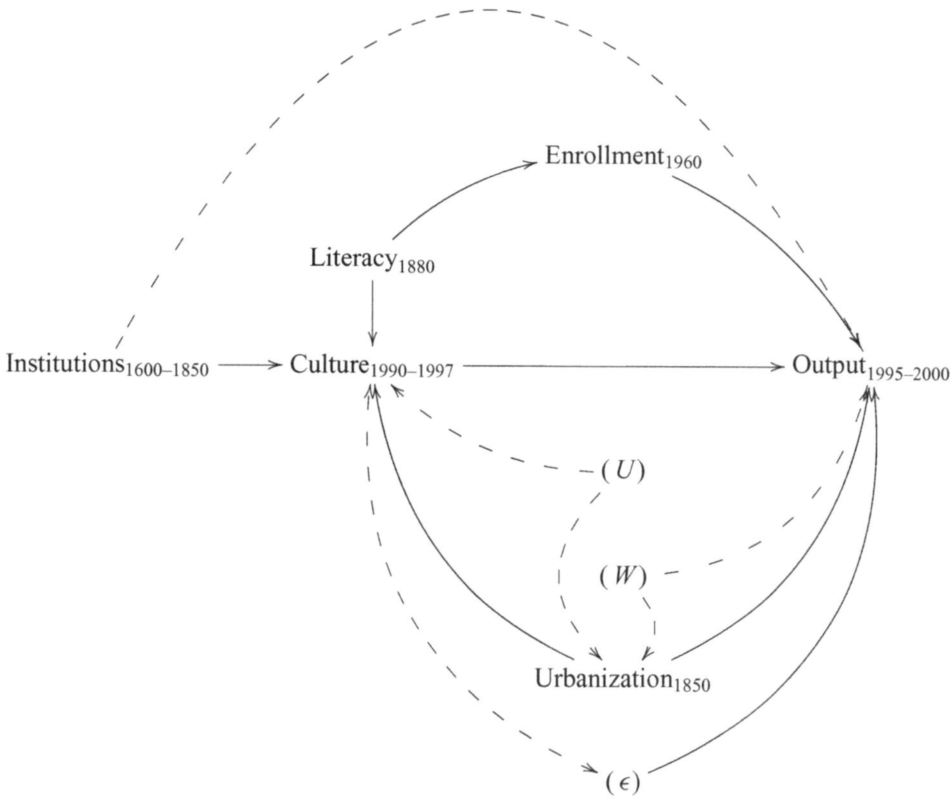

Figure 13.4 Relaxing the assumptions of Tabellini (2010)

should be no association between Z and Y after conditioning on D. If such a net association is detected, it would seem reasonable to conclude that the IV identifying assumption must be false.

It is certainly true that if the IV assumption is invalid, then Z and Y will be associated after conditioning on D. But the converse does not hold. Z and Y will always be associated after conditioning on D when the IV assumption is valid. This follows from the fact that D in Figure 13.2a is a collider that is mutually caused by Z and the unobserved common cause connecting D to Y. As we note in the previous section, conditioning on a collider induces an association between its causes. Thus conditioning on D in this DAG creates an association between Z and the unobserved common cause of D and Y, even though the IV identifying assumption is valid. Not finding a relationship between Z and Y after conditioning on D does not undermine the results of an IV analysis, since the only way not to observe an association in this regression is if the exclusion restriction is satisfied and there is no unobserved common cause confounding the relationship between D and Y (Elwert and Winship, 2014). But in this situation there is no need to use an IV to estimate the effect of D on Y. OLS will produce an unbiased and consistent estimate.[20]

OLS versus IV

The results reproduced in Tables 13.1–13.3 have an important commonality: since the IV estimator uses only a portion of the covariation between the treatment, D, and the outcome, Y, the IV estimates have far larger standard errors than the OLS estimates. If D is uncorrelated with ϵ, then the OLS estimates will be consistent and, according to the Gauss–Markov theorem, efficient, yielding smaller standard errors than the IV estimates. How, then, should we decide whether to use OLS or IV? Recall once more the causal regression setup introduced in equation (13.1),

$$Y = \alpha + \delta D + \epsilon, \tag{13.13}$$

as well as what is known as the first-stage equation predicting the treatment, D, using the instrumental variable Z,

$$D = \gamma + \beta Z + v. \tag{13.14}$$

If D is uncorrelated with ϵ, both the OLS and IV estimators will consistently estimate δ in equation (13.13). Our first intuition thus might be simply to compare the OLS and IV estimates. Different results would suggest that D is endogenous, leading us to prefer the IV estimator. If the estimates were similar, we would prefer OLS because it has smaller standard errors. The problem is that due to sampling error our estimates will never be exactly the same. Thus, we need a more formal test. This test is known as a Hausman test (Hausman, 1978).

Equation (13.14) allows us to segment D into two components: one, $\hat{D}$, that is a function of Z and therefore uncorrelated with ϵ; and another, $\hat{v}$, that may be associated with ϵ. If $\hat{v}$ is associated with Y, we have evidence that D is endogenous. We can test this formally by simply regressing Y on D and $\hat{v}$ and calculating a standard t test of the significance of the coefficient, λ, on $\hat{v}$:

$$Y = \alpha + \delta D + \lambda \hat{v} + w. \tag{13.15}$$

The estimate of δ in this equation will be equivalent to the IV estimate. In estimating the effect of D in equation (13.15), we are controlling for the component of D that may be associated with the error term, $\hat{v}$. This solves the endogeneity problem. If the effect of $\hat{v}$ is zero, however, then there is no need to control for it. In this case, OLS estimation will be both consistent and typically much more efficient.

IV estimation as LATE estimation

Following the adoption of a counterfactual perspective, a group of econometricians and statisticians clarified what IVs identify when unit-level causal effects are heterogeneous. In this section, we discuss the connections between traditional IV estimators and potential-outcome-defined treatment effects (Imbens and Angrist, 1994; Angrist and Imbens, 1995; Angrist et al., 1996; Imbens and Rubin, 1997). The key innovation here is the definition of a new treatment effect parameter: the local average treatment effect (LATE).

Until this point, we have assumed that the effect of D on Y is a constant structural effect. In this section, we relax this assumption and consider a different interpretation of the Wald estimator.[21] Recall the definition of Y in the fundamental problem of causal inference:

$$\begin{aligned} Y &= Y^0 + (Y^1 - Y^0)D \\ &= Y^0 + \delta D \\ &= \mu^0 + \delta D + v^0, \end{aligned} \tag{13.16}$$

where $\mu^0 \equiv E[Y^0]$ and $v^0 \equiv Y^0 - E[Y^0]$. Note that δ is now defined as $Y^1 - Y^0$, unlike its structural representation in equations (13.1) and (13.7) where δ was implicitly assumed to be constant across all units.

To understand when an IV estimator can be interpreted as an average causal effect estimator, Imbens and Angrist (1994) developed a counterfactual framework to classify units into those that respond positively to an instrument, those that remain unaffected by an instrument, and those that 'rebel' against an instrument. Their innovation was to define potential treatment assignment variables, $D^{Z=z}$, for each state z of the instrument Z. When D and Z are binary variables, there are four possible groups of units in the population.[22] These can be summarized by a four-category latent variable $\tilde{C}$ for compliance status:

$$\begin{aligned}
\text{compliers } (\tilde{C} = c): &\quad D^{Z=0} = 0 \text{ and } D^{Z=1} = 1,\\
\text{defiers } (\tilde{C} = d): &\quad D^{Z=0} = 1 \text{ and } D^{Z=1} = 0,\\
\text{always takers } (\tilde{C} = a): &\quad D^{Z=0} = 1 \text{ and } D^{Z=1} = 1,\\
\text{never takers } (\tilde{C} = n): &\quad D^{Z=0} = 0 \text{ and } D^{Z=1} = 0.
\end{aligned}$$

Although this terminology is best suited for describing experiments on individuals, it can accommodate natural experiments affecting aggregate units. Take Banerjee and Iyer (2005), for instance. Districts that adopted non-landlord systems only if their revenue control was taken over by the British between 1820 and 1856 can be defined as compliers ($\tilde{C} = c$). Districts that adopted non-landlord systems only if their revenue control was *not* taken over between 1820 and 1856 are defiers ($\tilde{C} = d$). Districts that would have adopted a non-landlord system irrespective of their date of revenue control are always takers ($\tilde{C} = a$). And districts that would never have adopted non-landlord systems are never takers ($\tilde{C} = n$).[23]

Analogous to the definition of the observed outcome, Y, the observed treatment indicator variable D can then be defined as

$$\begin{aligned}
D &= D^{Z=0} + (D^{Z=1} - D^{Z=0})Z \\
&= D^{Z=0} + \kappa Z,
\end{aligned} \tag{13.17}$$

where $\kappa \equiv D^{Z=1} - D^{Z=0}$. The parameter κ in equation (13.17) is the unit-level causal effect of the instrument on D. If the instrument represents encouragement to take the treatment, such as the adoption of non-landlord tenure systems in colonial India, then κ can be interpreted as the unit-level compliance inducement effect of the instrument. Accordingly, $\kappa = 1$ for compliers and $\kappa = -1$ for defiers. For always takers and never takers, $\kappa = 0$ because neither group responds to the instrument.

Given these definitions of potential outcome variables and potential treatment variables, a valid instrument Z for the causal effect of D on Y must satisfy three assumptions in order to identify a LATE:

$$\text{independence assumption: } (Y^1, Y^0, D^{Z=1}, D^{Z=0}) \perp\!\!\!\perp Z; \tag{13.18}$$

$$\text{nonzero effect of instrument assumption: } \kappa \neq 0 \text{ for all } i; \tag{13.19}$$

$$\text{monotonicity assumption: either } \kappa \geq 0 \text{ for all } i \text{ or } \kappa \leq 0 \text{ for all } i. \tag{13.20}$$

The independence assumption (13.18) is analogous to the assumption that $\text{Cov}(Z, \epsilon) = 0$ in the traditional IV literature. It stipulates that the instrument must be independent of the potential outcomes and potential treatments. Knowing the value of the instrument for unit i must not yield any information about the potential outcome of unit i under either treatment state. Moreover, knowing the realized value of the instrument for unit i must not yield any information about the probability of being in the treatment under the alternative hypothetical values of the instrument. A valid instrument predicts observed treatment status (D), but it does not predict potential treatment status ($D^{Z=z}$).

Assumptions (13.19) and (13.20) are about unit responses to shifts in the instrument. The assumption of a non-zero effect of Z on D is a stipulation that the instrument must predict treatment assignment for at least some units. There must be at least some compliers or some defiers in the population of interest. The monotonicity assumption further specifies that the effect of Z on D must be either weakly positive or weakly negative for all individuals i. Thus, there may be either defiers or compliers in the population, but not both.

If these three assumptions obtain, then an instrument Z identifies the LATE: the average causal effect of the treatment for the subset of the population whose treatment selection is induced by the instrument. If $\kappa \geq 0$ for all i, then the Wald estimator from equation (13.5) converges to a particular LATE:

$$\hat{\delta}_{\text{IV, Wald}} \xrightarrow{p} E[\delta|\tilde{C} = c], \tag{13.21}$$

which is equal to $E[Y^1 - Y^0|D^{Z=1} = 1, D^{Z=0} = 0]$ and is therefore the average causal effect among compliers. In contrast, if $\kappa \leq 0$ for all i, then the Wald estimator from equation (13.5) converges to the opposite of LATE:

$$\hat{\delta}_{\text{IV, Wald}} \xrightarrow{p} E[\delta|\tilde{C} = d], \tag{13.22}$$

which is equal to $E[Y^1 - Y^0|D^{Z=1} = 0, D^{Z=0} = 1]$ and is therefore the average causal effect among defiers. In either case, the treatment effects of always takers and never takers are not informed in any way by the IV estimate. Returning to Banerjee and Iyer (2005), we can thus interpret the IV estimates reproduced in Table 13.1 as the average causal effect for districts whose system of land tenure was set by the date when the British took control of revenue collection.

Two-stage least squares estimation

Most statistical software features packages that will automatically generate IV estimates. Identical point estimates can also be generated using what is known as two-stage least squares estimation. In the first stage, the analyst regresses the treatment, D, on the instrument, Z, all exogenous covariates, X, and a summary random variable, v, representing all other causes of D:

$$D = \gamma + \beta_1 Z + \beta_2 X + v. \tag{13.23}$$

In the second stage, the analyst uses the predicted values from the first-stage equation, $\hat{D}$, to predict the outcome conditional on the same exogenous covariates:

$$Y = \alpha + \delta_1 \hat{D} + \delta_2 X + \epsilon. \tag{13.24}$$

Often two-stage least squares is described as purging D of the component that is associated with v and then estimating the effect of D on Y using the purged D, or $\hat{D}$.

Because of its intuitive interpretation, we use a two-stage least squares framework to introduce our example below. However, in general it is not advisable to perform each stage manually, as the resulting standard errors and test statistics will be incorrect.[24]

EXAMPLE ANALYSIS

One of the primary theoretical motivations for the analysis presented in Tabellini (2010) is the claim, first advanced by Banfield (1958) and later studied by Putnam (1993), that differences

Table 13.4 Italian regions

Tabellini (2010)	ESS
Abruzzo – Molise – Basilicata	Abruzzo
	Molise
	Basilicata
Calabria	Calabria
Campania	Campania
Emilia-Romagna	Emilia-Romagna
Lazio	Lazio
Liguria	Liguria
Lombardia	Lombardia
Piemonte – Valle d'Aosta	Piemonte
	Valle d'Aosta
Puglia	Puglia
Sicilia – Sardegna	Sicilia
	Sardegna
Toscana	Toscana
Trentino-Alto Adige – Veneto – Fruili-Venezia Giulia	Trentino-Alto Adige
	Veneto
	Friuli-Venezia Giulia
Umbria – Marche	Umbria
	Marche

in economic development across the regions of Italy can be partially attributed to differences in citizens' generalized trust. In this didactic example, we use data from the European Social Survey (ESS) to test this hypothesis at the individual level. Our analysis thus provides a loose theoretical replication of the results reported in Tabellini (2010).

Despite a well-developed theoretical literature, the exact mechanisms by which Italian citizens' generalized trust might affect regional economic development remain incompletely understood. One possible way a lack of trust in others might hamper economic progress is by impeding borrowing and lending practices. Fortunately, the ESS enables us to test this hypothesis. All ESS survey respondents were asked to answer the following question: 'If for some reason you were in serious financial difficulties and had to borrow money to make ends meet, how difficult or easy would that be?' Respondents were given five choices, coded as follows: 'Very difficult' (1); 'Quite difficult' (2); 'Neither easy nor difficult' (3); 'Quite easy' (4); 'Very easy' (5). We treat these responses as continuous and run linear models throughout our analysis.

To predict citizens' beliefs about the ease or difficulty of borrowing money in times of need, we use the same trust measures used by Tabellini (2010). Both the World Values Survey and the ESS ask respondents: 'Generally speaking, would you say that most people can be trusted, or that you can't be too careful in dealing with people? Please tell me on a score of 0 to 10, where 0 means you can't be too careful and 10 means that most people can be trusted.' We treat this, our causal variable, as continuous as well.

Because the relationship between generalized trust and beliefs about borrowing money might be confounded by an unobserved common cause, we condition our estimates on several additional variables measured at the individual level. Specifically, we control for respondents' age in years, gender (coded as 1 for men and 0 for women), nativity (1 if native; 0 otherwise), and education (divided into five ascending categories).

Next we merge our data set with that of Tabellini (2010), linking individuals to information about the historical region in which they reside. There are 20 Italian regions in the ESS data set, each of which falls into one of the 13 historical regions identified by Tabellini (2010). Table 13.4 shows how Italy's current regions were organized historically.

We follow Tabellini (2010) and use a measure of historical institutions at the regional level as an instrument for individual respondents' levels of trust.[25] Tabellini (2010) argues that because these historical institutions have no modern equivalent, they should not have a direct effect on development. The author uses country fixed effects to block the effect of national historical institutions; likewise we restrict our analysis to a single country. We emphasize that the identifying assumption here is a strong one: as shown by Banerjee and Iyer (2005), it is possible for historical institutions at the regional level to have a direct effect on development. It might be that in our data historical regional institutions affected both individuals' level of generalized trust and their expectations about how easy or difficult it would be to borrow money in times of need. If so, the exclusion restriction would be violated. It is also important to bear in mind that despite our large sample of individuals, our design effectively leaves us only 13 observations at the regional level.[26]

We carry out our IV analysis using a traditional two-stage least squares approach. In the first stage, we write respondents' (i) degree of generalized trust as a function of the quality of the institutions governing the historical region (r) in which they reside and a vector of individual-level covariates, X:

$$\text{trust}_{i,r} = \alpha_1 + \beta \text{institutions}_r + \gamma_1 X_{i,r} + \nu_{i,r}. \tag{13.25}$$

We then substitute the predicted values ($\widehat{\text{trust}}$) from equation (13.25) for individuals' actual survey responses:

$$y_{i,r} = \alpha_2 + \delta \widehat{\text{trust}_{i,r}} + \gamma_2 X_{i,r} + \epsilon_{i,r}, \tag{13.26}$$

where y is an individual's belief about the ease or difficulty of borrowing money in times of need. We adjust the standard errors in all models to account for the clustering of individuals in regions.

Table 13.5 reports parameter estimates from equations (13.25) and (13.26), as well the OLS equations that correspond to them. Column 1 reports the bivariate OLS relationship between an individual's belief that others can generally be trusted and their belief that borrowing money in times of need is easy. As expected, we observe a positive relationship between these beliefs. The more people subscribe to the belief that 'most people can be trusted', the easier they believe it is to borrow money in times of need. Column 2 reports the bivariate IV relationship. Both the coefficient and standard error increase by more than 10 times and the relationship remains statistically significant. We can interpret this coefficient as the local effect of the portion of the total variation in trust that is attributable to historical institutions. Columns 3 and 4 show that both the OLS and the IV estimates are robust to the inclusion of individual-level controls. The first-stage results indicate that historical institutions are a strong predictor of individual respondents' levels of generalized trust, as measured both by the significance of the coefficient on institutions and by the F-statistic testing the strength of the instrument.[27] Provided the exclusion restriction is satisfied, these results provide additional support for Tabellini's (2010) thesis that trust affects economic development. People in regions with stronger historical institutions have more faith that people in general can be trusted, and this leads them to have greater confidence that they will be able to borrow money in times of need.

PITFALLS OF TRADITIONAL IV ESTIMATION: WEAK INSTRUMENTS

By using only a portion of the covariation in the causal variable and the outcome variable, IV estimators use only a portion of the information in the data. This represents a direct loss in statistical power, and as a result IV estimates typically have substantially more expected

Table 13.5 Regression of borrowing beliefs on generalized trust

Independent variables$^{\alpha}$	OLS	IV	OLS	IV
Trust	0.063***	0.682**	0.040**	0.682**
	(0.011)	(0.165)	(0.010)	(0.169)
Age			0.005***	0.005
			(0.001)	(0.002)
Male			0.198***	0.211*
			(0.025)	(0.079)
Native			–0.030	0.079
			(0.159)	(0.144)
Education			0.234***	–0.042
			(0.032)	(0.072)
Intercept	2.670***	–0.061	1.869***	–0.365
	(0.100)	(0.761)	(0.160)	(0.616)
First stage				
Institutions		0.243***		0.225***
		(0.048)		(0.040)
Age				0.000
				(0.003)
Male				–0.018
				(0.103)
Native				–0.110
				(0.265)
Education				0.415***
				(0.038)
Intercept		4.715***		3.775***
		(0.065)		(0.311)
F - Statistic		25.76		31.76
Number of observations	2522	2522	2516	2516

$^{\alpha}$Standard errors, clustered by region, in parentheses. $^{*}p < 0.05$; $^{**}p < 0.01$; $^{***}p < 0.001$.

sampling variance than other estimators. The problem can be especially acute in some cases. It has been shown that instruments that only weakly predict the causal variable of interest should be avoided entirely, even if they generate point estimates with acceptably small estimated standard errors (Bound et al., 1995). In brief, the argument here is fourfold. (A) In finite samples, IV point estimates can always be computed because sample covariances are never exactly equal to zero. (B) As a result, an IV point estimate can be computed even for an instrument that is invalid because it does not predict the endogenous variable in the population (i.e. even if $\text{Cov}(D, Z) = 0$ in the population, rendering equation (13.11) underdefined because its denominator is equal to 0). (C) At the same time, the formulas for calculating the standard errors of IV estimates fail in such situations, giving artificially small standard errors (when in fact the true standard error for the undefined parameter is infinity); and (D) the bias imparted by a small violation of the assumption that the IV affects the outcome variable only by way of the causal variable can explode if the instrument is weak.[28] To see this last result, consider equation (13.6), which depicts the expected bias in the Wald estimator for a binary IV as the term

$$\frac{E[\epsilon|Z=1]-E[\epsilon|Z=0]}{E[D|Z=1]-E[D|Z=0]}. \tag{13.27}$$

When the identifying assumption is violated, the numerator of equation (13.27) is non-zero because Z is associated with Y through ϵ. The bias is then an inverse function of the strength of the instrument; the weaker the instrument, the smaller the denominator and the larger the bias. If the denominator is close to zero, even a tiny violation of the identifying assumption can generate a large amount of bias. And, unfortunately, this relationship is independent of the sample size. Thus, even though a weak instrument may suggest a reasonable point estimate, and one with an acceptably small estimated standard error, the IV estimate may contain no genuine information whatsoever about the true causal effect of interest (see Hahn and Hausman, 2003; Staiger and Stock, 1997; Wooldridge, 2002).[29]

CONCLUSION

In this chapter we have provided an introduction to instrumental variables regression. We began by using directed graphs to demonstrate the logic underlying IVs. In order to show the broad applicability of IVs for applied researchers, we examined a number of examples of IV analysis in economic history. We noted the strong assumptions required for IVs to give consistent estimates of a causal effect and showed why a common approach to testing these assumptions is flawed. We introduced readers to a simple test that can be used to determine whether to prefer OLS or IV estimates and demonstrated that when the treatment effect is heterogeneous across units, IV analysis only estimates the treatment effect for compliers – those units that are affected by the instrument. We then introduced two-stage least squares estimation and presented a simple didactic example using the European Social Survey. Code to replicate this example in Stata and R is available in this chapter's online appendix. We closed with a brief discussion of some important limitations of IV regression.

FURTHER READING

The interested reader should consult more thorough and technical discussions in Angrist and Pischke (2009), Wooldridge (2002), Morgan and Winship (2007), Bollen (2012), and Murray (2006).

NOTES

1 Specifically, the chapter includes lightly edited material from pages 61–67, 187–190, 193–194, and 196–203 of Morgan and Winship (2007), but adds new discussions of the use of IVs in economic history.
2 Bollen (2012) provides an excellent discussion of many of the technical aspects of IV regression, including a broader treatment of the utility of instrumental variables in structural equation modeling.
3 In Pearl's framework, each variable is assumed to have an implicit probability distribution net of the causal effects reprented by the directed edges. This position is equivalent to assuming that background causes of each variable that exist are independent of the causes explicitly represented in the graph by directed edges.
4 For IV estimation, however, one requires what Pearl labels a linearity assumption. What this assumption means depends on the assumed distribution of the variables. For example, it would be satisfied in Figure 13.2 if the causal effect of Z on D is linear and the causal effect of D on Y is linear.
5 Selecting on the dependent variable and conditioning on an endogenous variable are two typical cases of the same fundamental problem: conditioning on the common outcome of two causes.
6 We have drawn the two bidirected edges separately for Figure 13.2b for simplicity. The same reasoning holds in more complex situations in which some of the common causes of Z and ϵ are also common causes of D and/or Y (and so on).
7 Equation (13.2) is generated in the following way. First, write $E[Y] = \alpha + \delta\, E[D] - E[\epsilon]$ conditional on the two values of Z, yielding $E[Y|Z = 1] = \alpha + \delta\, E[D|Z = 1] + E[\epsilon \mid Z = 1]$ and $E[Y \mid Z = 0] = \alpha + \delta\, E[D \mid Z = 0] +$

$E[\epsilon \mid Z = 0]$. Note that α and δ are constant structural effects and thus do not vary across individuals. Now, from $E[Y|Z = 1] = \alpha + \delta\ E[D|Z = 1] + E[\epsilon \mid Z = 1]$ subtract $E[Y|Z = 0] = \alpha + \delta\ E[D|Z = 0] + E[\epsilon \mid Z = 0]$. The parameter α is eliminated by the subtraction, and δ can be factored out of its two terms, resulting in equation (13.2).

8 As for the origin of the Wald estimator, it is customarily traced to Wald (1940) by authors such as Angrist and Krueger (1999). As we discuss later, the Wald estimator is only consistent, not generally unbiased, in a finite sample.

9 The reasons may well be clear to the reader already. In moving from equation (13.1) to equation (13.2), covariation in Y and D within levels of Z was purged from the equation. But if D has a causal effect on Y that varies across all individuals, it may be that whatever contrast is identified through Z may not be relevant to all members of the population.

10 All cultivable land in British India was governed by one of three systems: the *zamindari* or *malguzari* system, in which a class of landlords collected and set the terms of revenue for peasants under their jurisdiction; the *raiyatwari* system, under which revenue was collected directly from cultivators and usually varied according to annual outputs; and the *mahalwari* system, in which revenue was collected from collective village bodies.

11 In other specifications, they construct an indicator scoring one for districts mostly under non-landlord systems.

12 The authors include controls for the length of time under British rule to block a possible causal path from Z to Y.

13 Notice that substituting d_i for z_i in equation (13.9) results in the least squares regression estimator in equation (13.8). Thus, the OLS estimator implicitly treats D as an instrument for itself.

14 In order to be considered for trade with Europeans, societies needed to have passed a minimal threshold of institutional development. Societies with highly developed institutions tended to be densely populated and economically prosperous (Acemoglu et al., 2002). Thus, the societies most likely to enter the slave trade with Europeans were least likely to suffer from historical economic underdevelopment.

15 We describe a simple test of the difference between OLS and IV coefficients later in this chapter.

16 The authors attribute the larger IV estimates to measurement error in the institutions variables. See Card (2001) for a discussion of why IV estimates are consistently larger than OLS estimates in the literature on returns to schooling.

17 Urbanization is a standard proxy for historical economic development in economic history (see Acemoglu et al., 2005).

18 Indeed, we provide a loose theoretical replication of these results later in the chapter.

19 See Pearl (2009, pp. 274ff.) for an alternative test of the IV identifying assumption.

20 For completeness, consider what the test reveals when the IV assumption is invalid. Suppose that Figure 13.2a is augmented by an unobserved cause E and then two edges $E \rightarrow Z$ and $E \rightarrow \epsilon$. In this case Z and Y would be associated within levels of D for two reasons: (a) conditioning on the collider D generates a net association between Z and ϵ (and hence Z and Y); and (b) the common cause of E of Z and ϵ generates an unconditional association between Z and Y.

21 Here we denote potential outcome random variables for a binary cause D as Y^1 and Y^0. Accordingly, $Y = Y^1$ if $D = 1$ and $Y = Y^0$ if $D = 0$.

22 These four groups are considered principal strata in the framework of Frangakis and Rubin (2002).

23 In the IV regressions reported in Banerjee and Iyer (2005), the date of revenue control is actually used to predict the non-landlord *proportion* – not the dichotomous non-landlord indicator used elsewhere in the analysis.

24 See Gelman and Hill (2007, p. 223) for R code to correct two-stage least squares standard errors generated manually.

25 Tabellini (2010, pP. 597) uses a measure of 'constraints on the executive' to measure historical political institutions. This variable was constructed using information from historical sources.

26 This situation is akin to having only 13 group-level observations in a multilevel framework (Gelman and Hill, 2007) or an independent variable with low variance.

27 We discuss problems with weak instruments in the following section.

28 Complications (A)–(D) are all closely related. Situation (D) can be considered a less extreme version of the three-part predicament depicted in (A)–(C).

29 There is no consensus about how large an association between an IV and a treatment variable must be before an analysis can proceed safely. Many researchers rely on the rule of thumb that first-stage F-statistics testing the significance of the instrument should be around 10 or higher (Staiger and Stock, 1997; Stock et al., 2002).

REFERENCES

Acemoglu, D., Johnson, S. and Robinson, J. A. (2001). The colonial origins of comparative development: An empirical investigation. *American Economic Review*, 91, 1369–1401.

Acemoglu, D., Johnson, S. and Robinson, J. A. (2002). Reversal of fortune: Geography and institutions in the making of the modern world income distribution. *Quarterly Journal of Economics*, 117, 1231–1294.

Acemoglu, D., Johnson, S. and Robinson, J. A. (2005). The rise of Europe: Atlantic trade, institutional change, and economic growth. *American Economic Review*, 95, 546–579.

Angrist, J. D. and Imbens, G. W. (1995). Two-stage least squares estimation of average causal effects in models with variable treatment intensity. *Journal of the American Statistical Association*, 90, 431–442.

Angrist, J. D., Imbens, G. W. and Rubin, D. B. (1996). Identification of causal effects using instrumental variables. *Journal of the American Statistical Association*, 91, 444–455.

Angrist, J. D. and Krueger, A. B. (1999). Empirical strategies in labor economics. In O. C. Ashenfelter and D. Card (eds), *Handbook of Labor Economics*, volume 3 (pp. 1277–1366). Amsterdam: Elsevier.

Angrist, J. D. and Pischke, J.-S. (2009). *Mostly Harmless Econometrics*. Princeton, NJ: Princeton University Press.

Banerjee, A. and Iyer, L. (2005). History, institutions, and economic performance: the legacy of colonial land tenure systems in India. *American Economic Review*, 95, 1190–1213.

Banfield, E. C. (1958). *The Moral Basis of a Backward Society*. New York: Free Press.

Bollen, K. A. (1989). *Structural Equations with Latent Variables*. New York: Wiley.

Bollen, K. A. (1995). Structural equation models that are nonlinear in latent variables: a least-squares estimator. *Sociological Methodology*, 25, 223–251.

Bollen, K. A. (1996a). An alternative two-stage least squares (2sls) estimator for latent variable equations. *Psychometrika*, 61, 109–121.

Bollen, K. A. (1996b). A limited-information estimator for lisrel models with or without heteroscedastic errors. In G. A. Marcoulides and R. E. Schumacker (eds), *Advanced Structural Equation Modeling: Issues and Techniques* (pp. 227–241). Mahwah, NJ: Lawrence Erlbaum.

Bollen, K. A. (2001). Two-stage least squares and latent variable models: simultaneous estimation and robustness to misspecifications. In R. Cudeck, S. du Toit and D. Sörbom (eds), *Structural Equation Modeling: Present and Future – a Festschrift in Honor of Karl Jöreskog* (pp. 119–138). Lincolnwood, IL: Scientific Software International.

Bollen, K. A. (2012). Instrumental variables in sociology and the social sciences. *Annual Review of Sociology*, 38, 37–72.

Bound, J., Jaeger, D. A. and Baker, R. M. (1995). Problems with instrumental variables estimation when the correlation between the instruments and the endogenous variable is weak. *Journal of the American Statistical Association*, 90, 443–450.

Bowden, R. J. and Turkington, D. A. (1984). *Instrumental Variables*. Cambridge: Cambridge University Press.

Card, D. (2001). Estimating the return to schooling: progress on some persistant econometric problems. *Econometrica*, 69, 1127–1160.

Diamond, J. and Robinson, J. A. (eds) (2010). *Natural Experiments of History*. Cambridge, MA: Harvard University Press.

Duncan, O. D. (1975). *Introduction to Structural Equation Models*. New York: Academic Press.

Elwert, F. and Winship, C. (2014). Endogenous selection bias. *Annual Review of Sociology*, 40, 31–53.

Fox, J. (1984). *Linear Statistical Models and Related Methods: With Applications to Social Research*. New York: Wiley.

Frangakis, C. E. and Rubin, D. B. (2002). Principal stratification in causal inference. *Biometrics*, 58, 21–29.

Gelman, A. and Hill, J. (2007). *Data Analysis Using Regression and Multilevel/Hierarchical Models*. Cambridge: Cambridge University Press.

Goldberger, A. S. (1972). Structural equation methods in the social sciences. *Econometrica*, 40, 979–1001.

Haavelmo, T. (1942). The statistical implications of a system of simultaneous equations. *Econometrica*, 11, 1–12.

Hahn, J. and Hausman, J. (2003). Weak instruments: diagnosis and cures in empirical economics. *American Economic Review*, 93, 118–125.

Hausman, J. (1978). Specification tests in econometrics. *Econometrica*, 46, 1251–1271.

Heckman, J. J. (2000). Causal parameters and policy analysis in economics: a twentieth century retrospective. *Quarterly Journal of Economics*, 115, 45–97.

Hood, W. C. and Koopmans, T. (eds) (1953). *Studies in Econometric Method*. New York: Wiley.

Imbens, G. W. and Angrist, J. D. (1994). Identification and estimation of local average treatment effects. *Econometrica*, 62, 467–475.

Imbens, G. W. and Rubin, D. B. (1997). Estimating outcome distributions for compliers in instrumental variables models. *Review of Economic Studies*, 64, 555–574.
Koopmans, T. C. and Reiersol, O. (1950). The identification of structural characteristics. *Annals of Mathematical Statistics*, 21, 165–181.
Morgan, S. L. and Winship, C. (2007). *Counterfactuals and Causal Inference*. New York: Cambridge University Press.
Murray, M. (2006). Avoiding invalid instruments and coping with weak instruments. *Journal of Economic Perspectives*, 20, 111–132.
Nunn, N. (2008). The long-term effects of Africa's slave trades. *Quarterly Journal of Economics*, 123, 139–176.
Pearl, J. (2000). *Causality: Models, Reasoning, and Inference*. Cambridge: Cambridge University Press.
Pearl, J. (2009). *Causality: Models, Reasoning, and Inference*. Cambridge: Cambridge University Press.
Putnam, R. (1993). *Making Democracy Work: Civic Traditions in Modern Italy*. Princeton, NJ: Princeton University Press.
Reiersol, O. (1941). Confluence analysis by means of lag moments and other methods of confluence analysis. *Econometrica*, 9, 1–24.
Rosenzweig, M. R. and Wolpin, K. I. (2000). Natural 'natural experiments' in economics. *Journal of Economic Literature*, 38, 827–874.
Schultz, H. (1938). *The Theory and Measurement of Demand*. Chicago: University of Chicago Press.
Staiger, D. and Stock, J. H. (1997). IV regression with weak instruments. *Econometrica*, 65, 557–586.
Stock, J. H., Wright, J. H., and Yogo, M. (2002). A survey of weak instruments and weak identification in generalized method of moments. *Journal of Business and Economic Statistics*, 20, 518–529.
Tabellini, G. (2010). Culture and institutions: economic development in the regions of Europe. *Journal of European Economic Association*, 8, 677–716.
Wald, A. (1940). The fitting of straight lines if both variables are subject to error. *Annals of Mathematical Statistics*, 11, 284–300.
Wilson, N. H. (2011). From reflection to refraction: State administration in British India, circa 1770-1855. *American Journal of Sociology*, 116, 1437–1477.
Wooldridge, J. M. (2002). *Econometric Analysis of Cross Section and Panel Data*. Cambridge, MA: MIT Press.
Working, E. J. (1927). What do statistical 'demand curves' show? *Quarterly Journal of Economics*, 41, 212–235.
Working, H. (1925). The statistical determination of demand curves. *Quarterly Journal of Economics*, 39, 503–545.
Wright, S. (1921). Correlation and causation. *Journal of Agricultural Research*, 20, 557–585.
Wright, S. (1925). *Corn and Hog Correlations*. Washington, DC: U.S. Department of Agriculture.
Wright, S. (1934). The method of path coefficients. *Annals of Mathematical Statistics*, 5, 161–215.

14

Regression discontinuity designs in social sciences

David S. Lee and Thomas Lemieux[*]

INTRODUCTION

Regression discontinuity (RD) designs were initially introduced by Thistlethwaite and Campbell (1960) as a way of estimating treatment effects in a non-experimental setting where treatment is determined by whether an observed 'assignment' variable (also referred to in the literature as the 'forcing' variable or the 'running' variable) exceeds a known cutoff point. Thistlethwaite and Campbell (1960) analyzed the impact of merit awards on future academic outcomes in their original study, using the fact that the allocation of these awards was based on an observed test score. The main idea behind the research design was that individuals with scores just below the cutoff (who did not receive the award) were good comparisons to those just above the cutoff (who did receive the award). Although this evaluation strategy has been around for over 50 years, it only attracted limited attention in economics, and social sciences more generally, until relatively recently.

Since the late 1990s, a burgeoning literature in economics has relied on RD designs to estimate program effects in a wide variety of contexts. Like Thistlethwaite and Campbell (1960), early studies by Van der Klaauw (2002) and Angrist and Lavy (1999) exploited threshold rules often used by educational institutions to estimate the effect of financial aid and class size, respectively, on educational outcomes. Following these early papers in the area of education, there has been a rapid growth over the last ten years in the number of studies using RD designs to examine a range of other questions. Examples include: the labor supply effect of welfare, unemployment insurance, and disability programs; the effects of Medicaid on health outcomes; the effect of remedial education programs on educational achievement; the empirical relevance of median voter models; and the effects of unionization on wages and employment.

One important impetus behind this recent flurry of research is a recognition, formalized by Hahn et al. (2001), that RD designs require seemingly mild assumptions compared to those needed for other non-experimental approaches. Another reason for the recent wave of research is the realization that the RD design is not 'just another' evaluation strategy, and that causal inferences from RD designs are potentially more credible than those from typical 'natural experiment' strategies (e.g. difference-in-differences or instrumental variables), which have been heavily employed in applied research in recent decades. This notion has a theoretical justification: Lee (2008) formally shows that one need not *assume* the RD design isolates treatment variation that is 'as good as randomized'; instead, such randomized variation is a *consequence* of agents' inability to precisely control the assignment variable near the known cutoff.

So while the RD approach was initially thought to be 'just another' program evaluation method with relatively little general applicability outside of a few specific problems, recent work in economics has shown quite the opposite.[1] In addition to providing a highly credible and transparent way of estimating program effects, RD designs can be used in a wide variety of contexts covering a large number of important economic and social questions. These two facts likely explain why the RD approach is rapidly becoming a major element in the toolkit of empirical economists and empirical social science researchers more generally.

The goal of this chapter is twofold. First, it seeks to provide the conceptual framework underneath RD designs – what assumptions they require, and their strengths and weaknesses. Second, it discusses the 'nuts and bolts' of implementing RD designs in practice. Most of the issues discussed in this chapter are also covered in related pieces by Van der Klaauw (2008), Imbens and Lemieux (2008) and especially Lee and Lemieux (2010). Readers interested in learning more about conceptual and methodological issues should consult these studies, as we only briefly discuss these issues in this chapter.

The rest of the chapter is organized as follows. In the next section we introduce RD designs and discuss their main advantages and disadvantages. We introduce an important theme that we stress throughout the paper, namely that RD designs are particularly compelling because they are close cousins of randomized experiments. We then go through the main 'nuts and bolts' involved in implementing RD designs and provides a 'guide to practice' for researchers interested in using the design. We also provide a summary 'checklist' highlighting our key recommendations. These implementation issues are illustrated using an example from US House elections. After discussing caveats and frequent errors, we conclude by suggesting some further readings.

BACKGROUND AND CONCEPTUAL FRAMEWORK

In this section we set the stage for the rest of the chapter by discussing the origins and the conceptual framework underlying the RD design, beginning with the classic work of Thistlethwaite and Campbell (1960) and moving to the recent interpretation of the design using modern tools of program evaluation in economics and the potential outcomes framework. We show how RD designs can be viewed as local randomized experiments and discuss their generalizability. A key feature of RD designs is that they provide a very transparent way of graphically showing how the treatment effect is identified. We thus end the section by discussing how to graph the data in an informative way.

Origins and the potential outcomes approach

RD designs were first introduced by Thistlethwaite and Campbell (1960) in their study of the impact of merit awards on the future academic outcomes (career aspirations, enrollment in postgraduate programs, etc.) of students. The study exploited the fact that these awards were allocated on the basis of an observed test score. Students with test scores, X, greater than or equal to a cutoff value c received the award, and those with scores below the cutoff were denied the award. This generated a sharp discontinuity in the 'treatment' (receiving the award) as a function of the test score. Let the receipt of treatment be denoted by the dummy variable $D \in \{0, 1\}$, so that we have $D = 1$ if $X \geq c$, and $D = 0$ if $X < c$.

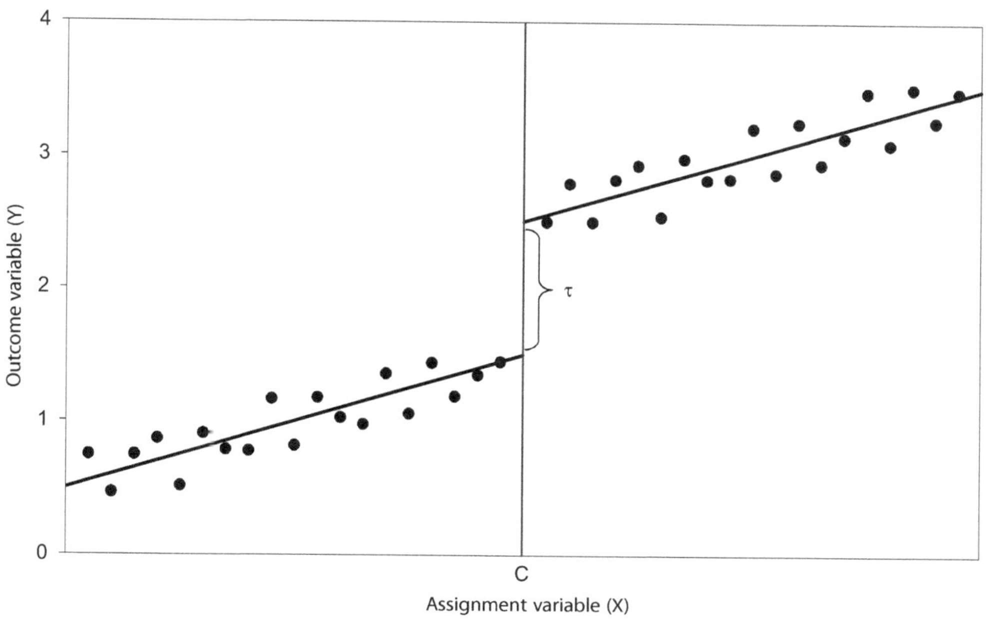

Figure 14.1 Simple linear RD setup

Importantly, there appears to be no reason, other than the merit award, for future academic outcomes, Y, to be a discontinuous function of the test score. This simple reasoning suggests attributing the discontinuous jump in Y at c to the causal effect of the merit award. Assuming that the relationship between Y and X is otherwise linear, a simple way of estimating the treatment effect τ is by fitting the linear regression

$$Y = \alpha + D\tau + X\beta + U, \tag{14.1}$$

where U is the usual error term that can be viewed as a purely random error generating variation in the value of Y around the regression line. This case is depicted in Figure 14.1, which shows both the true underlying function and numerous realizations of U.

While this simple regression approach is intuitively appealing, it is useful to analyze RD designs more formally to illustrate the key assumptions that need to be satisfied for the design to be valid. A key contribution in this regard is the work of Hahn et al. (2001), who used the approach developed in the treatment effects literature to analyze RD designs. Hahn et al. (2001) noted the key assumption of a valid RD design was that 'all other factors' were 'continuous' with respect to X, and suggested a non-parametric procedure for estimating τ that did not assume underlying linearity, as we have assumed in the simple example above.

The necessity of the continuity assumption is seen more formally using the 'potential outcomes framework' of the treatment effects literature, with the aid of a graph. It is typically imagined that for each individual i, there exists a pair of 'potential' outcomes: $Y_i(1)$ for what would occur if the individual were exposed to the treatment and $Y_i(0)$ if not exposed. The causal effect of the treatment is represented by the difference $Y_i(1) - Y_i(0)$. The fundamental problem of causal inference is that we cannot observe the pair $Y_i(0)$ and $Y_i(1)$ simultaneously. We therefore typically focus on average effects of the treatment, that is, averages of $Y_i(1) - Y_i(0)$ over (sub)populations, rather than on unit-level effects.

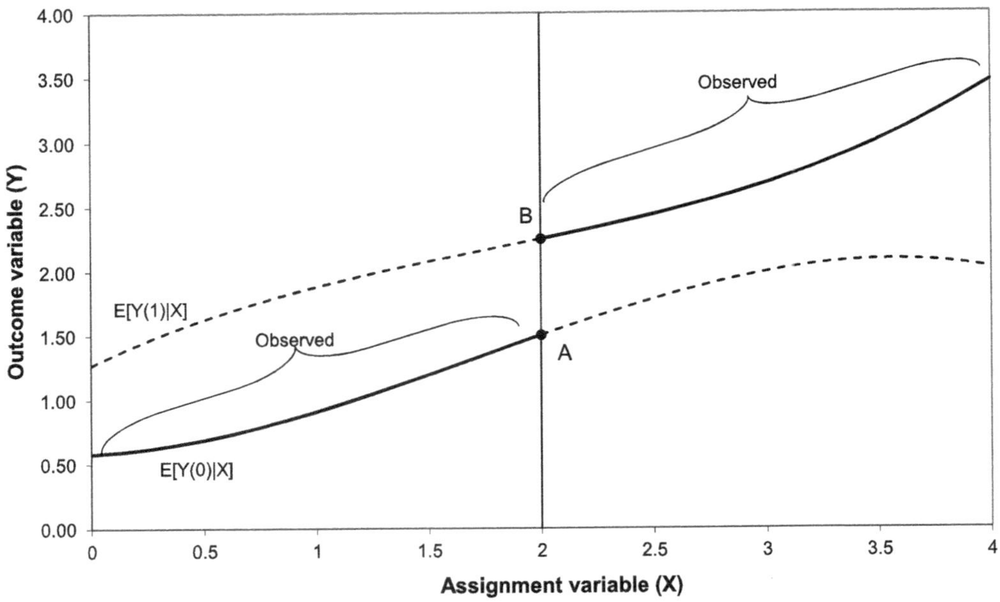

Figure 14.2 Non-linear regression discontinuity

In the RD setting, we can imagine there are two underlying relationships between average potential outcomes and X, represented by $E[Y_i(1)|X]$ and $E[Y_i(0)|X]$, as in Figure 14.2. But by definition of the RD design, all individuals to the right of the cutoff ($c = 2$ in this example) are exposed to treatment, and all those to the left are denied treatment. Therefore, we only observe $E[Y_i(1)|X]$ to the right of the cutoff and $E[Y_i(0)|X]$ to the left of the cutoff, as indicated in the figure.

It is easy to see that with what is observable, we could try to estimate the quantity

$$B - A = \lim_{\epsilon \downarrow 0} E[Y_i|X_i = c + \epsilon] - \lim_{\epsilon \uparrow 0} E[Y_i|X_i = c + \epsilon],$$

which would equal

$$E[Y_i(1) - Y_i(0)|X = c].$$

This is the 'average treatment effect' at the cutoff c. Note that this particular treatment effect is different from the conventional average treatment effect (ATE) one typically seeks to estimate using a randomized experiment. For example, in Figure 14.2 we see that the treatment effect (the difference between the two potential outcome curves) depends on the assignment variable X. Therefore, the treatment effect identified at $X = c$ may not be generalizable over the entire population (i.e. over the whole distribution of X).

Generalizability aside, inference is possible here because of the continuity of the underlying functions $E[Y_i(1)|X]$ and $E[Y_i(0)|X]$.[2] In essence, this continuity condition enables us to use the average outcome of those right below the cutoff (who are denied the treatment) as a valid counterfactual for those right above the cutoff (who received the treatment).

A key question is under what circumstances we expect this continuity assumption to hold. As it turns out, continuity is a direct consequence of the fact that, under the weak assumptions discussed below, in a RD design we have local randomization around the cutoff point. From that point of view, RD designs are more closely related to randomized experiments, the 'gold standard' of program evaluation methods, than to other commonly used methods such as matching on observables or instrumental variables (IV) methods.[3] We next explore the connection between RD designs and randomized experiments, and argue that RD designs can be analyzed and treated like randomized experiments.

Regression discontinuity design and local randomization

We consider a highly simplified example to illustrate the close connection between RD designs and randomized experiments. As we will explain later, the key results on local randomization can also be obtained in a much more general setting. More specifically, we assume that the treatment effect, τ, is constant for all individuals, and that potential outcomes are a linear function of baseline covariates, W, and an error term U:

$$Y(0) = W\delta_1 + U, \qquad (14.2)$$
$$Y(1) = \tau + W\delta_1 + U,$$

where we have omitted the subscript i to simplify the notation. Under these simplifying assumptions, we have a simple linear regression model for the observed outcome Y:

$$Y = (1 - D) \cdot Y(0) + D \cdot Y(1) = D\tau + W\delta_1 + U. \qquad (14.3)$$

The assignment variable, X, is assumed to depend linearly on the baseline covariates and a random component V,

$$X = W\delta_2 + V, \qquad (14.4)$$

and treatment assignment is given by

$$D = 1[X \geq c] = 1[W\delta_2 + V \geq c],$$

where $1(\cdot)$ is the indicator function.

Interestingly, a randomized experiment can be viewed as a special case of this model where $\delta_2 = 0$ and V is a randomly generated number used to divide individuals into treatments ($V \geq c$) and controls ($V < c$). Since treatment is randomly assigned, there are no systematic differences between the covariates W and the error term U between the treatment and control groups. In other words, W and U are 'balanced' between treatments and controls in the sense that:

$$E[W|D=0] = E[W|D=1] = E[W],$$
$$E[U|D=1] = E[U|D=0] = E[U].$$

It follows that

$$E[Y|D=1] = \tau + E[W]\delta_1 + E[U],$$
$$E[Y|D=0] = E[W]\delta_1 + E[U],$$

and

$$\tau = E[Y|D=1] - E[Y|D=0].$$

The treatment effect τ can, therefore, be estimated as a simple difference between the mean outcomes for treatments ($E[Y|D = 1]$) and controls ($E[Y|D = 0]$). As is well known, one does not need to control for baseline covariates since those are not systematically different for treatment and controls. In the context of the simple regression model in equation (14.3), this means that failing to include W in a regression of Y on D does not result in an omitted variable bias since W is uncorrelated with D.

Now consider the RD design. To make the above equations more concrete, we work with a case similar to Thistlethwaite and Campbell (1960) where the assignment variable X is a test score that depends on both intrinsic ability, W, and luck, V. Since future outcomes Y (earnings, choice of major, etc.) also likely depend on ability, we do not expect students above and below the cutoff c to be comparable. This means that, unlike in a randomized experiment, we have $E[W|D=1] \neq E[W|D=0]$ and $E[Y|D=1] - E[Y|D=0] \neq \tau$. But provided that the luck component, V, follows a continuous distribution $f(\cdot)$, randomization will hold locally around the cutoff, and the potential outcomes will be continuous functions of the assignment variable X.

To see this formally, consider a further simplification where W is a dummy variable indicating whether the student is high ($W=1$) or low ($W=0$) ability. Since $X = W\delta_2 + V$, for any given value x of the test score (assignment variable) X, high-ability students have a luck term $V = x - \delta_2$, while $V = x$ for low-ability students. Using a few manipulations, it follows that

$$\begin{aligned} E[W|X=x] &= \text{Prob}[W=1|X=x] \\ &= \frac{P_w \cdot \text{Prob}[X=x|W=1]}{P_w \cdot \text{Prob}[X=x|W=1] + (1-P_w) \cdot \text{Prob}[X=x|W=0]} \\ &= \frac{P_w \cdot f(x-\delta_2)}{P_w \cdot f(x-\delta_2) + (1-P_w) \cdot f(x)}, \end{aligned}$$

where $P_w = \text{Prob}[W=1]$ is the fraction of students who are high ability, $f(\cdot)$ is the probability density function of V, and we have used the fact that $\text{Prob}[X=x|W] = \text{Prob}[V=x-W\delta_2] = f(x-W\delta_2)$. While it is clear that $E[W|X=x]$ is now a function of the assignment variable X, the function is also *continuous* in X since the probability density function of V, $f(\cdot)$, is itself continuous. To simplify the notation we introduce the function $g(x)$ defined as

$$g(x) \equiv E[W|X=x] = \frac{P_w \cdot f(x-\delta_2)}{P_w \cdot f(x-\delta_2) + (1-P_w) \cdot f(x)}$$

When luck on the test, V, is unrelated to the error term U, it follows that

$$\begin{aligned} E[Y(0)\,|X] &= g(X)\,\delta_1 + E[U], \\ E[Y(1)\,|X] &= \tau + g(X)\,\delta_1 + E[U]. \end{aligned} \tag{14.5}$$

Since $g(X)$ is a continuous function, the expected value of the potential outcomes is also continuous in X, thereby satisfying the condition in Hahn et al. (2001). This simple example shows that continuity of the potential outcome functions illustrated in Figure 14.2 is a consequence of the assumption that there is a random and continuously distributed component V in the assignment variable X.

Local randomization is also a direct consequence of this assumption. In a randomized experiment where 50% of individuals are assigned to the treatment and control groups, respectively, each individual is equally likely to be a treatment or a control. In the simple RD design discussed above, we also get that individuals are randomly split in a 50–50 way right around the cutoff point. To see this, consider the probabilities that $X = c + \varepsilon$ and $X = c - \varepsilon$, where ε is a small number. Since the density $f(V)$ is continuous in V, it follows that

$$\lim_{\varepsilon \to 0} \frac{\text{Prob}(X = c+\varepsilon)}{\text{Prob}(X = c-\varepsilon)} = \lim_{\varepsilon \to 0} \frac{f(c+\varepsilon-W\delta_2)}{f(c-\varepsilon-W\delta_2)} = 1.$$

Since this holds regardless of W and U, it follows that W and U are balanced on each side of the cutoff, that is,

$$\begin{aligned} \lim_{\varepsilon \to 0} E[W|X = c + \varepsilon] &= \lim_{\varepsilon \to 0} E[W|X = c - \varepsilon], \\ \lim_{\varepsilon \to 0} E[U|X = c + \varepsilon] &= \lim_{\varepsilon \to 0} E[U|X = c - \varepsilon], \end{aligned}$$

and, therefore,[4]

$$\lim_{\varepsilon \to 0} E[Y|X = c + \varepsilon] - \lim_{\varepsilon \to 0} E[Y|X = c - \varepsilon] = \tau.$$

The difference between randomized experiments and the RD design is that while randomization holds globally (for any value of X) in a randomized experiment, it only holds locally in a RD design. Therefore, while the treatment effect can be computed as simple difference of mean outcomes in a randomized experiment, regression methods have to be used to estimate local means right around the cutoff point in a RD design. In the simple model above, equation (14.5) yields the following model for observed outcomes:[5]

$$Y = D\tau + g(X)\,\delta_1 + U, \tag{14.6}$$

which can be estimated by running a regression of Y on D where X is controlled in a flexible way to account for the function $g(X)$. In the next section, we explain in detail how such flexible regressions can be estimated in practice.

But besides the need to use regression methods instead of comparisons of means, RD designs can be analyzed using the same set of standard procedures that are commonly used in the case of randomized experiments. This includes, for example, checking whether baseline covariates W are balanced on the two sides of the cutoff point. As in a randomized experiment, one also does not need to include the covariates W in a regression model since the mean value of W is locally the same on each side of the cutoff.[6]

This simplified example can be easily generalized to a much richer setting where individuals have some control over the assignment variable, as shown in Lee (2008). To see this, consider again the test-taking example. When students know that scoring above a certain threshold (say 80%) will give them a scholarship benefit, we expect them to study harder and double-check their answers more thoroughly than in a lower-stake exam. Effort may well depend both on observed covariates and on the error term U in the outcome. For instance, high-ability students (high value of W) may have much better chances of scoring above 80%, which gives them a stronger incentive to try to score above 80%. Likewise, a student with a high value of U in the outcome equation may particularly benefit from the scholarship in terms of the program he/she will then be able to afford, etc.[8] Lee (2008) shows that the RD design remains valid in this setting as long as there is still a continuously distributed random component V in the assignment variable that remains beyond the control of the student. This is highly plausible in the test-taking example since students cannot perfectly control the grade they will get on an exam. More generally, one must have some knowledge about the mechanism generating the assignment variable, beyond knowing that if it crosses the threshold, the treatment is 'turned on'. It is 'folk wisdom' in the literature to judge whether the RD is appropriate based on whether individuals could manipulate the assignment variable and *precisely* 'sort' around the discontinuity threshold. The key word here is 'precise', rather than 'manipulate'. After all, in the above example, individuals do exert some control over the test score. And indeed in virtually every known application of the RD design, it is easy to tell a plausible story that the assignment variable is to some degree influenced by *someone.*

The main takeaway points from our discussion of local randomization are the following:

- *RD designs can be invalid if individuals can precisely manipulate the 'assignment variable'.* When there is a payoff or benefit to receiving a treatment, it is natural to consider how an individual may behave to obtain such benefits. For example, if students could effectively 'choose' their test score X through effort, those who chose a score c (and hence received the merit award) could be somewhat different from those who chose scores just below c. The important lesson here is that the existence of a treatment that is a discontinuous function of an assignment variable is *not* sufficient to justify the validity of a RD design. Indeed, if anything, discontinuous rules may generate incentives, causing behavior that would *invalidate* the RD approach.
- If *individuals – even while having some influence – are unable* to precisely *manipulate the assignment variable,* a consequence *of this is that the variation in treatment near the threshold is randomized as though from a randomized experiment.* This is a crucial feature of the RD design, since it is the reason why RD designs are often so compelling. Intuitively, when individuals have imprecise control over the assignment variable, even if some are especially likely to have values of X *near* the cutoff, *every* individual will have approximately the same probability of having an X that is just above (receiving the treatment) or just below (being denied the treatment) the cutoff – similar to a coin-flip experiment. This result clearly differentiates the RD and IV approaches. When using IV for causal inference, one must *assume* the instrument is exogenously generated as if by a coin-flip. Such an assumption is often difficult to justify – except when an actual lottery was run (as in Angrist, 1990), or if there were some biological process, such as gender determination of a baby, mimicking a coin-flip. By contrast, the variation that RD designs isolates is randomized *as a consequence* of individuals having imprecise control over the assignment variable.
- *RD designs can be analyzed – and tested – like randomized experiments.* This is the key implication of the local randomization result. If variation in the treatment near the threshold is approximately randomized, then it follows that all baseline characteristics – all those variables determined prior to the realization of the assignment variable – should have the same distribution just above and just below the cutoff. If there is a discontinuity in these baseline covariates, then, at a minimum, the underlying identifying assumption of individuals' inability to precisely manipulate the assignment variable is unwarranted. Thus, the baseline covariates are used to *test* the validity of the RD design. By contrast, when employing an IV or a matching/regression-control strategy, assumptions typically need to be made about the relationship of these other covariates to the treatment and outcome variables.[8]

Fuzzy regression discontinuity designs

The above discussion is based on what is called a 'sharp' RD design, where all individuals above the cutoff receive the treatment, while none of those below the cutoff get treated. However, in many interesting settings, treatment is only determined partly by whether the assignment variable crosses a cutoff point. This situation is very important in practice for a variety of reasons, including cases of imperfect take-up by program participants or when factors other than the threshold rule affect the probability of program participation. Starting with Trochim (1984), this setting has been referred to as a 'fuzzy' RD design. In the 'sharp' RD design the probability of treatment jumps from 0 to 1 when X crosses the threshold c. The fuzzy RD design allows for a smaller jump in the probability of assignment to the treatment at the threshold and only requires

$$\lim_{\epsilon \downarrow 0} \text{Prob}[D = 1|X = c + \epsilon] \neq \lim_{\epsilon \uparrow 0} \text{Prob}[D = 1|X = c + \epsilon].$$

Since the probability of treatment jumps by less than one at the threshold, the jump in the relationship between Y and X can no longer be interpreted as an average treatment effect. As in an instrumental variable setting, however, the treatment effect can be recovered by dividing the jump in the relationship between Y and X at c by the fraction induced to take up the treatment at the threshold – in other words, the discontinuous jump in the relation between D and X. In this setting, the treatment effect can be written as

$$\tau_F = \frac{\lim_{\epsilon \downarrow 0} E[Y|X = c + \epsilon] - \lim_{\epsilon \uparrow 0} E[Y|X = c + \epsilon]}{\lim_{\epsilon \downarrow 0} E[D|X = c + \epsilon] - \lim_{\epsilon \uparrow 0} E[D|X = c + \epsilon]},$$

where the subscript 'F' refers to the fuzzy RD design.

There is a close analogy between how the treatment effect is defined in the fuzzy RD design and in the well-known 'Wald' formulation of the treatment effect in an instrumental variables setting. Hahn et al. (2001) were the first to show this important connection and to suggest estimating the treatment effect using two-stage least squares (2SLS) in this setting. We discuss estimation of fuzzy RD designs in greater detail in the next section.

Hahn et al. (2001) furthermore pointed out that the interpretation of this ratio as a causal effect requires the same assumptions as in Imbens and Angrist (1994). That is, one must assume 'monotonicity' (i.e. X crossing the cutoff cannot simultaneously *cause* some units to take up and others to reject the treatment) and 'excludability' (i.e. X crossing the cutoff cannot impact Y except through impacting receipt of treatment). When these assumptions are made, it follows that[9]

$$\tau_F = E[Y(1) - Y(0) \,|\text{unit is complier}, X = c],$$

where 'compliers' are units that receive the treatment when they satisfy the cutoff rule ($X_i \geq c$), but would not otherwise receive it.

In summary, if there is local random assignment (e.g. due to the plausibility of individuals' imprecise control over X), then we can simply apply all of what is known about the assumptions and interpretability of instrumental variables. The difference between the 'sharp' and 'fuzzy' RD design is exactly parallel to the difference between the randomized experiment with perfect compliance and the case of imperfect compliance, when only the 'intent to treat' is randomized.

For example, in the case of imperfect compliance, even if a proposed binary instrument Z is randomized, it is necessary to rule out the possibility that Z affects the outcome, outside of its influence through treatment receipt, D. Only then will the instrumental variables estimand – the ratio of the reduced form effects of Z on Y and of Z on D – be properly interpreted as a causal effect of D on Y. Similarly, supposing that individuals do not have precise control over X, it is necessary to assume that whether X crosses the threshold c (the instrument) has no impact on Y except by influencing D. Only then will the ratio of the two RD gaps in Y and D be properly interpreted as a causal effect of D on Y.

In the same way that it is important to verify a strong first-stage relationship in an IV design, it is equally important to verify that a discontinuity exists in the relationship between D and X in a fuzzy RD design.

Furthermore, in this binary-treatment/binary-instrument context with unrestricted heterogeneity in treatment effects, the IV estimand is interpreted as the average treatment effect 'for the subpopulation affected by the instrument' (or local average treatment effect (LATE)). Analogously, the ratio of the RD gaps in Y and D (the 'fuzzy design' estimand) can be interpreted as a *weighted* LATE, where the weights reflect the ex-ante likelihood that the individual's X is near the threshold. In both cases, an exclusion restriction and monotonicity condition must hold.

Generalizability

As we pointed out while discussing Figure 14.2, in an RD design we can only identify the treatment effect right at the cutoff point c. In the fuzzy RD design, this means we can only estimate a local average treatment effect for individuals who are both marginally affected by the instrument (the usual LATE issue) and are right at the cutoff.

Depending on the context, this may be an overly simplistic and pessimistic assessment of how informative the treatment effect estimated using a RD design is, for at least two reasons. First, the treatment effect 'right at the cutoff' is often the parameter of policy interest. Going back to the test-score example, let us say that students with a GPA of at least 85 are offered a generous scholarship, and that a RD design is used to analyze its impact on future outcomes such as college attendance and earnings. A relevant policy question may be whether it is worth investing more into the program by allowing students with a GPA of 83 or 84 to also get the scholarship. In such a case, the average treatment effect for these students would likely be very close to the RD estimates obtained using the cutoff at a GPA of 85. In such a situation, the average treatment effect estimated using the RD design would be more policy-relevant than the average treatment effect for the whole population.

A second point, discussed in more detail in Lee and Lemieux (2010), is that the treatment effect estimated using a RD design is a weighted average of the individual treatment effect over the whole population. To see this, remember the treatment assignment rule introduced above: $D = 1[W\delta_2 + V \geq c]$. Since V is random, individuals right around the cutoff point c will have different values of the covariates W depending on the value of V they draw. In particular, individuals drawing a high value of V will tend to have a low value of $W\delta_2$, and vice versa. The treatment effect estimated using the RD design is, therefore, a weighted average of individual treatment effects where the weights are proportional to the conditional probability density function of X given W and U. While it is not possible to know how close the resulting RD gap is from the overall average treatment effect, it remains the case that the treatment effect estimated using a RD design is averaged over a larger population than one would have anticipated from a purely 'cutoff' interpretation.

Graphical presentation

A major advantage of the RD design over competing methods is its transparency, which can be illustrated using graphical methods. A standard way of graphing the data is to divide the assignment variable into a number of bins, making sure there are two separate bins on each side of the cutoff point (to avoid having treated and untreated observations mixed together in the same bin). Then the average value of the outcome variable can be computed for each bin and graphed against the mid-points of the bins.

More formally, for some bandwidth h, and for some number of bins K_0 and K_1 to the left and right of the cutoff value, respectively, the idea is to construct bins $(b_k, b_{k+1}]$, for $k = 1, \ldots, K = K_0 + K_1$, where

$$b_k = c - (K_0 - k + 1) \cdot h.$$

The average value of the outcome variable in the bin is

$$\overline{Y}_k = \frac{1}{N_k} \cdot \sum_{i=1}^{N} Y_i \cdot 1\{b_k \leq X_i < b_{k+1}\}.$$

It is also useful to calculate the number of observations in each bin,

$$N_k = \sum_{i=1}^{N} 1\{b_k \leq X_i < b_{k+1}\},$$

to detect a possible discontinuity in the assignment variable at the threshold, which would suggest manipulation (see the next section).

There are several important advantages in graphing the data this way before performing regressions to estimate the treatment effect. First, the graph provides a simple way of visualizing what the functional form of the regression function looks like on either side of the cutoff point. Since the mean of Y in a bin is a non-parametric kernel regression estimate, evaluated at the bin mid-point using a rectangular kernel, the set of bin means literally represents non-parametric estimates of the regression function. Seeing what the non-parametric regression looks like can then provide useful guidance in choosing the functional form of the regression models.

A second advantage is that comparing the mean outcomes just to the left and right of the cutoff point provides an indication of the magnitude of the jump in the regression function at this point (i.e. of the treatment effect). Since a RD design is 'as good as a randomized experiment' right around the cutoff point, the treatment effect could be computed by comparing the average outcomes in 'small' bins just to the left and right of the cutoff point. If there is no visual evidence of a discontinuity in a simple graph, it is unlikely the formal regression methods discussed below will yield a significant treatment effect.

A third advantage is that the graph also shows whether there are unexpected comparable jumps at other points. If such evidence is clearly visible in the graph and cannot be explained on substantive grounds, this calls into question the interpretation of the jump at the cutoff point as the causal effect of the treatment. We discuss below several ways of testing explicitly for the existence of jumps at points other than the cutoff.

Note that the visual impact of the graph is typically enhanced by also plotting a relatively flexible regression model, such as a polynomial model, which is a simple way of smoothing the graph. The advantage of showing both the flexible regression line and the unrestricted bin means is that the regression line better illustrates the shape of the regression function and the size of the jump at the cutoff point, and laying this over the unrestricted means gives a sense of the underlying noise in the data.

Of course, if bins are too narrow, the estimates will be highly imprecise. If they are too wide, the estimates may be biased, as they fail to account for the slope in the regression line (negligible for very narrow bins). More importantly, wide bins make the comparisons on both sides of the cutoff less credible, as we are no longer comparing observations just to the left and right of the cutoff point.

This raises the question of how to choose the bandwidth (the width of the bin). In practice, this is typically done informally by trying to pick a bandwidth that makes the graphs look informative in the sense that bins are wide enough to reduce the amount of noise, but narrow enough to compare observations 'close enough' on both sides of the cutoff point. While it is certainly advisable to experiment with different bandwidths and see how the corresponding graphs look, in Lee and Lemieux (2010) we also discuss formal procedures for selecting the bandwidth.

ESTIMATION AND INFERENCE

In this section we systematically discuss the nuts and bolts of implementing RD designs in practice. We first discuss what is, arguably, the most important issue in implementing an RD design: the choice of the regression model. We address this by presenting the various possible specifications, discussing how to choose among them, and showing how to compute the standard errors.

We then move on to a number of other practical issues that often arise in RD designs. Examples of questions discussed include whether one should control for other covariates and how to assess the validity of the RD design. We then summarize our recommendations for implementing the RD design.

Regression methods: Parametric or non-parametric regressions?

When we introduced the RD design in the previous section, we used a simple example where the resulting regression model is a non-linear function in the assignment variable X:

$$Y = \alpha + D\tau + g(X)\delta_1 + U,$$

where we have also added an intercept α to the model. Finding a good approximation for the functional form is fairly critical in RD designs since misspecification of the functional form typically generates a bias in the estimated treatment effect, τ.[10] Accordingly, the estimation of RD designs has generally been viewed as a non-parametric estimation problem. In particular, Hahn et al. (2001) suggest running local linear regressions to reduce the importance of the bias. As in many non-parametric estimation problems, one has to choose a particular kernel function. Following Imbens and Lemieux (2008) and Lee and Lemieux (2010), we only look at the case of a rectangular kernel. In practice, this means we can simply run standard linear regressions within a given bin on both sides of the cutoff point to better predict the value of the regression function right at the cutoff point.

The other important implementation issue in non-parametric estimation is the choice of the bandwidth (bin size). With a small bandwidth, the linear approximation will be highly accurate and the bias in the estimated treatment effect will be small. However, the drawback of a small bandwidth is that it yields more imprecise estimates. Therefore, we face a tradeoff between precision and bias, and optimal bandwidth selection procedures seek to find a balance between these two factors.

This being said, applied papers using the RD design often just report estimates from parametric models. Does this mean that these estimates are incorrect? Should all studies use non-parametric methods instead? As we discuss in more detail in Lee and Lemieux (2010), we think that the distinction between parametric and non-parametric methods has sometimes been a source of confusion to practitioners. In practice, it is typically more important to explore how RD estimates are robust to the inclusion of higher-order polynomial terms (the series or polynomial estimation approach) and to changes in the window width around the cutoff point (the local linear regression approach), than seeking to formally determine what is the 'best' specification to use for implementing the RD design. With this in mind, we next explain how to estimate these various regression models.

Estimating the regression

A simple way of implementing RD designs in practice is to estimate two separate regressions on each side of the cutoff point. In terms of computations, it is convenient to subtract the cutoff value from the assignment variable (i.e. transform X to $X - c$), so the intercepts of the two regressions yield the value of the regression functions at the cutoff point.

The regression model on the left-hand side of the cutoff point ($X < c$) is

$$Y = \alpha_l + g_l(X - c) + \epsilon,$$

while the regression model on the right-hand side of the cutoff point ($X \geq c$) is

$$Y = \alpha_r + g_r(X - c) + \epsilon,$$

where $g_l(\cdot)$ and $g_r(\cdot)$ are functional forms that we discuss later. The treatment effect can then be computed as the difference between the two regression intercepts, α_r and α_l, on the two sides of the cutoff point. A more direct way of estimating the treatment effect is to run the pooled regression on both sides of the cutoff point:

$$Y = \alpha_l + \tau \cdot D + g(X - c) + \epsilon,$$

where $\tau = \alpha_r - \alpha_l$ and $g(X - c) = g_l(X - c) + D \cdot [g_r(X - c) - g_l(X - c)]$. One advantage of the pooled approach is that it directly yields estimates and standard errors of the treatment effect τ. Note, however, that it is recommended to let the regression function differ on both sides of the cutoff point by including interaction terms between D and X. For example, in the linear case where $g_l(X - c) = \beta_l \cdot (X - c)$ and $g_r(X - c) = \beta_r \cdot (X - c)$, the pooled regression would be

$$Y = \alpha_l + \tau \cdot D + \beta_l \cdot (X - c) + (\beta_r - \beta_l) \cdot D \cdot (X - c) + \epsilon.$$

If we were to constrain the slope to be identical on both sides of the cutoff ($\beta_r = \beta_l$), this would amount to using data on the right-hand side of the cutoff to estimate α_l, and vice versa. Remember from the previous section that in an RD design, the treatment effect is obtained by comparing conditional expectations of Y when approaching from the left ($\alpha_l = \lim_{x \uparrow c} E[Y_i|X_i = x]$) and from the right ($\alpha_r = \lim_{x \downarrow c} E[Y_i|X_i = x]$) of the cutoff. Constraining the slope to be the same would thus be inconsistent with the spirit of the RD design.

In practice, however, estimates where the regression slope or, more generally, the regression function $g(X - c)$ are constrained to be the same on both sides of the cutoff point are often reported. One possible justification for doing so is that if the functional form is indeed the same on both sides of the cutoff, then more efficient estimates of the treatment effect τ are obtained by imposing that constraint. Such a constrained specification should only be viewed, however, as an additional estimate to be reported for the sake of completeness. It should not form the core basis of the empirical approach.

Local linear regressions and bandwidth choice

As discussed above, local linear regressions provide a non-parametric way of consistently estimating the treatment effect in a RD design (Hahn et al., 2001; Porter, 2003). Following Imbens and Lemieux (2008), we focus on the case of a rectangular kernel which amounts to estimating a standard regression over a window of width h on both sides of the cutoff point. While other kernels (triangular, Epanechnikov, etc.) could also be used, the choice of kernel typically has little impact in practice (Imbens and Lemieux, 2008). As a result, the convenience of working with a rectangular kernel compensates for efficiency gains that could be achieved using more sophisticated kernels.

The regression model on the left-hand side of the cutoff point is

$$Y = \alpha_l + \beta_l \cdot (X - c) + \epsilon, \quad \text{where } c - h \leq X < c,$$

while the regression model on the right-hand side of the cutoff point is

$$Y = \alpha_r + \beta_r \cdot (X - c) + \epsilon, \quad \text{where } c \leq X < c + h.$$

As before, it is also convenient to estimate the pooled regression

$$Y = \alpha_l + \tau \cdot D + \beta_l \cdot (X - c) + (\beta_r - \beta_l) \cdot D \cdot (X - c) + \epsilon, \quad c - h \leq X \leq c + h,$$

since the standard error of the estimated treatment effect can be directly obtained from the regression.

While it is straightforward to estimate the linear regressions within a given window of width h around the cutoff point, a more difficult question is how to choose this bandwidth. In general, choosing a bandwidth in non-parametric estimation involves finding an optimal balance between precision and bias. As the number of observations available increases, it becomes possible to use an increasingly small bandwidth since linear regressions can be estimated relatively precisely over even a small range of values of X. As it turns out, Hahn et al. (2001) show the optimal bandwidth is proportional to $N^{-1/5}$, which corresponds to a fairly slow rate of convergence to zero. [12] In practice, however, knowing at what rate the bandwidth should shrink in the limit does not really help since only one actual sample with a given number of observations is available. The importance of undersmoothing only has to do with a thought experiment on how much the bandwidth should shrink if the sample size were larger so that one obtains asymptotically correct standard errors, and does not help one choose a particular bandwidth in a particular sample.

In the econometrics and statistics literature, there are two 'benchmark' approaches commonly considered for choosing optimal bandwidths. The first procedure consists of characterizing the optimal bandwidth in terms of the unknown joint distribution of all variables. The relevant components of this distribution (such as the curvature of the regression function) can then be estimated and plugged into the optimal bandwidth function. Imbens and Kalyanaraman (2012) derive such an optimal bandwidth in the case of the RD design.

The second approach is based on a cross-validation procedure. In the case considered here, Ludwig and Miller (2007) and Imbens and Lemieux (2008) have proposed a 'leave one out' procedure aimed specifically at estimating the regression function at the boundary. The basic idea is to see how well a regression estimated over a window of width h fits data points (X_i, Y_i) just to the right or to the left of the window.[12] Repeating the exercise for each and every observation, we get a whole set of predicted values of Y that can be compared to the actual values of Y. The optimal bandwidth can be picked by choosing the value of h that minimizes the mean square of the difference between the predicted and actual value of Y.

Let $\widehat{Y}(X_i)$ represent the predicted value of Y obtained using these regressions. The cross-validation criterion is defined as

$$\mathrm{CV}_Y(h) = \frac{1}{N} \sum_{i=1}^{N} \left(Y_i - \widehat{Y}(X_i)\right)^2, \tag{14.7}$$

with the corresponding cross-validation choice for the bandwidth

$$h_{\mathrm{CV}}^{\mathrm{opt}} = \arg\min_h \mathrm{CV}_Y(h).$$

Imbens and Lemieux (2008) discuss this procedure in more detail and point out that since we are primarily interested in what happens around the cutoff, it may be advisable to only compute $\mathrm{CV}_Y(h)$ for a subset of observations with values of X close enough to the cutoff point.

Order of polynomial in local polynomial modeling

In the case of polynomial regressions, for a given bandwidth (typically a large one) one needs to choose the order of the polynomial regressions. As in the case of local linear regressions, it is advisable to try and report a number of specifications to see to what extent results are sensitive to the order of the polynomial. For the same reason mentioned earlier, it is also preferable to estimate separate regressions on the two sides of the cutoff point.

The simplest way of implementing polynomial regressions and computing standard errors is to run a pooled regression. For example, in the case of a third-order polynomial regression, we would have

$$
\begin{aligned}
Y = & \alpha_l + \tau \cdot D + \beta_{l1} \cdot (X - c) + \beta_{l2} \cdot (X - c)^2 + \beta_{l3} \cdot (X - c)^3 \\
& + (\beta_{r1} - \beta_{l1}) \cdot D \cdot (X - c) + (\beta_{r2} - \beta_{l2}) \cdot D \cdot (X - c)^2 + (\beta_{r3} - \beta_{l3}) \cdot D \cdot (X - c)^3 + \epsilon.
\end{aligned}
$$

While it is important to report a number of specifications to illustrate the robustness of the results, it is often useful to have some more formal guidance on the choice of the order of the polynomial. Starting with Van der Klaauw (2002), one approach has been to use a generalized cross-validation procedure suggested in the literature on non-parametric series estimators.[13] One special case of generalized cross-validation used by Black et al. (2007) that we also use in our empirical example is the well-known Akaike information criterion (AIC) of model selection. In a regression context, the AIC is given by

$$
AIC = N \ln(\widehat{\sigma}^2) + 2p,
$$

where $\widehat{\sigma}$ is the standard error of the regression, and p is the number of parameters in the regression model (order of the polynomial plus one for the intercept). Note that while this procedure is useful for choosing one polynomial specification over another, it does not provide a direct indication of how well a particular polynomial model fits the data. For instance, even if a cubic model is preferred to a quadratic model, this does not necessarily mean that the cubic model fits the data well right around the cutoff point. Lee and Lemieux (2010) address this issue by proposing a goodness-of-fit type test that compares how well a given regression model fits the data compared to a fully non-parametric model based on local averages of the outcome variable within a rich set of narrow bins (in X).

Estimation in the fuzzy RD design

As discussed earlier, in both the 'sharp' and the 'fuzzy' RD designs, the probability of treatment jumps discontinuously at the cutoff point. Unlike the case of the sharp RD where the probability of treatment jumps from 0 to 1 at the cutoff, though, the probability jumps by less than one in the fuzzy RD case. In other words, treatment is not solely determined by the strict cutoff rule in the fuzzy RD design. For example, even if eligibility for a treatment solely depends on a cutoff rule, not all the eligibles may get the treatment because of imperfect compliance. Similarly, program eligibility may be extended in some cases even when the cutoff rule is not satisfied. For example, while Medicare eligibility is mostly determined by a cutoff rule (age 65 or older), some disabled individuals under the age of 65 are also eligible.

Since we have already discussed the interpretation of estimates of the treatment effect in a fuzzy RD design in the previous section, here we focus on estimation and implementation issues. The key message to remember from the earlier discussion is that, as in a standard IV framework, the estimated treatment effect can be interpreted as a local average treatment effect provided monotonicity holds.

In the fuzzy RD design, we can write the probability of treatment as

$$
\text{Prob}(D = 1 | X = x) = \gamma + \delta T + g_D(x - c),
$$

where $T = 1[X \geq c]$ indicates whether the assignment variable exceeds the eligibility threshold c.[14] Note that the sharp RD is a special case where $\gamma = 0$, $g_D(\cdot) = 0$, and $\delta = 1$. It is advisable to draw a graph for the treatment dummy D as a function of the assignment variable X using the same procedure discussed at the beginning of this section. This provides an informal way of seeing how large the jump in the treatment probability δ is at the cutoff point, and what the functional form $g_D(\cdot)$ looks like.

Since $D = \text{Prob}(D = 1|X = x) + \nu$, where ν is an error term independent of X, the fuzzy RD design can be described by the following two-equation system:

$$Y = \alpha + \tau D + g(X - c) + \epsilon, \tag{14.8}$$

$$D = \gamma + \delta T + g_D(X - c) + \nu. \tag{14.9}$$

Looking at these equations suggests estimating the treatment effect τ by instrumenting the treatment dummy D with T. Note also that substituting the treatment determining equation into the outcome equation yields the reduced form

$$Y = \alpha_r + \tau_r T + g_r(X - c) + \epsilon_r, \tag{14.10}$$

where $\tau_r = \tau \cdot \delta$. In that setting, τ_r can be interpreted as an 'intent-to-treat' effect.

Estimation in the fuzzy RD design can be performed using either the local linear regression approach or polynomial regressions. Since the model is exactly identified, 2SLS estimates are numerically identical to the ratio of the reduced-form coefficients, τ_r/δ, provided that the same bandwidth is used for equations (14.9) and (14.10) in the local linear regression case, and that the same order of polynomial is used for $g_D(\cdot)$ and $g(\cdot)$ in the polynomial regression case.

How to compute standard errors?

As discussed above, for inference in the sharp RD case, we can use standard least squares methods. It is recommended to use heteroscedasticity-robust standard errors (White, 1980) instead of standard least squares standard errors, as usual.[15] One additional reason for doing so in the RD case is to ensure the standard error of the treatment effect is the same when either a pooled regression or two separate regressions on each side of the cutoff are used to compute the standard errors. As just discussed, it is also straightforward to compute standard errors in the fuzzy RD case using 2SLS methods, although robust standard errors should also be used in this case.

Implementing empirical tests of regression discontinuity validity and using covariates

In this subsection, we describe how to implement tests of the validity of the RD design and how to incorporate covariates in the analysis.

Inspection of the histogram of the assignment variable

Recall that the underlying assumption that generates the local random assignment result is that each individual has imprecise control over the assignment variable, as defined in the previous section. We cannot test this directly (since we will only observe one observation on the assignment variable per individual at a given point in time), but an intuitive test of this assumption is whether the *aggregate* distribution of the assignment variable is discontinuous, since a mixture of individual-level continuous densities is itself a continuous density.

McCrary (2008) proposes a simple two-step procedure for testing whether there is a discontinuity in the density of the assignment variable. In the first step, the assignment variable is partitioned into equally spaced bins and frequencies are computed within those bins. The second step treats the frequency counts as a dependent variable in a local linear regression. See McCrary (2008), who adopts the non-parametric framework for asymptotics, for details on this procedure for inference.

As McCrary (2008) points out, this test can fail to detect a violation of the RD identification condition if for some individuals there is a 'jump' up in the density, offset by jumps down for others, making the aggregate density continuous at the threshold. McCrary (2008) also notes it is possible the RD estimate could remain unbiased, even when there is important manipulation of the assignment variable causing a jump in the density. It should be noted, however, that in order to rely upon the RD estimate as unbiased, one needs to invoke other identifying assumptions and cannot rely upon the mild conditions we focus on in this chapter.[16]

Inspecting baseline covariates

An alternative approach for testing the validity of the RD design is to examine whether the observed baseline covariates are 'locally' balanced on either side of the threshold, which should be the case if the treatment indicator is locally randomized.

A natural thing to do is conduct both a graphical RD analysis and a formal estimation, replacing the dependent variable with each of the observed baseline covariates in W. A discontinuity would indicate a violation in the underlying assumption that predicts local random assignment. Intuitively, if the RD design is valid, we *know* that the treatment variable cannot influence variables determined prior to the realization of the assignment variable and treatment assignment; if we observe it does, something is wrong in the design.

Lee and Lemieux (2010) also discuss how to jointly test if the data are consistent with no discontinuities for any of the observed covariates. With many covariates, some discontinuities will be statistically significant by random chance. Lee and Lemieux (2010) suggest estimating a seemingly unrelated regression (SUR) system where each equation represents a different baseline covariate, and then performing an χ^2 test for the discontinuity gaps in all equations being zero.

Incorporating covariates in estimation

If the RD design is valid, the other use for the baseline covariates is to reduce the sampling variability in the RD estimates, just as in the case of randomized experiments. The simplest way to do so is to add the covariates W to the regression. Unlike the case of X, where we have to be careful when choosing the right functional form for the regression equation, here we can simply include W linearly in the regression. Intuitively, including or not including W in the regression does not affect the consistency of the estimates of τ since W is continuous at the cutoff. So, for the purpose of variance reduction, one can simply use a linear specification in W.

Lee and Lemieux (2010) also suggest a second procedure for incorporating covariates in the estimation based on 'residualizing' the dependent variable – subtracting from Y a prediction of Y based on the baseline covariates W – and then conducting a RD analysis on the residuals. Intuitively, this procedure nets out the portion of the variation in Y we could have predicted using the predetermined characteristics, making the question one of whether the treatment variable can explain the remaining residual variation in Y. The important thing to keep in mind is that if the RD design is valid, this procedure provides a consistent estimate of the same RD parameter of interest. Indeed, any combination of covariates can be used, and abstracting from functional form issues, the estimator will be consistent for the same parameter, just as the estimator for the treatment effect in a randomized experiment will be consistent for the same parameter, no matter what combination of covariates is included. Importantly, this two-step approach also allows one to perform a graphical analysis of the residual. See Lee and Lemieux (2010) for more detail.

A recommended 'checklist' for implementation

Below is a brief summary of our recommendations for the analysis, presentation, and estimation of RD designs. To make the 'checklist' more concrete, we refer to specific tables and figures that we discuss in the empirical example of the next section.

1. *To assess the possibility of manipulation of the assignment variable, show its distribution.* The most straightforward thing to do is to present a histogram of the assignment variable, using a fixed number of bins (see Figure 14.4). The bin widths should be as small as possible, without compromising the ability to visually see the overall shape of the distribution. The bin-to-bin jumps in the frequencies can provide a sense of whether any jump at the threshold is 'unusual'. For this reason, we recommend *against* plotting a smooth function comprised of kernel density estimates. A more formal test of a discontinuity in the density can be found in McCrary (2008).
2. *Present the main RD graph using binned local averages.* As with the histogram, we recommend using a fixed number of non-overlapping bins, as described in the previous section and illustrated in Figure 14.3. The non-overlapping nature of the bins for the local averages is important; we recommend against simply presenting a continuum of non-parametric estimates (with a single break at the threshold), as this will naturally tend to give the impression of a discontinuity even if there does not exist one in the population. We recommend generally 'undersmoothing', while at the same time avoiding 'overly narrow' bins that produce a scatter of data points, from which it is difficult to see the shape of the underlying function. Indeed, we recommend against simply plotting the raw data without a minimal amount of local averaging.
3. *Graph a benchmark polynomial specification.* Superimpose onto the graph the predicted values from a low-order polynomial specification (see Figure 14.3). One can often informally assess, by comparing the two functions, whether a simple polynomial specification is an adequate summary of the data. If the local averages represent the most flexible 'non-parametric' representation of the function, the polynomial represents a 'best-case' scenario in terms of the variance of the RD estimate, since if the polynomial specification is correct, under certain conditions, the least squares estimator is efficient.
4. *Explore the sensitivity of the results to a range of bandwidths, and a range of orders to the polynomial.* For an example, see Table 14.1. It is useful to supplement the table with information on optimal bandwidths selected using a plug-in or cross-validation procedure for local linear regression, as well as the AIC-implied optimal order of the polynomial. A useful graphical device for illustrating the sensitivity of the results to bandwidths is to plot the local linear discontinuity estimate against a continuum of bandwidths. For an example of such a presentation, see Figure 18 in Lee and Lemieux (2010).
5. *Conduct a parallel RD analysis on the baseline covariates.* If the assumption that there is no precise manipulation or sorting of the assignment variable is valid, then there should be no discontinuities in variables that are determined prior to the assignment (see Figure 14.5).

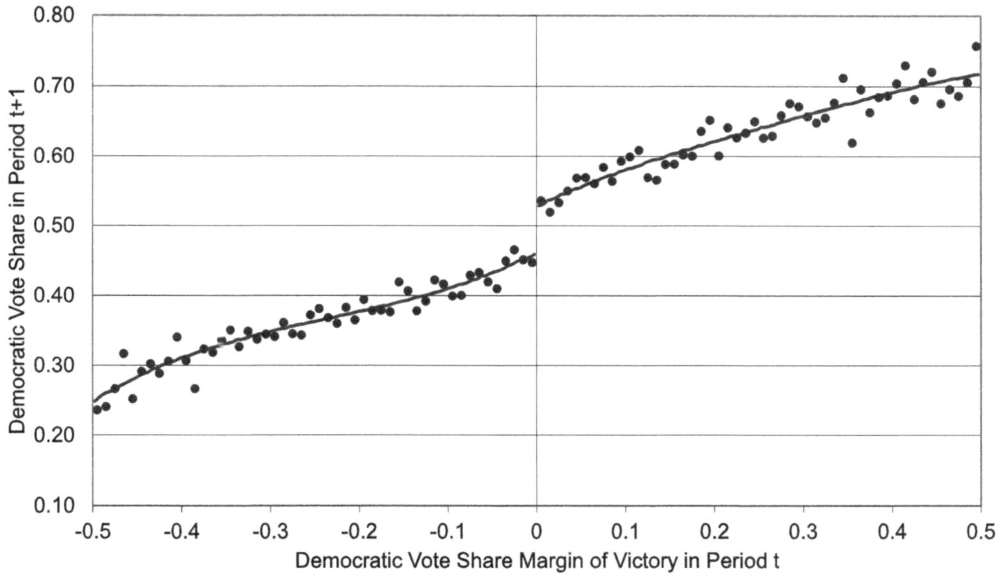

Figure 14.3 Share of vote in next election, bandwidth of 0.01 (100 bins)

6. *Explore the sensitivity of the results to the inclusion of baseline covariates.* The inclusion of baseline covariates – no matter how highly correlated they are with the outcome – should not affect the estimated discontinuity, if the no-manipulation assumption holds. If the estimates do change in an important way, it may indicate a potential sorting of the assignment variable that may be reflected in a discontinuity in one or more of the baseline covariates. In terms of implementation, in the previous subsection we suggest, after choosing a suitable order of polynomial, simply including the covariates directly. An alternative 'residualizing' procedure could also be used.

We recognize that due to space limitations, researchers may be unable to present every permutation of presentation within an published article. Nevertheless, we do believe that documenting the sensitivity of the results to an array of tests and alternative specifications – even if they only appear in unpublished, online appendices – is an important component of a thorough RD analysis.

EMPIRICAL EXAMPLE

In this section we illustrate the various concepts discussed above using an empirical example from Lee (2008) who uses a RD design to estimate the causal effect of incumbency in US House elections. We use a sample of 6558 elections over the 1946–1998 period (see Lee, 2008, for more detail). The assignment variable in this setting is the fraction of votes awarded to Democrats in the previous election. When the fraction exceeds 50%, a Democrat is elected and the party becomes the incumbent party in the next election. The outcome variable is the share of votes in the next election. In the presence of 'incumbency effects', we should observe a jump in the outcome variable for Democrats who barely won the previous election – and are just above the 50% cutoff – relative to those who barely lost it. The data and Stata programs used for this empirical example are provided along with this chapter.

Table 14.1 RD estimates of the effect of winning the previous election on the share of votes in the next election

	Bandwidth									
	1.00	0.50	0.25	0.15	0.10	0.05	0.04	0.03	0.02	0.01
Polynomial of order:										
Zero	0.347	0.257	0.179	0.143	0.125	0.096	0.080	0.073	0.077	0.088
	(0.003)	(0.004)	(0.004)	(0.005)	(0.006)	(0.009)	(0.011)	(0.012)	(0.014)	(0.015)
	[0.000]	[0.000]	[0.000]	[0.000]	[0.003]	[0.047]	[0.778]	[0.821]	[0.687]	
One	0.118	0.090	0.082	0.077	0.061	0.049	0.067	0.079	0.098	0.096
	(0.006)	(0.007)	(0.008)	(0.011)	(0.013)	(0.019)	(0.022)	(0.026)	(0.029)	(0.028)
	[0.000]	[0.332]	[0.423]	[0.216]	[0.543]	[0.168]	[0.436]	[0.254]	[0.935]	
Two	0.052	0.082	0.069	0.050	0.057	0.100	0.101	0.119	0.088	0.098
	(0.008)	(0.010)	(0.013)	(0.016)	(0.020)	(0.029)	(0.033)	(0.038)	(0.044)	(0.045)
	[0.000]	[0.335]	[0.371]	[0.385]	[0.458]	[0.650]	[0.682]	[0.272]	[0.943]	
Three	0.111	0.068	0.057	0.061	0.072	0.112	0.119	0.092	0.108	0.082
	(0.011)	(0.013)	(0.017)	(0.022)	(0.028)	(0.037)	(0.043)	(0.052)	(0.062)	(0.063)
	[0.001]	[0.335]	[0.524]	[0.421]	[0.354]	[0.603]	[0.453]	[0.324]	[0.915]	
Four	0.077	0.066	0.048	0.074	0.103	0.106	0.088	0.049	0.055	0.077
	(0.013)	(0.017)	(0.022)	(0.027)	(0.033)	(0.048)	(0.056)	(0.067)	(0.079)	(0.063)
	[0.014]	[0.325]	[0.385]	[0.425]	[0.327]	[0.560]	[0.497]	[0.044]	[0.947]	
Optimal order of the polynomial	6	3	1	2	1	2	0	0	0	0
Observations	6558	4900	2763	1765	1209	610	483	355	231	106

Notes: Standard errors in parentheses. p-values from the goodness-of-fit test in square brackets. The goodness-of-fit test is obtained by jointly testing the significance of a set of bin dummies included as additional regressors in the model. The bin width used to construct the bin dummies is 0.01. The optimal order of the polynomial is chosen using Akaike's criterion (penalized cross-validation)

Starting with the graphical representation of the data, Figure 14.3 shows the bin means using a bandwidth of 0.01, along with the fitted values from a quartic regression model estimated separately on each side of the cutoff point. Note that the assignment variable is normalized as the difference between the share of vote to Democrats and Republicans in the previous election. This means that a Democrat is the incumbent when the assignment variable exceeds zero. We also limit the range of the graphs to winning margins of 50% or less (in absolute terms) as data become relatively sparse for larger winning (or losing) margins. The graph shows clear evidence of a discontinuity at the cutoff point.

Turning to the local linear regressions, an important question is how to choose the bandwidth. Using the same voting data, Lee and Lemieux (2010) show that the optimal bandwidth chosen using a cross-validation procedure is equal to 0.282. The plug-in procedure of Imbens and Kalyanaraman (2012) yield an optimal bandwidth in the 0.26–0.29 range (for the same data), which is similar to the bandwidth selected using the cross-validation procedure.

Table 14.1 shows the estimates of the treatment effect for a rich set of specifications (up to a quartic) and bandwidths. Local linear regression estimates of the treatment effect are reported in the second row of the table. As expected, the precision of the estimates declines quickly as we approach smaller and smaller bandwidths. Notice also that estimates based on very wide bandwidths (0.5 or 1) are slightly larger than those for the smaller bandwidths (in the 0.05–0.25 range) that are still large enough for the estimates to be reasonably precise. The fact that the estimates from bandwidths around 0.25 (near the calculations of the 'optimal bandwidth') are quite similar is consistent with the graphical evidence in Figure 14.3, which suggest relatively little curvature in the regression function. Since the linear approximation is accurate over a relatively wide range of values of X, bias is not much of an issue and it is sensible to use a relatively large bandwidth to increase precision.

As in Lee (2008), the estimates reported in Table 14.1 suggest large incumbency effects in the 0.05–0.10 range. When Democrats barely win a congressional election they get, on average, an additional 5–10% share of the vote in the next election relative to Democrats who barely lose an election. This large causal effect of incumbency may come from a variety of channels such as more exposure in the media and community, more ability to raise money, etc.

This example also illustrates the importance of first graphing the data before running regressions and trying to choose the optimal bandwidth. When the graph shows a more or less linear relationship – as is the case here – it is natural to expect different bandwidths to yield similar results and the bandwidth selection procedure not to be terribly informative. But when the graph shows substantial curvature, it is natural to expect the results to be more sensitive to the choice of bandwidth and that bandwidth selection procedures will play a more important role in selecting an appropriate empirical specification.

In the case of polynomial regressions, the order of the polynomial selected using the AIC for each bandwidth is presented at the bottom of Table 14.1. The p-values of Lee and Lemieux's (2010) goodness-of-fit tests are reported in square brackets. Broadly speaking, the goodness-of-fit tests do a very good job ruling out clearly misspecified models, like the zero-order polynomials (simple comparison of means on each side of the cutoff) with large bandwidths that yield upward biased estimates of the treatment effect. Estimates of τ from models that pass the goodness-of-fit test mostly fall in the 0.05–0.10 range.

Looking informally at the fit of the model (goodness-of-fit test) and the precision of the estimates (standard errors) suggests the following strategy: use higher-order polynomials for large bandwidths of 0.50 and more, lower-order polynomials for bandwidths between 0.05 and 0.50, and zero-order polynomials (comparisons of means) for bandwidths of less than 0.05, since the latter specification passes the goodness-of-fit test for these very small bandwidths. Interestingly, this informal approach more or less corresponds to what is suggested by the AIC. In this specific example, it seems that given a specific bandwidth, the AIC provides reasonable suggestions on which order of the polynomial to use.

Turning to possible evidence of manipulation, Figure 14.4 shows a graph of the raw densities computed over bins with a bandwidth of 0.005 (200 bins in the graph), along with a smooth second-order polynomial model. Consistent with McCrary (2008) who also uses Lee's (2008) data in his paper, the graph shows no evidence of discontinuity at the cutoff. McCrary also shows that a formal test fails to reject the null hypothesis of no discontinuity in the density at the cutoff.

There is also no evidence of discontinuity in baseline covariates in the voting data. For instance, Figure 14.5 considers the case where the Democratic vote share in the election prior to the one used for the assignment variable (four years prior to the current election) is used as baseline covariate. Consistent with Lee (2008), there is no indication of a discontinuity at the cutoff. The actual RD estimate using a quartic model is −0.004 with a standard error of 0.014.

CAVEATS AND FREQUENT ERRORS

We now comment on two implementation issues that seem innocuous at a superficial level, but pose problems in correctly interpreting evidence from a regression discontinuity design. The first is the issue of relying on a single or *the* optimal bandwidth in computing RD estimates. While it is tempting to think there is a single, best bandwidth formula that 'should' be used

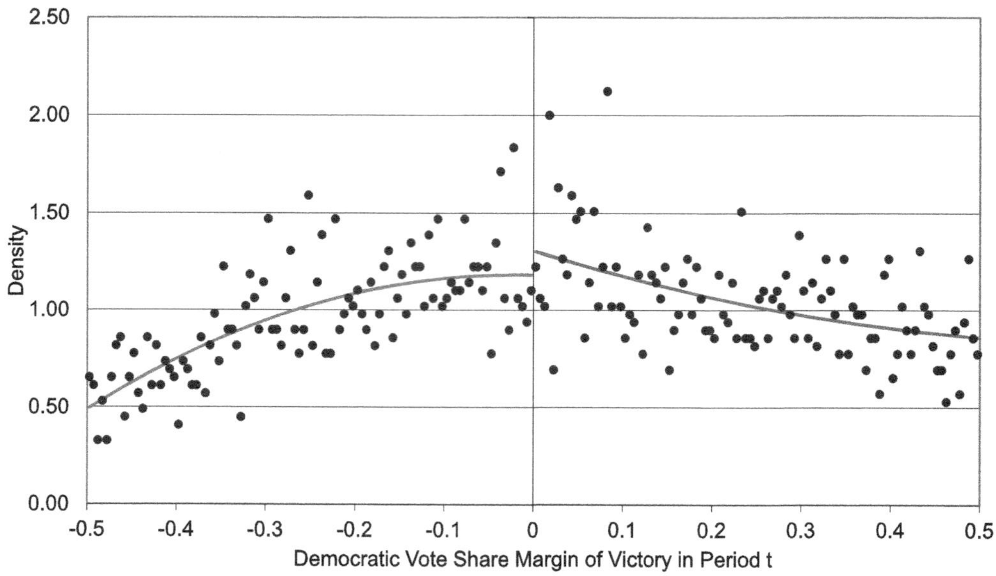

Figure 14.4 Density of the assignment variable (vote share in previous election)

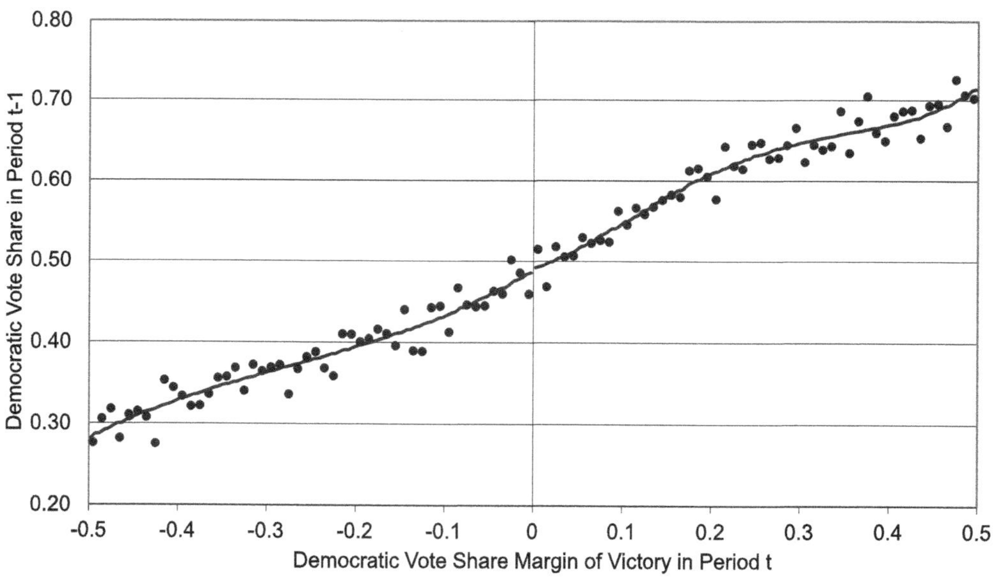

Figure 14.5 Discontinuity in baseline covariate (share of vote in prior election)

in all RD analyses, the fact is that 'optimality' can be defined in so many different ways (see the discussion in Imbens and Kalyanaraman, 2012), and different notions of optimality will lead to different recommended bandwidth formulas. In our view, it is a mistake in empirical analyses to only report the estimate from a single bandwidth, no matter its claim to be optimal. At some level, there is no escaping that one limitation of RD designs is that the underlying functional form is unknown, and for every problem there will be a number of different and equally sensible specifications (or bandwidths), and – especially with relatively smaller sample sizes – those specifications may lead to somewhat different answers. Our recommendation is, in all cases, to report estimates using a range of bandwidths (or specifications) in addition to computing 'benchmark' bandwidths commonly computed in the literature for reference. In a sense, exploring the sensitivity of the empirical analysis to equally plausible specifications is common practice in quality empirical work. The case of RD analysis is no exception. The only difference is that the question of the 'right' specification does not have anything to do with what variables to include; instead it has everything to do with how best to model the shape of the relation between the outcome and the assignment variable.

A second, rather pernicious problem has to do with sample selection and missing values. For example, suppose there is missing data in the voting data that we use in this chapter, and that whether the data is missing is related to whether the district was won in period t by a Democrat or a Republican. Then selecting only data for which there are non-missing values for outcomes in $t + 1$ (the outcome variable) will necessarily induce a classic sample selection problem. So when one examines, for example, the baseline characteristics, one might be tempted to interpret a discontinuity as evidence of precise sorting or manipulation of the assignment variable. Unfortunately, this is potentially an erroneous inference, since a discontinuity in the baseline covariates could be completely consistent with a valid RD design (i.e. local randomization), with the discontinuity being driven by sample selection. To see why this is the case, it is helpful to consider the analogy of the randomized experiment. Clearly, if treatment status affects attrition propensities, then for the *selected* sample, the treatment and control no longer have a similar composition, even if the treatment was perfectly randomized at the outset. Another way such non-random sample selection can occur is through the merging of extra variables from other data sets. As an example, suppose one wanted to add campaign expenditure data to the election data used in this chapter, but the availability of such data was partially influenced by whether a district was won by a Democrat or Republican. After merging the data, the researcher is tempted to simply select the data for which there are no missing values for all of the variables. This seemingly innocuous merging of data can also generate sample selection bias for the same reasons discussed above.

Unfortunately, just as in a randomized experiment with non-random sample selection, there are no assumption-free, bullet-proof methods to adjust for this. In the case where missing values are only a problem for the outcome variable, one could impute the missing values (e.g. with the mean of the outcome variable) to ensure that the full sample (not a selected one) is being utilized in the analysis. This may not eliminate any bias in the estimated causal effect of the treatment on the outcome, but at least a full sample would ensure that the test of continuity of the baseline covariates would still be informative about the imprecise control/manipulation assumption, rather than a test that confounds an invalid RD design and sample selection.

One diagnostic to gauge whether sample selection is a problem is to perform a RD analysis of the dummy variable that is equal to 1 if the outcome data is non-missing, and 0 otherwise. Inspecting whether there is a discontinuity in the probability of sample selection would at least give some evidence on whether the treatment did impact selection into the sample. Failing to reject continuity would not prove an absence of sample selection, but a rejection would strongly suggest a problem. As an example, Caughey and Sekhon (2011) use the US House election data for 1942–2008, and for years that overlap use a subsample of the data used in Lee (2008). They find, in contrast to Lee (2008), discontinuities in key baseline covariates, as well as discontinuities in new covariates that were merged on to the original Lee (2008) data. They interpret this as evidence of sorting or manipulation of the assignment variable. As mentioned above, a simpler explanation is that the data selection process and/or merging of new variables with missing

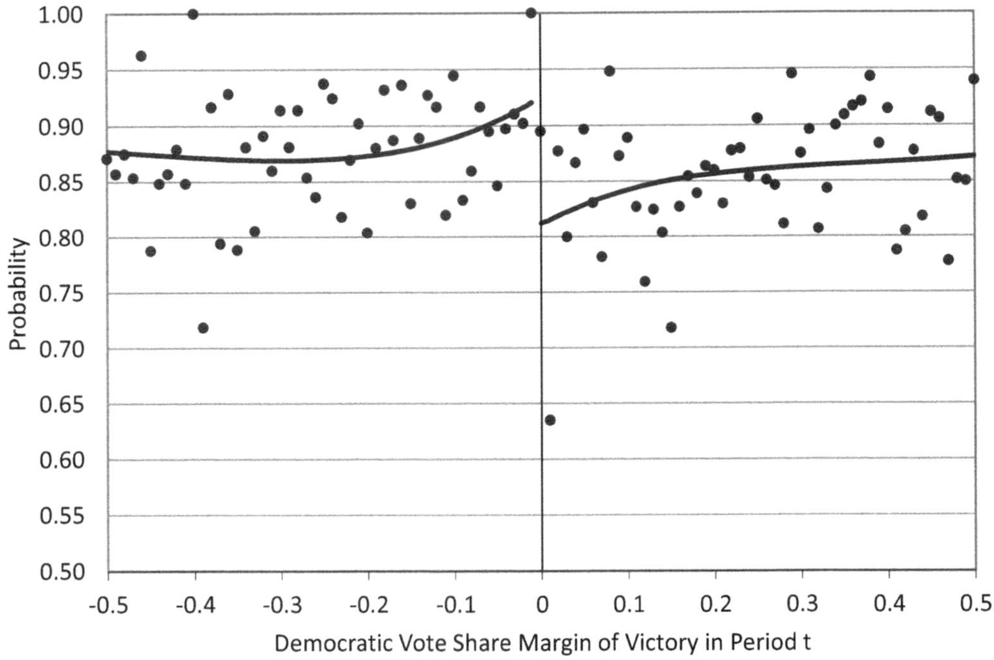

Figure 14.6 Discontinuity in the probability that an observation is non-missing in Caughey and Sekhon (2011) data

values induced sample selection. Figure 14.6 plots the estimated probability that an observation in the Lee (2008) data has a non-missing Democratic vote share (period $t-1$) in the Caughey and Sekhon (2011) data. As the figure shows, over the same time frame, Caughey and Sekhon (2011) use a strict subset of the data. More importantly, the selection (into the Caughey and Sekhon data) probability drops discontinuously from about 0.9 to about 0.8 (the point estimate is -0.114 with a standard error of 0.035): the data do seem to indicate a sample selection problem of the sort described above.

FURTHER READING

Most of the issues discussed in this chapter are addressed in more detail in Lee and Lemieux (2010) who also provide an extensive survey of recent applications of the RD design in economics. Cook (2008) provides some complementary coverage on the history of the RD design in other disciplines. Other recent surveys include Van der Klaauw (2008) and Imbens and Lemieux (2008). Finally, a more thorough coverage of the methodological issues discussed in the second section of this chapter is provided by Hahn et al. (2001) and Lee (2008).

NOTES

* This chapter is an abridged and modified version of Lee and Lemieux (2010). We thank David Autor, David Card, John DiNardo, Guido Imbens, and Justin McCrary for suggestions, as well as for numerous illuminating discussions on the various topics we cover in this review. We also thank Henning Best, Christof Wolf, and Thornsten Kneip for their constructive comments on a first draft of this paper. Diane Alexander, Emily

Buchsbaum, Mingyu Chen, Elizabeth Debraggio, Enkeleda Gjeci, Ashley Hodgson, Andrew Langan, Yan Lau, Steve Mello, Pauline Leung, Xiaotong Niu, and Zhuan Pei provided excellent research assistance.

1 See Cook (2008) for an interesting history of the RD design in education research, psychology, statistics, and economics. Cook argues the resurgence of the RD design in economics is unique as it is still rarely used in other disciplines.

2 The continuity of both functions is not the minimum that is required, as pointed out in Hahn et al. (2001). For example, identification is still possible even if only $E[Y_i(0)|X]$ is continuous, and only continuous at c. Nevertheless, it may seem more natural to assume that the conditional expectations are continuous for all values of X, since cases where continuity holds at the cutoff point but not at other values of X seem peculiar.

3 In the survey of Angrist and Krueger (1999), RD is viewed as an IV estimator, thus having essentially the same potential drawbacks and pitfalls. Here we argue that the assumptions required for RD designs to be valid are much weaker than what has to be imposed in the case of IVs.

4 This last results follows from the fact that $\lim_{\varepsilon \to 0} E[Y|X = c + \varepsilon] = \tau + \lim_{\varepsilon \to 0} E[W|X = c + \varepsilon]\delta_1 + \lim_{\varepsilon \to 0} E[U|X = c + \varepsilon]$, $\lim_{\varepsilon \to 0} E[Y|X = c - \varepsilon] = \lim_{\varepsilon \to 0} E[W|X = c - \varepsilon]\delta_1 + \lim_{\varepsilon \to 0} E[U|X = c - \varepsilon]$, and, thus, $\lim_{\varepsilon \to 0} E[Y|X = c + \varepsilon] - \lim_{\varepsilon \to 0} E[Y|X = c - \varepsilon] = \tau$.

5 From equation (14.5), it follows that $Y(0) = E[Y(0)|X] + U = g(X)\delta_1 + U$ and $Y(1) = E[Y(1)|X] + U = \tau + g(X)\delta_1 + U$. Thus, $Y = (1 - D) \cdot Y(0) + D \cdot Y(1) = D\tau + g(X)\delta_1 + U$.

6 Note that in the simple example we use here, since $g(X)$ is the fraction of high-ability types ($W = 1$), it would be fully captured by simply controlling for W in the regression model. But in a more realistic setting, the assignment variable X would also depend on other unobserved factors (e.g. unobserved ability) that are not captured by the covariates W, and are also likely correlated with the error term U. But since the above argument about the continuity of the potential outcomes in W and in X holds regardless of whether W is observed or not, the RD design remains valid and the treatment effect can still be estimated using a flexible regression model.

7 In a more realistic setting we would expect the error term to take on different values $U(0)$ and $U(1)$ in the two potential outcome equations. Since individuals with higher values of $U(1) - U(0)$ gain more from the treatment (higher treatment effect), they would likely put more effort into trying to score high enough to indeed receive the treatment. Lee (2008) shows that local randomization still holds in that setting provided, once again, that individuals have imperfect control over the assignment variable (some randomness in the test score in the example considered here).

8 Typically, one assumes that *conditional on the covariates*, the treatment (or instrument) is essentially 'as good as' randomly assigned.

9 See Imbens and Lemieux (2008) for a more formal exposition.

10 By contrast, when one runs a linear regression in a model where the true functional form is non-linear, the estimated model can still be interpreted as a linear predictor that minimizes specification errors. But since specification errors are only minimized globally, we can still have large specification errors at specific points including the cutoff point and, therefore, a large bias in RD estimates of the treatment effect.

11 For technical reasons, however, it would be preferable to undersmooth by shrinking the bandwidth at a faster rate requiring that $h \propto N^{-\delta}$ with $1/5 < \delta < 2/5$, in order to eliminate an asymptotic bias that would remain when $\delta = 1/5$. In the presence of this bias, the usual formula for the variance of a standard least squares estimator would be invalid. See Hahn et al. (2001) and Imbens and Lemieux (2008) for more details.

12 In order to mimic the fact that RD estimates are based on regression estimates at the boundary, the regression is estimated using only observations with values of X on the left of X_i ($X_i - h \leq X < X_i$) for observations on the left of the cutoff point ($X_i < c$). For observations on the right of the cutoff point ($X_i \geq c$), the regression is estimated using only observations with values of X on the right of X_i ($X_i < X \leq X_i + h$).

13 See Blundell and Duncan (1998) for a more general discussion of series estimators.

14 Although the probability of treatment is modeled as a linear probability model here, this does not impose any restrictions on the probability model since $g_D(x - c)$ is unrestricted on both sides of the cutoff c, while T is a dummy variable. So there is no need to write the model using a probit or logit formulation.

15 One small complication that arises in the non-parametric case of local linear regressions is that the usual (robust) standard errors from least squares are only valid provided that $h \propto N^{-\delta}$ with $1/5 < \delta < 2/5$. This is not a very important point in practice, and the usual standard errors can be used with local linear regressions.

16 McCrary (2008) discusses an example where students who barely fail a test are given extra points so that they barely pass. The RD estimator can remain unbiased if one assumes that those who are given extra points were chosen randomly from those who barely failed.

REFERENCES

Angrist, J. D. (1990). Lifetime earnings and the Vietnam era draft lottery: Evidence from social security administrative records. *American Economic Review*, 80(3), 313–336.

Angrist, J. D. and Krueger, A. B. (1999). Empirical strategies in labor economics. In O. Ashenfelter and D. Card (eds), *Handbook of Labor Economics*, volume 3A. Amsterdam: Elsevier.

Angrist, J. D. and Lavy, V. (1999). Using Maimonides' rule to estimate the effect of class size on scholastic achievement. *Quarterly Journal of Economics*, 114(2), 533–575.

Black, D. A., Galdo, J. and Smith, J. A. (2007). Evaluating the worker profiling and reemployment services system using a regression discontinuity approach. *American Economic Review*, 97(2), 104–107.

Blundell, R. and Duncan, A. (1998). Kernel regression in empirical microeconomics. *Journal of Human Resources*, 33(1), 62–87.

Caughey, D. and Sekhon, J. S. (2011). Elections and the regression discontinuity design: Lessons from close U.S. House races, 1942–2008. *Political Analysis*, 19, 385–408.

Cook, T. (2008). 'Waiting for life to arrive': A history of the regression-discontinuity design in psychology, statistics and economics. *Journal of Econometrics*, 142(2), 636–654.

Hahn, J., Todd, P. and der Klaauw, W. V. (2001). Identification and estimation of treatment effects with a regression-discontinuity design. *Econometrica*, 69(1), 201–209.

Imbens, G. and Kalyanaraman, K. (2012). Optimal bandwidth choice for the regression discontinuity estimator. *Review of Economic Studies*, 79(3), 933–960.

Imbens, G. and Lemieux, T. (2008). Regression discontinuity designs: A guide to practice. *Journal of Econometrics*, 142(2), 615–635.

Imbens, G. W. and Angrist, J. D. (1994). Identification and estimation of local average treatment effects. *Econometrica*, 62(2), 467–475.

Lee, D. S. (2008). Randomized experiments from non-random selection in U.S. House elections. *Journal of Econometrics*, 142(2), 675–697.

Lee, D. S. and Lemieux, T. (2010). Regression discontinuity designs in economics. *Journal of Economic Literature*, 48(2), 281–355.

Ludwig, J. and Miller, D. (2007). Does Head Start improve children's life chances? Evidence from a regression discontinuity design. *Quarterly Journal of Economics*, 122(1), 159–208.

McCrary, J. (2008). Manipulation of the running variable in the regression discontinuity design: A density test. *Journal of Econometrics*, 142(2), 698–714.

Porter, J. (2003). Estimation in the regression discontinuity model. *Department of Economics, University of Wisconsin*.

Thistlethwaite, D. L. and Campbell, D. T. (1960). Regression-discontinuity analysis: An alternative to the ex-post facto experiment. *Journal of Educational Psychology*, 51, 309–317.

Trochim, W. M. K. (1984). *Research Design for Program Evaluation: The Regression-Discontinuity Approach*. Beverly Hills, CA: Sage.

Van der Klaauw, W. (2002). Estimating the effect of financial aid offers on college enrollment: A regression-discontinuity approach. *International Economic Review*, 43(4), 1249–1287.

Van der Klaauw, W. (2008). Regression-discontinuity analysis: A survey of recent developments in economics. *Labour*, 22(2), 219–245.

White, H. (1980). A heteroskedasticity-consistent covariance matrix estimator and a direct test for heteroskedasticity. *Econometrica*, 48(4), 817–838.

15

Fixed-effects panel regression

Josef Brüderl and Volker Ludwig*

INTRODUCTION

Fixed-effects (FE) regression is a method that is especially useful in the context of causal inference (Gangl, 2010). While standard regression models provide biased estimates of causal effects if there are unobserved confounders, FE regression is a method that can (if certain assumptions are valid) provide unbiased estimates in this situation (other methods are instrumental variables regression and regression discontinuity; see Chapters 13 and 14 in this volume). Since unobserved confounders are ubiquitous in social science applications, FE regression should be standard in the toolkit of modern social research.

FE regression is most often used with panel data, and therefore the focus of this chapter will be on FE regression with panel data. However, before we begin, we want to place FE regression in a wider context. FE regression is not only applicable with panel data, it can be used with any kind of multi-level data. These are data where lower-level units are nested within higher-level groups (sometimes also called 'clusters'). A multi-level regression (see Chapter 7 in this volume) has to assume that there is neither unit-specific nor group-specific unobserved heterogeneity.[1] With non-experimental data, this assumption is often violated due to self-selection on the group level. However, the assumption of no unobserved heterogeneity can be weakened if the researcher uses FE models. A fixed-effects regression is specified on the level of the units and includes group-specific constants (the so-called 'fixed effects'). Because group-specific fixed effects wipe out all group-specific unobserved heterogeneity, FE models only require the assumption of no unit-specific unobserved heterogeneity. Thus, compared with standard regression models, FE models allow a causal effect to be identified under weaker assumptions. Obviously, this makes FE models attractive for social researchers undertaking causal analysis.

There are many examples of multi-level data where FE models have been used:

- Pupils nested within classes nested within schools: School class fixed effects wipe out unobserved heterogeneity on the class level (Legewie, 2012).
- Workers nested within firms (matched employer-employee data): Firm fixed effects control for unobservables on the firm level (Hinz and Gartner, 2005).
- Siblings nested within families: Family fixed effects wipe out unobservables on the family level (Arránz Becker et al., 2013).

- Individuals nested within countries: Country fixed effects control for all country-level heterogeneity.
- Repeated observations within individuals: Individual fixed effects control for all person-level heterogeneity.

The latter case describes an FE model for panel data: here we observe persons repeatedly over several panel waves. Panel data are especially useful for applying FE models because, due to their richness, they allow many relevant social science questions to be investigated. Therefore, the combination of panel data and FE modeling is especially promising.

The basic fixed-effects framework

Panel data are typically set up in long format. That is, the observations of each subject are ordered chronologically, and the time series (panels) of subjects are stacked below each other (pooled data). We refer to a setting where there are short time series, but many cross-sections ($N \to \infty$, T fixed). This is the typical situation when estimating an effect of some causal variable using data from an annual panel survey conducted in years $t = 1, \ldots, T$ for individuals $i = 1, \ldots, N$. In this case a single observation from a subject i is called a person-year. A leading case is estimation of the causal effect of an event that occurs during the life course (e.g. marriage). We assume that the outcome variable Y is continuous, while the K regressors $X_1, \ldots, X_K$ may be measured on any scale.

FE estimation builds on the error components model,

$$y_{it} = \mathbf{x}_{it}\boldsymbol{\beta} + \alpha_i + \epsilon_{it}. \tag{15.1}$$

Here, y_{it} denotes the observed outcome of person i at time t, $\mathbf{x}_{it}$ is the $(1 \times K)$ vector of covariates of this person measured contemporaneously, and $\boldsymbol{\beta}$ is the corresponding $(K \times 1)$ vector of parameters to be estimated. The error term of this model is split into two components. The α_i are stable person-specific characteristics which not only are often unobserved by the researcher (e.g. cognitive ability, genetic disposition, personality), but also are very often related to the covariates. Hence, the α_i are unobserved effects capturing time-constant individual heterogeneity. The second component ϵ_{it} is an idiosyncratic error that varies across subjects and over time. The intercept α that is standard in regression models is dropped here due to collinearity with the person-specific errors α_i (in fact, these can be seen as person-specific intercepts).

Formally, such an error term decomposition is always possible. However, the two terms can only be identified if panel data are available (or, more generally, multi-level data). The reason is that we can infer the person-specific characteristics only from repeated observations. With cross-sectional data the error components model is not identified.

The easiest way to estimate the parameters of the model is by pooling the data and running ordinary least squares (pooled OLS, POLS). Consistency of POLS requires exogeneity of time-constant individual heterogeneity and idiosyncratic errors. POLS estimation does not distinguish between the two error components, which are replaced by the composite error $\nu_{it} = \alpha_i + \epsilon_{it}$. The condition for consistency is $E(\mathbf{x}'_{it}\nu_{it}) = \mathbf{0}$ which is equivalent to assuming $E(\mathbf{x}'_{it}\alpha_i) = \mathbf{0}$ and $E(\mathbf{x}'_{it}\epsilon_{it}) = \mathbf{0}$. The second condition requires that idiosyncratic errors are contemporaneously exogenous, an assumption that is often reasonable. But the first condition imposes exogeneity of stable characteristics, which very often is not a reasonable assumption. In case of self-selection into treatment, the assumption is violated and POLS estimates are biased and inconsistent.

On the other hand, the key assumption for consistency of the FE estimator is the strict exogeneity condition imposed on the idiosyncratic errors,

$$E(\epsilon_{it} | \mathbf{x}_{i1}, \ldots, \mathbf{x}_{iT}, \alpha_i) = E(\epsilon_{it} | \mathbf{x}_{it}, \alpha_i) = 0. \tag{15.2}$$

Strict exogeneity implies $E(\mathbf{x}'_{is}\epsilon_{it}) = \mathbf{0}$, for all $s, t = 1, \dots, T$. This rules out not only contemporaneous correlation of regressors and the idiosyncratic errors, but also correlation of past and future values of covariates and errors. This is why the assumption is called 'strict' exogeneity. Obviously this assumption is somewhat stronger than the contemporaneous exogeneity assumption required for POLS. The key advantage, however, of the FE framework is that no assumption is needed concerning the relation of stable characteristics and regressors. In fact, α_i and $\mathbf{x}_{it}$ can be related in arbitrary ways which will not result in biased estimates of the coefficients.

FE estimation applies POLS to transformed data where the transformation (called 'demeaning' or 'within transformation') extracts the variation within subjects over time, but discards variation across units. Averaging equation (15.1) over time gives

$$\bar{y}_i = \bar{\mathbf{x}}_i\boldsymbol{\beta} + \alpha_i + \bar{\epsilon}_i, \qquad (15.3)$$

where $\bar{y}_i = T^{-1}\sum_{t=1}^{T} y_{it}$, $\bar{\mathbf{x}}_i = T^{-1}\sum_{t=1}^{T}\mathbf{x}_{it}$ and $\bar{\epsilon}_i = T^{-1}\sum_{t=1}^{T}\epsilon_{it}$ are person-specific means. Because α_i is time-constant for each person it is identical to the mean for that person. Therefore, subtracting equation (15.3) from equation (15.1) eliminates α_i and any bias that might result from its association with regressors.

The demeaned regression is given by

$$y_{it} - \bar{y}_i = (\mathbf{x}_{it} - \bar{\mathbf{x}}_i)\boldsymbol{\beta} + \epsilon_{it} - \bar{\epsilon}_i. \qquad (15.4)$$

The conventional FE estimator is the POLS estimator applied to these demeaned data. As mentioned above and shown in the statistical framework section below, the strict exogeneity assumption is sufficient for consistency of the FE estimator. The main point is that the assumption of no person-specific unobserved heterogeneity $E(\mathbf{x}'_{it}\alpha_i) = \mathbf{0}$ is no longer needed. The FE estimator is consistent even if $E(\mathbf{x}'_{it}\alpha_i) \neq \mathbf{0}$. Thus, with panel data and the FE estimator it is possible to identify a causal effect under weaker assumptions (compared to cross-sectional OLS or POLS).

A didactic example

To strengthen the intuition on FE methodology we now demonstrate with stylized data how the FE estimator works. The data are plotted in Figure 15.1.[2] For didactic reasons we imagine that these are the wage careers of four men ($N = 4$) over six panel waves ($T = 6$). The outcome variable is the monthly wage in euros. The treatment variable is a marriage dummy that is zero before marriage (black dots) and unity afterwards (grey triangles). As can be seen, two of the men are low earners who do not marry during the observation window. The other two men are high earners and they marry between panel waves (time) 3 and 4. The wage careers are constructed so that with marriage there is a wage increase of €500. Thus we built a causal marriage effect of plus €500 into these data (marital wage premium). Further, we built self-selection into these data: it is the high earners who marry (perhaps because they are the more attractive marriage partners). This has the consequence that these data are plagued by person-specific heterogeneity.

Next we want to compare the results of three wage regressions that are specified as follows:

$$w_{it} = \beta m_{it} + \alpha_i + \epsilon_{it}. \qquad (15.5)$$

Here w_{it} is the monthly wage and m_{it} is the marriage dummy. β is then an estimate for the marital wage premium.

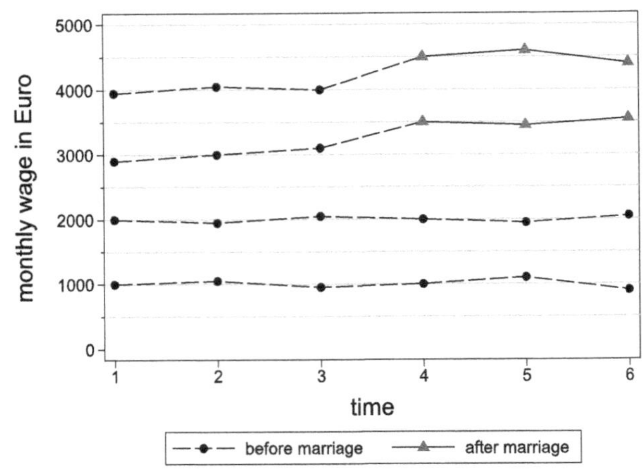

Figure 15.1 Fictional wage careers of four men

First, we estimate a *cross-sectional OLS regression* at $t = 4$ (see Table 15.1, column (1)). Because the regression coefficient of a dummy is simply the difference of the group means, we obtain an estimate of the marriage effect of plus €2500 (it can easily be seen from Figure 15.1 that at $t = 4$ the wage average of those married is €4000 and of those not married is €1500). Obviously, this estimate of the marital wage premium is strongly biased upwards. The statistical reason for this bias is that we have person-specific unobserved heterogeneity in these data. The person-specific error term α_i is correlated with the marriage dummy m_{it}.

More intuitively, the reason is that a cross-sectional regression compares outcomes of different persons: the wage of those married with the wage of those not married. We term this a *between comparison*. A between comparison only works if the assumption of unit (here, person) homogeneity holds: the persons we compare must not differ in anything relevant but the treatment (in other words, there is no unobserved heterogeneity). However, in our data this assumption is violated: men differ not only by marriage but also on other relevant characteristics that affect their wage and marriage propensity. In fact, in most non-experimental social science research the assumption of unit homogeneity will not be tenable: units differ in so many respects that it is impossible to control for all of them.

Would capitalizing on the panel structure of our data remedy the problem? To investigate this, in a next step we estimate an *OLS regression with the pooled data* (POLS, see Table 15.1, column (2)). POLS uses all 24 observations and estimates a marital wage premium of plus €1833. This is the mean of the six wage observations after marriage minus the mean of the 18 wage observations before marriage. Again, the statistical reason for this heavy upward bias is that we have unobserved heterogeneity.

More intuitively, the reason is that POLS, once again, does a between comparison. With panel data, we can distinguish *between* and *within variation*. Between variation is generated by person differences (more technically, it is the variation of the person-specific means $\bar{y}_i$). Within variation is the variation generated by changes over time within the persons (more technically, it is the variation of the demeaned data $y_{it} - \bar{y}_i$). POLS uses both components of variation to estimate the marriage effect. But, as we already know, the between variation is 'contaminated' (Allison, 2009, p. 3) by unobserved heterogeneity due to self-selection. Therefore, the POLS estimate is so strongly biased. Compared with cross-sectional OLS the bias is lower, because POLS also uses the within variation (which is not contaminated).

Table 15.1 Comparing three regression methods: the effect of marriage on wages

	(1) Cross-sectional OLS	(2) POLS	(3) FE
Marriage	2500	1833	500
Constant	1500	2167	–
Number of persons	4	4	2
Number of person-years	4	24	12

Source: own computations from fabricated data.

Obviously, panel data *per se* do not remedy the problem of unobserved heterogeneity. Part of the variation in panel data – namely between variation – is contaminated if there is self-selection on the person level. Therefore, estimation techniques that recur on between variation will be biased. However, if the within variation is exogenous, a solution to the problem of unobserved person heterogeneity would be to base the estimation on within variation alone.

A *fixed-effects regression* does exactly this: it discards the between variation and infers the causal effect from the within variation only. In Table 15.1, column (3), we see that FE regression in fact provides the correct estimate of the marital wage premium: plus €500. By demeaning the data all between variation has been eliminated. Only within variation is left and estimates are based on within variation only. Therefore, person-specific heterogeneity no longer disturbs estimation. FE estimation is not biased by person-specific unobserved heterogeneity.

To get more intuition on how FE estimation 'works', we plot the demeaned data in Figure 15.2 (data are jittered for better visibility). Marriage (demeaned) has always value zero for those who never marry. Their 12 person-years are therefore in the middle of the plot. Marriage (demeaned) is −0.5 for the six person-years before marriage and +0.5 after marriage. Now, what determines the FE regression line? The FE regression line is estimated by OLS applied to the demeaned data. First, because demeaned variables have a mean of zero, the FE regression line passes through (0, 0). This makes clear that those persons who never marry contribute nothing to the FE estimation. The reason is that they have no within variation on the treatment variable. Therefore, the FE regression line is based only on the 12 observations of those two men who eventually marry. Second, the slope of the regression line is determined by the mean of the (demeaned) wage of the six person-years after marriage minus the mean wage of the six person-years before marriage. This difference is €500.

Thus we see that the FE estimator does a pure *within comparison*: the wages after a marriage are compared with the wages before a marriage. And this within comparison is not biased by any kind of person heterogeneity. The intuition on within comparison becomes even clearer when one describes FE estimation in a slightly different (but equivalent) way. Compute for each man who married the within wage difference (average wage after marriage minus average wage before). This is the marital wage premium estimated for each man separately. Then the FE estimator is the (weighted) average of the individual wage premia.

The main point is that FE estimation does not infer the causal effect from a comparison of different persons, but by comparing within person change that is induced by a treatment event. Thus, FE estimation no longer needs the strong and in many situations untenable assumption of unit homogeneity. It needs the assumption of 'temporal homogeneity': nothing relevant changes over time, only the treatment. This assumption usually is also strong, because many things change over time. However, the assumption is valid in our stylized data and therefore the FE estimator works well here.

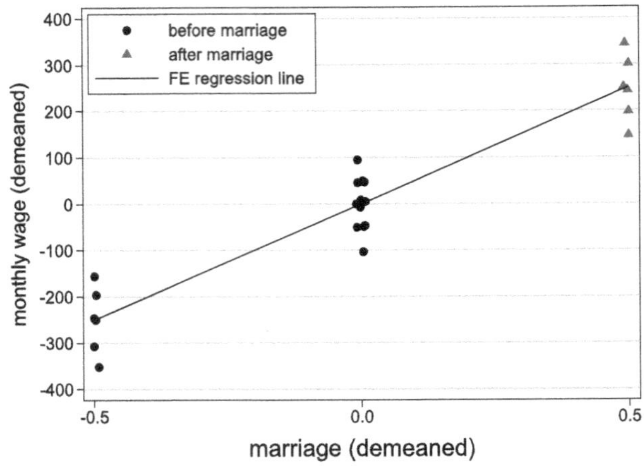

Figure 15.2 The 'mechanics' of FE estimation

More generally, the assumption of temporal homogeneity can be weakened easily by including a control group of non-treated units. In our stylized data these would be the men who never marry. By including time dummies in addition to the marriage dummy they also contribute to FE estimation by providing an estimate of the time trend. This trend is then differenced out from the within comparison in the treatment group. One no longer has to assume that nothing relevant changes over time, but only that the time trends are parallel in the treatment and control group (parallel trends assumption, see below). More formally, this is the strict exogeneity assumption introduced above. In many social science research contexts, this assumption is weaker than a unit homogeneity assumption. This is the reason why panel data and FE estimation allow a causal effect to be identified under weaker assumptions than standard regression with cross-sectional data.

STATISTICAL FRAMEWORK OF FIXED-EFFECTS REGRESSION AND ALTERNATIVE PANEL ESTIMATORS

The basic fixed-effects framework continued

In the previous section we introduced the FE estimator by applying POLS to the demeaned data. We can write the transformed estimation equation (15.4) as

$$\ddot{y}_{it} = \ddot{\mathbf{x}}_{it}\boldsymbol{\beta} + \ddot{\epsilon}_{it}, \tag{15.6}$$

where the dots indicate that respective variables have been demeaned. Consistency of FE estimation is preserved because strict exogeneity is sufficient for consistency of POLS applied to the demeaned data. Assumption $E(\ddot{\mathbf{x}}_{it}'\alpha_i) = \mathbf{0}$ holds because α_i has gone after demeaning. The assumption $E(\ddot{\mathbf{x}}_{it}'\ddot{\epsilon}_{it}) = \mathbf{0}$ holds because $E(\mathbf{x}_{is}'\epsilon_{it}) = \mathbf{0}$, for all $s, t = 1, \ldots, T$. Hence, $E(\ddot{\mathbf{x}}_{it}'\ddot{\epsilon}_{it}) = E[(\mathbf{x}_{it} - \bar{\mathbf{x}}_i)'(\epsilon_{it} - \bar{\epsilon}_i)] = \mathbf{0}$. Neither ϵ_{it} nor $\bar{\epsilon}_i$ is related to $\mathbf{x}_{it}$ or $\bar{\mathbf{x}}_i$ by assumption. Like any assumption, strict exogeneity may fail. Two important violations may arise if time-constant unobservables produce heterogeneous growth in the outcome and if current outcomes are determined by prior outcomes. These problems are discussed later in this section.

The conventional FE estimator can be derived by the analogy principle, replacing expectations by sample moments. Write (15.6) as a system of T OLS equations

$$\ddot{\mathbf{y}}_i = \ddot{\mathbf{X}}_i\boldsymbol{\beta} + \ddot{\boldsymbol{\epsilon}}_i, \tag{15.7}$$

where $\ddot{\mathbf{y}}_i$ and $\ddot{\boldsymbol{\epsilon}}_i$ are $T \times 1$, $\ddot{\mathbf{X}}_i$ is $T \times K$ and $\boldsymbol{\beta}$ is $K \times 1$. Premultiplying (15.7) by $\ddot{\mathbf{X}}_i'$, taking expectations and solving for $\boldsymbol{\beta}$ gives

$$\begin{aligned} \ddot{\mathbf{X}}_i'\ddot{\mathbf{y}}_i &= \ddot{\mathbf{X}}_i'\ddot{\mathbf{X}}_i\boldsymbol{\beta} + \ddot{\mathbf{X}}_i'\ddot{\boldsymbol{\epsilon}}_i \\ E(\ddot{\mathbf{X}}_i'\ddot{\mathbf{y}}_i) &= E(\ddot{\mathbf{X}}_i'\ddot{\mathbf{X}}_i\boldsymbol{\beta}) + E(\ddot{\mathbf{X}}_i'\ddot{\boldsymbol{\epsilon}}_i) = \boldsymbol{\beta}E(\ddot{\mathbf{X}}_i'\ddot{\mathbf{X}}_i) \\ \boldsymbol{\beta} &= \left[E(\ddot{\mathbf{X}}_i'\ddot{\mathbf{X}}_i)\right]^{-1}E(\ddot{\mathbf{X}}_i'\ddot{\mathbf{y}}_i). \end{aligned} \tag{15.8}$$

Note that a rank condition must hold for $E(\ddot{\mathbf{X}}_i'\ddot{\mathbf{X}}_i)$ to be invertible (i.e. nonsingular). The rank of the matrix must be equal to the number of regressors, rk $\sum_{n=1}^{N} E(\ddot{\mathbf{X}}_i'\ddot{\mathbf{X}}_i) = K$, which imposes linear independence among the demeaned covariates.

Demeaning not only wipes out unobserved time-constant variables α_i, but along with them all observed time-constant covariates. The effect of variables that do not change for any person over time (e.g. sex, birth cohort) thus cannot be estimated. This is often seen as a shortcoming of FE estimation, but it actually is a major strength of the method because the number of potential confounders that need be measured and included in the model is reduced tremendously. For example, one never has to worry about bias induced by genetic differences between persons, a potential confounder that is virtually impossible to measure in large surveys.

Given that strict exogeneity and the rank condition are satisfied, the FE estimators are identified as

$$\hat{\boldsymbol{\beta}}_{\mathrm{FE}} = \left(\sum_{n=1}^{N} \ddot{\mathbf{X}}_i'\ddot{\mathbf{X}}_i\right)^{-1} \left(\sum_{n=1}^{N} \ddot{\mathbf{X}}_i'\ddot{\mathbf{y}}_i\right). \tag{15.9}$$

Inference

For tests of hypotheses concerning the FE coefficients further assumptions are necessary. The idiosyncratic errors must be homoscedastic and serially uncorrelated to obtain consistency of the variance–covariance matrix from which to get the standard errors of coefficients. Both conditions hold if

$$E(\boldsymbol{\epsilon}_i\boldsymbol{\epsilon}_i'|\mathbf{x}_i, \alpha_i) = E(\boldsymbol{\epsilon}_i\boldsymbol{\epsilon}_i') = \sigma_\epsilon^2\mathbf{I}_T. \tag{15.10}$$

Since FE regression is based on POLS for the demeaned data, however, the above condition must guarantee that POLS conditions of homoscedasticity and no serial correlation hold for the transformed errors. The POLS conditions are (Wooldridge, 2010, p. 192)

$$E(\epsilon_{it}^2\mathbf{x}_{it}'\mathbf{x}_{it}) = \sigma_\epsilon^2 E(\mathbf{x}_{it}'\mathbf{x}_{it}), \tag{15.11}$$

$$E(\epsilon_{it}\epsilon_{is}\mathbf{x}_{it}'\mathbf{x}_{is}) = \mathbf{0}, \quad \text{for all } t \neq s. \tag{15.12}$$

Assumption (15.11) is the homoscedasticity condition for each of the T cross-sections. Assumption (15.12) states that the composite errors must be uncorrelated over time.

The conventional asymptotic variance of the FE estimator is

$$\widehat{\mathrm{Var}}(\hat{\boldsymbol{\beta}}_{\mathrm{FE}}) \approx^{\mathrm{asy}} \hat{\sigma}_\epsilon^2 \left(\sum_{i=1}^{N} \ddot{\mathbf{X}}_i'\ddot{\mathbf{X}}_i\right)^{-1}, \tag{15.13}$$

where the standard errors are the square roots of the diagonal elements. After FE regression, the residuals are given by $\hat{\ddot{\epsilon}}_{it} = \ddot{y}_{it} - \ddot{\mathbf{x}}_{it}\hat{\boldsymbol{\beta}}$. These are squared and then summed over t and i to estimate the error variance $\hat{\sigma}_\epsilon^2 = \sum_{i=1}^{N}\sum_{t=1}^{T}\hat{\ddot{\epsilon}}_{it}^2/(NT - N - K)$, where $NT - N - K$ are the correct degrees of freedom. (The usual standard errors obtained when running OLS on the demeaned equation are not valid. They have to be adjusted for the fact that N means have to be calculated for demeaning.)

Assumption (15.10) guarantees that (15.11) holds because the variance of the demeaned errors of period t is $E(\ddot{\epsilon}_{it}^2) = E[(\epsilon_{it} - \bar{\epsilon}_i)^2] = E(\epsilon_{it}^2) + E(\bar{\epsilon}_i^2) - 2E(\epsilon_{it}\bar{\epsilon}_i) = \sigma_\epsilon^2 + \sigma_\epsilon^2/T - 2\sigma_\epsilon^2/T = \sigma_\epsilon^2(1 - 1/T)$, which does not vary over t. However, assumption (15.12) is violated. Even if the idiosyncratic errors are serially uncorrelated in the untransformed data so that condition (15.10) holds, the demeaned errors are negatively correlated because $E(\ddot{\epsilon}_{it}\ddot{\epsilon}_{is}) = E[(\epsilon_{it} - \bar{\epsilon}_i)(\epsilon_{is} - \bar{\epsilon}_i)] = E(\epsilon_{it}\epsilon_{is}) - E(\epsilon_{it}\bar{\epsilon}_i) - E(\epsilon_{is}\bar{\epsilon}_i) + E(\bar{\epsilon}_i^2) = 0 - \sigma_\epsilon^2/T - \sigma_\epsilon^2/T + \sigma_\epsilon^2/T = -\sigma_\epsilon^2/T$. Although this term usually is small in practice (especially as T gets larger), in many applications there is substantial serial correlation already before transformation. It is therefore recommended to use panel-robust standard errors by default.

The panel-robust standard errors correct for arbitrary clustering of time series and heteroscedasticity. They are obtained from

$$\widehat{\mathrm{Var}}(\hat{\boldsymbol{\beta}}_{\mathrm{FE}}) \overset{\mathrm{asy}}{\approx} (\ddot{\mathbf{X}}'\ddot{\mathbf{X}})^{-1}\left(\sum_{i=1}^{N}\ddot{\mathbf{X}}_i'\hat{\ddot{\boldsymbol{\epsilon}}}_i\hat{\ddot{\boldsymbol{\epsilon}}}_i'\ddot{\mathbf{X}}_i\right)(\ddot{\mathbf{X}}'\ddot{\mathbf{X}})^{-1}. \tag{15.14}$$

For large N and small T, the degrees of freedom are $NT - K$, which is smaller than the adjustment of the usual standard errors. It is therefore quite often the case that the panel-robust standard errors are actually smaller. Angrist and Pischke (2009) recommend reporting the larger of the two standard errors to be on the conservative side.

Panel-robust standard errors are biased if there are only few clusters. Kézdi (2004) shows for simulated data that the clustered standard errors perform very well if there are 50 or more clusters. Stock and Watson (2008) present simulation results showing little bias of clustered standard errors in the event of heteroscedasticity and 100 or more clusters. This result also holds when errors are at the same time serially correlated (modeled as a first-order moving average). Nevertheless, as Angrist and Pischke (2009, Chapter 8) argue, standard errors will always be biased in finite samples. However, it is not the prime task of social researchers to get the standard errors right. Much more important is to get the coefficient estimates right. Therefore, we recommend following their advice: 'Your standard errors probably won't be quite right, but they rarely are. Avoid embarrassment by being your own best skeptic, and especially, DON'T PANIC!' (Angrist and Pischke, 2009, p. 327).

Other basic within estimators

There are three other within estimators which yield results that are identical or similar to the FE estimator. The first, least squares dummy variable (LSDV) regression, is conceptually simple and makes the intuition of within estimation very clear. Instead of demeaning the data one could include $N - 1$ dummy variables in a POLS regression, one indicator variable for each person (leaving one out as the reference category). That is, one 'fixes' the time-constant differences between subjects by replacing the common constant of the POLS model with individual-specific constants. The term 'fixed-effects' regression actually stems from this approach where the unobserved effects are viewed as fixed quantities and individual constants as parameters that need to be estimated.[3] However, in large samples it is more appropriate to view them as random variables, which is the modern approach to FE regression. The LSDV and FE estimators are identical.

The second alternative to FE regression is first differencing (FD). Instead of demeaning the data, FD builds on differences between ensuing observations from the same subject. This is another within transformation that wipes out time-constant unobservables (since these are the same at t and $t-1$). FD is the POLS regression of

$$\Delta y_{it} = \Delta \mathbf{x}_{it}\boldsymbol{\beta} + \Delta \epsilon_{it}, \tag{15.15}$$

where $\Delta y_{it} = y_{it} - y_{it-1}, \Delta \mathbf{x}_{it} = \mathbf{x}_{it} - \mathbf{x}_{it-1}, \Delta \epsilon_{it} = \epsilon_{it} - \epsilon_{it-1}$. FD and FE yield identical estimates for $\boldsymbol{\beta}$ when $T = 2$. For panels that are longer than two periods, estimates differ in general. They will differ substantially if there is very strong or very weak serial correlation in the untransformed data. Both estimators require strict exogeneity. However, while FE builds on the assumption of no serial correlation prior to demeaning (see condition (15.10)), FD relies on no serial correlation in the differenced errors. The latter assumption is equivalent to very strong correlation in the untransformed errors. FE and FD therefore rely on assumptions that are opposite extremes.

The third well-known alternative is the difference-in-differences (DiD) estimator. The DiD approach is similar to LSDV in that both explicitly model differences in outcome levels. However, DiD is more parsimonious because only mean differences between groups are modeled, not individual differences. For example, if one is interested in the effect of a training program that takes place between years $t = 1$ and $t = 2$ and uses DiD for evaluation of earnings in $t = 2$, one may estimate by POLS the equation

$$y_{it} = \alpha_0 + \delta_1 p_{it} + \alpha_1 d_i + \delta_2 p_{it} d_i + \mathbf{x}_{it}\boldsymbol{\beta} + \epsilon_{it}. \tag{15.16}$$

This regression contains a dummy for the second (post-treatment) year, p_{it}, a time-constant dummy indicating that a person belongs to the treated group, d_i, the interaction of these two, $p_{it}d_i$, and possibly control variables, $\mathbf{x}_{it}$. The overall constant α_0 gives us the mean earnings of the control group in the first year (pre-treatment). α_1 tells us by how much the mean earnings of the treated differed from those of the non-treated before treatment. The coefficient on the year dummy, δ_1, is the change in mean earnings of the non-treated from year 1 to year 2, and δ_2 captures by how much the change in mean earnings across the two years (i.e. from before to after treatment) differed in the treated group. δ_2 is thus an estimate of the treatment effect. Since only group mean differences are modeled, the DiD estimator essentially builds on aggregate data. This is why it does not necessarily require panel data, but can be applied also to a pseudo-panel of pooled independent cross-sections. This property makes DiD a very useful tool for policy evaluation when only pseudo-panel data are available.

Any of the four basic within estimators is appropriate for dealing with time-constant confounders. Nevertheless, FE regression has some important practical advantages over the others. While LSDV is conceptually very simple, it is computationally impracticable with large N because any statistical software will refuse to estimate a model containing several thousand regressors. As noted above, FD might be preferable in the presence of strong serial correlation. However, besides that advantage, it has the disadvantage of being inefficient because the initial period is dropped in any case. Moreover, the inefficiency can be very large in the presence of missing data, because first differences can be built only on ensuing observations. For example, if one person is observed at $t = 1, 3, 5$, then FE would use the three person-years, but in FD the person would be dropped completely. With (balanced) panel data for $T = 2$, the DiD estimator is identical to FE and FD. For longer panels, however, it differs in general. In fact, it can give misleading answers because all variables enter the regression in levels. If there are control variables (which usually is the case) their effect is likely to be biased, which may also induce bias on the treatment effect. It is therefore recommended to use FE (or FD) where all variables are transformed (Wooldridge, 2010, p. 321).

Extending the fixed-effects approach to account for heterogeneous growth

As mentioned above, strict exogeneity may fail if some unobservable produces heterogeneity with regard to individual trajectories of the outcome. More precisely, the assumption must fail if heterogeneous growth is related to the covariate(s) of interest. Strict exogeneity implies that individual outcome trajectories of treated subjects would have developed parallel to trajectories of non-treated subjects *had they not chosen treatment.* In this section, we show how conventional within estimators fail if this assumption of 'parallel trends' is not met, and how failure of FE can be repaired by extending the FE approach to allow for individual-specific slopes.

Any of the four basic within estimators can be extended to allow for heterogeneous growth. Suppose there are $T = 3$ observations per person and consider the FD model with one regressor x_{it} (say, a binary treatment indicator) and individual-specific slope parameters of (process) time. The structural model can be written as a system of three equations:

$$\text{at } t, \quad y_{it} = \alpha_{1i} + \alpha_{2i}t + \beta x_{it} + \epsilon_{it}, \tag{15.17}$$

$$\text{at } t-1, \quad y_{it-1} = \alpha_{1i} + \alpha_{2i}(t-1) + \beta x_{it-1} + \epsilon_{it-1}, \tag{15.18}$$

$$\text{at } t-2, \quad y_{it-2} = \alpha_{1i} + \alpha_{2i}(t-2) + \beta x_{it-2} + \epsilon_{it-2}. \tag{15.19}$$

Here, α_{1i} are individual-specific constants capturing any differences in outcome *levels* between individuals produced by stable characteristics. In addition, α_{2i} are individual-specific slopes that capture individual differences in the *growth* of outcomes over time. Technically, α_{2i} are stable characteristics that interact with time to produce outcome trajectories specific to individuals. These may be observable or unobservable characteristics. For example, the steepness of wage trajectories might differ by cohort, because younger cohorts face better labor market opportunities over the life course. Or they might differ by unobserved career orientation, because the motivated are likely to get promotions at a faster rate.

Individual heterogeneity in the outcome levels, i.e. in α_{1i}, is eliminated by first differencing. Subtract (15.18) from (15.17), and (15.19) from (15.18), to get

$$\text{at } t, \quad (y_{it} - y_{it-1}) = \alpha_{2i} + \beta(x_{it} - x_{it-1}) + (\epsilon_{it} - \epsilon_{it-1}), \tag{15.20}$$

$$\text{at } t-1, \quad (y_{it-1} - y_{it-2}) = \alpha_{2i} + \beta(x_{it-1} - x_{it-2}) + (\epsilon_{it-1} - \epsilon_{it-2}). \tag{15.21}$$

Estimation of the FD model will nevertheless be biased if α_{2i} is related to x_{it}. Strict exogeneity will hold only if the individual trends do not differ in the treated and control group, on average. Hence, it must be true that $E(\alpha_{2i}|x_{it}, \alpha_{1i}) = 0$. In words, the outcome trajectories of the two groups would have developed parallel to each other had there been no treatment. There might have been differences in outcome levels prior to treatment, and there might have been a treatment effect for the treated. But conditioning on α_{1i} and x_{it}, there should be no difference in the mean outcomes.

How can we cure the bias? A simple way to get rid of α_{2i} is to apply differencing once more to get

$$\text{at } t, \quad (y_{it} - 2y_{it-1} + y_{it-2}) = \beta(x_{it} - 2x_{it-1} + x_{it-2}) + (\epsilon_{it} - 2\epsilon_{it-1} + \epsilon_{it-2}). \tag{15.22}$$

This extension of FD, second differencing (SD), provides an unbiased estimate of the treatment effect even if there is heterogeneity with respect to individual growth that is systematically related to treatment, that is, if the parallel trends condition is violated.

The LSDV and DiD estimators can also be extended to eliminate bias that is due to α_{2i}. Instead of just including dummy variables in a POLS regression (LSDV), one could additionally include interactions of these dummies with time. The coefficients on the interaction terms would then give estimates for individual growth. Similarly, when following the DiD approach, one would add the interaction of time and the time-constant indicator of the treatment group.

As explained above, conventional FE estimation is based on within transformation of the data, also known as demeaning. The idea was to subtract the individual mean as a time-constant estimate of individuals' outcome trajectory to eliminate individual differences in outcome levels. The extension of this approach now subtracts a time-varying estimate instead (i.e. an estimate of individual growth curves), to eliminate the individual-specific slopes α_{2i} along with the individual-specific constants α_{1i}. Intuitively, this idea leads to the following estimation procedure:

(1) Estimate for each subject i the individual POLS regression $y_{it} = \alpha_{1i} + \alpha_{2i}t + \xi_{it}$ and get predicted values $\hat{y}_{it} = \hat{\alpha}_{1i} + \hat{\alpha}_{2i}t$.
(2) From the actual outcome values y_{it}, subtract the estimated values $\hat{y}_{it}$ from step (1) to get detrended outcomes $\tilde{y}_{it} = y_{it} - \hat{y}_{it}$.
(3) Repeat steps (1) and (2) for the independent variable(s) to get, for any variable x_j, the detrended variable $\tilde{x}_{jit} = x_{jit} - \hat{x}_{jit}$.
(4) Pool the transformed data from steps (2) and (3) and run a POLS regression.

The predicted values $\hat{y}_{it}$ from step (1) are time-varying predicted values of the expected outcome of individual i. These estimated values are used in step (2) to purge individual i's outcomes from the expected individual-specific trend. Hence, the $\tilde{y}_{it}$ are just the residuals of the individual time-series regression from step (1). Detrending all variables of the model in this way thus makes clear what the extended within transformation actually does to the model. The estimate of the treatment effect is based only on variation within subjects over time that cannot be predicted from the (initial) level and slope of individual trajectories.

The fixed-effects model with individual-specific constants and slopes (FEIS), introduced by Polachek and Kim (1994) and generalized by Wooldridge (2010, pp. 377–381), can be described as follows.[4] If the data are in long format, we can write the structural model for unit i as

$$\mathbf{y}_i = \mathbf{Z}_i\boldsymbol{\alpha}_i + \mathbf{X}_i\boldsymbol{\beta} + \boldsymbol{\epsilon}_i, \tag{15.23}$$

where $\mathbf{y}_i$ and $\boldsymbol{\epsilon}_i$ are $T \times 1$, $\mathbf{Z}_i$ is $T \times J$, $\boldsymbol{\alpha}_i$ is $J \times 1$, $\mathbf{X}_i$ is $T \times K$ and $\boldsymbol{\beta}$ is $K \times 1$.

Now define a 'residual maker' matrix for unit i: $\mathbf{M}_i \equiv \mathbf{I}_T - \mathbf{Z}_i(\mathbf{Z}_i'\mathbf{Z}_i)^{-1}\mathbf{Z}_i'$. Premultiply equation (15.23) through by $\mathbf{M}_i$. This gives $\mathbf{M}_i\mathbf{y}_i = \mathbf{M}_i\mathbf{Z}_i\boldsymbol{\alpha}_i + \mathbf{M}_i\mathbf{X}_i\boldsymbol{\beta} + \mathbf{M}_i\boldsymbol{\epsilon}_i$. Since $\mathbf{M}_i\mathbf{Z}_i = \mathbf{Z}_i - \mathbf{Z}_i(\mathbf{Z}_i'\mathbf{Z}_i)^{-1}\mathbf{Z}_i'\mathbf{Z}_i = \mathbf{0}$, $\boldsymbol{\alpha}_i$ is eliminated. Conventional FE is a special case of this model where $\mathbf{Z}_i \equiv \mathbf{1}$, that is, the model includes only individual constants. Another special case is the model including additionally individual-specific linear time trends such that $\mathbf{Z}_i \equiv (\mathbf{1}, \mathbf{t})$. As discussed above, this model could be estimated by POLS on second differences. However, FEIS is more efficient, and this advantage can be very important if panels are short and/or if panels have gaps due to missing data (analogous to the pros and cons of FE and FD; see the previous subsection). FEIS is also much more general than SD because individual slopes can be assumed for variables other than calendar time. For example, one could assume that unobserved career orientation produces steeper wage trajectories over labor market experience (and perhaps experience squared). Given there are enough observations per subject (at least $J+1$ are needed to remove $\boldsymbol{\alpha}_i$), the model is easily extended to include further variables that interact with unobservables. These variables do not have to be a function of time. For example, returns to schooling might be greater for very motivated people. If the motivated people are also more likely to participate in further education, the effect of a training program on wages would be biased in an FE regression model, but the bias could be removed using an FEIS model with individual slopes for schooling.

The general detrended model for unit i at time t can be written as

$$\tilde{y}_{it} = \tilde{\mathbf{x}}_{it}\boldsymbol{\beta} + \tilde{\epsilon}_{it}. \tag{15.24}$$

Because detrending eliminated $\boldsymbol{\alpha}_i$ from the model (along with $\tilde{\mathbf{z}}_{it}$) the FEIS estimators are obtained by running POLS on the detrended data. Consistency of $\boldsymbol{\beta}_{\text{FEIS}}$ requires

$$E(\epsilon_{it}|\mathbf{z}_{it}, \mathbf{x}_{it}, \boldsymbol{\alpha}_i) = 0, \tag{15.25}$$

which is a weaker form of the strict exogeneity condition required for conventional FE because idiosyncratic errors are expected to be unrelated to regressors only conditional on individual slopes. Hence, the parallel trends assumption is no longer needed.

Despite this obvious advantage, the FEIS model has not yet been widely used in the social sciences (for an exception, see Ludwig and Brüderl, 2011). Morgan and Winship (2007) are even skeptical about the usefulness of the model, because 'substantial amounts of data are needed to estimate it, and certainly from more than just one pretreatment time period and one posttreatment time period' (pp. 264–265). While it is true that the method is 'data hungry', it is also the case that there are some long-running panel surveys around that provide long enough panels.[5] The main practical reason why researchers do not use FEIS models seems to be that it is not implemented in standard statistical software. Therefore, a Stata ado file (xtfeis.ado) is available for download from the website accompanying this volume.

Random-effects models

A very popular class of panel estimators is based on the random-effects (RE) approach. The conventional RE model starts from the error components model given in (15.1). RE requires the same strict exogeneity assumption as FE. As for POLS, however, an assumption is needed about stable unobserved characteristics. The orthogonality condition $E(\alpha_i|\mathbf{x}_i) = E(\alpha_i) = 0$ states that stable unobserved confounders may not be related to any of the regressors. Because this assumption is so restrictive, RE often will fail to identify the causal effect of interest. If both conditions hold and the conditional mean is modeled correctly, the RE estimator is consistent because $E(\nu_{it}|x_{it}) = 0$.

We can write the model as

$$\mathbf{y}_i = \mathbf{X}_i\boldsymbol{\beta} + \boldsymbol{\nu}_i, \tag{15.26}$$

where $\boldsymbol{\nu}_i = \boldsymbol{\alpha}_i\mathbf{j}_T + \boldsymbol{\epsilon}_i$ and $\mathbf{j}_T$ is a $(T \times 1)$ vector of ones.

RE estimation exploits the time structure of the composite errors, which is why strict exogeneity is required. Define the $(T \times T)$ matrix $\boldsymbol{\Omega} \equiv E(\boldsymbol{\nu}_i\boldsymbol{\nu}_i')$. This is the variance–covariance matrix of the composite errors. In order to estimate it, further assumptions on the error structure are needed. Standard assumptions of constant variance of the idiosyncratic errors across t, $E(\epsilon_{it}^2|\mathbf{x}_i, \alpha_i) = E(\epsilon_{it}^2) = \sigma_\epsilon^2$ and $E(\epsilon_{it}\epsilon_{is}|\mathbf{x}_i, \alpha_i) = E(\epsilon_{it}\epsilon_{is}) = 0$ for all $t \neq s$, as well as constant variance of the stable components across i, $E(\alpha_i^2|\mathbf{x}_i) = \sigma_\alpha^2$, are sufficient (Wooldridge, 2010, p. 294). The two assumptions together imply equality of the conditional and unconditional variance–covariance matrix, $E(\boldsymbol{\nu}_i\boldsymbol{\nu}_i'|\mathbf{x}_i) = E(\boldsymbol{\nu}_i\boldsymbol{\nu}_i')$. The variance–covariance matrix

$\mathbf{\Omega}$ then has a special form. Elements on the diagonal are $\sigma_\alpha^2 + \sigma_\epsilon^2$ and off-diagonal elements are σ_α^2. This special form arises because under strict exogeneity $E(\alpha_i \epsilon_{it}) = 0$, hence, $E(\nu_{it}^2) = E(\alpha_i^2) + E(\epsilon_{it}^2) + 2E(\alpha_i \epsilon_{it}) = \sigma_\alpha^2 + \sigma_\epsilon^2$, and given the additional assumptions on the error terms, $E(\nu_{it}\nu_{is}) = E(\alpha_i^2) = \sigma_\alpha^2$.

The RE estimator is given by

$$\hat{\boldsymbol\beta}_{\text{RE}} = \left(\sum_{n=1}^{N} \mathbf{X}_i'\hat{\mathbf{\Omega}}_i^{-1}\mathbf{X}_i\right)^{-1}\left(\sum_{n=1}^{N} \mathbf{X}_i'\hat{\mathbf{\Omega}}_i^{-1}\mathbf{y}_i\right) \tag{15.27}$$

$$= \left(\sum_{n=1}^{N}\sum_{t=1}^{T} \check{\mathbf{x}}_{it}'\check{\mathbf{x}}_{it}\right)^{-1}\left(\sum_{n=1}^{N}\sum_{t=1}^{T} \check{\mathbf{x}}_{it}'\check{y}_{it}\right), \tag{15.28}$$

where $\check{y}_{it} = y_{it} - \theta\bar{y}$ and $\check{\mathbf{x}}_{it} = \mathbf{x}_{it} - \theta\bar{\mathbf{x}}_i$ indicate that the data have been quasi-demeaned.

To derive θ, write the variance–covariance matrix in full matrix notation as $\mathbf{\Omega} \equiv \sigma_\epsilon^2\mathbf{I}_T + \sigma_\alpha^2\mathbf{j}_T\mathbf{j}_T'$. This can also be written as $\mathbf{\Omega} = \sigma_\epsilon^2[\mathbf{Q} + ((\sigma_\epsilon^2 + T\sigma_\alpha^2)/\sigma_\epsilon^2)(\mathbf{I}_T - \mathbf{Q})]$, where $\mathbf{Q} = \mathbf{I}_T - T^{-1}\mathbf{j}_T\mathbf{j}_T'$. This expression results because $\sigma_\epsilon^2(\mathbf{I}_T - T^{-1}\mathbf{j}_T\mathbf{j}_T') + (\sigma_\epsilon^2 + T\sigma_\alpha^2)\mathbf{I}_T - (\sigma_\epsilon^2 + T\sigma_\alpha^2)(\mathbf{I}_T - T^{-1}\mathbf{j}_T\mathbf{j}_T') = \sigma_\epsilon^2\mathbf{I}_T + \sigma_\alpha^2\mathbf{j}_T\mathbf{j}_T'$.

Note further that $\mathbf{Q} = \mathbf{Q}\mathbf{Q}'$ and $\mathbf{I}_T = \mathbf{I}_T\mathbf{I}_T'$. Therefore, $\mathbf{\Omega}^{-1/2} = 1/\sigma_\epsilon^2[\mathbf{Q} + (\sigma_\epsilon^2/(\sigma_\epsilon^2 + T\sigma_\alpha^2))^{1/2}(\mathbf{I}_T - \mathbf{Q})]$. Premultiplying the outcome by $\mathbf{\Omega}^{-1/2}$ gives the quasi-demeaned outcome $\check{\mathbf{y}}_i = \mathbf{y}_i - [1 - (\sigma_\epsilon^2/(\sigma_\epsilon^2 + T\sigma_\alpha^2))^{1/2}]\bar{\mathbf{y}}_i$, which is just the compact form of $\check{y}_{it} = y_{it} - \theta\bar{y}_i$ with $\theta = 1 - (\sigma_\epsilon^2/(\sigma_\epsilon^2 + T\sigma_\alpha^2))^{1/2}$.

The RE estimator therefore is the POLS estimator on the quasi-demeaned data. The model is usually estimated by feasible generalized least squares (FGLS). The procedure is called 'feasible' because consistent estimators for the variance of the two error components are available. A consistent estimator for σ_ϵ^2 is obtained by running a within (FE) regression and calculating the residual variance as $\hat{\sigma}_\epsilon^2 = \sum_{n=1}^{N}\sum_{t=1}^{T}[\hat{\epsilon}_{it}^2/(NT - N - K)]$. Using this result and residuals from a between regression of equation (15.3), one gets $\hat{\sigma}_\alpha^2 = \sum_{n=1}^{N}[\hat{\nu}_{it}^2/(N - K) - (\hat{\epsilon}_{it}^2/T)]$. The variance of the unobserved effect results because the between regression leaves variation to the residuals that is due to α_i and ϵ_{it}, whereas the FE regression residuals contain only variation due to ϵ_{it}.

From the expression for θ, it follows that the RE estimates are in between POLS and FE estimates. The POLS estimator results when all the error variance is due to the idiosyncratic errors ($\theta = 0$). The FE estimator results when all the variance is due to the stable unobserved effects ($\theta = 1$). The advantage of RE over POLS is in terms of greater efficiency at the price of a stronger exogeneity assumption. The advantage of RE over FE is also greater efficiency given the orthogonality condition holds. If it does not hold, RE is inconsistent and FE should be preferred.

Panel regression as multi-level models

As mentioned in the introduction to this chapter, models for panel data can be formulated as multi-level models. In the case of individuals observed over time, there are two levels: person-years on the lower level (level 1) which are nested in persons (level 2). Note that multi-level models allow for individual-specific coefficients α_{ki}. Different specifications are possible. Each α_{ki} may be specified as a 'fixed coefficient' that does not vary over persons, $\alpha_{ki} = \gamma_k$, as a coefficient varying non-randomly across persons, $\alpha_{ki} = \mathbf{w}_{ki}\boldsymbol\gamma_k + \xi_{ki}$, or as a coefficient with variation that is purely random, $\alpha_{ki} = \gamma_k + \xi_{ki}$ (Cameron and Trivedi, 2005, pp. 846–847).[6]

Usually, models contain both fixed and varying coefficients. Such models are often classified as 'mixed-effects models', but 'mixed-coefficient models' would be a better term. In these models, there is a vector of variables $\mathbf{x}_{it}$ for which coefficients $\boldsymbol{\beta}$ are homogeneous across units, while for some variables $\mathbf{z}_{it}$ coefficients $\boldsymbol{\alpha}_i$ are individual-specific.[7] We specify a linear regression model on the lower level (level 1 equation):

$$y_{it} = \mathbf{x}_{it}\boldsymbol{\beta} + \mathbf{z}_{it}\boldsymbol{\alpha}_i + \epsilon_{it}. \tag{15.29}$$

For the individual-specific coefficients, we specify a general model on the upper level (level 2 equation):

$$\alpha_{ki} = \mathbf{w}_{ki}\boldsymbol{\gamma}_k + \xi_{ki}. \tag{15.30}$$

Conventional FE and RE models are special cases. The FE model assumes an individual-specific intercept that is non-random, $\alpha_{1i} = \gamma_{1i}$, while all other coefficients are 'fixed' (homogeneous across units). Plugging this into equation (15.29) gives $y_{it} = \mathbf{x}_{it}\boldsymbol{\beta} + \gamma_{1i} + \epsilon_{it}$, which is just the model of equation (15.1) (albeit using different notation). The conventional RE model assumes a random individual intercept $\alpha_{1i} = \gamma_1 + \xi_{1i}$ (keeping all other coefficients fixed). Therefore, $y_{it} = \mathbf{x}_{it}\boldsymbol{\beta} + \gamma_1 + \xi_i + \epsilon_{it}$, where γ_1 is a common constant that can be subsumed in $\mathbf{x}_{it}$. We see that this is the RE model.[8]

Using the multi-level framework, an obvious extension of the RE approach is to allow for individual variation not only in the individual intercepts, but also in coefficients. This idea leads to the random-slopes (RS) model, where the individual slope on time-varying regressors is modeled as a random coefficient, $\alpha_{2i} = \gamma_2 + \xi_{2i}$, and $\alpha_{1i} = \gamma_1 + \xi_{1i}$ is the random intercept (as in the RE model).

As an example, suppose we are interested in the effect of parental divorce (x_{it}) on children's cognitive ability (measured repeatedly by some IQ test, y_{it}). We envisage that children differ with respect to their level and growth in ability. For simplicity, we specify a linear trend over age (z_{it}). We end up with the model $y_{it} = \gamma_1 + \gamma_2 z_{it} + \beta x_{it} + \xi_{1i} + \xi_{2i} z_{it} + \epsilon_{it}$, where $\gamma_1 + \gamma_2 z_{it}$ reflect a common constant and trend and $\xi_{1i} + \xi_{2i} z_{it}$ are child-specific deviations from these. A further extension of the model would be to include a covariate such as parental education (w_i) in the level 2 equation to model group-specific differences in levels or growth. This way, we would specify a 'cross-level' interaction effect.

Raudenbush (2001) argues that the RS model should be used to estimate the effect of an event because it controls for unobserved heterogeneity regarding individual growth. However, this is not correct. The RS model imposes exogeneity assumptions on ξ_{1i} and ξ_{2i}. In fact, if the model is estimated by maximum likelihood, as it usually is, full distributional assumptions are needed. For consistency of RS estimates, ξ_{1i} and ξ_{2i} then have to be distributed according to a bivariate normal distribution which implies zero conditional mean. The model therefore is inconsistent if the parallel trends condition needed for consistency of FE is not met. Unlike FEIS, RS estimation cannot handle violations of this assumption.

Lagged dependent variable models

Panel models as considered so far induce correlation of the outcome over time. For instance, in the error components model (15.1) the stable person-specific unobservables (α_i) produce a correlation in y over time. The mechanism behind this correlation is simply that stable unobservables affect the outcome all the time and thus tend to produce similar outcomes over time. This outcome correlation over time produced by stable unobservables is sometimes called 'spurious state dependence'. However, often researchers argue that there might be a second source of correlation over time: 'true state dependence'. Here it is supposed that outcomes affect each other causally over time. The mechanism behind this is that outcomes tends to reproduce themselves.

Usually true state dependence is modeled by including a lagged dependent variable (LDV) on the right-hand side of a panel model:

$$y_{it} = \rho y_{it-1} + \mathbf{x}_{it}\boldsymbol{\beta} + \alpha_i + \epsilon_{it}, \tag{15.31}$$

where $|\rho| < 1$ (stationary outcome process) and everything else is as defined before. In this model ρ captures state dependence. A ρ close to one would indicate that the outcome process strongly tends to reproduce itself. In fact, it can be shown that such a model includes two sources of correlation over time: stable unobservables and true state dependence (Cameron and Trivedi, 2005, p. 763). It is important to understand that LDV estimators, unlike within estimators, are *not* designed to handle unobserved heterogeneity. Nor are they able to ameliorate it, let alone solve the problems associated with it, under general conditions. Instead, they aim to tackle the problem of true state dependence where the outcome in any period is determined by its past values.

Introducing an LDV in a panel regression seems intuitively appealing to most social scientists. Further, LDV models are termed often 'dynamic panel models', a term that signals superiority over the 'static panel models' that we have discussed so far. However, estimation of LDV models is inherently problematic. Most researchers naively estimate the LDV model (15.31) with POLS (POLS-LDV). Examples are given by Halaby (2004). However, POLS-LDV provides inconsistent estimators of both ρ *and* $\boldsymbol{\beta}$. Intuitively, this is because the regressor y_{it-1} is related to α_i by definition (α_i affects all outcomes of a person). So the estimate of ρ is biased. This is true even if α_i is purely random, that is, not related to any of the regressors in $\mathbf{x}_{it}$. The exogeneity condition required for POLS is violated by construction of the model. And even worse, this bias transfers to the estimate of $\boldsymbol{\beta}$. More exactly, whenever $\alpha_i \neq 0$ for some i and lagged outcomes y_{it-1} are correlated with $\mathbf{x}_{it}$, the POLS estimates of $\boldsymbol{\beta}$ are inconsistent as well (for more details see Angrist and Pischke, 2009, Section 5.4). In many applications, y_{it-1} and x_{it} will be correlated because covariates are trended. In fact, if we are interested in the effect of a binary treatment dummy (the effect of an event), this will necessarily be the case. In this setting, POLS-LDV is biased if $\boldsymbol{\beta} \neq 0$.

An idea that has intuitive appeal is to build on the within approach to eliminate the selection bias and construct an LDV estimator for the transformed data to deal with state dependence (FE-LDV). This idea has stimulated much methodological research, and several estimators have been proposed. However, an estimator that is consistent under general forms of selection into treatment is not available. Research has shown that FE estimation including a lagged dependent variable produces inconsistent estimators. This inconsistency has become known in the econometric literature as the 'Nickell bias' (Nickell, 1981). More recently, Phillips and Sul (2007) generalized Nickell's findings to cover the FE model with individual-specific time trends. Phillips and Sul give the formula for the bias of ρ and β and also show by simulations that detrending the data often increases the bias dramatically (p. 166). Clearly, within transformation of the data alone does not help to reduce the bias of the LDV model.

Therefore, several approaches that use instrumental variables have been suggested. Arellano and Bond (1991) suggested using further lags of the outcome variable as instruments in an FD model (AB-LDV). The FD model is

$$(y_{it} - y_{it-1}) = \rho(y_{it-1} - y_{it-2}) + (\mathbf{x}_{it} - \mathbf{x}_{it-1})'\boldsymbol{\beta} + (\epsilon_{it} - \epsilon_{it-1}). \tag{15.32}$$

AB-LDV uses y_{it-2} as an instrument for $y_{it-1} - y_{it-2}$. To increase efficiency AB-LDV uses additional lags as instruments (if available). The rationale for this is that lagged outcomes are unrelated to the error term in first differences ($\epsilon_{it} - \epsilon_{it-1}$) if sequential exogeneity holds. This assumption is weaker than the 'strict exogeneity' assumption from FE. Both strict and sequential exogeneity assume that past outcome levels are (conditionally) independent of the contemporaneous idiosyncratic error term. However, only strict exogeneity implies that this is also valid for future outcomes.

As N goes to infinity (keeping T fixed), the AB-LDV estimator becomes consistent if unobserved heterogeneity is purely random. Arellano and Bond (1991) show by simulation that the model works reasonably well even in small samples ($N = 100, T = 7$). However, if time-constant unobserved heterogeneity is related to covariates the estimator is inconsistent unless further conditions hold. Most importantly, it is required that the idiosyncratic errors are serially uncorrelated prior to first differencing (cf. Halaby, 2004). As we have noted earlier, this will often not be the case. Individual outcomes often exhibit strong persistence over time even after controlling for covariates. Then y_{it-2} will correlate with $\epsilon_{it} - \epsilon_{it-1}$ and thus will be an invalid instrument. Moreover, the estimator is known to suffer from bias due to weak instruments if $|\rho|$ gets large (close to one) or if $\sigma_\alpha^2/\sigma_\epsilon^2$ gets large (unobserved heterogeneity is very important) (Bun and Windmeijer, 2010; Blundell and Bond, 1998). Therefore, AB-LDV is also highly likely to fail in practice.

In conclusion, we see that LDV models pose inherent estimation problems that can hardly be solved with real-world data. Therefore, we would argue that LDV models are not useful at all. Fortunately, we also see theoretical reasons why researchers do not have to bother with LDV models. We would argue that the mechanism behind a correlation over time in many cases is due to stable unobservables, not to some kind of reproduction. That is, state dependence is 'spurious', not 'true'. For instance, stable income levels over time are produced by (observed and unobserved) human capital levels, personality traits, and so on. We see no reason why one would want to invoke some mysterious 'wage reproduction'. Therefore, we would argue that in many cases careful theoretical reasoning would lead to the conclusion that there is no need for an LDV specification.

Monte Carlo simulations comparing panel estimators

In this subsection, we will present simulation results in order to demonstrate the strengths and shortcomings of FE regression *vis-á-vis* alternative estimators. We compare FE to POLS and RE. Because we will also simulate scenarios with individual-specific slopes, we include FEIS and RS. Further, we will simulate situations with true state dependence and therefore also include POLS-LDV and AB-LDV.

The general setup of the simulations is as follows. We draw random samples with $N = 200$ and $T = 10$. We create artificial data based on the data generating process

$$y_{it} = \rho y_{it-1} + \beta x_{it} + \alpha_{1i} + \alpha_{2i}t + \epsilon_{it}. \tag{15.33}$$

Variable x_{it} is a binary treatment indicator. At $t = 0$, the treatment dummy is zero for all subjects. Half of the subjects are treated in later period $t = 5$ (all at the same time). From this point onward, the treatment indicator equals one for the treated.

The data generating process allows for unobserved heterogeneity and true state dependence. Heterogeneity will be modeled by including individual-specific intercepts α_{1i} and individual-specific slopes α_{2i}. We consider heterogeneity with respect to α_{1i} that is either purely random or systematically related to treatment (selection). We also model situations where there is selection into treatment with respect to not only α_{1i} but also α_{2i}. True state dependence will be modeled by setting $\rho \neq 0$, thereby allowing for individual outcomes that are 'self-reproducing' over time. The simulations cover cases with state dependence and either random heterogeneity or selection into treatment. We simulate data under several conditions (described more precisely

Table 15.2 Consistency of panel estimators under different scenarios (*N* large, *T* small)

Scenario	$\hat{\beta}_{POLS}$	$\hat{\beta}_{RE}$	$\hat{\beta}_{FE}$	$\hat{\beta}_{FEIS}$	$\hat{\beta}_{RS}$	$\hat{\beta}_{POLS\text{-}LDV}$	$\hat{\beta}_{AB\text{-}LDV}$
Heterogeneity w.r.t. α_{1i}	con.	con.	con.	con.	con.	inc.	con.
Selection w.r.t. α_{1i}	inc.	inc.	con.	con.	inc.	inc.	inc.
Selection w.r.t. α_{2i}	inc.	inc.	inc.	con.	inc.	inc.	inc.
TSD, het. w.r.t. α_{1i}	inc.	inc.	inc.	inc.	inc.	inc.	con.
TSD, sel. w.r.t. α_{1i} or α_{2i}	inc.	inc.	inc.	inc.	inc.	inc.	inc.

α_{1i}, individual-specific intercepts; α_{2i}, individual-specific slopes; TSD, true state dependence; con., consistent; inc., inconsistent.
Source: own compilation.

below) to investigate how unobserved heterogeneity and state dependence affect the estimates of our seven models. We focus on the estimate of the treatment effect β only.

Table 15.2 shows what we expect from the simulations. If unobserved heterogeneity is purely random, all of the estimators considered are consistent, except the POLS-LDV estimator. Of the consistent estimators, the RE estimator should be the most efficient. The RE estimator should, however, be inconsistent if unobserved heterogeneity is related to treatment (selection with respect to α_{1i}). This is the strength of FE regression. The conventional FE estimator should be consistent when the parallel trends condition holds. If there is self-selection with respect to α_{2i} the parallel trends assumption is violated and FE should be inconsistent. Of all estimators considered here, only FEIS should be able to deal with non-parallel trends. In the case with true state dependence, the AB-LDV estimator should be consistent provided unobserved heterogeneity is not systematically related to treatment. If there is both state dependence and selection with respect to α_{1i} or α_{2i}, all estimators considered should be inconsistent (even AB-LDV because it strongly depends on the condition of no serial correlation).

In the following, we describe the simulation scenarios more precisely and discuss the results of the simulations. For each of eight scenarios we draw 1000 samples and compute the mean of each of the seven estimates of β as well as the mean of their standard errors.[9] The treatment effect β is set to one, except in scenario (2), where it is set to zero (no treatment effect). In all simulations (except for the last scenario), ϵ_{it} is a Gaussian random variable. In scenario (8), however, we allow for serial correlation in the idiosyncratic errors. The results are shown in Table 15.3.

In scenarios (1)–(5) we model unobserved heterogeneity and assume that there is no true state dependence ($\rho = 0$). In scenarios (1) and (2) we assume that time-constant heterogeneity is purely random, that is, not related to treatment (α_{1i} is a standard normal random variable). Even in this arguably innocuous case without self-selection into treatment, there is one model that provides strongly biased estimates. The POLS-LDV estimator is consistent only if there is no treatment effect (column (2)), but inconsistent whenever the treatment effect is different from zero (column (1)). All other estimators provide correct answers most of the time. They all are consistent, but standard errors differ. In terms of efficiency, the random-effects models (RE or RS) would be the preferred choice as the standard errors are slightly smaller than those obtained by FE.

In scenario (3) we assume that there is selection into treatment with respect to time-constant unobservables. Here, we model α_{1i} as a normally distributed random variable with variance 1, but mean 1 for the treated and mean 0 for the untreated subjects. In this case of parallel trends, the FE and FEIS models provide the true value. POLS is heavily biased. The RE model provides an estimate that is in between POLS and FE, but is still substantially biased. The same holds for the RS model. POLS-LDV is biased as well. The AB-LDV estimator seems to perform very well,

Table 15.3 Simulation results for eight scenarios: mean of $\hat{\beta}$ for $N = 200$, $T = 10$ (1000 replications)

	(1)	(2)	(3)	(4)	(5)	(6)	(7)	(8)
	het. α_1	het. α_1	sel. α_1	sel. α_2	sel. α_2	het. α_1	sel. α_1	sel. α_1
	$\beta = 1$	$\beta = 0$	$\beta = 1$	$\beta = 1$	$\beta = 1$	$\beta = 1$	$\beta = 1$	$\beta = 1$
	$\rho = 0$	$\rho = 0$	$\rho = 0$	$\rho = 0$	$\rho = 0$	$\rho = 0.5$	$\rho = 0.5$	$\rho = 0.5$, $\gamma = 0.2$
POLS	1.005	−0.006	1.896	2.495	1.257	1.557	3.314	3.297
	(0.139)	(0.139)	(0.140)	(0.890)	(0.892)	(0.269)	(0.272)	(0.280)
RE	1.004	−0.001	1.139	1.482	0.649	1.506	1.864	1.860
	(0.077)	(0.077)	(0.077)	(0.598)	(0.599)	(0.127)	(0.130)	(0.147)
RS	1.005	−0.001	1.131	0.917	0.904	1.303	1.239	1.202
	(0.078)	(0.078)	(0.078)	(0.126)	(0.126)	(0.103)	(0.104)	(0.111)
FE	1.004	−0.000	1.002	1.391	0.594	1.502	1.771	1.760
	(0.080)	(0.080)	(0.080)	(0.572)	(0.574)	(0.125)	(0.127)	(0.145)
FEIS	1.002	0.003	1.003	0.994	1.001	1.051	0.853	0.847
	(0.128)	(0.128)	(0.128)	(0.128)	(0.128)	(0.141)	(0.141)	(0.153)
POLS-LDV	0.644	−0.003	1.053	0.093	0.040	0.585	0.775	0.713
	(0.082)	(0.077)	(0.081)	(0.065)	(0.063)	(0.059)	(0.065)	(0.066)
AB-LDV	1.007	0.003	1.002	0.993	0.908	0.993	1.011	0.884
	(0.147)	(0.149)	(0.128)	(0.205)	(0.206)	(0.146)	(0.127)	(0.133)

Mean of panel-robust standard errors in parentheses. het. α_1, random unobserved heterogeneity; sel. α_1, selection into treatment with respect to α_{1i}; sel. α_2, selection into treatment with respect to α_{1i} and α_{2i}. For exact definition of simulation scenarios (1)–(8), see text and corresponding table in the online appendix on the volume's website.

Source: Simulated data.

but this is not true in general, for example, in settings with serially correlated errors (as shown below). Unlike AB-LDV, the FE and FEIS point estimates are consistent and even unbiased in the presence of serial correlation which affects only the standard errors. Hence, FE or FEIS would clearly be preferable. In fact, FE is the best choice because it is more efficient than FEIS.

In scenarios (4) and (5) there is selection into treatment due to both α_{1i} and α_{2i}. α_{1i} is modeled as in scenario (3). In scenario (4) we model α_{2i} as a normally distributed random variable with variance 1, but mean 0.1 for the treated and mean 0 for the untreated subjects. This results in diverging trends between treatment and control group. In scenario (5), α_{2i} is normally distributed with variance 1, but mean −0.1 for the treated and mean 0 for the untreated subjects. Hence, trends of the treatment and control group converge over time. Both scenarios simulate unobserved heterogeneity with respect to the outcome levels and growth curves, which both are related to treatment. In these cases, FE estimates are biased, as are all other estimators except FEIS. FEIS is the only model that provides estimates that are reasonably close to the true value.[10] Note that the RS model is substantially biased, though according to Raudenbush (2001) it is perfect for this situation. The reason is that RS assumes that the conditional means of the individual growth curves are equal across treatment groups. In our simulations, however, means differ systematically between treatment groups. In many real social science data it can be expected that the latter condition holds, and then the RS model provides biased estimates. Clearly, FEIS is the preferred choice if the parallel trends assumption does not hold.

Finally, in scenarios (6)–(8) we allow for state dependence. We set $\rho = 0.5$ so that there is substantial time-series dependence (but not strong enough to induce severe problems due to weak instruments. Scenario (6) assumes in addition that there is only random unobserved heterogeneity, defined as in scenarios (1) and (2). In scenario (7) there is selection into treatment due to α_{1i} (as in scenario (3)). Theoretically, if unobserved heterogeneity is purely random, the AB-LDV estimator is the only estimator that is consistent. This can be seen from the results of our simulations (column (6)). AB-LDV is the only estimator that is reasonably close to one. The estimator is even close to the true parameter if heterogeneity is related to treatment (column (7)). However, this result strongly depends on the assumption of no serial correlation in the idiosyncratic errors. To show this, we alter the setup of scenario (7) to allow for such serial correlation. In scenario (8) we assume a first-order autoregressive process $\epsilon_{it} = \gamma\epsilon_{it-1} + \nu_{it}$, where $\gamma = 0.2$ and ν_{it} is a Gaussian random variable. Serial correlation of this or even greater magnitude is often found with panel data. The results show that the AB-LDV model provides estimates that are heavily biased in such situations.

We have already emphasized that the POLS-LDV estimator is biased under all conditions. It is obvious that this estimator should not be used (nevertheless, it has been used very often in psychological and sociological research). However, that even the acclaimed AB-LDV estimator is biased under the very realistic scenario (8) is dramatic. This leads to the conclusion that hitherto no estimator has existed that allows a treatment effect to be estimated consistently if both true state dependence and self-selection are present in the data. Fortunately, as argued above, for theoretical reasons there is often no need to model true state dependence anyway.

AN APPLICATION: MARRIAGE AND HAPPINESS

In this section we want to apply the most important models discussed so far to real world panel data. We will use data from the German Socio-Economic Panel 1984–2009 (SOEP, 2010). The SOEP is a household panel with annual waves (for a description, see Wagner et al., 2007). Thus, for our analyses panels covering up to 26 waves are available. It is rare to have such long panels. However, it is obvious that long panels make the task of identifying the detailed nature of a causal effect easier. We will make ample use of this advantage.

Our research question is how marriage affects life satisfaction (happiness). The common-sense hypothesis would be that marriage makes people happy. However, there is the potential for self-selection: it is very plausible that it is the happier people who marry, because it is easier for them to find a partner (cf. Stutzer and Frey, 2005). So our working hypothesis is that POLS estimates of the marriage effect should be too high. An FE approach should provide lower estimates that perhaps are even close to zero.

This is a question on the effects of an event (marriage). It is in such questions that a within approach can unfold its full potential (cf. Allison, 1994). FE will compare happiness before and after a marriage within persons. Therefore, self-selection into marriage will not bias results.

Preparing the data

Every year since 1984 the SOEP questionnaire has concluded with the happiness question: 'We would like to ask you about your satisfaction with your life in general.' Respondents can answer on an 11-point scale from 0 ('completely dissatisfied') to 10 ('completely satisfied'). This will be our outcome variable. It is an ordinal variable; however, we follow the standard approach in happiness research and treat it as a metric variable.[11]

Our treatment variable is marriage. We restrict our analysis to first marriages. An advantage of panel analysis is that it is possible to model and identify time-varying causal effects. Therefore, the analyst should think about how to model the time path of the causal effect. Simply adding an event dummy (as we did in the first section of this chapter) assumes that the causal effect is immediate and permanent. In most situations, however, one would want to use a more flexible modeling by adding an event time clock. One creates a new variable that measures time since the event has occurred and includes power terms thereof in the model (or, even more flexibly, splines or dummies). In our case we include 'years since marriage'.

Further, we include three controls. First, with panel data it is possible to separate age effects from cohort effects (see below). Therefore, including these variables in panel models should be standard. For didactic reasons, in a first step we include age only linearly. More detailed age/cohort analyses follow later. Second, we include household income as a time-varying variable (natural logarithm). Third, we include sex (time-constant).

Panel data preparation generally is much more complex than cross-sectional data preparation. Here we do not have the space to dwell on this in more detail.[12] However, we want to emphasize one important point, because it deviates from what one is used to from cross-sectional data preparation: *one should restrict the estimation sample to those persons who can potentially experience the treatment during the observation window*. In our case, treatment is first marriage. Thus, we restrict our estimation sample to those who are 'never married' when entering the SOEP. Those who are already married (or divorced, widowed, etc.) when entering the SOEP are dropped. Second, we exclude all person-years after the end of a marriage (separation, divorce, widowhood). Third, we exclude all persons who have only one person-year. In the SOEP (v26) there are 422,734 happiness person-years from 51,543 persons. In our estimation sample that is restricted as described above we are left with 122,919 person-years from 14,634 persons. Thus our estimation sample comprises only 29% of all available person-years.

This may seem strange because, coming from a between approach, social scientists are accustomed to 'hunt for each observation'. The standard approach is to base the between comparison on as many observations as available (sometimes missing data are even imputed). From a within approach, however, it seems reasonable to restrict the estimation sample to those person-years that potentially contribute to within estimation of the treatment effect. In the current context, these are person-years from persons who are 'never married' at the beginning and who have more than one person-year. Person-years of those never marrying are included because they provide the control group for estimating the common age effect. Furthermore, person-years after a marriage has dissolved contribute nothing to estimating the marriage effect and are therefore discarded.

Comparing different regression models

In this section we will compare the results of five regression models. POLS, RE and FE are canonical. We supplement the set by an extreme between approach: a regression that is run on the person-specific means is called 'between regression' (BE). The model is given in equation (15.3) and uses between variation only. Finally, we add another within approach: first differences regression (FD).[13]

For didactic reasons we simplify our modeling strategy and only include a marriage dummy. Before running a within panel regression, one should investigate whether there is enough within variation on the regressors. With 3793 marriages observed in our estimation sample, there is certainly enough within variation on the variable of main interest. It is no surprise that this is also the case for age and household income.

In Table 15.4 we present the results of our model comparison. Our estimates are based on 14,634 persons. BE regression is based on the person-specific means. The other regressions utilize the panel structure of the data and are based on the 121,919 person-years provided by these persons. However, FD is based on fewer cases, because we lose one person-year per person and another one for each gap in the panels (see above). There are 2614 gaps in the data. We lose 123 persons because they do not have any consecutive person-years.

To assess model fit we report overall R^2 for BE and POLS. For the other three models we report the within R^2. Our BE and POLS models are not very successful in explaining the overall happiness variation in the data. Too many other factors affect happiness. For RE, FE and FD it makes sense to report the within R^2, because it is the happiness variation within the persons that we want to explain with our time-varying regressors. The FE model, for instance, succeeds in explaining 1.6% of the within person happiness variation. More precisely, changes in marital status, age and household income can explain 1.6% of the happiness variation of a person over time. The remaining 98.4% is produced by other factors that change over time (health, labor market status, mood, weather, etc.). The strict exogeneity assumption states that these factors must not be correlated with marriage. Otherwise the within marriage effect estimate will be biased.

In all models (with the natural exception of BE) we estimate panel-robust standard errors. These are in most cases larger than conventional standard errors. This results in lower t-values. For instance, the conventional marriage effect t-value is 9.95, whereas in Table 15.4 we report a t-value of 7.37 (for FE). In any case, the marriage effect is highly significant.

Looking at the estimated marriage effect, we see that BE, in particular, provides a quite strong marriage effect: married people are happier by 0.34 scale points. However, the within estimates show that this is an overestimate: the happiness increase with marriage is much lower, namely only 0.14 scale points (FD). The most plausible explanation for this result is self-selection of the happy into marriage.[14]

A subtle, but important point concerning the interpretation of regression coefficients from within models has to be made here. Whereas with standard between regression we compare different people, with within regression we investigate changes over time within the same people. Thus, standard regression coefficients tell us how people differ. For instance, the marriage coefficient in the BE model tells us that married people are on average happier by 0.34 points. However, regression coefficients from within models tell us how the outcome changes when treatment status changes. Thus, the FE estimate tells us that after a marriage, happiness increases by 0.17 points on average. It is obvious that such a change coefficient is closer to what we mean by a treatment effect than a coefficient that reports a difference between people only.

Concerning the age effect, we see that the within estimators are largest (in absolute terms): in 10 years happiness declines by 0.41 scale points (FE). BE, POLS and RE seem to underestimate the negative age effect. Below we will demonstrate that this underestimation is due to not controlling for cohort (older cohorts are happier) and to self-selection (the happy live longer).

Concerning the effect of household income, we see that between estimators strongly overestimate the effect. FE (and especially FD) provide much lower estimates. Again the most plausible reason for this are confounders: unobservables that increase both happiness and income. Nevertheless even the most conservative FD estimate predicts that with an income gain happiness increases significantly.

Finally, the effect of sex cannot be estimated by the within estimators because sex is time-constant. The other three models provide an estimate of the sex effect that is very similar across models: women are (marginally) happier by 0.05 scale points (RE). This effect is statistically significant at the 5% level.

The conclusion of our model comparison is that BE and POLS (and to a lower extent also RE) provide biased estimates. Obviously, using between regression models in happiness research has

Table 15.4 The effect of marriage on happiness: comparing five regression models

	(1) BE	(2) POLS	(3) RE	(4) FE	(5) FD
Marriage	0.343*** (8.80)	0.190*** (7.34)	0.098*** (4.98)	0.167*** (7.37)	0.143*** (5.07)
Age	–0.009*** (–9.25)	–0.014*** (–11.42)	–0.027*** (–24.78)	–0.041*** (–23.96)	–0.062*** (–23.09)
Household income (ln)	0.480*** (25.84)	0.325*** (23.18)	0.162*** (15.28)	0.125*** (10.16)	0.048** (3.22)
Woman	0.058** (2.84)	0.060** (2.65)	0.051* (2.49)	–	–
(Within) R^2	0.024	0.027	0.014	0.016	0.015
Persons	14,634	14,634	14,634	14,634	14,511
Person-years	–	121,919	121,919	121,919	104,671

t-values in parentheses (based on panel-robust standard errors).
$^{*}\ p \leq 0.05$; $^{**}\ p \leq 0.01$; $^{***}\ p \leq 0.001$.

Source: own computations

Data: SOEP 1984–2009 (v26)

the potential to yield grossly misleading results. Experience tells that this can be generalized: trying to infer causal effects from regression models that use variation between individuals leads to misleading conclusions in most cases. The reason is that people tend to self-select into (or out of) treatment. Therefore, it is generally not a good idea to compare different people to identify a causal effect. It is much better to try to identify the causal effect by looking at within person changes.

Presenting the results from a fixed-effects regression

In the previous subsection we modeled the causal effect very simply using an event dummy. We now want to use a more flexible modeling strategy. Instead of a single marriage dummy, we add a separate dummy for every year after marriage (cf. Allison, 1994). Altogether we add 25 'years since marriage dummies' to the FE model. Such a modeling strategy allows us to identify the time path of the causal effect in a very differentiated way. Further, to investigate whether there is an anticipation effect we also add a dummy for the year before marriage.

Presenting the results of such a regression model in a table would not be very helpful, because there are so many regression coefficients. In fact, some people would even argue that this is a waste of paper. It is much better to present the results in a graph. In principle there are three helpful graphs for interpreting regression models (see Chapter 10 in this volume): (i) one could plot the predicted values (profile plot); (ii) one could plot the regression coefficients (or more generally: the average marginal effects, AMEs) (effect plot); or (iii) one could plot the AMEs of X conditional on the values of Z (conditional effect plot). To show in a graph how the marriage effect evolves over time, we plot the regression coefficients of the marriage year dummies over time (with Stata's `marginsplot` command). We plot only over the first 10 years of marriage, because afterwards the estimates become very imprecise due to small numbers of cases.

As can be seen in Figure 15.3, in the year of marriage (year 0) happiness is higher by 0.38 scale points. The reference group is the average happiness in all person-years at least 2 years

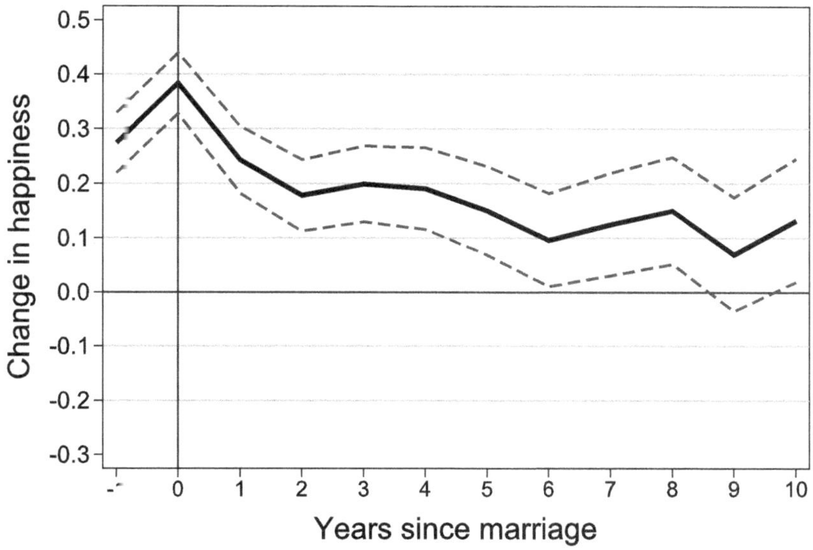

Figure 15.3 The time path of the marriage effect

before marriage. This marriage effect is much larger than that reported above (0.17). The reason is that the dummy marriage effect averages over all the (after) marriage effects we see in Figure 15.3, and many of these are much smaller. As we can see, the marriage effect quickly declines to 0.15 after 5 years. In the ninth year it becomes even insignificant, as indicated by the confidence interval (dashed line) that overlaps zero. In summary, this means that the marriage effect is immediate and strong. However, it is not persistent. It declines quickly and vanishes after about 10 years. Furthermore, we see a clear anticipation effect: in the year before marriage happiness is higher by 0.27 points.

We conclude that the modeling strategy suggested here allows us a very detailed investigation of the time path of a causal effect, including possible anticipation effects.

Modeling individual growth

So far we have emphasized that one advantage of panel data is that they can help in identifying causal effects. In the previous subsection, we saw that panel data even help in identifying the time path of a causal effect. Closely related is the second big advantage of panel data: panel data allow us to model individual dynamics, that is to say, with panel data we can model the development of the outcome over time. In most cases time will be 'age', but it could also be labor market experience, etc. Curves that describe the development of the outcome with age are called 'growth curves'.

A growth curve is easily modeled by including age terms in a regression model. Age dummies are most flexible, but age splines or age polynomials are also possible. Wunder et al. (2013) give a detailed discussion on the different modeling possibilities.

When modeling growth curves it is important to control for birth cohort (if one has panel data that were collected from several cohorts). As sociologists know, cohort effects are ubiquitous and therefore might spoil the age effects found. Cross-sectional data offer no solution here, because it is impossible to separate age effects from cohort effect. However, with panel data we can separate both effects due to repeated measurement of the outcome at different ages.

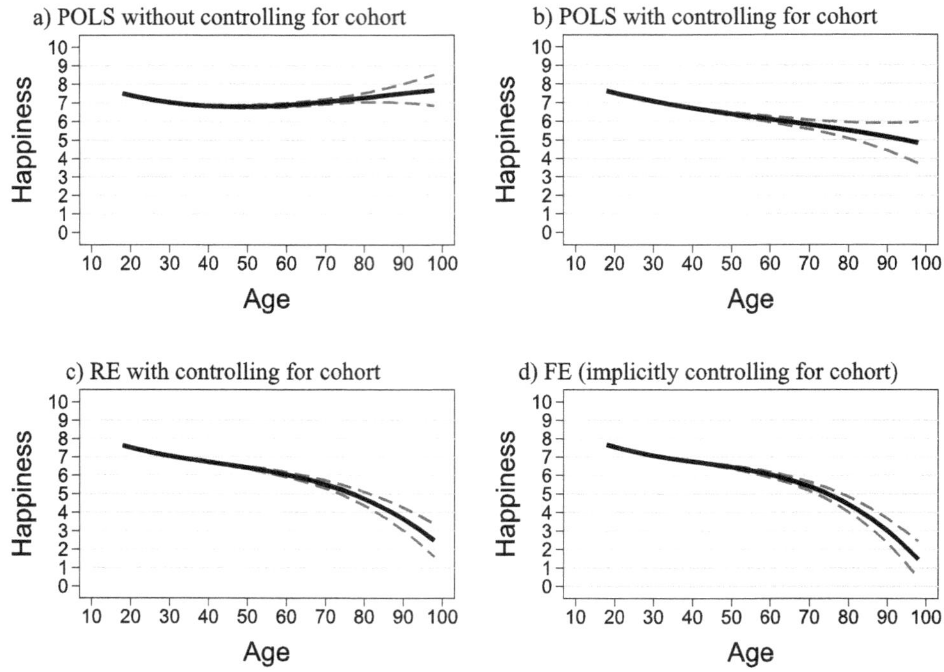

Figure 15.4 Comparing cubic growth curves obtained with different methods

When using FE to estimate a growth curve, cohort is automatically controlled for (because it is time-constant).

Growth curves are often estimated by RE modeling or structural equation modeling (latent growth curves). However, these modeling strategies potentially provide biased growth curves if there is self-selection: people are selected into different age groups according to their value on the outcome. This might happen because of differential mortality, or also because of differential panel attrition. Therefore, we recommend FE methodology for estimating growth curves. FE growth curves have the advantage that they provide a within age effect: the age effect is not estimated by comparing people of different ages, but by looking how the outcome changes when persons grow 1 year older. This avoids self-selection bias. Furthermore, Frijters and Beatton (2012) demonstrate that biased estimates of other regressors also might bias growth curves. In so far as FE helps in identifying unbiased effects of regressors, it will also help to estimate an unbiased growth curve.

To demonstrate growth curve estimation we estimate happiness regressions that include a linear, quadratic and cubic age term (cubic growth curve). As controls we include the marriage dummy and household income. After estimation we plot predicted values by age (profile plots using Stata's `marginsplot` command). Figure 15.4 presents the results of four different estimation strategies. In Figure 15.4a we present the happiness growth curve obtained with POLS. As can be seen, we get a U-shaped growth curve. Happiness declines slightly until age 50 and then increases again. Such a pattern has been reported in numerous studies, especially in the economics literature. However, common sense tells us this is a quite surprising result: it seems implausible that old people are happier than middle-aged people. Many studies show that happiness declines with deteriorating health. Therefore, because we do not control for health, happiness should decline with age. Meanwhile it has been shown by several studies (e.g. Frijters and Beatton, 2012; Wunder et al., 2013) that a U-shaped happiness growth curve is an artifact of not controlling for cohort and of not taking regard of potential self-selection.

Therefore, in Figure 15.4b we estimate the growth curve after controlling for cohort (by including 90 birth year dummies). We see that the growth curve now declines monotonically. This striking change in the growth curve pattern happens because there is a cohort effect in the data. As further inspection showed us, younger cohorts are on average less happy. In particular, the cohorts born after the Second World War are two full scale points less happy than the oldest cohorts born before the First World War.

However, we still do not see a pronounced old age decline in happiness. This might be due to self-selection at higher ages. Taking regard of self-selection, we further estimate RE (Figure 15.4c) and FE models (Figure 15.4d). In fact, we now see that there is a steep decline in happiness starting at about age 70. The decline estimated with FE is even steeper than that estimated with RE. Such a decline seems very plausible because, as Gerstorf et al. (2010) show, 3–5 years before death there is a 'terminal decline' in happiness.

The age–period–cohort problem

Until now we have ignored period effects. It is well known that models which do not include a period variable are potentially misspecified: age (and cohort) effects are potentially biased. Period effects (i.e. idiosyncratic events shortly before the panel interview took place and that affect the outcome) are ubiquitous. Therefore, it is generally recommended to include period effects in panel regression models. Including period effects in addition to age and cohort effects will not work, however. Age (in years), period (interview year), and cohort (birth year) are linearly dependent due to the relation $age = period - cohort$. Therefore, it is not possible to include all three terms in one regression model (the so-called age–period–cohort (APC) problem). In an FE model the problem already occurs if we include age and period, because cohort is implicitly controlled for.

As a solution one has to impose some kind of non-linear restrictions (APC restrictions). For instance, if one tries to estimate an FE model including age and interview year dummies, Stata automatically drops the last year dummy (in addition to the first year dummy that is dropped as a base). With this APC restriction perfect collinearity is broken and the model can be estimated. Unfortunately, this provides reasonable results only if the restriction is plausible (i.e. the mean outcomes are very similar in the first and last year). Otherwise, the growth curve will be misleading. Therefore, one should not use automatic APC restrictions. The researcher has to think very carefully about which APC restriction makes sense.

Therefore, we closely inspected the interview year dummies in models with a cohort restriction (cohort is missing): happiness in the years 1984, 1985, 1986, 1990, and 1991 turned out to be highest and at a very similar level. Hence, for our final model that includes period effects we define those five years as base category. Now we can estimate an FE growth curve that controls for cohort (implicitly) and period (with APC restrictions). In addition, we want to go beyond a parametric specification of the age growth curve. This is why we now include a full set of age dummies in the FE model instead of a cubic specification. The result can be seen in Figure 15.5.

Compared to the FE growth curve without controlling for period effects (Figure 15.4d), we see an important change: happiness stays essentially constant up to age 65. Only then does the decline in happiness begin. In addition, due to our dummy specification, we see two interesting details. There is a little 'happiness hump' around age 60. Wunder et al. (2013) argue that this is due to people anticipating and/or entering retirement. Further, there is a sharp happiness drop before age 20. This is due to a panel-conditioning effect, where respondents in their first three

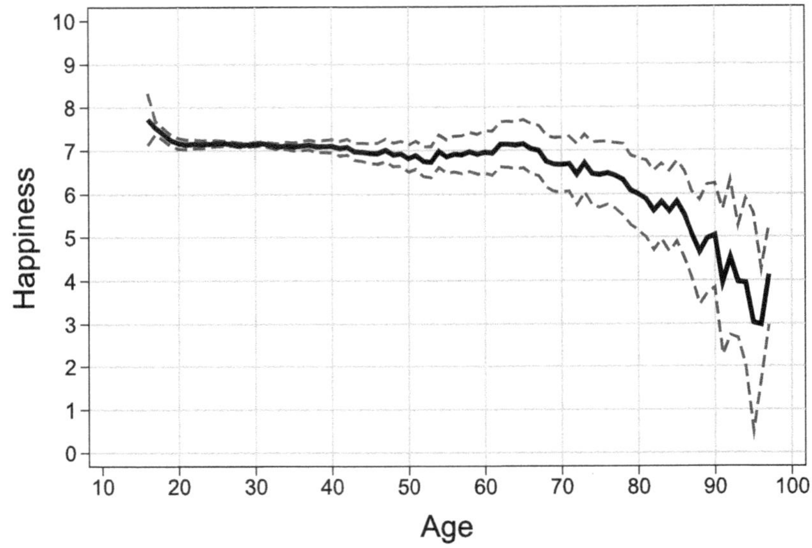

Figure 15.5 A non-parametric FE growth curve (base category years 1984, 1985, 1986, 1990 and 1991).

SOEP years report systematically higher happiness values. This can be confirmed by including first year dummies (not shown here).

One has to be cautious, however, in interpreting these results. We cannot be sure that our APC restriction 'works'. Therefore, a better solution to the APC problem could be to impose substantive restrictions. Cohort and period dummies are only proxies for specific circumstances and events. For instance, people who have grown up during a war might have higher happiness levels throughout their lives after the end of the war. Another example is that Germans were very happy in 1990 and 1991 because of German unification and because their soccer team won the World Cup. Therefore, instead of the proxies one could introduce the substantive relevant variables directly. It seems difficult to get a handle on all variables that produce cohort effects. It might be easier to identify those (large-scale) events that produce period effects. Hence, we recommend including age and cohort terms, dropping the period terms, and instead including event dummies. Further important period variables are usually gross national product and the unemployment rate.

Finally, age is a proxy variable, too. It captures increasing human capital, maturation, declining health, and so on. Therefore, growth curves do not have a causal interpretation. Growth curves are a description of how the outcome develops over time. We hope that the reader of this section has been convinced that such a description is of value in itself. However, the mechanisms that produce the growth curve can also be uncovered by pushing the analysis further: including indicators for the mechanisms should flatten the growth curve. For instance, if the happiness growth curve is mainly due to the development of health, then the inclusion of health measures should make the growth curve flat (this is not the case with the SOEP data; cf. Wunder et al., 2013). Thus, the growth curve methodology as presented in this section can also be very helpful for causal analysis.

CAVEATS AND FREQUENT ERRORS

The main point with FE estimation is that it discards potentially 'contaminated' between variation. It uses only within variation to estimate the causal effect of an event. Thus, to identify the causal effect FE only requires the assumption that the within variation is exogenous. This is a weaker assumption, compared to POLS or RE, where both between and within variation have to be exogenous. Several implications follow:

1 Usually it is not a good idea to use also the between variation for causal inference (i.e. to use POLS or RE).
2 However, for descriptive purposes it makes sense to also use the between variation.
3 Because the FE estimator uses only the within variation, it is only generalizable to those units that show within variation.
4 If the within variation is not exogenous the FE estimator will yield biased estimates.

Concerning (1), in most non-experimental social science settings, between variation will be endogenous due to self-selection. Thus, for the purpose of causal analysis one should avoid models that rest on the assumption of exogenous between variation. However, some researchers are uncomfortable with discarding so much variation and therefore use RE or POLS. Two arguments are put forward in this context. (i) One should use RE estimation because it is more efficient. It is argued that estimators should be judged not only by bias, but also by their standard error (e.g. according to the mean square error). This argument certainly makes sense from a statistical perspective. Nevertheless, experience tells us that the bias introduced by using the between variation is so huge that the lower standard error is bought at an exorbitant price: estimates are so far off the true value that the whole analysis is misleading. (ii) The effects of time-constant variables are interesting, and therefore one should use RE estimation (or the hybrid model; cf. Firebaugh et al., 2013). Obviously, this argument is motivated by the common practice of reporting big regression tables with many coefficients. Again, this might spoil the whole analysis. If one is interested in the effects of time-constant variables, it makes much more sense to estimate group-specific growth curves (by including group–age interactions).

Nevertheless, and turning now to implication (2), in the social sciences not only causal questions are of value. Descriptive questions are also of prime interest. Some even argue that we should first have good descriptions of the social world, before we start analyzing causal effects. For descriptive purposes the between variation is important. Therefore, for descriptive purposes one should use POLS (not RE, because RE is 'biased' toward FE). For instance, if one wants to investigate the causal wage effect of migrating from West Germany to East Germany, then one should use only the within variation produced by those who actually migrate. However, if one simply wants to describe what the wage difference is between West and East Germans, one clearly should use all variation in the data (i.e. estimate the wage difference by POLS). In fact, panel data are suboptimal for such descriptive purposes, because between and within variation are mixed. Trend data from independent cross-sections are much better suited for descriptive analyses.

Concerning (3), by using only the variation of those who show within variation in treatment status, FE estimators cannot be generalized to the whole population. This point confuses many social researchers because from cross-sectional research they are used to generalizing to the whole population (given that we have a random sample from the whole population). For instance, if we compare the wages of non-married and married people, we would generalize the wage differential found to the whole population. This makes sense for a descriptive research question.

However, a within estimator provides a treatment effect. And this can be generalized only to those who potentially experience such a treatment. For instance, the FE estimator on the marital wage premium uses only the information from those who have married during the observation period. Therefore, we can generalize the result found only to those in the population who marry. In the counterfactual literature this is called an 'average treatment effect on the treated' (ATT). For instance, those who do not marry might have quite different marriage effects (effect heterogeneity). Thus, the within marriage effect found cannot be generalized to them. However, this is not a shortcoming of FE estimation. It merely reflects the real-world fact that not everybody marries. And therefore we are satisfied with an ATT that only makes a statement on the causal effect of those who experience the treatment. Note that with an experimental design we would 'force' some people to marry, who in the real world would never marry. This would provide an 'average treatment effect' (ATE) that could be generalized to the whole population. From a substantive point of view, such an ATE does not seem to be preferable.

Policy recommendations, however, should be based on an ATE. For example, if we use an FE model and find a strong positive effect of a training program for some unemployed people on their subsequent wages, it is not clear whether we would find a similar effect for all other unemployed people who did not participate. Even after controlling for unobserved ability, participants might benefit more than other people would. In this case of 'differential treatment effects', generalizations to the whole population would be misleading (Morgan and Winship, 2007).

Finally, concerning (4), within estimation fails if strict exogeneity does not hold. Generally there are three sources of endogeneity (cf. Wooldridge, 2010, p. 321). (i) There may be time-varying confounders (unobservables that affect both the outcome and the treatment). For instance, in the context of our marital wage premium example, people might undergo cosmetic surgery and as a consequence have a higher probability of marriage and earn more money (beauty premium). Then what in fact is a beauty premium would erroneously be attributed to marriage. (ii) There may be simultaneity, that is, a change in the outcome may affect treatment (reverse causality). For instance, men might react to a wage increase with a higher probability of marriage. Again, FE would erroneously estimate a marital wage premium. (iii) There might be measurement errors in the treatment indicator. It is well known that this might also bias estimates.

If at least one of these mechanisms is at work, then the within variation used by FE to estimate the effect is no longer exogenous and FE fails to identify the true causal effect. All three of these sources of endogeneity are potentially present in most social science applications. Therefore, one should always be critical in the face of the FE estimates found and look for arguments why the strict exogeneity assumption might be violated. Only if no such arguments are found, one should provisionally accept the FE results.

Different methods have been suggested to deal with endogeneity in FE models (instrumental variables, structural equation modeling). Generally these methods rest on untestable assumptions and do not provide robust results. Research fields where these methods have been used abundantly are full of contradictory results. For example, Mouw (2006) gives an overview of longitudinal studies on the effects of social capital and drives this point home very nicely. Therefore, we do not recommend these methods. Instead of investing in complicated statistical modeling it is recommended to invest in 'shoe leather' – that is, to go out and collect better data that, for instance, include information on the supposed time-varying confounder or simply are measured with greater precision.

As we saw in the second section of this chapter, strict exogeneity can be weakened by allowing for individual-specific trends (FEIS). In many applications it seems likely that the parallel trends assumption does not hold, because there is self-selection not only on the outcome level but also on the outcome growth. For instance, Ludwig and Brüderl (2011) argue not only that there is self-selection of high earners into marriage, but also that those on a steeper wage profile self-select into marriage. If this is the case, FE overestimates the marital wage premium (see the simulations above). They actually find that the marital wage premium in Germany and the US vanishes if one uses FEIS. They interpret this as evidence that there is self-selection on wage growth and that no causal marital wage premium exists.

FURTHER READING

The main reference for FE regression models for panel data is Allison (2009). He discusses not only linear FE models, as we have in this chapter, but also non-linear FE regression models (fixed-effects logistic, count data, and Cox models). Halaby (2004) gives a short but excellent introduction to FE modeling, including a discussion of LDV models. In addition, he dissects the erroneous arguments that sociologists often put forward against FE regression (Rogosa, 1988, does the same with psychologists' arguments). Firebaugh et al. (2013) discuss panel models with a focus on the hybrid model. Econometric details on panel regression methods can be found, for instance, in Baltagi (2008), Cameron and Trivedi (2005) and Wooldridge (2010). Finally, introductions into the modern methods of causal inference – of which FE regression is one – can be found in Angrist and Pischke (2009) and Morgan and Winship (2007).

NOTES

* We thank Henning Best, Klaus Pforr, Patrick Riordan and an anonymous reviewer for careful reading and helpful comments on an earlier version of this chapter.

1 By 'unobserved heterogeneity' we mean unobserved heterogeneity that is correlated with the regressors (i.e. endogenous). In the following we will often use this somewhat loose wording which, however, is common in the social sciences.

2 The data (panelanalysis stylized.dta) and the Stata do-file (panelanalysis stylized.do) are available on the volume's website.

3 If one happens to be interested in estimates of inter-individual differences then one should apply LSDV. The estimates of the individual constants are unbiased under strict exogeneity, but inconsistent as $N \to \infty$ because, with T fixed, adding one parameter for each individual prevents convergence in probability. In general, one would like to have many observations per subject to get reliable estimates of the constants. Furthermore, for time-dependent processes, where process time might be age, tenure or marriage duration, each individual should be observed from the same point in time onward. (In event history parlance, this rules out left-censoring.) Otherwise the estimates of individual constants might be very misleading.

4 The procedure actually is an application of the Frisch–Waugh–Lovell theorem for partitioned regression. See Baltagi (2008) for a detailed exposition of the basic FE estimator using this framework.

5 The model actually is not as hungry as Morgan and Winship suggest. If we just want to allow for individual-specific constants and linear time trends, three observations are sufficient.

6 What we term 'fixed coefficients' here is usually called 'fixed effects'. This terminology is obviously confusing. What is meant is that the effect is assumed homogeneous across subjects. It is not a fixed effect in the sense used in this chapter.

7 We deviate here from the usual notation in the multi-level literature where the index of the lower level (t in this case) comes first.

8 When motivated by a multi-level framework, the conventional RE model is usually called the 'random-intercept' (RI) model and is estimated by maximum likelihood (instead of FGLS). In practice, maximum likelihood and FGLS often yield very similar coefficients in large samples.

9 The Stata do-file (panelsim.do) used to run the simulations is available for download on the volume's website.

10 The AB-LDV provides reasonable estimates in scenario (4) merely by coincidence, as further simulations with different values for the conditional means of α_{2i} showed us.

11 An anonymized version of the data file that we extracted from the SOEP (Happiness Anonym.dta) and the Stata do-file (Happiness Regressions.do) are available on the volume's website.

12 A tutorial on panel data preparation can be found on our 'Teaching Materials' page under the heading 'Working with panel data': `http://www.ls3.soziologie.uni-muenchen.de/teach-materials`.

13 Another potential candidate would be the 'hybrid model'. This model provides the FE estimates for time-varying regressors, and RE-type estimates for time-constant regressors (Allison, 2009; Firebaugh et al., 2013). However, the effect of time-constant regressors such as sex should not be interpreted as causal effects in this model. Generally, there is much confusion over how to interpret the estimates of this model. For causal analysis the model is therefore not useful.

14 The marriage effects estimated by BE, POLS, and RE are even biased downwards, because we do not control for cohort in these models (FE and FD do implicitly). This biases the age effect upwards (see below), which in turn biases the marriage effect downwards.

REFERENCES

Allison, P. D. (1994). Using panel data to estimate the effects of events. *Sociological Methods and Research*, 23, 174–199.

Allison, P. D. (2009). *Fixed Effects Regression Models*. Thousand Oaks, CA: Sage.

Angrist, J. D. and Pischke, J.-S. (2009). *Mostly Harmless Econometrics*. Princeton, NJ: Princeton University Press.

Arellano, M. and Bond, S. (1991). Some tests of specification for panel data: Monte Carlo evidence and an application to employment equations. *Review of Economic Studies*, 58, 277–297.

Arránz Becker, O., Salzburger, V., Lois, N. and Nauck, B. (2013). What narrows the stepgap? Closeness between parents and adult (step)children in Germany. *Journal of Marriage and Family*, 75(5), 1130–1148.

Baltagi, B. (2008). *Econometric Analysis of Panel Data*, 4th edn. Chichester: Wiley.

Blundell, R. and Bond, S. (1998). Initial conditions and moment restrictions in dynamic panel data models. *Journal of Econometrics*, 87, 115–143.

Bun, M. J. G. and Windmeijer, F. (2010). The weak instrument problem of the system GMM estimator in dynamic panel data models. *Econometrics Journal*, 13, 95–126.

Cameron, A. and Trivedi, P. (2005). *Microeconometrics: Methods and Applications*. New York: Cambridge University Press.

Firebaugh, G., Warner, C. and Massoglia, M. (2013). Fixed effects, random effects, and hybrid models for causal analysis. In S. L. Morgan (ed.), *Handbook of Causal Inference for Social Research* (pp. 113–132). New York: Springer.

Frijters, P. and Beatton, T. (2012). The mystery of the U-shaped relationship between happiness and age. *Journal of Economic Behavior & Organization*, 82, 525–542.

Gangl, M. (2010). Causal inference in sociological research. *Annual Review of Sociology*, 36, 21–47.

Gerstorf, D., Ram, N., Mayraz, G., Hidajat, M., Lindenberger, U. and Wagner, G. (2010). Late-life decline in well-being across adulthood. *Psychology and Aging*, 25, 477–485.

Halaby, C. (2004). Panel models in sociological research: theory into practice. *Annual Review of Sociology*, 30, 507–544.

Hinz, T. and Gartner, H. (2005). Geschlechtsspezifische Lohnunterschiede in Branchen, Berufen und Betrieben. *Zeitschrift für Soziologie*, 34, 22–39.

Kézdi, G. (2004). Robust standard error estimation in fixed-effects panel models. *Hungarian Statistical Review*, 9, 96–116.

Legewie, J. (2012). Die Schätzung von kausalen Effekten: Überlegung zu Methoden der Kausalanalyse anhand von Kontexteffekten in der Schule. *Kölner Zeitschrift für Soziologie und Sozialpsychologie*, 64, 123–153.

Ludwig, V. and Brüderl, J. (2011). *Is there a Male Marital Wage Premium? Resolving an Enduring Puzzle with Panel Data from Germany and the U.S.* Paper presented at RC 28 meeting in Essex.

Morgan, S. and Winship, C. (2007). *Counterfactuals and Causal Inference*. New York: Cambridge University Press.

Mouw, T. (2006). Estimating the causal effect of social capital: A review of recent research. *Annual Review of Sociology*, 32, 79–102.

Nickell, S. J. (1981). Biases in dynamic models with fixed effects. *Econometrica*, 49, 1417–1426.

Phillips, P. C. B. and Sul, D. (2007). Bias in dynamic panel estimation with fixed effects, incidental trends and cross section dependence. *Journal of Econometrics*, 137, 162–188.

Polachek, S. and Kim, M.-K. (1994). Panel estimates of the gender earnings gap: Individual-specific intercept and individual-specific slope models. *Journal of Econometrics*, 61, 23–42.

Raudenbush, S. W. (2001). Comparing personal trajectories and drawing causal inferences from longitudinal data. *Annual Review of Psychology*, 52, 501–525.

Rogosa, D. (1988). Myths about longitudinal research. In K. W. Schaie (ed.), *Methodological Issues in Aging Research* (pp. 171–209). New York: Springer.

SOEP (2010). *Data for years 1984-2009, version 26*. doi:10.5684/soep.v26.

Stock, J. H. and Watson, M. W. (2008). Heteroskedasticity-robust standard errors for fixed effects panel data regression. *Econometrica*, 76, 155–174.

Stutzer, A. and Frey, B. S. (2005). Does marriage make people happy or do happy people get married? *Journal of Socio-Economics*, 35, 326–347.

Wagner, G. G., Frick, J. R. and Schupp, J. (2007). The German Socio-Economic Panel Study (SOEP) – scope, evolution and enhancements. *Schmollers Jahrbuch*, 127, 139–169.
Wooldridge, J. (2010). *The Econometrics of Cross-Section and Panel Data*, 2nd edn. Cambridge, MA: MIT Press.
Wunder, C., Wiencierz, A., Schwarze, J. and Küchenhoff, H. (2013). Well-being over the life span. *Review of Economics and Statistics*, 95, 154–167.

16

Event history analysis

Hans-Peter Blossfeld and Gwendolin J. Blossfeld

INTRODUCTION TO THE METHOD

The empirical investigation of causal relationships is an important but difficult scientific endeavor. In the social sciences, two understandings of causation have guided the empirical analysis of causal relationships (see Goldthorpe, 2001): 'causation as robust dependence' and 'causation as consequential manipulation'. Both approaches clearly have strengths and weaknesses for the social sciences which will be shortly discussed. Based on this elaboration, a third understanding of 'causation as generative process', proposed by David Cox (1990, 1992), is then further developed. This idea seems to be particularly valuable as a framework for event history analysis because it directs the attention of causal analysis towards the study of time lags between causes and their effects as well as the different temporal shapes of effects.

Using data from the National Educational Panel Study (NEPS), the usefulness of the approach of 'causation as generative process' is demonstrated in an event history analysis of entry into first motherhood (as the dependent process) explained by various other parallel and interdependent processes in the life courses of women (such as educational careers and job trajectories).

Models of causal inference

The goal of finding scientifically based empirical evidence for causal relationships leads to design questions, such as which inference model is appropriate to specify the relationship between cause and effect and which statistical procedures can be used to determine the strength of that relationship (Schneider et al., 2007). Two different models of causal inference have dominated the work of practitioners in the social sciences in recent decades: 'causation as robust dependence' and 'causation as consequential manipulation'. The former approach – which is also known as the 'control variable' or 'partialing' approach (Duncan, 1966; Kerlinger and Pedhazur, 1973; Blalock, 1970) – starts from the presumption that correlation does not necessarily imply causation but causation must in some way or the other imply correlation. In this view, the key problem of causal inference is to determine whether an observed correlation of variables X and Y, where X is temporally prior to Y, can be established as a 'genuine causal relationship'.

The advocates of the 'causation as robust dependence' approach call X a 'genuine' cause of Y in so far as the dependence of Y on X cannot be eliminated through additional variables being introduced into the statistical analysis. Thus, in this approach causation is established essentially through the elimination of spurious (or non-causal) influences. Although this approach has dominated the social sciences for several decades, sociologists now consider it as too limited an approach. First, they think that causal inference should not be limited entirely to a matter of statistical predictability but should include predictability in accordance with theory (Goldthorpe, 2001, p. 3). Second, since scientists rarely know all of the causes of observed effects or how they relate to one another, it is impossible to be sure that all other important variables have in fact been controlled for (Shadish et al., 2002). A variable X can therefore never be regarded as having causal significance for Y in anything more than a provisional sense (Goldthorpe, 2001, p. 5).

The second understanding of 'causation as consequential manipulation' seems to have emerged as a reaction to the limitations of 'causation as robust dependence'. Instead of 'establishing the causes of effects', Holland (1986, 1988) and Rubin (1974, 1978, 1980) are concerned with 'establishing the effects of causes'. They propose that it is more to the point to take causes simply as given, and to concentrate on the question of how their effects can be reliably determined. According to this approach, causes can only be those factors that could serve as treatments or interventions in well-designed controlled experiments or quasi-experiments. Thus, given appropriate experimental controls, if a causal factor X is manipulated, then a systematic effect is produced on the response variable Y. The particular strength of this design is that 'while statements in the form "X is a cause of Y" are always likely to be proved wrong as knowledge advances, statements in the form "Y is an effect of X", once they have been experimentally verified, do not subsequently become false: "Old, replicable experiments never die, they just get reinterpreted" ' (Goldthorpe, 2001, p. 5).

Understood in this way, causation is always relative in the sense that the specific treatment and its observed outcome are compared with what would have happened to the same unit if it had not been exposed to this treatment (counterfactual account of causality). Referring to a particular unit, u, we denote $X(u)$ as the treatment variable. If there is a treatment, then $X(u) = 1$ and we observe the outcome $Y_1(u)$; otherwise $X(u) = 0$ and we observe $Y_0(u)$. However, since it is not possible in the same experiment for a unit to be both exposed and not exposed to the treatment, the conception of 'causation as consequential manipulation' leads to what Holland (1986) has called the 'fundamental problem of causal inference'. For example, a student who completes one mathematics program cannot go back in time and complete a different program so that we can compare the outcomes of the two conditions. Thus, the question arises of how we make sure that one gets convincing measurements for something that is in fact impossible to measure: the outcome of $Y_0(u)$ in a situation where $X(u) = 1$.

In sociology, economics, and demography, strict experimental controls are often hard to apply, in particular, if we study long-term processes such as life histories (see Mayer and Huinink, 1990). In addition, randomization is often practically or socially unacceptable (e.g. it is morally and legally impossible to assign twins at birth randomly to families of different social classes in order to study the impact of various social environments on school success). Thus, well-designed randomized controlled experiments or quasi-experiments are rarely carried out by practitioners in the social sciences and most demographic, economic and sociological causal inference is based on non-experimental observations of social processes.

Since these observational data are often highly selective, Rubin, Holland and others subscribing to the approach of 'causation as consequential manipulation' recommend that in their empirical work social scientists should make the process of unit assignment itself a prime concern of the inquiry. In particular, social scientists should attempt to identify, and then to represent through covariates in their data analyses, all unobserved and observed influences on the response variable that could conceivably be involved in or follow from this unit assignment process (Goldthorpe, 2001). Again, as in the 'causation as robust dependence' approach, the question immediately arises: have all relevant variables been included and adequately measured and controlled?

A whole battery of statistical techniques – as demonstrated in various chapters of this book – has been developed to help to approximate randomized controlled experiments with observational data (Schneider et al., 2007). These methods include fixed effect models (i.e. the adjustment for time-constant unobserved individual characteristics), random effect models (i.e. the so-called variance components approach where the residual variance is partitioned into two components, the between-group variance and the within-group (or between-individual) variance), instrumental variables (i.e. a method to correct for omitted variables bias due to unobserved characteristics), propensity score matching (an approach where individuals are matched on the basis of their observed aggregate characteristics), and regression discontinuity designs (where samples and comparisons between groups are restricted to individuals who fall just above or below a specific cut-off point and, at the same time, are likely to be similar on a set of unobserved variables). Despite how valuable these techniques might be, 'it is still difficult to avoid the conclusion that, in non-experimental social research, attempts to determine the effects of causes will lead not to results that "never die" but only to ones that have differing degrees of plausibility' (Goldthorpe, 2001, p. 6).

A major shortcoming of the approaches of 'causation as robust dependence' and 'causation as consequential manipulation' is that they neglect the dynamic relationships between causes and their effects over time t. After a causal event $\Delta X(t)$, it may take some time (perhaps weeks, months or even years) until an effect $\Delta Y(t)$ starts to appear (time lag) or the effect $\Delta Y(t)$ changes its strength over time t. If dynamic relationships between events of social processes are studied over longer time-spans, a third understanding of 'causation as generative process' seems therefore to be helpful. According to Cox (1990, 1992) it is crucial to the claim of a causal link that there is a detailed theoretical elaboration of the underlying, generative process existing in space and time which then can be empirically tested. We will develop this idea further when we discuss the analysis of parallel and interdependent processes in event history analysis in the next section.

MATHEMATICAL FOUNDATIONS AND ADVANCED ASPECTS

Social scientists often study substantive processes which can be characterized as follows (Coleman, 1981): (1) there is a collection of units (which may be individuals, organizations, societies, etc.), each moving among a finite (usually small) number of states; (2) these changes (or events) may occur at any point in time (i.e. they are not restricted to predetermined points in time); and (3) there are time-constant and/or time-varying factors influencing the timing of these events. Examples are workers who move forth and back between unemployment and employment or men and women who make transitions from being single to a consensual union or first marriage or from having no child to first parenthood.

The most restricted event history model is based on a process with only a single episode and two states: one origin and one destination state (see Blossfeld et al., 2007, pp. 38ff.). An example is the duration (or episode) of first marriage until the end of the marriage (for whatever reason). In this case, each individual who entered into first marriage (origin state) started an episode, which could be terminated by a transition to the destination state 'not married anymore'. If more than one destination state exists, these models are called multistate models. They are also referred to as models with competing events or risks. For example, the first marriage might be terminated by the event 'death' or the event 'divorce'. If there are repeated events over the life course, these models are called multiepisode models. For example, we might analyze not just first marriages but all marriages of individuals. The individual then moves repeatedly between different states. Thus, in event history analysis, we have often a sample of $i = 1, \ldots, N$ multistate–multiepisode data. A complete description of the data is given by

$$(u_i, m_i, o_i, d_i, s_i, t_i, x_i), \quad i = 1, \ldots, N \tag{16.1}$$

where u_i is the identification number of the individual or any other unit of analysis the ith episode belongs to; m_i is the serial number of the episode; o_i is the origin state, the state held during the episode until the ending time of the episode; d_i is the destination state defined as the state reached at the end of the episode; s_i and t_i are the starting and ending times, respectively. In addition, there is a covariate vector x_i with time-constant or time-changing factors associated with the episode. We always assume that the starting and ending times are coded such that the difference $t_i - s_i$ is the duration of the episode, which is positive and greater than zero.

Observations of event histories are often censored. Censoring occurs when the information about the duration in the origin state is incompletely recorded. If the length of time a subject has already spent in the origin state is unknown, the episode is censored on the left. This kind of censoring should be avoided because it is not easy to handle in statistical terms (see Blossfeld et al., 2007, pp. 39ff.). In this chapter, we assume that we can observe all processes right from the beginning. Right censoring on the other hand, is very common. This type of censoring typically occurs in life course studies at the time of the retrospective interview, or in panel studies at the time of the last panel wave. For example, if one is interested in entry into first motherhood, women who are still in their fertile years and do not have a first baby at the time of interview have a right-censored episode. This means that these women might have a first baby later, but we do not know. Because the timing of the interview is normally independent of the timing of the substantive processes under study, this type of right censoring is unproblematic. It can easily be handled with event history methods. For example, we can easily compute the probability that a woman in our example has not yet had her first baby up to the time of the interview. This information about having no event up to a certain point in time (survivor function) can then be utilized in the estimation.

Given such an event history data set, the typical problem of the social scientist is to apply appropriate statistical models to analyze the time-dependent causal relationships among various independent and dependent events over time.

The dependent variable in continuous-time and discrete-time event history analysis

Event history models can be formulated in continuous-time and discrete-time. For each point in time (continuous-time models) or for each time interval (discrete-time models), they predict future levels or changes of the transition rate of the dependent process on the basis of states and/or events of other processes in the past. The central concept of event history analysis is the transition rate. Because of the various origins of event history analysis in the different disciplines, the transition rate is also called the hazard rate, intensity rate, failure rate, transition intensity, risk function, or mortality rate. The transition rate describes in detail how the dependent process evolves over time.

If time can be considered to be (approximately) continuous (i.e. when events, at least in principle, can happen at any point in time), the transition rate $r(t)$ can be interpreted as the propensity (or intensity) to change from an origin state j to a destination state k, at time t (see Blossfeld et al., 2007, p. 36):

$$r(t) = \lim_{t' \to t} \Pr(t \leq T < t' | T \geq t) / (t' - t), \quad \text{with } t < t'. \tag{16.2}$$

It is important to note that this propensity $r(t)$ is defined in relation to a risk set $(T \geq t)$ at t, that is, the set of units that still can experience the event because they have not yet had the event before t. In other words, the transition rate is a conditional density function, that is, the density function $f(t)$ divided by the survivor function $G(t)$:

$$r(t) = f(t) / G(t)$$

where

$$f(t) = \lim_{t' \to t} \Pr(t \leq T < t') / (t' - t) \quad \text{with } t < t' \tag{16.3}$$

and

$$G(t) = \Pr(T \geq t).$$

If events can only happen within fixed time intervals, the continuous-time axis is arbitrarily split into a series of time intervals $\tau_0 < \tau_1 < \tau_2 < \ldots < \tau_q$, with $\tau_0 = 0$. The number of the time interval then becomes a discrete random variable $T^* = t \Leftrightarrow T \in [\tau_{t-1}, \tau_t)$, with $t = 1, \ldots, q$. It denotes the time interval where an event happens. The discrete-time transition rate is then a (conditional) probability

$$r^*(t) = \Pr(T^* = t | T^* \geq t), \quad \text{with } 0 \leq r^*(t) \leq 1. \tag{16.4}$$

$r^*(t)$ is the probability that the dependent variable changes from origin state j to a destination state k in time interval t under the condition that the event did not yet happen until the beginning of that interval. In the discrete-time model the transition rate $r^*(t)$ is therefore a conditional probability function, where the probability function $f^*(t)$ is divided by the survivor function $G^*(t)$,

$$r^*(t) = f^*(t) / G^*(t).$$

where

$$f^*(t) = \Pr(T^* = t) \quad \text{with } t = 1, \ldots, q \tag{16.5}$$

and

$$G^*(t) = \Pr(T^* > t).$$

Statistical models

The central idea in event history analysis is to make the continuous-time transition rate $r(t)$ or the discrete-time transition rate $r^*(t)$ dependent on concepts of time (e.g. time t) and on a set of time-constant x or time-dependent $x(t)$ covariates:

$$r(t) = g(t, \mathbf{X}, \mathbf{X}(t)) \quad \text{or} \quad r^*(t) = g(t, \mathbf{X}, \mathbf{X}(t)). \tag{16.6}$$

The causal interpretation of the transition rate requires that we take the temporal order in which the processes evolve very seriously. In the continuous-time mode, at any given point in time t, the transition rate $r(t)$ can be made dependent on conditions that happened to occur in the past (i.e. before t), but not on what is the case at t or in the future after t. Equivalently, in the discrete-time model, at any given time interval t, the discrete-time transition rate $r^*(t)$ can be made dependent on conditions that happened to occur before the beginning of interval t, but not on what is the case at time interval t or after time interval t.

There are several possibilities for specifying the functional relationship $g(\cdot)$ between covariates and the transition rate in continuous-time event history analysis (see Blossfeld et al., 2007). First, the 'exponential model', which normally serves as a baseline model, assumes that the transition rate can vary with different constellations of time-constant covariates, so that the rates are time-constant too (see Blossfeld et al., 2007, pp. 87ff.):

$$r(t) = \exp(\beta_0 + \beta_1 x_1 + \beta_2 x_2 + \ldots + \beta_n x_n) = \exp(\mathbf{X}\boldsymbol{\beta}) \tag{16.7}$$

where β_0 is a constant, $\mathbf{X}$ is the row vector of covariates and $\boldsymbol{\beta}$ is a corresponding column vector of coefficients.

The exponential model, like all parametric transition rate models described below, is estimated using the maximum likelihood method. To explain the general setup of the likelihood for parametric transition rate models, we consider the case of a single transition from one origin to one destination state. The notation to set up the likelihood is as follows: N is the set of all episodes, Z is the set of all censored episodes, and E is the set of all uncensored episodes. The likelihood may then be written as

$$L = \prod_{i \in E} f(t_i) \prod_{i \in Z} G(t_i) = \prod_{i \in E} r(t_i) \prod_{i \in N} G(t_i), \tag{16.8}$$

where $f(t)$ is the density function and $G(t)$ is the survivor function for the parametric model. The contribution to the likelihood of an episode with an event at t_i is given by the density function, evaluated at the ending time t_i and with appropriate covariate values. The contribution of a censored episode is given by the survivor function evaluated at the censoring ending time t_i, but possibly depends on covariates changing their values during the episode The likelihood can be expressed, then, by using only the transition rate and the survivor function. The exponential model and the following parametric models can be estimated with Stata (see Blossfeld et al., 2007).

In most applications of transition rate models, the assumption that the forces of change are constant over time is, however, not justified. This is particularly true if we are interested in causal relationships which relate changes (or events) in some (explaining) processes to changes (or events) in a dependent process. In our view, the most important step forward in event history analysis has been to explicitly measure and include time-dependent covariates in transition rate models. In such cases, covariates can change their values over process time. Time-dependent covariates can be qualitative or quantitative, and may stay constant for finite periods of time or change continuously (see Blossfeld et al., 2007, p. 128). Time-dependent covariates can be included by using a piecewise constant exponential model, by applying the method of episode splitting in parametric or semi-parametric transition rate models, and by specifying the distributional form of the time-dependence and directly estimating its parameters using the maximum likelihood method.

A first simple and very useful generalization of the 'exponential model' is the so-called 'piecewise constant exponential model'. This allows the transition rate to vary across fixed time periods with period-constant or period-specific effects of covariates (see Blossfeld et al., 2007, pp. 116ff.). The piecewise constant model is particularly helpful when researchers are not in a position to measure and include important causal factors (time-dependent covariates) explicitly or when they do not have a clear idea about the form of the time-dependence of the process, after controlling for important covariates. The basic idea of the piecewise constant exponential model is to split the duration into L time periods

$$I_t = \{t \mid \tau_l < t \leq \tau_{l+1}\}, \quad l = 1, \ldots, L, \tag{16.9}$$

based on arbitrary split points on the time axis $0 = \tau_1 < \tau_2 < \ldots < \tau_L$, with $\tau_L = \infty$. The transition rate can then vary over the $l = 1, \ldots, L$ time periods based on the changing period-specific constants β_l (baseline hazard rate):

$$r(t) = \exp\{\beta_l + \mathbf{X}\boldsymbol{\beta}\}, \quad \text{if } t \in I_l, \tag{16.10}$$

where $\mathbf{X}$ is a (row) vector of covariates, and $\boldsymbol{\beta}$ is an associated vector of coefficients. Note that in this model there is no additional constant in $\mathbf{X}$.

Since causal relationships relate changes (or events) in explaining processes to changes (or events) in a dependent process, it is very important to be able to represent the changes of the causal forces in event history models directly (see the next subsection). An easy way to do this is to include time-varying covariates $x(t)$ in the exponential model via the method of episode splitting:

$$\begin{aligned} r(t) = \exp(&\beta_1 x_1 + \beta_2 x_2 + \ldots + \beta_n x_n \\ &+ \beta_{n+1} x_{n+1}(t) + \beta_{n+2} x_{n+2}(t) + \ldots + \beta_{n+p} x_{n+p}(t)) \\ = \exp(&\mathbf{X}(t)\,\boldsymbol{\beta}), \end{aligned} \tag{16.11}$$

where β_1 is the regression constant ($x_1 = 1$). The transformation of the person-spell file into a person-period file through episode splitting will be described in detail in the application example below (see Blossfeld et al., 2007, pp. 135ff.):

Another model is the so-called Cox model (see Cox, 1972), where the baseline hazard rate $h(t)$ is left unspecified:

$$\begin{aligned} r(t) = h(t)\exp(&\beta_1 x_1 + \beta_2 x_2 + \ldots + \beta_n x_n \\ &+ \beta_{n+1} x_{n+1}(t) + \beta_{n+2} x_{n+2}(t) + \ldots + \beta_{n+p} x_{n+p}(t)) \\ = h(t)\exp(&\mathbf{X}(t)\,\boldsymbol{\beta}) \end{aligned} \tag{16.12}$$

Thus, it is only possible to estimate the effects of the time-constant and/or time-varying covariates, controlling for an unknown baseline hazard rate. This model is also called a semi-parametric model, since only part of the rate function is parametrically specified, or partial likelihood model, because only part of the likelihood function is maximized (see Blossfeld et al., 2007, pp. 216ff.). Note that in this model there is again no regression constant in $\mathbf{X}$.

However, in some applications in the social sciences, substantive theory or previous empirical research may suggest a specific shape of time-dependence of the transition rate. However, time itself is no causal factor. Rather measures of time may serve as proxies for time-changing causal factors that could not be observed directly (see Blossfeld et al., 2007, pp. 182ff.). For example, duration might serve as a proxy for 'the changing amount of marriage-specific investments' in divorce studies or 'the changing stock of job-specific labor force experience' in mobility analyses. There are different parametric models which are based on specific shapes of the time-dependence. For example, the Gompertz(–Makeham) and Weibull models are able to specify monotonically increasing or monotonically decreasing shapes of the hazard rate. For example, if labor force experience x_{lfx} can be assumed to change linearly with job duration ($x(t)_{lfx} \approx t$) and if it is hypothesized that the job exit rate declines with job-specific investments (e.g. according to the human capital theory), then this hypothesis can be tested with the following Gompertz model ($H_0 : \beta_{lfx} < 0$):

$$\begin{aligned} r(t) &= \exp(\mathbf{X}(t)\,\boldsymbol{\beta} + \beta_{lfx} x(t)_{lfx}) \\ &= \exp(\mathbf{X}(t)\,\boldsymbol{\beta} + \beta_{lfx} t) \\ &= \exp(\mathbf{X}(t)\,\boldsymbol{\beta}) \exp(\beta_{lfx} t), \end{aligned} \tag{16.13}$$

where $r(t)$ is the job exit rate.

However, if it can be justified that the labor force experience $x(t)_{lfx}$ changes only as a logarithmic function of duration in a job ($x_{lfx} \approx \log(t)$) and if it can be hypothesized again that job-specific investments decrease the job exit rate, a Weibull model might be estimated ($H_0 : \beta_{lfx} < 0$):

$$\begin{aligned} r(t) &= \exp(\mathbf{X}(t)\,\boldsymbol{\beta} + \beta_{lfx} x(t)_{lfx}) = \exp(\mathbf{X}(t)\,\boldsymbol{\beta} + \beta_{lfx} \log(t)) \\ &= \exp(\mathbf{X}(t)\,\boldsymbol{\beta}) \exp(\beta_{lfx} \log(t)) \\ &= \exp(\mathbf{X}(t)\,\boldsymbol{\beta})\, t_{lfx}^{\beta}, \end{aligned} \tag{16.14}$$

where $r(t)$ is again the job exit rate.

Further parametric models such as the sickle, the log-logistic and the log-normal models allow an (at first increasing and then decreasing) non-monotonic time-dependency to be estimated. Since these models are rarely used in causal analysis, they will not be discussed here (for more details, see Blossfeld et al., 2007, pp. 204ff.).

Finally, an important problem of event history analysis is the issue of unobserved heterogeneity. In this case, the transition rate that is estimated for a population can be the result (a mixture) of quite different transition rates in the subpopulations (see Blossfeld et al., 2007, pp. 247ff.). There have been several proposals to deal with unobserved heterogeneity in time-continuous transition rate models. The basic idea is to incorporate an 'error term' into the model specification. For example, the continuous-time transition rate can be made dependent on an exponential model, the observed (time-dependent) covariates $x(t)$ and a stochastic error term v, which is gamma distributed (see Blossfeld et al., 2007, pp. 256ff.):

$$r(t|v) = \exp(\mathbf{X}(t)\,\boldsymbol{\beta})\, v, \quad \text{with } v \geq 0. \tag{16.15}$$

A likelihood ratio test can then be used to test whether the transition rate models with and without this unobserved heterogeneity term differ. The issue and the application of models with unobserved heterogeneity will be discussed in more detail below.

In the case of discrete-time models, the estimation of covariate parameters is normally achieved by preparing a period-person data file, which will be explained in the application example below, and then estimating a simple logit model:

$$r^*(t) = \frac{\exp(\alpha_1 x_1 + \alpha_2 x_2 + \ldots + \alpha_n x_n) \exp(\beta(t))}{1 + \exp(\alpha_1 x_1 + \alpha_2 x_2 + \ldots + \alpha_n x_n) \exp(\beta(t))}$$

or

$$r^*(t) = \frac{ab(t)}{1 + ab(t)}, \tag{16.16}$$

$$\text{with } a = \exp(\alpha_1 x_1 + \alpha_2 x_2 + \ldots + \alpha_n x_n),\ b(t) = \exp(\beta(t)).$$

In this model, if $b(t) = \exp(\beta_0)$ the logit model estimates an exponential model, if $b(t) = \exp(\beta_0 + \beta_1 t)$ it estimates a Gompertz model, and if $b(t) = \exp(\beta_0 + \beta_1 \ln t)$ it estimates a Weibull model. A piecewise constant model with n periods $(P_1, P_2, \ldots, P_n)$ is estimated for $b(t) = \exp(\beta_1 P_1 + \beta_2 P_2 + \ldots + \beta_n P_n)$, and a flexible time-dependent model (suggested by Coale, 1971) is estimated for $b(t) = \exp[\beta_0 + \beta_1 \ln(t - t_{min}) + \beta_1 \ln(t_{max} - t)]$ (see the application example below).

If we have estimated the discrete-time transition rate $r^*(t)$ based on a specified event history model, it is easy to compute the survivor $G^*(t)$ function for various constellations of the covariates based on

$$G^*(t) = \prod_{l=0}^{t-1} (1 - r^*(l)). \tag{16.17}$$

Parallel and interdependent processes

In causal analysis, researchers are interested in the extent to which changes (or events) in one process influence changes (or events) in a dependent process. Thus, the study of how parallel or interdependent processes influence each other is one of the most important advances of event history analysis (Blossfeld et al., 2007; Willekens, 1991; Courgeau and Lelièvre, 1992). Parallel or interdependent processes can operate at a variety of different levels. There may be interdependent or parallel processes at the level of:

- different domains of an individual's life. For instance, one may ask how educational career or upward and downward moves in a woman's job history influence her entry into marriage and motherhood (e.g. Blossfeld and Huinink, 1991).
- various individuals interacting with each other such as 'interdependent or linked lives' (Elder, 1987). One might study the effect of the career of the husband on his wife's labor force participation (Blossfeld and Drobnic, 2001).
- intermediate organizations, such as how the changing household structure determines women's labor force participation.
- macro processes, where the researcher may be interested, for instance, in the effect of changes in the business cycle or unemployment rate on family formation (e.g. Blossfeld and Huinink, 1991).
- any combination of the aforementioned processes. For example, in the study of life course, cohort, and period effects, time-dependent covariates measured at different levels must be included simultaneously (Blossfeld, 1986; Mayer and Huinink, 1990). Such an analysis combines processes at the individual level (life course change) with two kinds of processes at the macro level: variations in structural conditions across successive (birth, marriage, etc.) cohorts; and changes in particular historical conditions affecting all cohorts in the same way.

In event history analysis, time-dependent covariates are used to include the changing conditions of explanatory processes into the transition rate models. In the literature, however, only two types of time-dependent covariates have been described as not being subject to reverse causation (e.g. Kalbfleisch and Prentice, 1980; Tuma and Hannan, 1984; Blossfeld et al., 1989; Yamaguchi, 1991; Courgeau and Lelièvre, 1992). The first are defined time-dependent covariates whose total time path (or functional form of change over time) is determined in advance in the same way for all subjects under study. For example, process time, such as age or duration in a state (e.g. duration of marriage in divorce studies), is a defined time-dependent covariate because its values are predetermined for all subjects. It is the predefined onset of the process when the individual becomes 'at risk' in the event history model. Thus, by definition, the values of these time-dependent covariates cannot be affected by the dependent process under study. The second type are ancillary time-dependent covariates whose time path is the output of a stochastic process that is external to the units under study. Again, by definition, the values of these time-dependent covariates are not influenced by the dependent process itself. Examples of time-dependent covariates that are approximately external in the analysis of individual life courses are variables that reflect changes at the macro level of society (unemployment rates, occupational structure, etc.) or the population level (composition of the population in terms of age, sex, race, etc.), provided that the contribution of each unit is small and does not really affect the structure in the population (Yamaguchi, 1991).

In contrast to defined or ancillary time-dependent covariates are internal time-dependent covariates, which are often referred to as being problematic for causal analysis in event history models (e.g. Kalbfleisch and Prentice, 1980; Tuma and Hannan, 1984; Blossfeld et al., 1989; Yamaguchi, 1991; Courgeau and Lelièvre, 1992). An internal time-dependent covariate $X_1(t)$ describes a stochastic process, considered in a causal model as being the cause, which in turn is affected by another stochastic process $X_2(t)$, considered in the causal model as being the effect. Thus, there are direct effects in which the processes autonomously affect each other ($X_1(t)$ affects $X_2(t)$ and $X_2(t)$ affects $X_1(t)$), and there are 'feedback' effects in which these processes are affected by themselves via the respective other processes ($X_2(t)$ affects $X_2(t)$ via $X_1(t)$ and $X_1(t)$ affects $X_1(t)$ via $X_2(t)$). In other words, such processes are interdependent and form what has been called a dynamic system (Tuma and Hannan, 1984). Interdependence is typical at the individual level of processes in different domains of life and at the level of a few individuals interacting with each other – for example, career trajectories of partners (see Blossfeld and Drobnic, 2001). For example, the empirical literature suggests that the employment trajectory of an individual is influenced by his/her marital history and that the marital history is dependent on the employment trajectory.

Blossfeld and Rohwer (2002) have suggested a causal approach for analyzing interdependent processes. Based on theoretical reasoning, the researcher focuses on one of the interdependent processes and considers it as the dependent one. The future changes of this process are linked to the present state and history of the entire dynamic system as well as to other exogenous variables (see Blossfeld and Huinink, 1991). Thus, the history and the present state of the system are seen as a condition for change in (any) one of its processes. To find an empirical approach to examine longitudinal causal relations, Blossfeld and Rohwer (2002) suggested the examination of conditions which actually do change in time, controlling for other factors. These changes are characterized as events or transitions. More formally, an event is specified as a change in a variable, and this change must happen at a specific point in time or time interval. In other words, the role of a time-dependent covariate in this approach is to indicate that a (qualitative or metric) causal factor has changed its state at a specific time or time interval and that the unit under study is exposed to another causal condition. From this point of view, it seems somewhat misleading to regard whole processes as causes. Rather, only events, or changes in state space can sensibly be viewed as possible causes.

Consequently, we do not suggest that process $X_1(t)$ is a cause of process $X_2(t)$, but that a change in $X_1(t)$ could be a cause of (or provide a new condition for) a change in $X_2(t)$. Or, more formally, $\Delta X_1(t) \rightarrow \Delta X_2(t')$ (with $t > t'$), meaning that a change in variable $X_1(t)$ at an earlier time point t (or earlier time interval t) is a cause of a change in variable $X_2(t')$ at a later point in time (or a later time interval). Of course, it is not implied that $X_1(t)$ is the only cause which might affect $X_2(t')$. We speak of causal conditions to stress that there might be, and normally is, a quite complex set of causes (see Marini and Singer, 1988). Thus, if causal statements are studied empirically, they must intrinsically be related to time, which relates to three important aspects of 'causation as generative process':

- First, to speak of a change in variables necessarily implies reference to a time axis. Therefore, we use the following symbols to refer to changes in the values of the time-dependent covariate $\Delta X_1(t)$ and the state variable $\Delta X_2(t)$ at time t. This leads to the important point that causal statements relate to changes in two (or more) variables, if we think in terms of 'causation as generative process'.
- Second, we must consider time ordering, time intervals and apparent simultaneity. Time ordering assumes that cause must precede the effect in time, $t < t'$, in the formal representation given above, an assumption which is generally accepted (Eells, 1991, Ch. 5). As an implication, the 'causation as generative process' approach must specify a temporal interval between the change in the variable representing a cause and the corresponding effect (Kelly and McGrath, 1988). This time interval may be very short or very long, but can never be zero or infinity. In other words, also in time-continuous event history models there can never be simultaneity of the causal event and its effect event. Some effects take place almost instantaneously. They may occur in a time interval that requires small time units (e.g. microseconds) or may be too small to be measured by any given methods, so that cause and effect seem to occur at the same point in time. Apparent simultaneity is often the case where temporal intervals are relatively crude, as in yearly data. For example, the events 'first marriage' and 'first childbirth' may be interdependent, but whether these two events are observed simultaneously or successively depends on the degree of temporal refinement of the scale used in making the observations. Other effects need a long time until they start to occur. Thus, there might be sometimes a delay or lag between cause and effect (see Figure 16.1) that must be specified in an appropriate model of 'causation as generative process'. Unfortunately, in most of the current social science theories and interpretations of research findings, this interval is left unspecified.
- This leads to the third point of 'causation as generative process': temporal shapes of the unfolding effect. This means that there might be different shapes of how the causal effect $X_2(t)$ unfolds over time (see Figure 16.1). While the problem of time lags is widely recognized in the social science literature, only little attention has been given to the temporal shapes of effects in the social sciences (Kelly and McGrath, 1988). Researchers (using experimental or observational data) often seem to either ignore or be ignorant about the fact that causal effects could be highly time-dependent, which, of course, is an important aspect of 'causation as generative process'. For instance, in Figure 16.1a, there may be an immediate impact of change that is then maintained (this obviously is the idea underlying the approaches of 'causation as robust dependence' and 'causation of consequential manipulation' because there is no explicit notion of an underlying generative process present in these models). Or the effect could occur with a lengthy time lag and then become time-invariant (see Figure 16.1b). The effect could start almost immediately and then gradually increase (see Figure 16.1c) or there may be an almost all-at-once increase which reaches a maximum after some time and then decreases (see Figure 16.1d). Finally, there could exist a cyclical effect pattern over time (see Figure 16.1e).

Thus, based on these examples, it is clear that we cannot rely on the assumption of eternal, timeless laws but have to recognize that the causal effect may change during the development of social processes. Since the approaches of 'causation as robust dependence' and 'causation of consequential manipulation' do not have an explicit idea of an underlying generative process in

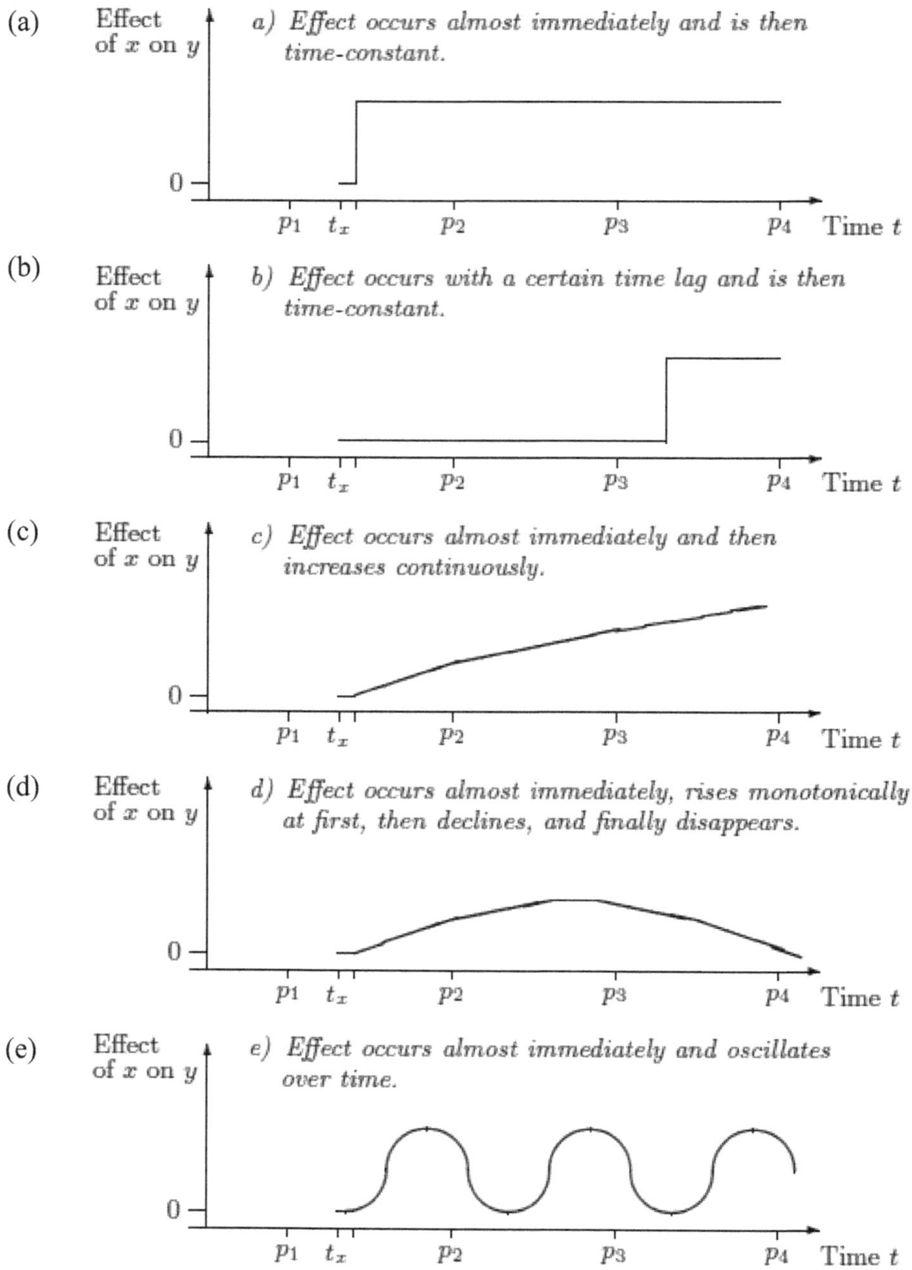

Figure 16.1 Hypothetical temporal lags and effect shapes (Blossfeld and Rohwer, 2002, p. 17)

time and space, it might happen that the timing of observations in observational or experimental studies (see, for example, the arbitrarily chosen observation times p_2, p_3 or p_4 in Figure 16.1) lead to completely different empirical evidence for causal relationships.

We consider here only interdependent processes that are not just an expression of another underlying process so that it is meaningful to assess the properties of the two processes without regarding the underlying one (control variable approach). This means, for instance, that what happens next to $X_1(t)$ should not be directly related to what happens to $X_2(t)$, at the same point in time, and vice versa. This condition, which we call 'local autonomy' (see Pötter and Blossfeld, 2001), can be formulated in terms of the uncorrelatedness of the prediction errors of both processes, $X_1(t)$ and $X_2(t)$, and excludes stochastic processes that are functionally related.

Combining the ideas above, a causal view of parallel and interdependent processes becomes easy, at least in principle. Given two parallel processes, $X_1(t)$ and $X_2(t)$, a change in $X_1(t)$ at any (specific) point in time t may depend on the history of both processes up to, but not including t. Or stated in another way: what happens with $X_1(t)$ at any point in time t is conditionally independent of what happens with $X_2(t)$ at t, conditional on the history of the joint process $X(t) = (X_1(t), X_2(t))$ up to, but not including, t. Of course, the same reasoning can be applied if one focuses on $X_2(t)$ instead of $X_2(t)$ as the 'dependent variable.' This is the principle of conditional independence for parallel and interdependent processes.

The same idea can be developed more formally. Beginning with a transition rate model for the joint process, $X(t) = (X_1(t), X_2(t))$, and assuming the principle of conditional independence, the likelihood for this model can be factorized into a product of the likelihoods for two separate models: a transition rate model for $X_1(t)$ which is dependent on $X_2(t)$ as a time-dependent covariate, and a transition rate model for $X_2(t)$ which is dependent on $X_1(t)$ as a time-dependent covariate. The effects of time-dependent (qualitative and metric) processes on the transition rate can be easily estimated by applying the method of episode splitting (Blossfeld and Rohwer, 2002). For example, with this method, the original episodes (or durations) of the dependent process $X_1(t)$ are split into two subepisodes at the time, when the time-dependent covariate of $X_2(t)$ changes its value. The first subepisode ending at the time of the split is censored (i.e. the indicator variable gets the value 0). The second subepisode starting at the time of the split gets the indicator variable of the original episode.

This result has important implications for the modeling of event histories. From a technical point of view there is no need to distinguish between defined, ancillary, and internal covariates because all of these time-dependent covariate types can be treated equally in the estimation procedure. However, a distinction between defined and ancillary covariates on the one hand and internal covariates on the other makes sense from a theoretical perspective, because only in the case of internal covariates does it make sense to examine whether parallel processes are independent, whether one of the parallel processes is endogenous and the other ones are exogenous, or whether parallel processes form an interdependent system (i.e. they are all endogenous).

The principle of conditional independence implies that the prediction errors (or residuals) of the correlated processes $X_1(t)$ and $X_2(t)$ are uncorrelated, given the history of each process up to t and the covariates. In practice, however, there may be time-invariant unmeasured characteristics that affect both $X_1(t)$ and $X_2(t)$, leading to a residual correlation between both processes. In that case, we say that the two processes are jointly determined by some unmeasured influences. Suppose, for example, that we are interested in studying the relationships between employment transitions and fertility among women. We might expect that a woman's chance of making an employment transition at t would depend on her childbearing history up to t (e.g. the presence and age of children), and that her decision on whether to have a(nother) child at t would depend on her employment history up to t. There may be unobserved individual characteristics, fixed over time, that affect the chances of both an employment and a fertility transition at t. For example, more 'career-minded' women may delay childbearing and have fewer children than less 'career-minded' women. In the absence of suitable measures of 'career-mindedness,' this variable would be absorbed into the residual terms of both processes, leading to a cross-process residual correlation. If the residual correlation cannot be explained by time-dependent and time-invariant covariates, the two processes should be modeled simultaneously, and multiprocess models (Lillard and Waite, 1993) have been developed for this purpose.

Unobserved heterogeneity

Unfortunately, we are not always able to include all important factors in the event history analysis. One reason is the limitation of available data. We would like to include some important variables, but we simply do not have the information. Furthermore, we often do not know what is important. So what are the consequences of this situation? Basically, there are two aspects to be taken into consideration. The first one is well known from 'causation as robust dependence'. Because our covariates are often correlated, the parameter estimates depend on the specific set of covariates included in the model. Every change in this set is likely to change the parameter estimates of the variables already included in previous models. Thus, as in the 'causation as robust dependence' approach, the only way to proceed is to estimate a series of models with different specifications and then to check whether the estimation results are stable or not. Since our theoretical models are normally weak, this procedure can provide additional insights into what may be called context sensitivity of causal effects in the social world.

Second, changing the set of covariates in a transition rate model will very often also lead to changes in the time-dependent shape of the transition rate. A similar effect occurs in traditional regression models: depending on the set of covariates, the empirical distribution of the residuals changes. But, as opposed to regression models, where the residuals are normally only used for checking model assumptions, in transition rate models the residuals become the focus of modeling. In fact, if transition rate models are reformulated as regression models, the transition rate becomes a description of the residuals, and any change in the distribution of the residuals becomes a change in the time-dependent shape of the transition rate (see Blossfeld et al., 2007). Consequently, the empirical insight that a transition rate model provides for the time-dependent shape of the transition rate more or less depends on the set of covariates used to estimate the model.

The transition rate that is estimated for a population can be the result (a mixture) of quite different transition rates in the subpopulations. What are the consequences? First, this result means that one can 'explain' an observed transition rate at the population level as the result of different transition rates in subpopulations. Of course, this will only be a sensible strategy if we are able to identify important subpopulations. To follow this strategy one obviously needs observable characteristics to partition a population into subpopulations. Although there might be unobserved heterogeneity (and we can usually be sure that we were not able to include all important variables), just making more or less arbitrary distributional assumptions about unobserved heterogeneity will not lead to better models. On the contrary, the estimation results will be more dependent on assumptions than would be the case otherwise (Lieberson, 1985). Therefore, we would like to stress our view that the most important basis for any progress in model building is better theoretical models as well as sufficient and appropriate data.

There remains the problem of how to interpret a time-dependent transition rate from a causal view. The question is: can time be considered as a proxy for an unmeasured variable producing a time-dependent rate, or is it simply an expression of unobserved heterogeneity, which does not allow for any substantive interpretation? There have been several proposals to deal with unobserved heterogeneity in continuous-time transition rate models, which cannot be developed here in detail (see Blossfeld et al., 2007; Tuma and Hannan, 1984). Furthermore, models with unobserved heterogeneity in the discrete-time event history have been developed (Yamaguchi, 1986; Allison, 1996; Steele, 2003; Zhang and Steele, 2004). In our example analysis below, we will use a model that was suggested by Steele (2003):

$$r^*(t) = \frac{\exp(\alpha_1 x_1 + \alpha_2 x_2 + \ldots + \alpha_n x_n)\exp(\beta(t))\exp(u_\nu)}{1 + \exp(\alpha_1 x_1 + \alpha_2 x_2 + \ldots + \alpha_n x_n)\exp(\beta(t))\exp(u_\nu)} \quad (16.18)$$

with $u_\nu \sim N(0, \sigma_\nu^2)$ representing unobserved factors across $\nu = 1, \ldots, n$ individuals. σ_ν^2 is then a measure of unobserved heterogeneity or frailty. Thus, we can test with a likelihood ratio test whether the standard deviation of unobservables σ_ν is statistically significant (for further discussion, see Steele, 2003).

All these models for unobserved heterogeneity broadly enrich the spectrum of models and can be quite helpful in separating robust estimation results (i.e. estimation results that are to a large degree independent of a specific model specification) and 'spurious' results, which might be defined by the fact that they heavily depend on a specific type of model.

EXAMPLE ANALYSIS

Data

In order to demonstrate the strength of the 'causation as generative process' approach with event history models in a substantive context, we use data from the National Educational Panel Study.[1] The NEPS is a project collecting longitudinal data in a multicohort sequence design in Germany. One of the six NEPS cohorts is a representative sample of adults aged 23–65. For this cohort the NEPS offers retrospective life histories for 11,649 respondents born between 1944 and 1986. The NEPS data collection took place from November 2009 until June 2010. The example analysis in this chapter uses a subsample of this data set including 600 women born in West Germany. The example data set `rrdat1_1.dta` and the Stata do-file `event_history_analysis.do` can be downloaded from the online supplement of this edited volume. For a detailed description of the NEPS project and the design of the NEPS study we refer the reader to Blossfeld et al. (2011).

Causal modeling strategies and dependent variable

In our example, we analyze entry into first motherhood (as the dependent process) explained by various other parallel and interdependent processes in the life courses of women (such as educational careers and job trajectories). The event of a child birth to a woman in her fertile years could happen at any point in time; however, the NEPS only collects dates of transitions and events on a monthly basis. For our analysis, we define for each woman a spell starting at age 16. This spell ends when the woman gives birth to a first child. However, since it is not the date of birth that is theoretically important but the conditions under which women get pregnant, we use in our analysis the time of conception leading to the first birth as the event terminating a woman's spell. Based on the NEPS data, the time of conception is defined as date of child birth minus 9 months. For women who do not give birth to a first child, we censor the spells at age 45. In other words, right censoring takes place when a woman has not given birth to her first child by age 45 or when the interview has taken place before this age. This leads to a person-oriented spell data set, where each woman in the sample has exactly one record of data, including a series of time-constant covariates and information when time-varying covariates change their values. We have then transformed this person-level data set into a person-period data set using the method of episode splitting (see Blossfeld et al., 2007, p. 137), in which each woman has multiple records – one for each month. The idea of this method can be described as follows. At the end of each month, the original episode is split into subepisodes of one month length. For each of these monthly subepisodes a new record is created containing:

1 Information about the origin state of the original episode.
2 The values of all covariates at the beginning of the subepisodes. In other words, such a person-period data set allows an easy integration of time-varying covariates since the covariates can change their values across each of the person-month records.
3 The starting and ending times of the subepisodes.
4 Information indicating whether the subepisode ends with the destination state of the original episode or is censored. All the monthly subepisodes, apart from the last one, are regarded as right-censored. Only the last subepisode is given the same destination state as the original episode.

Splitting the original episode into a series of $l = 1, \ldots, L$ monthly subepisodes does not change the likelihood. Obviously, only the survivor function $G(t)$ is influenced by the process of episode splitting. However, $G(t)$ can be written as a product of the conditional survivor functions for each split of the episode (Blossfeld et al., 2007, p. 137):

$$G(t) = \prod_{l=1}^{L} G(t_l | s_l) . \tag{16.19}$$

It is the responsibility of the user to do any episode splitting in such a way that the splits add up to a sample of meaningful episodes.

Because we are studying the occurrence of a single non-repeatable event (pregnancy leading to a first birth), for each woman i, chronology of event occurrence can be conveniently recorded using a sequence of dummy variables Y_{ij} whose values y_{ij} are defined as (Singer and Willett, 1993, p. 168):

$$y_{ij} = \begin{cases} 0 & \text{if a woman } i \text{ does not conceive in month } j \text{ (event leading to a first birth),} \\ 1 & \text{if a woman } i \text{ does conceive in month } j \text{ (event leading to a first birth).} \end{cases} \tag{16.20}$$

y_{ij} is the dependent variable in discrete-time event history models and the variable containing the censoring information in continuous-time event history models. We estimate both continuous-time and discrete-time models. To control for unobserved heterogeneity, we eventually run a continuous-time model with a stochastic error term and a discrete-time model with a term for unobserved heterogeneity as described above.

Causal modeling, substantive hypotheses and definition of explanatory variables

There seems to be a consensus that causal inferences cannot simply be made from empirical data. Causal statements are based preliminarily on substantive hypotheses that the researcher develops about the social world. In this sense, causal inference is always theoretically driven, specifying a particular mechanism of how a causal event produces an effect event. The following substantive mechanisms and explanatory variables are included in our example analysis:

Age dependency (time-dependent covariate). In the literature, the rate of entry into first birth is considered to have a non-monotonic age pattern in modern societies (Blossfeld, 1995). As women's age increases, the rate of entry into first motherhood initially rises, reaches a peak, and then decreases. This bell-shaped baseline hazard rate can be explained by two mechanisms: the readiness to enter into motherhood, which is influenced by participation in education and the job market; and the dynamics of the marriage market, influencing the probability of meeting a single partner in a specific age interval. For example, younger people get ready for motherhood step by step when they leave the education system and enter the job market. At this age, there are also plenty of unmarried singles available on the marriage market in the relevant age range. Then, some of these young men and women marry, and with increasing age it is more and more difficult to meet an unmarried individual of the opposite sex in the relevant age band. Thus, the marriage rate at first increases with age, reaches a peak, and then decreases. Given the importance of this pattern over the life course for entry into first motherhood, it seems reasonable to include age dependency in our models of first pregnancy. To model the non-monotonic shape of entry into first birth across age, we include two variables, log(current age – 15.9) and log(45.1 – current age), which allow us to flexibly test different baseline shapes, also including the bell-shaped pattern across the life course (Blossfeld and Huinink, 1991). If there is a bell-shaped pattern, we expect the β-coefficients of these two variables to be positive in our event history analysis, but the first one to be smaller than the second one.

Women's enrollment in education (time-dependent covariate). In many studies, sociologists expect the effect of women's enrollment in the education system to be negative (e.g. Blossfeld, 2011). The reason is that enrollment in education is connected with financial dependence on the family or the welfare state. Another reason is that educational activities tend to be incompatible with adult family roles such as child care. This means that participation in education should be in conflict with entry into first motherhood. In our analysis, a time-dependent dummy variable is therefore included, with the value one if a woman is enrolled in education and zero otherwise. We expect that enrollment in school reduces the rate of entry into first motherhood.

Women's investments in education (time-dependent covariate). In the economic literature, differences in fertility behavior of women have often been attributed to women's investments in education. For example, according to Becker (1981), higher investments in education increase the productivity of individuals and therefore lead to higher incomes and wages. Because women are still mainly responsible for child care in West Germany, there is a conflict between spending time in the labor market (and earning money) and child care as well as child rearing. Better-qualified women have higher opportunity costs since their time is more valuable. Women's educational attainment level should therefore have a negative effect on entry into first motherhood, according to the economic theory of the family. Since Becker assumes that women can turn their investments in education into employment opportunities, this effect also means that there is a permanent reduction of qualified women's fertility over the life course. We model women's educational investments as a time-dependent covariate: whenever a woman attains a higher level of education, the educational attainment level will be adjusted. We distinguish nine educational attainment levels and express each degree as the number of years necessary to achieve it (see Blossfeld et al., 2007).

Women's socioeconomic status (time-dependent covariate). If women interrupt employment and take care of children, women's potential income is the opportunity cost according to the economic theory of the family (Becker, 1981). In other words, women's opportunity costs increase with income. Unfortunately, in our NEPS data set, individual income histories are not available. We therefore take women's prestige (International Socio-Economic Index, ISEI) as a proxy measure of how good their job is and how much they can earn. If women are not employed, we allocate the value zero. We expect a negative coefficient for women's socioeconomic status (ISEI) on their rate of entry into first motherhood.

Fertility pressure (time-dependent covariate). Women who are enrolled in school longer and attain higher qualifications are subject to increasing pressure not only from potential medical problems connected with having children later, but also from violating societal age norms ('women should have their first child at the latest by age 30'; see Yamaguchi, 1991). We model this increasing pressure by including a time-dependent covariate which increases linearly from age 16 onwards. We expect that this fertility pressure variable should increase the rate of entry into first motherhood when we have controlled for enrollment in education.

Father's educational attainment (time-constant covariate). In the sociological literature, it is well known that socioeconomic background has a strong effect on children's education. Based on different enrollments in education there is an indirect effect of family of origin on women's age at entry into first motherhood. Women from higher social strata participate longer in the education system, so they are ready at a later age to become mothers. We are using father's highest educational attainment level as social background information. To model father's highest educational attainment, we distinguish seven education levels. We then attach the number of years that are necessary to achieve these levels (see Blossfeld et al., 2007). We expect that father's educational attainment has a negative impact on women's rate of entry into first motherhood as long as women's enrollment in education is not included in the model. If women's enrollment in education is included, father's educational attainment should become insignificant.

Marital status (time-dependent covariate). Fertility and nuptiality are closely connected. In West Germany in particular, marriage is still an important normative setting for having a child. The rate of extramarital birth is therefore still quite low in West Germany compared to other modern societies. We include a time-dependent covariate indicating whether or not a woman is married in the respective month in our model. We expect that this covariate has a strongly positive effect on the rate of entry into first motherhood.

Historical period and unemployment rate (time-dependent covariates). The literature shows that variations in age at entry into first motherhood are also dependent on historical period and unemployment rates (Blossfeld and Huinink, 1991). In Germany this might be especially the case before and after German unification. To be able to take the most important political and economic historical phases into account, we have included period dummy variables distinguishing three historical periods in our analysis: the period before 1990 (reference), the period between 1990 and 2003, and the period after 2003. The start of the third period in the year 2003 was suggested by indications in the *Demography Report 2010* (European Commission, 2011) that fertility rates all over Europe have been declining since that year.

In addition, unemployment rates have changed drastically in Germany over the period of observation. These macro changes influence the economic and social conditions in which young women make their fertility decisions. Using annual unemployment rates from 1960 until the date of the interview for West Germany, we analyze whether macro structural effects such as increasing uncertainty in the course of rising unemployment have an effect on women's fertility behavior. We expect the uncertainty after German unification and increasing unemployment rates each to have a negative effect on the rate of entry into first motherhood.

Interpretation of results

Stata provides maximum likelihood estimates for the event history model coefficients and their standard errors. The standard errors are useful in order to assess the precision of the model parameters. In particular, one can check whether the estimated coefficients are significantly different from zero. In Tables 16.1–16.3 we attach * to the coefficient if it is significant at the 0.05 level, ** if it is significant at the 0.01 level and *** if it is significant at the 0.001 level.

If the coefficient is significant and has a negative sign, then the rate of change in the dependent process is reduced if the values of the time-dependent covariate increase. If the coefficient is significant and has a positive sign, then the rate of change in the dependent process is increasing if the value of the time-dependent covariate increases. Unfortunately, the current state of sociological knowledge only allows us to interpret the significance and direction of the influence of a covariate on the dependent process and not the concrete numeric values of the coefficients. We therefore do not compute and interpret the relative change of the rate if the covariate changes its value (see Blossfeld et al., 2007, p. 99). Instead, in the following causal inferences we focus only on the significance and direction of the estimated coefficients.

Stata also provides values of the log-likelihood function. Thus, one can compare each model with the exponential model without covariates using a likelihood ratio test. Under the null hypothesis that the additionally included covariates do not significantly improve the model fit, the likelihood ratio test statistic (LR) follows approximately a χ^2 distribution with m degrees of freedom, where m is the number of additionally included covariates (see Tables 16.1 and 16.2). These test statistics can be calculated as twice the difference of the log-likelihoods:

$$LR = 2(\,\text{LogLik}(\text{present model}) - \text{LogLik}(\text{reference model})\,)\,. \tag{16.21}$$

To estimate the rate of entry into first motherhood as the dependent variable, we use the person-period file (with time-constant and time-varying covariates) `rrdat1_1.dta` as described above. The Stata do-file `event_history_analysis.do`, which can be downloaded from the online supplement of this volume, demonstrates how to estimate a discrete-time event history model, a continuous-time event history model, a discrete model with unobserved heterogeneity, and a continuous model with a stochastic error term ν, which is gamma distributed.

We first interpret the discrete-time causal analysis with the logit model. In order to check whether women's age at entry into first motherhood initially rises, reaches a peak, and then decreases, we include in Model 1 of Table 16.1 two variables, log(current age − 15.9) and log(45.1 − current age). Both β-coefficients are positive and significant at the 0.001 level, which means that there is indeed a strong non-monotonic pattern of the observed fertility rate across age. Because the β-coefficient for log(45.1-current age) is greater than that for log(current age -15.9), the bell-shaped curve of entry into first motherhood is right-skewed across age.

In a second step (Model 2 in Table 16.1) we include women's social origin as an explanatory variable. As a measure of family background we use father's educational attainment level. We expect women from families with higher educated fathers to value education more and therefore to participate longer in the education system. Extended enrollment in and delayed entry into the labor market lead to a later readiness for motherhood. The estimates of Model 2 show a significant negative β-coefficient for father's education. In other words, women of higher social origins enter first motherhood much later than women of lower social origins.

In Model 3, we include the time-dependent covariate of women's enrollment in education. As expected, there is a strong and significant negative effect of enrollment in education on women's fertility behavior. This means that it is very difficult for women to balance enrollment in education and motherhood, for example because of time constraints related to child care and financial dependence on family of origin or the welfare state. The effect of social background is not significant. In other words, women's fertility is not directly influenced by social background, but by enrollment in education. This supports the importance of education for entry into first motherhood.

In Model 4, level of women's educational attainment is included. According to the economic theory of the family, women's educational attainment level should have a strong negative effect

Table 16.1 Discrete-time event history models for the conception of a first child

Variable	Model								
	1	2	3	4	5	6a	6b	7	8
Age dependency									
log(current age–15.9)	1.994***	2.004***	1.681***	1.702***	1.741***	0.484			
log(45.1–current age)	6.988***	7.022***	6.694***	6.699***	6.892***	32.835***	41.212***	31.302***	31.367***
Social background									
Father's educational attainment		–0.044*	–0.019	–0.015	–0.015	–0.015	–0.014	0.013	0.013
Education									
Enrollment in education			–0.884***	–0.876***	–1.078***	–1.078***	–1.105***	–0.806***	–0.805***
Women's educational attainment				–0.009	–0.005	–0.009	–0.008	0.012	0.012
Employment									
Socioeconomic status (ISEI)					–0.005*	–0.005*	–0.005*	–0.007**	–0.007**
Fertility pressure						0.810***	1.079***	0.781***	0.783***
Marital status									
Unmarried (ref.)									
Married								1.618***	1.619***
Historical periods									
Period before 1990 (ref.)									
Period 1990–2003									0.054
Period after 2003									0.023
Macro level insecurity									
Unemployment rate									–0.006
Constant	–35.320***	–34.945***	–33.133***	–33.122***	–33.775***	–148.190***	–185.182***	–142.034***	–142.297***
Number of events	405	405	405	405	405	405	405	405	405
Number of subepisodes	93,076	93,076	93,076	93,076	93,076	93,076	93,076	93,076	93,076
LR (χ^2)	197.59	202.68	248.02	248.43	253.04	271.05	268.21	502.85	503.06
Degrees of freedom	2	3	4	5	6	7	6	7	10
Log-likelihood									–2644.602

Note: $^{*}p < 0.05$; $^{**}p < 0.01$; $^{***}p > 0.001$

Source: Estimations based on NEPS data from the adult study

Table 16.2 Continuous-time event history models for the conception of a first child

Variable	Model								
	1	2	3	4	5	6a	6b	7	8
Age dependency									
log(current age–15.9)	1.986***	1.996***	1.674***	1.694***	1.733***	0.484			
log(45.1–current age)	6.960***	6.994***	6.665***	6.669***	6.860***	32.627***	41.006***	31.085***	31.151***
Social background									
Father's educational attainment		–0.043*	–0.018	–0.015	–0.015	–0.014	–0.014	0.013	0.013
Education									
Enrollment in education			–0.880***	–0.872***	–1.072***	–1.072***	–1.100***	–0.800***	–0.799***
Women's educational attainment				–0.009	–0.005	–0.009	–0.008	0.012	0.012
Employment									
Socioeconomic status (ISEI)					–0.005*	–0.005*	–0.005*	–0.007**	–0.007**
Fertility pressure						0.805***	1.073***	0.776***	0.778***
Marital status									
Unmarried (ref.)									
Married								1.603***	1.604***
Historical periods									
Period before 1990 (ref.)									
Period 1990–2003									0.053
Period after 2003									0.023
Macro level insecurity									
Unemployment rate									–0.006
Constant	–35.204***	–34.831***	–33.019***	–33.007***	–33.653***	–147.287***	–184.287***	–141.098***	–141.360***
Number of events	459	459	459	459	459	459	459	459	459
Number of subepisodes	93,076	93,076	93,076	93,076	93,076	93,076	93,076	93,076	93,076
LR (χ^2)	196.80	201.86	246.94	247.35	251.93	269.80	266.93	499.37	499.57
Degrees of freedom	2	3	4	5	6	7	6	7	10
Log-likelihood									–2647.48

Note: $^{*}p < 0.05$; $^{**}p < 0.01$; $^{***}p > 0.001$
Source: Estimations based on NEPS data from the adult study

on entry into first motherhood. However, the β-coefficient of women's educational attainment has no additional significant effect. Women's education only affects entry into first motherhood via educational participation, not via educational attainment level, since the variable for the educational attainment level is not significant. This means that higher qualified women only shift their fertility decisions to higher ages. So, there is only a temporary conflict between education and fertility over the life course.

Model 5 introduces women's socioeconomic status as a time-varying covariate and estimates the effect on entry into first motherhood. The economic theory of the family expects that women's opportunity costs increase with increasing income and status. Our results in Model 5 of Table 16.1 show that socioeconomic status has a significantly negative effect in this model. This effect supports the economic theory of the family which expects higher status jobs to have higher opportunity costs for women. So, these women have indeed a lower rate of entry into first motherhood.

In Models 6a and 6b of Table 16.1, fertility pressure is included as a linearly increasing measure over the life course, starting at age 16. Women who are enrolled in school longer and attain higher qualifications delay entry into first motherhood and are therefore subject to increasing pressure not only from potential medical problems connected with having children later, but also from violating societal age norms. Our estimates in Model 6a show indeed a strong effect of fertility pressure. In other words, the more women delay first motherhood because of enrollment in education, the stronger is the effect of fertility pressure. So, after completing school, higher qualified women have their first babies more quickly. However, fertility pressure in Model 6a is highly correlated with one of our age variables. Therefore, we run Model 6b and leave out the variable log (current age – 15.9). The coefficients of the covariates in Models 6a and 6b are very similar and the causal inferences are basically the same. To be on the safe side, we therefore estimate the remaining models in Tables 16.1–16.3 without the variable log (current age – 15.9).

Fertility and nuptiality are closely connected in West Germany. West German women still try to have their babies in a marital setting. We therefore include women's marital status as a time-varying covariate in Model 7 (see Table 16.1). As expected, being married has a strong positive and significant influence on entry into first motherhood.

Finally, in Model 8 of Table 16.1, we include macro-structural influences of historical periods and changing unemployment rates in estimating the rate of entry into first motherhood. None of the included variables is significant. In other words, the fertility behavior of women in West Germany is not influenced by historical periods (before and after unification) or changes in unemployment rates in this historical period.

The coefficients of Table 16.1 are hard to interpret in combination because the model includes many diverse and sometimes contradictory influences of time-varying covariates. It therefore represents a complex web of mechanisms and influences over the life course. In this situation it is helpful to use the β-coefficients of Table 16.1 and to compute the survivor function for certain constellations of the covariates. Figure 16.2 illustrates survival functions for women using estimates for Model 4 in Table 16.1. It shows the proportion of childless women for each age. The survivor functions show that women who leave the education system earlier (with lower secondary school qualifications or middle school qualifications with vocational training) have a very similar pattern of childlessness. Half of these women have already entered motherhood by the age of 25. In contrast, women with university degree at first tend to delay the birth of a first child until they leave the education system. However, these women then quickly catch up with the lower qualified women so that the difference in childlessness between the various educational groups almost disappears at the age of 45. So, education has mainly an effect on the time structure of entry into first motherhood over the life course of differently qualified women

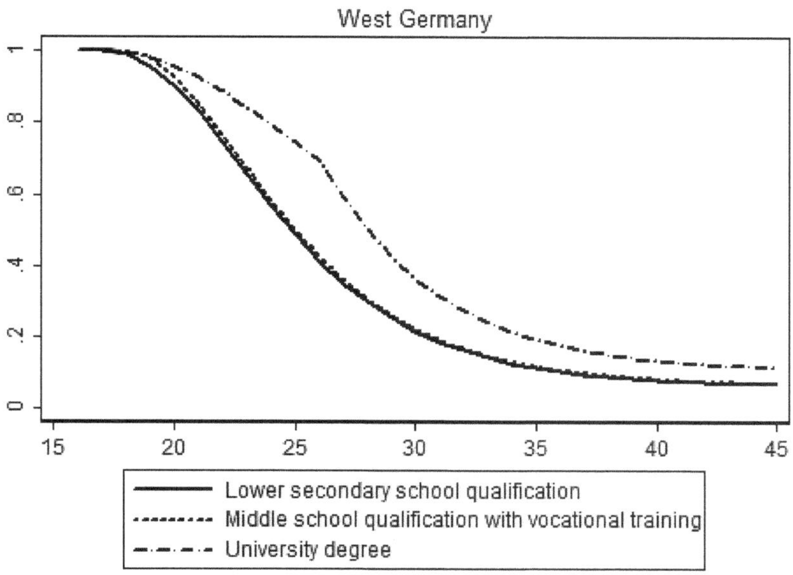

Figure 16.2 Survivor functions based on Model 4 of Table 16.1

but not so much on their final childlessness. This is of course only true if higher qualified women do not turn their better investments in education into better jobs (see Figure 16.3).

Figure 16.3 illustrates the survival function for the estimation of Model 6 in Table 16.1. The graph shows that women who can turn their higher educational attainment into better jobs are indeed less likely to enter into first motherhood than women of lower socioeconomic status. For example, in Figure 16.3 it can be seen that women who work as medical doctors have a higher childlessness at the age of 45 than women who work as chefs or office workers. Of course, this finding supports the economic theory of the family.

Since the discrete-time rate $r^*(t) = \Pr(t \leq T < t'|T \geq t)$ is an approximation of the continuous-time rate $r(t) = \lim \Pr(t \leq T < t'|T \geq t)/(t' - t)$ for small time intervals $(t' - t)$, that is, $\Pr(t \leq T < t'|T \geq t) \approx (t' - t)\, r(t)$, we can expect in our example very similar results for the continuous-time event history models in Table 16.2 (see Blossfeld et al., 2007, p. 37). Thus, using the same person-period file `rrdat1_1.dta`, we estimate all discrete-time models of Table 16.1 with continuous-time event history models (Table 16.2).[2] A comparison of Tables 16.1 and 16.2 demonstrates that the estimates are indeed very similar.

Finally, in Table 16.3 we add an unobserved heterogeneity term to the discrete-time Model 8 of Table 16.1 and the continuous-time Model 8 of Table 16.2. If we compare the various β-coefficients of Models 8 in Tables 16.1 and 16.2 with the β-coefficients of Table 16.3, it is obvious that the causal inferences reached in all four models are the same. Thus, estimations based on different model assumptions do not alter our substantive interpretations.

SUMMARY AND SOME PITFALLS

The aim of this chapter is to introduce the reader to the concepts of discrete and continuous event history analysis and to demonstrate with a practical example the estimation, interpretation and possible complications of event history models. We have particularly emphasized that the

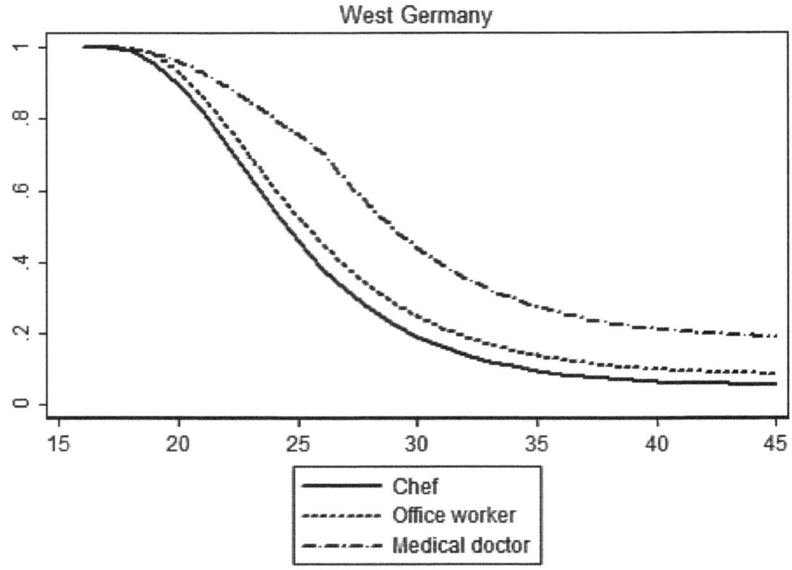

Figure 16.3 Survivor functions based on Model 6 of Table 16.1

application of event history models is closely related to a new understanding of causality, namely 'causation as generative process'. In substantive terms, the investigations of our example showed the existence of highly time-dependent causal processes. Events in women's educational and job careers have a clear time-dependent impact on entry into first-time motherhood.

One shortcoming of this approach is that our application example is only based on observed behavior. We do not have time-related information on individual decision processes. For example, we have analyzed the relationship between exiting education and entry into first motherhood. Yet the time order between decisions and observed behavior might be exactly the other way around. A woman could decide to have a baby and then leave education and become pregnant. Courgeau and Lelièvre (1992) have introduced the notion of 'fuzzy time' to represent this time span between decisions and behavior. Since the time between decisions and behavior is probably not random and differs across couples, examining observed behavior could lead to false causal inferences. This does not alter the key temporal issues embedded within the causal logic. However, we must admit that using the time order of only behavioral events without taking into account the timing of decisions could lead to serious misspecification. Thus, for studies aiming to model 'causation as a generative process' a combination of prospective panel observations of individuals' objectives and decisions and retrospective information on their behavioral events appears to be a very desirable design for causal inference.

FURTHER READING

We recommend the following books for further reading. Allison (1984) is a classic and introduces the reader to discrete-time event history analysis. Blossfeld et al. (2007) provide an introductory account of event history modeling techniques using the statistical package Stata. The specific emphasis is on the usefulness of continuous-time event history models for causal analysis in the social sciences. Courgeau and Lelièvre (1992) gives an informative introduction to event history analysis with many demographic applications. Lillard and Waite (1993) introduces the

Table 16.3 Discrete-time and continuous-time event history analyses with unobserved heterogeneity for the conception of a first child

Variable	Discrete		Continuous	
	γ	SE	γ	SE
Age dependency				
log(45.1–current age)	36.507***	5.386	31.150***	4.352
Social background				
Father's educational attainment	–0.024	0.035	0.013	0.021
Education				
Enrollment in education	–0.847***	0.201	–0.799***	0.173
Women's educational attainment	–0.005	0.023	0.012	0.014
Employment				
Socioeconomic status (ISEI)	–0.009**	0.003	–0.007**	0.002
Fertility pressure	1.031***	0.158	0.778***	0.115
Marital status				
Unmarried (ref.)				
Married	2.246***	0.180	1.604***	0.112
Historical periods				
Period before 1990 (ref.)				
Period 1990–2003	–0.102	0.177	0.053	0.128
Period after 2003	–0.128	0.293	0.023	0.220
Macro level insecurity				
Unemployment rate	–0.041	0.028	–0.006	0.020
Constant	–166.986***	23.822	–141.357***	19.025
ln (σ_v^2)	0.666*	0.334		
σ_v	1.395	0.233	–221.789	
ρ	0.372	0.078		
Log-likelihood	–2624.35		–2647.478	

Note: *$p < 0.05$; **$p < 0.01$; ***$p > 0.001$; $n = 405$
Source: Estimations based on NEPS data from the adult study

reader to the simultaneous modeling of parallel and interdependent processes. Vermunt (1997) embeds event history analysis into log-linear models. Yamaguchi (1991) introduces the reader to discrete-time event history analysis.

NOTES

1 NEPS Starting Cohort 6 – Adults (Adult Education and Lifelong Learning), doi:10.5157/NEPS:SC6:1.0.0. The NEPS data collection is part of the Framework Program for the Promotion of Empirical Educational Research, funded by the German Federal Ministry of Education and Research and supported by the federal states.

2 In Stata this is achieved with `streg` which then estimates an exponential model (see the Stata commands in do-file `event_history_analysis.do` to be found in the online supplement of this volume).

REFERENCES

Allison, P. D. (1984). *Event History Analysis. Regression for Longitudinal Event Data*. Beverly Hills, CA: Sage.
Allison, P. D. (1996). Fixed-effects partial likelihood for repeated events. *Sociological Methods & Research*, 25, 207–222.
Becker, G. (1981). *A Treatise on the Family*. Cambridge, MA: Harvard University Press.
Blalock, H. M. (1970). *Causal Methods in the Social Sciences*. Chicago: Aldine.

Blossfeld, G. J. (2011). *Die Vereinbarkeit von Ausbildung, Familie und Beruf bei Frauen. Langfristige Trend und neueste Entwicklungen in Ost- und Westdeutschland*. Leverkusen-Opladen: Budrich UniPress.

Blossfeld, H.-P. (1986). Career opportunities in the Federal Republic of Germany: a dynamic approach to the study of life-course, cohort, and period effects. *European Sociological Review*, 2, 208–225.

Blossfeld, H.-P. (1995). *The New Role of Women. Family Formation in Modern Societies*. Oxford: Westview.

Blossfeld, H.-P. and Drobnic, S. (eds) (2001). *Careers of Couples in Contemporary Societies*. Oxford: Oxford University Press.

Blossfeld, H.-P., Golsch, K. and Rohwer, G. (2007). *Event History Analysis with Stata*. Mahwah, NJ: Erlbaum.

Blossfeld, H.-P., Hamerle, A. and Mayer, K. U. (1989). *Event History Analysis*. Hillsdale, NJ: Erlbaum.

Blossfeld, H.-P. and Huinink, J. (1991). Human capital investments or norms of role transition? How women's schooling and career affect the process of family formation. *American Journal of Sociology*, 97, 143–168.

Blossfeld, H.-P. and Rohwer, G. (2002). *Techniques of Event History Modeling*. Mahwah, NJ: Erlbaum.

Blossfeld, H.-P., Roßbach, H.-G. and von Maurice, J. (2011). *Education as a Lifelong Process*. Wiesbaden: VS Verlag.

Coale, A. (1971). Age patterns of marriage. *Population Studies*, 25, 193–214.

Coleman, J. S. (1981). *Longitudinal Data Analysis*. New York: Basic Books.

Courgeau, D. and Lelièvre, E. (1992). *Event History Analysis in Demography*. Oxford: Clarendon Press.

Cox, D. R. (1972). Partial likelihood. *Biometrika*, 62, 269–276.

Cox, D. R. (1990). Role of models in statistical analysis. *Statistical Science*, 5, 169–174.

Cox, D. R. (1992). Causality: some statistical aspects. *Journal of the Royal Statistical Association, Series A*, 155, 291–301.

Duncan, O. D. (1966). Path analysis: sociological examples. *American Journal of Sociology*, 72, 1–16.

Eells, E. (1991). *Probabilistic Causality*. Cambridge: Cambridge University Press.

Elder, G. H. (1987). War mobilization and the life course. A cohort of World War II veterans. *Sociological Forum*, 2, 449–472.

European Commission (2011). *Demography Report 2010: Older, more numerous and diverse Europeans*. Technical report, Publications Office of the European Union, Luxembourg.

Goldthorpe, J. (2001). Causation, statistics, and sociology. *European Sociological Review*, 17(1), 1–20.

Holland, P. W. (1986). Statistics and causal inference. *Journal of the American Statistical Association*, 81(396), 945–960.

Holland, P. W. (1988). Causal inference, path analysis, and recursive structural equation models. *Sociological Methodology*, 18, 449–484.

Kalbfleisch, J. D. and Prentice, R. L. (1980). *The Statistical Analysis of Failure Data*. New York: Wiley.

Kelly, J. R. and McGrath, J. E. (1988). *On Time and Method*. Newbury Park, CA: Sage.

Kerlinger, F. N. and Pedhazur, E. (1973). *Multiple Regression in Behavioral Sciences*. New York: Holt, Rinehart and Winston.

Lieberson, S. (1985). *Making It Count. The Improvement of Social Research and Theory*. Berkeley: University of California Press.

Lillard, L. A. and Waite, L. J. (1993). A joint model of marital childbearing and marital disruption. *Demography*, 30, 653–681.

Marini, M. M. and Singer, B. (1988). Causality in the social sciences. *Sociological Methodology*, 347–409.

Mayer, K. U. and Huinink, J. (1990). Age, period, and cohort in the study of the life course: a comparison of classical A-P-C-analysis with event history analysis or farewell to Lexis? In D. Magnusson and L. R. Bergmann (eds), *Data Quality in Longitudinal Research* (pp. 211–232). Cambridge: Cambridge University Press.

Pötter, U. and Blossfeld, H.-P. (2001). Causal inferences from series of events. *European Sociological Review*, 17(1), 21–32.

Rubin, D. B. (1974). Estimating causal effects of treaments in randomized and nonrandomized studies. *Journal of Educational Psychology*, 66, 688–701.

Rubin, D. B. (1978). Bayesian inference for causal effects: the role of randomization. *Annals of Statistics*, 6, 34–58.

Rubin, D. B. (1980). Discussion of 'Randomization analysis of experimental data in the Fisher randomization test' by Basu. *Journal of the American Statistical Association*, 81, 961–962.

Schneider, B., Carnoy, M., Kilpatrick, J., Schmidt, W. H. and Shavelson, R. J. (2007). *Estimating Causal Effects Using Experimental and Observational Designs*. Washington, DC: American Educational Research Association.

Shadish, W. R., Cook, T. D. and Campbell, D. T. (2002). *Experimental and Quasi-experimental Designs for Generalized Causal Inference*. Boston: Houghton Mifflin.

Singer, J. D. and Willett, J. B. (1993). It's about time. Discrete-time survival analysis and the timing of events. *Journal of Educational Statistics*, 18, 155–195.

Steele, F. (2003). A multilevel mixture model for event history data with long-term survivors: An application to an analysis of contraceptive sterilisation in Bangladesh. *Lifetime Data Analysis*, 9, 155–174.
Tuma, N. B. and Hannan, M. T. (1984). *Social Dynamics: Models and Methods*. Orlando, FL: Academic Press.
Vermunt, J. K. (1997). *Log-Linear Models for Event Histories*. Newbury Park, CA: Sage.
Willekens, F. J. (1991). Understanding the interdependence between parallel careers. In J. J. Siegers (ed.), *Female Labour Market Behaviour and Fertility* (pp. 11–31). Berlin: Springer-Verlag.
Yamaguchi, K. (1986). Alternative approaches to unobserved heterogeneity in the analysis of repeatable events. *Sociological Methodology*, 16, 213–249.
Yamaguchi, K. (1991). *Event History Analysis*. Newbury Park, CA: Sage.
Zhang, W. and Steele, F. (2004). A semiparametric multilevel survival model. *Applied Statistics*, 53, 387–404.

17

Time-series cross-section

Jessica Fortin-Rittberger

INTRODUCTION

Social scientists have long investigated relationships between institutional, economic and social variables either by using comparisons between entities, such as schools, cities, regions, and countries, or through repeated observations in one geographic unit over time. Methodologies to examine these dimensions simultaneously, for instance time-series cross-section (TSCS) estimations, were only developed and applied in political science in the mid-1980s, and gained popularity with the publication of Stimson's (1985) seminal essay on the topic. Since then, the number of publications harnessing the strength of this technique has surged. The combination of cross-sections and time-series is a powerful analytical strategy to accommodate the interaction of the temporal and spatial dimensions in social science theories. It is therefore not surprising that TSCS models have become extremely popular among social scientists. This chapter offers an introduction to the methodology of TSCS analyses in the social sciences, highlighting its advantages, but most importantly focusing on the many empirical challenges brought forth by the combination of temporal and cross-sectional dimensions in a single analysis. While the challenges will at first seem daunting, they are not insurmountable; the inferential advantages of using pooled models over time largely outweigh the complexities.

The remainder of this introduction will introduce the methodology of TSCS, its logics, main applications, and advantages. The next section will present the technical aspects of the methodology as well as the main sources of violations, strategies to detect them, and possible solutions. The third section presents an empirical example using a simple model estimating the effects of levels of development on levels of democracy in 26 countries. The fourth section touches on the most common errors and demonstrates how to improve the application of the methodology, and the chapter concludes by suggesting useful further readings.

TSCS data are a class of longitudinal data consisting of repeated observations over time in multiple units, typically involving the interplay of several independent variables. Data of this type are often referred to as being 'pooled' since they combine N spatial units (e.g. nations, provinces, cities, institutions) and T time periods (e.g. years, since most data are measured annually), producing a set of $N \times T = NT$ observations. Pooled data thus marry both time and space and present units of analysis in 'place–time' or 'country–year' form rather than a single dimension taken separately, as shown in Figure 17.1. TSCS allows the researcher to analyze variables that vary over time together with variables that only vary across units, and not

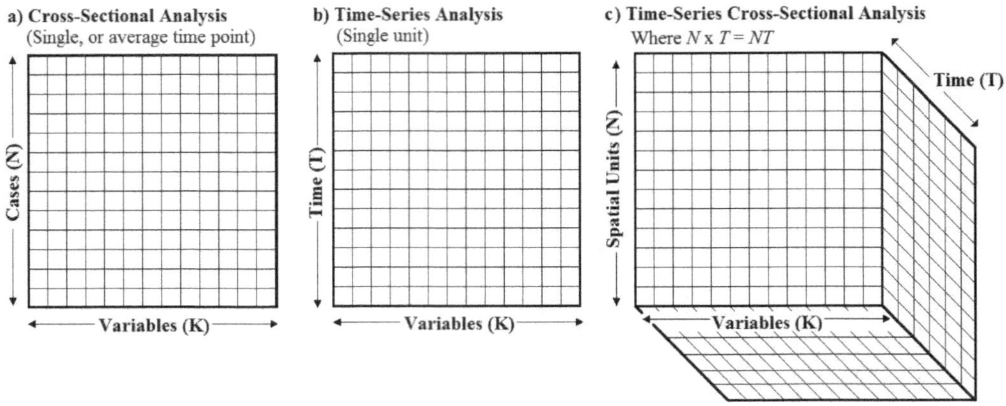

Figure 17.1 Data structure

over time, which could not be accomplished with either cross-sectional or time-series models considered separately.[1] The analytical leverage gained by moving from two-dimensional research designs such as cross-sections and time series to three-dimensional analyses, both comparative and dynamic, as shown in Figure 17.1, is tremendous, as will be highlighted in the ensuing paragraphs.

Panel and TSCS data are organized in analogous structures. In panel data, units are generally randomly sampled over a large number of individuals, whose features are assumed to be identical (see Chapter 15 in this volume). Because of the large number of units contained in such studies, the asymptotics are in N, while T is considered to be fixed. In other words, under asymptotic theory, we assume that T remains constant, while we assume that N could grow to infinity ($N \rightarrow \infty$). The crucial difference between panel and TSCS data is that in the latter, the asymptotics are in T, while N is fixed. These asymptotic properties are important to keep in mind since they provide limiting approximations of the probability distributions of sample statistics. In spite of the similarities between panel and TSCS, because of their different asymptotic properties, some techniques developed for the estimation of panel data are not necessarily suitable for TSCS data, although many are applicable (Hsiao, 2003; Beck, 2008).

In contrast with panel data, typical TSCS data tend to contain fewer units since data are most often pooled from geographic entities such as countries or regions. Moreover, it is not uncommon for units not to be randomly selected or sampled: hence the fixed N. In characteristic applications of TSCS in political science, units usually vary between 10 and 100.[2] Although a large N is not harmful, TSCS analysis does not necessarily require a large number of units to be performed appropriately. The units – their similarities and differences – therefore become much more interesting. Table 17.1 shows actual data stacked in a pool. These data are sorted by country, which provides the cross-sectional variation, and then by year, which provides variation over time; the other variables shown are potential X and Y variables – here GDP per capita, Freedom House scores, and the Cingranelli–Richards index of human rights protection – measured over N countries and T points in time.

TSCS data generally assume two types of patterns. The first pattern is where observations are dominated by a large number of units N relative to time periods T. For example, a study of the 50 states in the United States over 9–10 time periods would be considered 'cross-sectional dominant' or 'stacked' in the terminology of Stimson (1985). The second pattern is where the time component is larger than the number of units used (e.g. 20 developed countries observed

Table 17.1 TSCS data matrix, sorted by country and year

Case	Country	Year	GDP per capita	Freedom House	CIRI
2	Albania	1990	3019.77	6.5	3
3	Albania	1991	2189.24	4	4
4	Albania	1992	2048.06	3.5	6
5	Albania	1993	2275.35	3.5	6
⋮	⋮	⋮	⋮	⋮	⋮
126	Estonia	1990	8900.39	4.5	8
127	Estonia	1991	8230.33	2.5	7
128	Estonia	1992	6606.59	3	7
⋮	⋮	⋮	⋮	⋮	⋮
140	Estonia	2004	12831.76	1	7
⋮	⋮	⋮	⋮	⋮	⋮
289	Romania	1999	5769.04	2	6
290	Romania	2000	5894.13	2	6
291	Romania	2001	6317.64	2	6
292	Romania	2002	6739.99	2	6
⋮	⋮	⋮	⋮	⋮	⋮
378	Ukraine	2004	5895.07	3.5	4

over a period of 40 years). These are considered 'temporally dominant'. The analytical strategy adopted by researchers will be contingent on the number of N units at hand as well as the number of T observations of each unit. If T is limited to a few observations, then the investigation of temporal processes will be restricted. Although not committed to a 'hard-and-fast rule', Beck (2001) suggests a boundary where $T \geq 10$ for using TSCS methods, given how they are justified by their asymptotics in T.[3]

The first uses of TSCS models in political science were made in the fields of political economy of advanced industrial countries (Pampel and Williamson, 1988; Alvarez et al., 1991; Hicks and Swank, 1992; Garrett, 1998), and in the study of international relations, starting with Maoz and Russett's (1993) influential work on democratic peace. Since then, TSCSs have been used in an increasing variety of substantive applications, such as identifying the dynamics of public support for the European Union over time (Eichenberg and Dalton, 1993; Gabel, 1998), investigating voter turnout patterns in democracies and in the European Union (Jackman and Miller, 1995; Grey and Caul, 2000; Franklin and Hobolt, 2011), and explaining the emergence of new political parties (Tavits, 2008). TSCSs are also found in the literature on the determinants of democratization and quality of democracy (Burkhart and Lewis-Beck, 1994; Ross, 2001; Gerring et al., 2005; Fortin, 2012), in studies seeking to investigate the role of democracy on economic performance (Gasiorowski, 2000; Gerring et al., 2005), economic development (Leblang, 1997; Gerring et al., 2005), and the state (Bäck and Hadenius, 2008), among others.

The advantages of TCSC over cross-sectional models largely explain why this methodology rapidly gained popularity. First, using TSCS increases statistical leverage. An advantageous side effect of the combination of time and units is the multiplication of the number of observations available to the researcher. Pooled TSCS estimations are especially useful when cross-national and temporally comparable quantitative indicators are available only for a short time and are applicable only to a small group of countries (fixed, not sampled), as is often the case in political science where observations are available for a handful of countries, for a decade or two. For example, research designs centered on the study of the welfare state will be limited to about 20 countries over three or four decades (Hicks, 1994). TSCS analyses solve some of the classical problems of having too many explanatory variables for too few cases, and expand the degrees of freedom required to model complex relationships (Judge et al., 1985). TSCS thus gives researchers a large number of data points, increases degrees of freedom, and reduces the collinearity among explanatory variables which improves the efficiency of econometric estimates (Hsiao, 2003, p. 3).

In addition to solving the 'small-*N* problem', pooled data allow the study of 'variables' whose variability is limited either across space, or time. This is the case of important features of political regimes that tend to be temporally invariant. Institutions of power distribution such as parliamentary versus presidential arrangements, federal versus unitary constitutional structures, and electoral rules are features that display continuity over time, especially in advanced industrial democracies. Some attributes of nations such as age structures are disposed to inertia, while patterns of utilization of productive capacity are prone to cyclical movements (Hicks, 1994, p. 171). Regression with pooled data can accommodate these types of variables.

TSCS analyses allow the possibility of capturing variation of time and space simultaneously. This combination allows powerful inquiries into causal forces that vary both temporally and cross-sectionally. TSCS data can increase our theoretical leverage by allowing researchers to model the effects of a temporally variable characteristic such as changes in partisan government, with other traits such as types of electoral systems, which do not vary over time (Hicks and Swank, 1992). TSCS therefore lends researchers better footing to draw inferences about the dynamics of change than by solely using cross-sectional evidence.

With TSCS, researchers can therefore construct, and *test*, more complicated theories than with either cross-sectional or time-series data. The combination of time and multiple cross-sections allows some of the effects of missing variables to be overcome: 'By utilizing information on both the intertemporal dynamics and the individuality of the entities being investigated, one is better able to control in a more natural way for the effects of missing or unobservable variables' (Hsiao, 2003, p. 5). Moreover, TSCS design might bring researchers some steps closer to establishing causal claims by allowing the demonstration of covariation, the elimination of rival hypotheses, and determining time order: 'Many of the possible threats to valid inference are specific to either cross-sectional or time-serial design, and many of them can be jointly controlled by incorporating both space and time in the analysis' (Stimson, 1985, p. 916). Campbell and Stanley (1967) even refer to panel/TSCS as excellent quasi-experimental designs, and consider them to be perhaps the best of the most feasible designs.

While offering clear sample size and inferential advantages, the pooling of time-series from a number of cross-sectional units combines estimation difficulties from integrating different dimensions: 'what distinguishes the pooled case [from other types of designs] is that the opportunity to be wrong is considerably enhanced when the design is two-dimensional' (Stimson, 1985, p. 916). The following section elaborates on the technical aspects of TSCS, the frequent violations of pooled models estimated with ordinary least squares regression procedures, and some of the solutions available to researchers to remedy them.

MATHEMATICAL FOUNDATIONS AND ADVANCED ASPECTS

Since most applications of TSCS are made with the standard linear model, otherwise referred to as ordinary least squares (OLS), this chapter will focus on this type of estimation. Starting with a simple model, consider a classical cross-sectional model estimable with OLS;

$$y_i = \alpha + \mathbf{x}_i\boldsymbol{\beta} + e_i, \qquad (17.1)$$

where i refers to cross-sectional units $1, 2, 3, \ldots, N$ ($\mathbf{x}_i$ is a $1 \times K$ vector and $\boldsymbol{\beta}$ is a $K \times 1$ vector). Adding time to equation 17.1, consider a simple TSCS model here, also estimable via OLS;

$$y_{i,t} = \alpha + \mathbf{x}_{i,t}\boldsymbol{\beta} + e_{i,t}, \tag{17.2}$$

where i refers to cross-sectional units $1, 2, 3, \ldots, N$ and t refers to time periods $1, 2, 3, \ldots, T$. In this case, $y_{i,t}$ refers to the a continuous dependent variable for unit i and time t and $\mathbf{x}_{i,t}$ is a vector of exogenous variables. Note that equation 17.2 does not allow a distinction to be made between panel and TSCS data. This equation also makes the assumption of rectangular data, which means that units are observed during the same time points. As is the case in cross-sectional regressions, equation (17.2) is best estimated by OLS if it meets certain assumptions, that is, if the errors assume a spherical form, meaning that each of the $e_{i,t}$ from equation 17.2 are independent and identically distributed, just as assumed in the standard linear model.

However, in practice, when scholars estimate TSCS data using OLS, the errors can display five types of violations that can lead to inefficient or biased estimates (Sayrs, 1989; Hicks, 1994). First, errors can display temporal dependence (sometimes described as autocorrelated or autoregressive), that is, when errors are not independent from one time period to the next. Second, errors can be non-random as a consequence of heterogeneity between units (spatial, or temporal), which violates the assumption that all units are fitted by the same model (homogeneity). Third, errors can suffer from panel heteroscedasticity, where they display constant variance within units, but might vary from unit to unit, or within subsets of units. Fourth, errors can be correlated across units. Fifth, errors can contain both temporal and cross-sectional components. In sum, using Table 17.1 as an example, we assume that there should be no connection between the cross-sections 'Romania' and 'Ukraine' at the time point '2000'; 'Russia' in '1999' with 'Russia' in '2000'; or 'Romania' in '1999' with 'Russia' in '2000'. If these connections exist, but are not specified in the model, they are sources of contamination of the regression estimates (Sayrs, 1989).

Analysts using TSCS will rarely possess data that do not exhibit one or several violations of OLS assumptions. Given the data structure, violations of OLS assumptions are easier in pools of time series. This explains why most of the literature on TSCS has focused on establishing the consequences and producing 'fixes' for these expected violations, treating them as either nuisances, or features that can be modeled (Beck and Katz, 1996; Beck, 2001). While violations can occur along any dimension of pooled data: cross-sections, time, or both combined, the following deals with each process separately as much as possible. In practice, cross-sectional and dynamic issues are not technically separable since many specifications to correct issues related to cross-sections will also have implications for modeling dynamics (Beck and Katz, 2011).

Dealing with the dynamic properties of the data

Although there are no formal rules as to which potential source of violation should be addressed first, Beck and Katz (2001, 2011) suggest that it is generally advisable to tackle model dynamics before confronting cross-sectional issues.[4] One way to think about dynamics in TSCS is to depart from a classical time-series perspective: after all, for each of the units included in TSCS, we have a time-series. Although in cross-sectional analyses we make the assumption that the units are independent, in time-series in social science, data almost always display some degree of dependence over time.

Serial correlation

Violations of OLS assumptions occur when cases are not independent along the time dimension within units.[5] As mentioned earlier, observations within TSCS data are seldom independent along the time dimension: some degree of serial dependence is almost always to be expected when we think about the properties of the type of phenomena we study. Countries display a myriad of characteristics that are interdependent across time. For instance, it is reasonable to assume that the size of the population in country i at time t will be linked to what is observed at time $t + 1$, and so on. An autoregressive (AR) process is denoted by AR(0) when there is no dependence between time points, AR(1) when a dependent variable displays linear dependence on its previous value, AR(2) when the dependence is on the two previous terms, etc. Processes that involve dependence on more than a time point are referred to as higher-order autocorrelation; these are more likely when using weekly, monthly or quarterly data.

Many indicators used in economics and political science, attributing values to a particular unit from one time period to the next period, will be closely associated: the share of budgets apportioned to finance the welfare state, the degree of democracy, or the level of socioeconomic development observed in countries will be linked from one year to the next even if they tend to vary widely over long periods of time. As underlined by Gerring et al. (2012, p. 2), '[r]egimes do not begin again, de novo, with each calendar year. Where one is today depends critically upon where one has been.' These linkages, although substantively meaningful, will lead to error processes that are serially correlated (also autocorrelated or autoregressive) and such dynamics need to be accounted for in TSCS models. The data example in Table 17.1 reveals – by mere visual inspection– that some variables such as GDP per capita might contaminate errors with autoregression within a country. The value of GDP per capita in Romania in 1999 will be closely linked with the value observed in 2000, 2001, 2002, etc. since annual economic outputs are connected from one year to the next.

The dynamics in TSCS can assume many shapes, and overlooking the dynamic structure of data can lead to serious biases in the estimates (Adolph et al., 2005). Before proceeding with OLS, serial correlation must be removed. Temporal or serial autocorrelation can be detected with Lagrange multiplier tests.[6] A pooled Durbin–Watson statistic can also be used to detect autoregression, which is calculated for each cross-section, and then averaged over the pooled cases.[7] Since AR(1) processes are the most prevalent in applications, testing for this form of autocorrelation is a good point of departure. However, one should not exclude higher-order processes.

What to do in the face of serial correlation? Leading contributors (Beck and Katz, 1995, 1996; Beck, 2001) suggest that one of the simplest ways to model these dynamics is the addition of a lagged dependent variable (LDV) among the independent variables as shown in the following equation:

$$y_{i,t} = \lambda y_{i,t-1} + \alpha + \mathbf{x}_{i,t}\boldsymbol{\beta} + e_{i,t} \tag{17.3}$$

The LDV usually removes much serial correlation since the lagged term of the dependent variable includes lagged error terms, $\lambda y_{i,t-1}$, in the model.

Considering that most TSCS data are measured on an annual basis, a single LDV is often the most appropriate fix. However, an LDV does not always remove autocorrelation, and researchers ought to test for remaining autocorrelation even after adding an LDV. The consequences of not investigating this issue are significant; if the addition of an LDV does not remove the serial correlation, OLS estimates will be inconsistent (Keele and Kelly, 2006). In such cases, practitioners may turn to maximum likelihood or Cochrane–Orcutt methods (also called distributed lag models) to obtain consistent estimates for their LDV model (Hamilton, 1994; Beck and Katz, 2011).

Despite its benefits, this method was criticized for adding even more bias to estimations since it is possible that the effects of the LDV will be overestimated and consequently absorb all the predictive power of other independent variables we know to be important (Achen, 2000). In addition, the inclusion of LDVs is not recommended where T is small, especially since LDVs reduce the number of observations available to calculate estimates along the time dimension. Ultimately the decision to include an LDV should be driven by theory. If the researcher knows that successive values of a dependent variable are theoretically dependent on previous values (e.g. national budgets where most items are not re-discussed with each new budget), or that the past matters to explain current values of a variable, the LDV should probably be included.

Practitioners must keep in mind that the LDV approach is one of many potential ways to deal with autocorrelation. The classical time-series literature is replete with discussions of dynamics. Some practicable solutions to deal with serially correlated errors are to use feasible generalized least squares (FGLS) (using the Prais–Winsten transformation), non-linear regression, or, as already mentioned, maximum likelihood, or Cochrane–Orcutt methods when OLS estimates are known to be inconsistent.[8] Another way of dealing with AR(1) processes is by differencing either only the dependent variable, or all the variables, that is, by replacing the raw data $\{\mathbf{x}_{i,t}\}$ by the differenced series $\{\mathbf{x}_{i,t} - \mathbf{x}_{i,t-1}\}$. While these techniques, often referred to as first-difference models, are effective at removing first-order autoregressive processes, these types of corrections are performed at the cost of long-term trends: when using differenced data, analysts can only interpret the short-term effects of variables since long-term effects are effectively differenced out.

Stationarity and non-stationarity

TSCS models function under the assumption that data are stationary. A time-series process is said to be stationary when its statistical properties – mean, variance, autocorrelation, etc. – are all constant over time periods $T = 1, 2, 3, \ldots$. In other words, trends can fluctuate upward or downward over time but return to a mean value (e.g. white noise). Non-stationary (unit roots, trending, drifting, or integrated) series are those in which there is no tendency to return to the mean, and where the error term exhibits a permanent influence on the time series. Although there are different types of non-stationarity, a random walk is an illustrative example of a stochastic (non-stationary) process since it is unpredictable and impossible to forecast. Results obtained by using non-stationary data will lead to flawed hypothesis tests, and can be spurious (indicating a relationship when none exists), especially when estimating models containing either a LDV or serially correlated errors (Franzese, 2002; Beck, 2008). For example, if two variables are trending over time, performing a regression containing both could lead to a high R^2 even if these variables are not related. Stationarity is therefore a critical attribute to tackle.

Non-stationarity is typically detected through Dickey–Fuller or augmented Dickey–Fuller unit root testing (Dickey and Fuller, 1981). A large proportion of contributions dealing with non-stationarity testing was intended either for panels in which N is very large, or where T is large enough to use classical time-series tests on each of the units. For anything in-between (e.g. typical TSCS with 10–250 units and 25–250 time series) some argue that existing solutions are either not computationally feasible, or not powerful enough (Enders, 1995; Levin et al., 2002; Pesaran, 2007). In the face of controversial tests, Beck suggests verifying if the residuals of a dynamic model appear stationary by regressing residuals on their lags,[9] and if the coefficients on any LDV term are near one. Alternatively, researchers could investigate the time-series properties of individual countries/units for variables of interest using classical techniques, to avoid misleading inference driven by a few cases. When T is short, however, it is impossible to know whether or not the effects observed during this period are typical of what we would observe over a longer period.

What to do with non-stationary TSCS data? The short answer is to make the data stationary somehow. While this problem is simple to deal with in classical time series (typically univariate), the territory remains uncharted in TSCS data with numerous independent variables and hence multiple sources of non-stationarity (Beck and Katz, 2011). One solution for dealing with stochastic trends mentioned by Katz is to revert to first-difference models as described in the above discussion on autocorrelation. The other solution, better suited to dealing with trend-stationary data, is to include a time index component as an independent variable in a regression to model the non-stationarity and reduce the risk of spurious correlations.

Cross-sectional issues

Equation (17.2) assumes that all countries can be fitted by the same model. In OLS, geographic entities, although pooled, are treated as independent entities. Using country-level pooled data increases the likelihood that the assumption of independent errors is violated. For example, thinking about economic models, given the increasing level of financial integration of countries with one another, some degree of interdependence is predictable. This subsection elaborates on the most common unit-related sources of disturbances.

Heterogeneity

TSCS equations assume that the units (countries, states, institutions) are completely homogeneous, that is, that they differ only in the levels of their explanatory variables. In the context of TSCS, heterogeneity refers to unobserved variables that remain constant over time and are not explained by the independent variables included in a model. These differences are omitted variables that are specific to the units (Adolph et al., 2005). Unit homogeneity is not a realistic assumption in most cases of macro-level research using geographic entities such as countries. We observe heterogeneity in situations where some features remain relatively stable over time in a geographical unit; for example, attitudes such as the level of conservatism or liberalism tend to remain stable over time across different electoral constituencies. Even if the units considered in an analysis present large similarities in displaying certain properties, they can still differ on other characteristics. For example, taking the member countries of the European Union, all of which are advanced industrial democracies, these countries still would present many differences in size, history, culture and population, among many other potential attributes. Researchers should therefore investigate whether pooling is really the best option for all cases (Maddala, 1991; Baltagi et al., 2000).[10] When the assumption that cases can be pooled does not hold, researchers should consider reverting to pure time-series analysis, cross-sectional regressions or to seemingly unrelated regressions.

The consequences of using OLS on units that exhibit heterogeneity are illustrated in Figure 17.2, reproduced from Wilson and Butler (2007), using simulated data. Relationships between two variables in two fictitious countries were estimated with the OLS cross-section equation (17.1) and again pooling the two countries in a single model following equation (17.2). The model for each country estimated independently has the same slope coefficient $\beta = 1$, but differs on values of α and distributions of x. In (a) and (b) the pooled line overestimates the relationship between x and y, in (c) the relationship is underestimated, while in (d) the pooled line incorrectly estimates the sign of the slope parameter β. Figure 17.2 makes it clear that in the face of unit heterogeneity, the pooled OLS regression line can be misleading. Failure to take care of heterogeneity can have profound consequences on results.

There are several ways to detect heterogeneity. A first step is to gain preliminary visual perspective by producing box-plots of the dependent variables for each of the units, in order

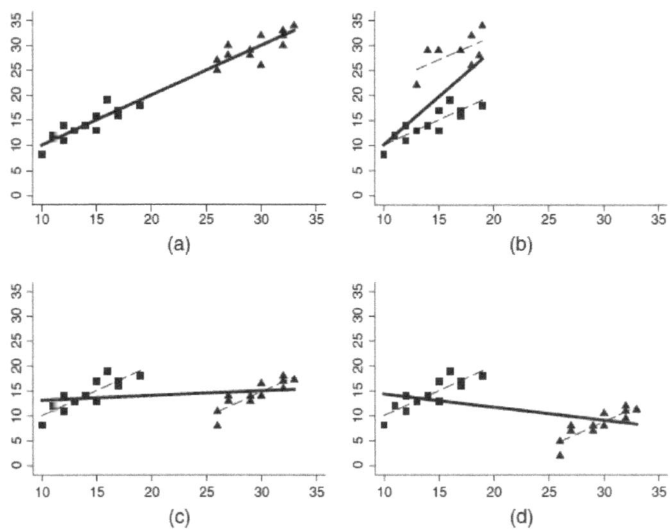

Figure 17.2 Pooled regression slopes versus unit slopes when units display heterogeneity (from Wilson and Butler, 2007, p. 105)

to establish whether or not unit effects are present and of what type they are. Box-plots of the dependent variable offer good clues as to which cases are potentially problematic. Alternatively, plotting each cross-section's fitted values against those of one of the regressors, along with the pooled regression line such as in Figure 17.2 yields concrete visual evidence as to whether the pooled sample presents problematic cases. Practitioners can also set up an F-test where the null hypothesis is that a restrictive model is the most appropriate – meaning that all of the units share the same intercept – against the alternative that intercepts vary across units.[11] Researchers can also turn to solutions based on the technique of cross-validation such as the one developed by Stone (1974).

What to do in the face of unit heterogeneity? Two main solutions are usually presented to researchers: fixed effects (FE), or random effects (RE), also called the random coefficient model (RCM). The FE approach models heterogeneity by making the assumption that each unit has its own intercept, as shown by the addition of the term v_i to equation (17.2):

$$y_{i,t} = \alpha + \mathbf{x}_{i,t}\boldsymbol{\beta} + v_i + e_{i,t} \tag{17.4}$$

This approach is called 'fixed effects' because the covariation is fixed in an intercept term instead of varying as a random variable (Sayrs, 1989; see also Chapter 15 in this volume).

Equation (17.4) includes least squares dummy variable (LSDV) estimators, where v_i is a dummy variable for each of the units. Dummy variables estimate a fixed effect α for each of the units using either $n - 1$ dummies, or n dummies and suppressing the constant. This LSDV will account for the possibility of intercept differences across units, but also by the same token account for variance from potentially influential variables that were left out of the model (Judge et al., 1985; Hicks, 1994).[12] In practice, the FE model allows the unobserved effects to be correlated with the included variables. The LSDV approach can be extended to include time-specific aspects as well, since time periods can have a systematic influence on the error in the same way as cross-sections can be influential. For instance, seasonal shifts can be modeled, such as consumer expenditures around Christmas time.

The use of the LSDV involves a trade-off in the cost of degrees of freedom and loss of efficiency on variables that are of interest. The problem is especially acute when other independent variables are highly correlated with the dummies. FEs are particularly problematic when some of the independent variables are stable over time or slow-moving. In such situations, not uncommon when analysts include features of political institutions in their models such as presidential versus parliamentary arrangements, or types of electoral rules, country dummies are collinear with these variables. In other words, 'unit dummies *completely absorb* differences in the level of independent variables across units' (Plümper et al., 2005, p. 331). Because FEs remove the average country effect, the coefficients represent a cross-country average of the longitudinal effect (Kittel and Winner, 2005, p. 272). One should therefore be careful when testing theories about the effects of *levels* of variables on others with FE models.

The main alternative to FEs is to consider random effects (the RCM), also used in analyses of multilevel data. While FE models are relatively straightforward to implement, the RCM comes in a variety of estimators, usually employing maximum likelihood, Bayesian, or generalized least squares estimates with different specifications and implementations (Sayrs, 1989; Beck and Katz, 2007). Instead of assuming the variation across units to be fixed, the RCM assumes that the variation is random and uncorrelated with the independent variables included in the model. This represents an important advantage when we have weak theoretical justifications for heterogeneity among units. Another important benefit of RCMs is that, by contrast with FE, they allow time-invariant variables to be modeled. The RE model is as follows:

$$y_{i,t} = \alpha + \mathbf{x}_{i,t}\boldsymbol{\beta} + u_{i,t} + e_{i,t}, \tag{17.5}$$

where $e_{i,t}$ represents the within-country error, and the new term $u_{i,t}$ encapsulates the between-country error. The term $u_{i,t}$ changes the interpretation of the regression coefficients since these will now include within- and between-country effects. In other words, coefficients represent the average effect of X on Y when X changes across time and between countries by one unit, and this shift makes substantive interpretations more difficult. Another drawback stems from the assumptions underpinning RCMs, namely that there is no correlation of random effects with regressors and no correlation between random components. Since these assumptions are controversial and rarely met in reality (Bartels, 2008), this topic has sparked a debate as to whether this approach is applicable to populations of countries across which heterogeneity cannot be considered random (Beck, 2001; Kristensen and Wawro, 2003).

The practitioner is left with the daunting question of which fix to choose in the face of heterogeneity.[13] How to model unobserved heterogeneity in TSCS is a hotly debated issue (Judge et al., 1985; Stimson, 1985; Beck, 2001; Green et al., 2001; Zorn, 2001; Wooldridge, 2002; Hsiao, 2003; Wilson and Butler, 2007; Baltagi, 2008; Bell and Jones, 2014). Wooldridge (2002) suggests that if the units are exchangeable (i.e. their names are irrelevant, as with respondents in a survey), then random effects often make sense. If the units are not exchangeable, they are of interest in their own right. As a consequence, the RCM is probably not the right model. Judge et al. (1985) suggest characterizing the problem in terms of the correlation between the independent variables and the error when the choice is not clear-cut. If the number of units is small, and there is a possibility that the dummies and the explanatory variables are correlated, then the FE approach is recommended. However, as T becomes larger, the differences between FE and RCM decrease (Plümper et al., 2005). In most cases, implementing a Hausman test, described in Greene (2012), might help reaching a decision, as will be shown in the example presented below.

Panel heteroscedasticity and contemporaneously correlated errors

Heteroscedasticity in the context of cross-sectional analyses refers to a situation 'in which – contrary to the assumption of homoscedasticity – the error term in a regression model does not have constant variance' (Berry and Feldman, 1985, p. 72). Panel heteroscedasticity in the context of TSCS refers to situations where errors display constant variance within units, but vary from unit to unit, or within subsets of units. When the variance of error processes differs from unit to unit, the Gauss–Markov assumptions are violated. Such violations have important repercussions for OLS estimations since they bias standard errors. An example of this is when cases display both high values and high variance on certain variables compared to others: the United States tends to have more volatile and higher unemployment rates than Switzerland (Hicks, 1994).

Contemporaneously correlated errors in TSCS are a form of time-specific heterogeneity. Observations from a unit may be correlated with another unit during the same period: 'We might expect TSCS errors to be contemporaneously correlated in that large errors for unit i at time t will often be associated with large errors for unit j at time t' (Beck and Katz, 1995, p. 636). Such contemporaneous correlations may differ by unit. Unlike heterogeneity between units, this type of heterogeneity cannot simply be modeled using time-specific dummy variables, since time dummies serve to control for events that affect all units at a given time point. Panel heteroscedasticity and contemporaneously correlated errors will affect the OLS standard errors, making them inefficient. Panel heteroscedasticity can be detected via modified Wald tests, while contemporaneously correlated errors can be detected via the Breusch–Pagan test for cross-sectional independence (Baum, 2001), or Pesaran's cross-sectional dependence (CD) test (De Hoyos and Sarafidis, 2006).

What to do in the face of panel heteroscedasticity and contemporaneously correlated errors? In early applications, a version of FGLS proposed by Parks (1967) and Kmenta (1971) was used to model data exhibiting panel heteroscedasticity and contemporaneously correlated errors. In the typical TSCS applications, however, the Parks–Kmenta approach often yields grossly downward-biased standard errors, and potentially flawed hypotheses tests: Beck and Katz (1995) even go as far as advising not to employ this technique. To reduce the risks associated with overconfidence in the performance of estimators, Beck and Katz developed panel-corrected standard errors (PCSEs) for OLS. Simulations show that PCSEs tend to be more accurate than alternative computations of OLS standard errors for TSCS data, and have the advantage of being easy to estimate with standard statistical software (Beck and Katz, 1995).

A related issue is that of spatial correlation among units, which is also generally treated as nuisance to be 'corrected' either by Parks' procedure (FGLS) or by resorting to PCSE. Spatial correlation among geographical units, especially when they are contiguous, is frequent. It is not unreasonable to assume that when two countries share a border, they will likely face some of the same problems. An example drawn from criminology illustrates spatial correlation very well. If there is an increase in homicides in one city, one could reasonably expect that this problem could affect surrounding cities (Worrall and Pratt, 2004). Spatial autocorrelation is the result of a non-random geographic clustering of values across observations, and is usually considered less of a problem when units are randomly sampled. When the stochastic element $e_{i,t}$ (from equation (17.2)) exhibits spatial correlation, the standard errors are pushed downwards, and efficiency is overstated. More than being a nuisance, some analysts have underlined that many of the phenomena of interest to political scientists are explicitly subject to spatial interdependence: policy and institutional diffusion across national governments, coups d'état, riots, revolutions, civil wars and democratization are often theorized to display spatial interference, which should explicitly be included in models rather than corrected for (Beck et al., 2006; Franzese and Hays, 2009).

Other important issues less frequently addressed

Missing data (unbalanced panels)

Sample size differences between units are sources of heterogeneity. Because of missing data, or the way data are recorded, panels in which the groups differ in size across time are not unusual. These are called unbalanced panels. Most analyses assume equal group sizes. Honaker and King (2010) frame the issue of unbalanced panels as one of missing data. The default technique employed by analysts and statistical software alike to deal with this problem is listwise deletion: entire observations are deleted from an analysis even if only one variable is missing, since most statistical procedures necessitate complete cases from all variables. Because TSCS requires a minimum of continuous panel information and assumes balanced panels to produce reliable estimates, missing cases and listwise deletion can be particularly damaging to practitioners. In such cases, data imputation should be considered (Allison, 2000; Honaker and King, 2010).

Time-invariant predictors

In practice, time-invariant characteristics of countries are perfectly collinear with country dummies (Kohler and Kreuter, 2008). Therefore, time-invariant predictors make the use of fixed effects impossible, while slow-moving predictors generate large standard errors (Plümper and Troeger, 2007; Breusch et al., 2011). Large standard errors raise the likelihood of researchers making Type II errors, that is, rejecting that a variable has an effect when in fact the effect should be significant. The issue of time-invariant and slowly changing covariates raises the question as to whether pooled models are appropriate at all in such situations, and what inferences in a pooled data set mean: 'if the frequency of observation is high enough, any unchanging variable can appear statistically significant' (Wilson and Butler, 2007, p. 120). While RCMs can handle time-invariant predictors, Plümper and Troeger (2007) have proposed a solution via vector decomposition to integrate such predictors in FE models. However. some have voiced reservations about the approach (Greene, 2012), or claim it is functionally equivalent to an instrumental variables approach (Breusch et al., 2011).

EXAMPLE ANALYSIS

Let us turn to an example testing a simplified modernization-based hypothesis (Lipset, 1959) proposing a relationship between countries' level of development and democratization.[14] In the example that follows, we will link an indicator of development, GDP per capita, with levels of democracy (Freedom House) in 26 former communist countries from 1989 to 2004. While panels are slightly unbalanced, a rectangular shape will be assumed (where observations are available the same time periods for all cases) to facilitate interpretation.

After examining the variables of interest for skewness, we proceed to verify if an OLS model would suffer from autocorrelation. Performing Wooldridge's test for autocorrelation yields a statistically significant result, which means the null hypothesis of no serial correlation is strongly rejected (displayed in Table 17.2). This indicates that we should consider the addition of an LDV in the following models to be performed.[15] The addition of the LDV is further justified by theory which suggests that current levels of development will impact current levels of democracy, but also that past levels of development will impact current levels of democracy. The LDV specification encapsulates this knowledge by suggesting that levels of democracy today are functions of past levels, as modified by new information on development (Keele and Kelly, 2006). In addition, we should pay attention to the assumption of stationarity by performing

Table 17.2 Diagnostic tests of a pooled model testing the effects of economic development on democracy in 26 post-communist countries, restrictive model (unless otherwise noted)

Issue	Test	H_0	p
Autocorrelation, first order	Wooldridge	No first-order autocorrelation	0.000
Unit root, 1 lag	Aug. Dickey–Fuller (Fisher)	All panels contain unit roots	0.000
OLS vs FE	F-test	No joint fixed effects	0.000
OLS vs RE	Lagrange multiplier test	No significant difference across units	0.000
FE vs RE	Hausman test	Difference in coefficients not systematic	0.047
Contemporaneously correlated errors*	Pesaran's CD test	No cross-sectional dependence	0.000
Groupwise heteroscedasticity*	Modified Wald test	Homoscedasticity (constant variance)	0.000
Necessity of time fixed effects	F-test	No joint fixed effects	0.255

*Calculated for FE model.

some tests. A series of statistically significant augmented Dickey–Fuller tests (at different lags) for each cross-section shown in Table 17.2 provides preliminary information that the panels do not seem to present unit roots, and can be considered stationary.

To verify whether our pooling of the 26 countries makes sense, let us look at box-plots of the dependent variable, Freedom House ratings in each of the units (Figure 17.3). While researchers should expect some differences among units, Figure 17.3 indicates that some units appear to be different and that pooling all 26 countries in a single model, assuming homogeneity, would be problematic. However, the clearest evidence is presented in Figure 17.4 which shows the countries' independent regression lines (dashed lines) against a pooled regression line (continuous line). Recalling that equation (17.2) assumes that all countries are fitted by the same model, the patterns of fitted values for each of the 26 cross-sections do not look as if they could all be summarized by the continuous line representing a restrictive pooled model shown in Figure 17.4. Although most country-specific slopes are in the expected direction, some slopes are steeper, while some are even in the opposite direction (albeit not all statistically significant). Figure 17.4 illustrates just how much information would be swept under the carpet by subsuming all the cases under a single regression line.

Now that we have visually established the presence of heterogeneity, let us look at options to deal with this issue in our model, as well as some of the additional diagnostics mentioned in the previous sections, also displayed in Table 17.2. Performing an F-test to verify if an FE model is preferable to a restrictive model reveals that the country dummies are jointly statistically significant, and hence that an FE model is preferable to the restrictive OLS. Next, a Lagrange multiplier test is used to verify if random intercepts are preferable to a restrictive model. The statistically significant test result leads us to reject the null hypothesis that there is no significant difference across units; an RE model would also be preferable to a restrictive model. Having established that a restrictive model is the least desirable option of the three, a Hausman test is performed to determine whether an FE or an RE model would be the most appropriate option to model our data. In this case, we reject (albeit weakly) the null hypothesis that the difference between coefficients is not systematic: an FE model appears to be the most suitable alternative. FEs, or country dummies, are a sensible option in the context of this exercise, given the small number of cases. Now turning to additional diagnostics displayed in Table 17.2, performing Pesaran's CD test reveals that there is cross-sectional dependence in the group of countries

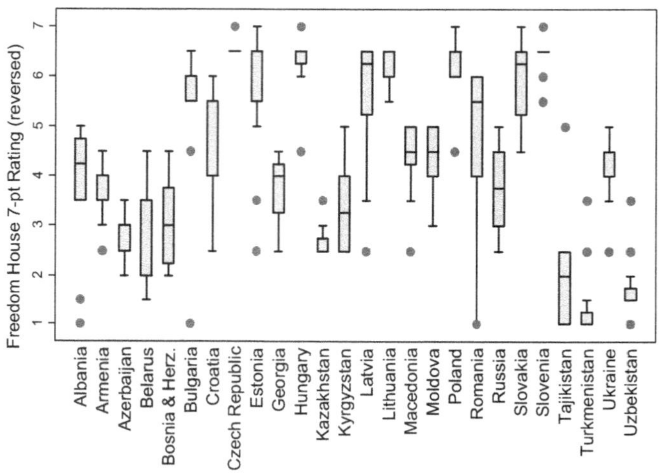

Figure 17.3 Box-plots of the dependent variable by country

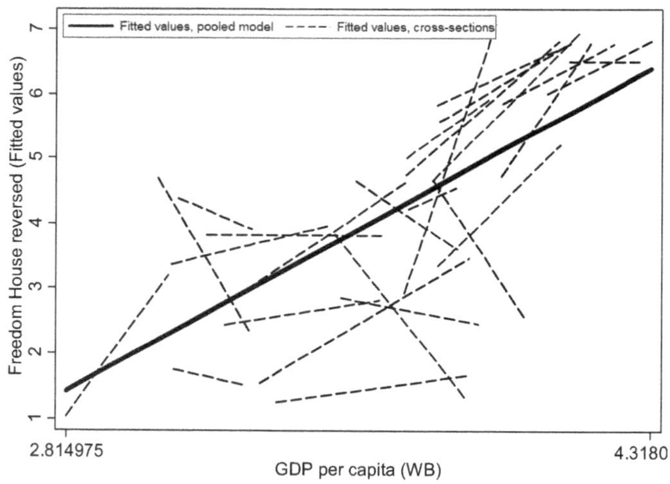

Figure 17.4 Effects of levels of development on democracy by cross-section vs pooled model in former communist countries

considered, as well as groupwise heteroscedasticity (modified Wald test). These results indicate either using Beck and Katz's PCSE routine or reverting to another form of robust standard errors.

Given what we already know about the data, Table 17.3 presents a series of alternative model specifications to estimate our relationship of interest. Model 1 is a restrictive OLS, Model 2 adds PCSEs, Model 3 introduces an LDV, Model 4 uses country dummies as fixed effects, Model 5 combines LDV and country dummies and Model 6 uses both country and year dummies to test the limits of our hypothesis. The difference between restrictive Models 1 and 2 is in the size of the standard errors. It may be argued, however, that Models 1 and 2 are not properly specified, because they do not include a control for the previous measurement of level of democracy since the results of our test revealed the presence of AR(1) autocorrelation. The use of the LDV is

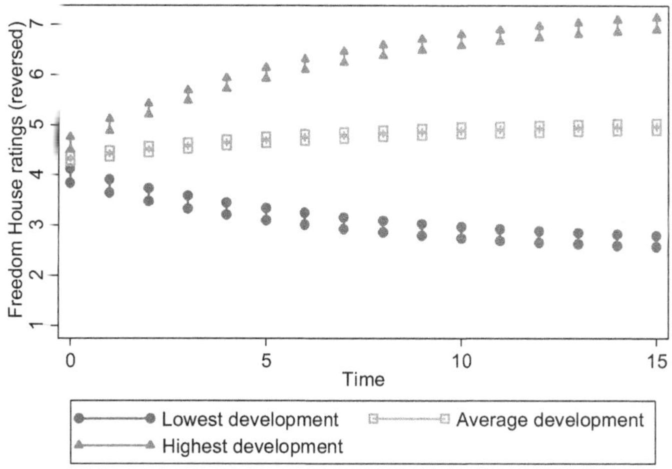

Note: bars depict 95% confidence intervals.

Figure 17.5 Dynamic simulations of the effects of levels of development on democracy

therefore needed to remove effects of autocorrelation which might otherwise have biased the effects of interest. Model 3 incorporates a measurement of level of democracy at $t-1$. As noted in the literature criticizing the use of LDVs (Achen, 2000; Plümper et al., 2005), the effect of GDP per capita is significantly reduced when the LDV is introduced in the model. However, since we know that theoretically, current measurements of democracy are largely driven by the level recorded in the previous year, the addition of a LDV is necessary in the present case. This addition changes the hypothesis in the following way: the model containing a LDV specifies the effect of levels of development after the effect of Freedom House scores (lagged one period) are taken into account: we are not modeling yearly changes in Freedom House scores.

The addition of country dummies in Model 4 reduces the size of the effect of GDP per capita estimated in Model 2, an issue raised by practitioners noticing how much cross-sectional variance was absorbed by unit dummies (Huber and Stephens, 2001). This is attributable to the fact that each dummy is absorbing the effects particular to each cross-section that were left out of the model. The addition of a dummy for each country estimates the net effects of levels of development on democracy, by controlling for the heterogeneity we have observed in Figures 17.3 and 17.4. Models 5 and 6 apply brute force to the hypothesis of modernization by overloading the models with both a LDV and country dummy variables, and finally with both country and year dummy variables. Despite these extreme model specifications, the variable depicting levels of development retains its statistically significant impact on the dependent variable. Although no other control variables are employed in this example, we can conclude that this variable has a robust impact on the dependent variable that is resistant to alternative specifications. The main difference across the six models, shown in Table 17.3, concerns the size of the effects of development on levels of democracy.

In addition to presenting tables, illustrating results with dynamic simulations allows drawing substantive inferences about whether two scenarios are significantly different from each other at any time period, and enhances result interpretation.[26] Instead of assuming that variables have the same effect at all levels, graphical representations enable researchers to make finer distinctions across levels of variables of interest, and over time. Dynamic simulations are powerful tools for making assessments about the long-term effects of variables. Figure 17.5 presents the results

Table 17.3 OLS, PSCE, LSDV and LDV estimates of the effects of economic development on democracy in 26 post-communist countries

	Model 1 OLS	Model 2 PCSE	Model 3 LDV	Model 4 LSDV	Model 5 LDV + LSDV	Model 6 LSDV country + year
GDP per capita	2.522***	3.315***	0.610***	1.992***	0.327*	3.228***
	(0.36)	(0.13)	(0.16)	(0.34)	(0.19)	(0.71)
Freedom House $t-1$	–	–	0.862***	–	0.565***	–
			(0.04)		(0.04)	
Unit effects (countries)						
Albania (omitted)				–	–	–
Armenia				0.295*	–0.094	0.456**
				(0.16)	(0.14)	(0.19)
Azerbaijan				–1.004***	–0.800***	–1.046***
				(0.26)	(0.17)	(0.25)
Belarus				–1.332***	–0.764***	–1.538***
				(0.26)	(0.18)	(0.27)
Bosnia & Herzegovina				–1.314***	–0.605***	–1.694***
...				...	...	...
Uzbekistan				–1.494***	–1.129***	–1.233***
				(0.28)	(0.14)	(0.33)
Time effects (years)						
1989 (omitted)						–
1990						1.523***
						(0.16)
1991						2.135***
						(0.13)
...						...
2002						1.592***
						(0.19)
Constant	–5.004***	–7.915***	–1.531***	–3.145***	0.772	–9.193***
	(1.33)	(0.49)	(0.42)	(1.20)	(0.61)	(2.47)
R^2	0.442	0.442	0.895	0.778	0.936	0.843
Observations	341	341	318	341	318	341

$^{*}\ p < 0.10$, $^{**}\ p < 0.05$, $^{***}\ p < 0.01$

of a simple dynamic simulation (using Model 3) holding the lagged value of the dependent variable at its mean, and letting levels of development vary from its lowest value, its mean and its highest value over a period of 15 time points, using 95% confidence intervals. Figure 17.5 shows that development seems to have a linear effect on levels of democracy: all three scenarios remain statistically different from each other in both the short and long term. In this group of countries at least, there is no larger or smaller effect at the highest or lowest levels of development, something we are not able to infer just from regression coefficients on their own.

CAVEATS AND FREQUENT ERRORS

The previous sections should have made clear that despite the huge advantages of TSCS models, there is an equivalent amount of estimation difficulties tied to them. There is no 'one-size-fits-all' way to deal with TSCS data (Adolph et al., 2005). While they acknowledge the importance of contributions made by Beck and Katz, Wilson and Butler (2007) warn against simply treating their advice – lagged dependent variable and panel-corrected standard errors – as a universal solution to the many problems researchers can face when using TSCS. For example, including a lagged dependent variable in a LSDV or FE model can lead to serious biases when there is remaining autocorrelation (Nickel, 1981; Judson and Owen, 1999; Beck and Katz, 2004; Keele and Kelly, 2006). While broadly appropriate, the approach proposed by Beck and Katz was not meant to be a universally applicable solution, and can introduce biases when used under certain situations. Studies of comparative politics usually focus on 10–20 nations; while at the other extreme, scholars of international relations often have thousands of dyads of countries as units in their analyses: the wide range in types of samples used in applied research indicates that different applications will face different sets of challenges.

Contributors have shown that TSCS estimates tend to be fragile to model specifications (Kittel and Winner, 2005; Wilson and Butler, 2007). Using the same variables, modifications in some of the model specifications proposed in this chapter (e.g. using cross-sectional and or time fixed effects, LDV, PCSE) have a strong impact on the estimates. Such fragility of results in the face of different – legitimate – specifications is the most challenging aspect of using TSCS. The number of potential problems that can be exhibited in the data, and the proliferation of estimators to deal with these problems, can often leave researchers at loss as to which procedure to choose and which errors to correct. Controlling for both spatial and temporal disturbances is problematic and costly in terms of degrees of freedom. Corrections for dynamic issues have repercussions for cross-sectional issues, so much that it might not be possible to correct for all types of violations to OLS assumptions without jeopardizing the efficiency of parameter estimates in one way or another. Perhaps more important is to remember is that there is no such thing as a general set of 'best practices' yet.

While correcting for the myriad of problems that can occur in TSCS is challenging, one possible strategy to increase robustness is to make additional diagnostics and robustness checks part of the analyses. As highlighted in this chapter, testing for potential sources of violations in order to choose an appropriate model specification is crucial. In their review of 195 published articles in political science employing TSCS, Wilson and Butler (2007) found that few of these considered the issue of unit heterogeneity and tested for alternative dynamic structures: 'Careful studies using TSCS should consider unit heterogeneity and alternative dynamic specification and test for autocorrelation' (Wilson and Butler, 2007, p. 109). Since TSCSs are particularly fragile to alternative specifications, one idea would be to follow Leamer's 'extreme bound analysis' (Leamer, 1985, 1983), and to verify what happens to the magnitude, the size and the statistical significance of findings when they are put under stress. Testing alternative specifications allows the sensitivity of results to be gauged, provides stronger tests, and raises the bar for confirming our theories.

Devoting effort to robustness checks should not be at the price of interpretation: researchers should make sure they get the most out of their results once they have selected suitable model specifications. Parameter estimates from TSCS models encompass the combined average partial effect of the cross-sectional dimension and time: their substantive meaning is less readily interpretable than in cross-section models (Firebaugh, 1980; Kittel, 1999). In other words, one cannot really know the relative contribution of time and cross-section in a TSCS parameter estimate, or if there are changes in this proportion because it is assumed to be constant. For instance, a relationship could only exist during a certain period and still show a significant relationship over the entire period. In short, Kittel (1999) argues that pooling 'averages over' a lot of information and might thus not be as advantageous as we think. Researchers should try to further evaluate the contributions of time and space by running additional models such as yearly cross-sections and time-series in each cross-sectional unit.

In their review of TSCS applications, Williams and Whitten (2012) noticed that most contributions have limited themselves to standard significance and parameter estimates interpretations: few contributions offered dynamic interpretations, and even fewer presented graphical depictions of inferences, such as those presented in the example section above, which they consider a missed opportunity to make substantive inferences. They suggest going further than simple interpretations of the short-term effects of a one-unit change in the independent variable on the dependent variable (when a LDV is present), and to present long-term effects of changes in independent variables accompanied by confidence intervals (or some estimation of uncertainty), as well as graphical depictions of effects of substantive interest.

FURTHER READING

At this point, the reader will have guessed that modeling TSCS data requires a solid grasp of econometrics. Most articles dealing with this topic in the social sciences assume that readers are familiar with a series of manuals, keeping in mind that most of the relevant literature is geared at users of panel data where $N > T$ (Wooldridge, 2002; Arellano, 2003; Hsiao, 2003; Baltagi, 2008; Greene, 2012). Lois W. Sayrs (1989) also provides a succinct introduction to TSCS. For recent methodological advances and applications in political science, the 2007 special issue of *Political Analysis* is a good point of departure. The series of articles by Beck and Katz have become some of the most cited articles in political science (e.g. Beck and Katz, 1996, 1995; Beck, 2008, 2007, 2001, 2012). While this chapter has been concerned with continuous dependent variables (and OLS applications), using categorical or binary dependent variables involves different types of considerations which are addressed in more details in Beck et al. (1998). Researchers using dyads as units of analyses should pay attention to some crucial issues not covered in this chapter; see Beck and Katz (2001), Green et al. (2001) and Zorn (2001). Researchers interested in using R should consult Bailey and Katz (2011).

NOTES

1 Like ordinary least squares regression, TSCS is generally used on continuous dependent variables, although models can accommodate categorical dependent variables with some modifications. The present chapter will concentrate on continuous dependent variable estimation, and present the appropriate literature to deal with categorical response variables in the final section.

2 In the fields of American politics and especially international relations, the number of units has sometimes been over 1000. See Adolph et al. (2005) for a more comprehensive review of TSCS uses in political science journals from 1970 to 2005.

3 Panels models are better equipped to deal with very small T given the large samples that characterize this type of data.

4 The reader must note that this order is contested. Sayrs (1989), for instance, argues that autoregression can only be detected once heteroscedasticity has been controlled for.

5 Data that are 'cross-sectionally dominant' usually present less concern for autocorrelation. Since only $N(t\text{-}1)$ cases can possibly be serially dependent, this concern decreases as N becomes larger and T smaller. As T grows, however, the importance of dynamic specification also rises.

6 Assuming the null hypothesis that errors are independent, a Lagrange multiplier test is performed by first running an OLS regression and saving the residuals. Second, the saved residuals are regressed on the independent variables of the model and the lagged residual. This step can include a lagged dependent variable. The null hypothesis can be rejected if the coefficient for the lagged residual is statistically significant. Alternatively, a user developed program in Stata (Drukker, 2003) produces a simple test for autocorrelation in panel data models based on Wooldridge (2002).

7 The closer the Durbin–Watson statistic is to 2, the less the autoregression. The regular Durbin–Watson statistic is not valid for models with a lagged dependent variable, and is only used for processes of order AR(1). The alternative Durbin–Watson test (`durbina` in Stata) allows lagged variables, and instances of higher-order autoregressive processes.

8 Other fixes proposed in the panel literature are not necessarily appropriate for typical usages in political science since they assume a large N (with all asymptotics in N), while in typical political science studies, N is small (with all asymptotics in T). See Wawro (2002) for an overview of a review of alternative techniques.

9 If the coefficient of the lag residual term is near one, it can be considered stationary.

10 An important contribution was made by Bartels (1996) on how to deal with disparate observations using fractional pooling, allowing degrees of pooling.

11 The way to test this is to compare the F-statistics from equation (17.1) and a model comprising n-1 dummies of the cross-sections.

12 An F-test indicated that the null hypothesis of no effects should be rejected. The addition of country dummy variables should account for the remaining possibility of intercept differences across units. It is important to note that the addition of these variables can bias downward the coefficients of those variables whose effects are partly cross-sectional. The intercepts themselves are not explanations for the between-unit variance, but a way to characterize the variance so that the bias in the 'true' explanation is minimized, what Maddala (1977) has termed 'specific ignorance'.

13 Some also propose using instrumental variables or generalized methods of moments (Anderson and Hsiao, 1982; Arellano and Bond, 1991).

14 The data set used has been made available by Pippa Norris (see Norris, 2009).

15 It is important to emphasize that one should perform further tests to establish whether a LDV removes the serial correlation.

16 See Williams and Whitten (2012) for details on how to perform dynamic simulations using Stata. While at the present time the commands developed cannot accommodate all statistical techniques dealing with some of the problems raised in this chapter, future developments in this direction are foreseeable.

REFERENCES

Achen, C. A. (2000). *Why lagged dependent variables can suppress the explanatory power of other dependent variables*. Paper presented at the Annual Meeting of the Political Methodology Section of the American Political Science Association, UCLA.

Adolph, C., Butler, D. M. and Wilson, S. E. (2005). *Which time-series cross-section estimator should I use now? Guidance from Monte Carlo experiments.* Paper presented at the Annual Meeting of the American Political Science Association, Washington, DC.

Allison, P. D. (2000). Multiple imputation for missing data: A cautionary tale. *Sociological Methods & Research*, 28(3), 301–309.

Alvarez, M., Garrett, G. and Lange, P. (1991). Government partnership, labour organization and macroeconomic performance 1967–1984. *American Political Science Review*, 85, 539–556.

Anderson, T. and Hsiao, C. (1982). Formulation and estimation of dynamic models using panel data. *Journal of Econometrics*, 18, 47–82.

Arellano, M. (2003). *Panel Data Econometrics*. Oxford: Oxford University Press.

Arellano, M. and Bond, S. (1991). Some tests of specification for panel data: Monte Carlo evidence and an application to employment equations. *Review of Economics Studies*, 58(2), 277–297.

Bäck, H. and Hadenius, A. (2008). Democracy and state capacity: Exploring a J-shaped relationship. *Governance*, 21(1), 1–24.

Bailey, D. and Katz, J. N. (2011). Implementing panel-corrected standard errors in R: The pcse package. *Journal of Statistical Software*, 42.

Baltagi, B. (2008). *Econometric analysis of panel data*. Chichester: Wiley.

Baltagi, B. H., Griffin, J. M. and Xiong, W. (2000). To pool or not to pool: Homogeneous versus heterogeneous estimators applied to cigarette demand. *Review of Economics and Statistics*, 82(1), 117–126.

Bartels, B. L. (2008). Beyond 'fixed versus random effects': a framework for improving substantive and statistical analysis of panel, time-series cross-sectional, and multilevel data. Prepared for the 2008 Annual Meeting of the Society for Political Methodology, Ann Arbor, MI.

Bartels, L. M. (1996). Pooling disparate observations. *American Journal of Political Science*, 40(3), 905–942.

Baum, C. F. (2001). Residual diagnostics for cross-section time series regression models. *Stata Journal*, 1(1), 101–104.

Beck, N. (2001). Time-series-cross-section data: What have we learnt in the past few years? *Annual Review of Political Science*, 4, 271–293.

Beck, N. (2007). From statistical nuisances to serious modeling: Changing how we think about the analysis of time-series cross-section data. *Political Analysis*, 15(2), 97–100.

Beck, N. (2008). Time-series cross-section methods. In J. M. Box-Steffensmeier, H. E. Brady and D. Collier (eds), *The Oxford Handbook of Political Methodology* (pp. 475–493). Oxford: Oxford University Press.

Beck, N. (2012). Sweeping fewer things under the rug: tis often (usually?) better to model than be robust. Prepared for the 2012 Annual Meeting of the Society for Political Methodology, Chapel Hill, NC.

Beck, N., Gleditsch, K. S. and Beardsley, K. (2006). Space is more than geography: Using spatial econometrics in the study of political economy. *International Studies Quarterly*, 50, 27–44.

Beck, N. and Katz, J. N. (1995). What to do (and not to do) with time-series cross-section data. *American Political Science Review*, 89(3), 634–647.

Beck, N. and Katz, J. N. (1996). Nuisance vs. substance: Specifying and estimating time-series-cross-section models. *Political Analysis*, 6(1), 1–36.

Beck, N. and Katz, J. N. (2001). Throwing out the baby with the bath water: A comment on Green, Kim and Yoon. *International Organization*, 55, 487–495.

Beck, N. and Katz, J. N. (2004). Time-series–cross-section issues: Dynamics, 2004. Working Paper, Society for Political Methodology.

Beck, N. and Katz, J. N. (2007). Random coefficient models for time-series–cross-section data: Monte Carlo experiments. *Political Analysis*, 15(2), 185–195.

Beck, N. and Katz, J. N. (2011). Modeling dynamics in time-series-cross-section political economy data. *Annual Review of Political Science*, 14, 331–352.

Beck, N., Katz, J. N. and Tucker, R. (1998). Taking time seriously: Time-series-cross-section analysis with a binary dependent variable. *American Journal of Political Science*, 42(4), 1260–1288.

Bell, A. and Jones, K. (2014). Explaining fixed effects: Random effects modelling of time-series cross-sectional and panel data. *Political Science Research and Methods*, to appear.

Berry, W. D. and Feldman, S. (1985). *Multiple Regression in Practice*. Newbury Park, CA: Sage.

Breusch, T., Ward, M. B., Nguyen, H. T. M. and Kompas, T. (2011). On the fixed-effects vector decomposition. *Political Analysis*, 19(2), 123–134.

Burkhart, R. E. and Lewis-Beck, M. S. (1994). Comparative democracy: the economic development thesis. *American Political Science Review*, 88(4), 903–910.

Campbell, D. and Stanley, J. (1967). *Experimental and Quasi-experimental Designs for Research*. Chicago: Rand McNally.

De Hoyos, R. E. and Sarafidis, V. (2006). Testing for cross-sectional dependence in panel-data models. *Stata Journal*, 6(4), 482–496.

Dickey, D. A. and Fuller, W. A. (1981). Likelihood ratio statistics for autoregressive time series with a unit root. *Econometrica*, 49(4), 1057–1072.

Drukker, D. M. (2003). Testing for serial correlation in linear panel-data models. *Stata Journal*, 3, 168–177.

Eichenberg, R. C. and Dalton, R. J. (1993). Europeans and the European Community: The dynamics of public support for European integration. *International Organization*, 47(4), 507–534.

Enders, W. (1995). *Applied Econometric Time Series*. New York: Wiley.

Firebaugh, G. (1980). Cross-national versus historical regression models: Conditions of equivalence in comparative research. In R. F. Tomasson (ed.), *Comparative Social Research*, Vol. 3 (pp. 333–344). Greenwich, CT: JAI Press.

Fortin, J. (2012). Is there a necessary condition for democracy? The role of state capacity in post-communist countries. *Comparative Political Studies*, 45(7), 903–930.

Franklin, M. N. and Hobolt, S. (2011). The legacy of lethargy: How elections to the European parliament depress turnout. *Electoral Studies*, 31(1), 67–76.

Franzese, R. J. (2002). *Macroeconomic Policies of Developed Democracies*. New York: Cambridge University Press.

Franzese, R. J. and Hays, J. C. (2009). Empirical modeling of spatial interdependence in time-series cross-sections. In S. Pickel, G. Pickel, H.-J. Lauth and D. Jahn (eds), *Neuere Entwicklungen und Anwendungen auf dem Gebiet der Methoden der vergleichenden Politikwissenschaft, Band II* (pp. 233–261). Wiesbaden: Westdeutscher Verlag.
Gabel, M. (1998). Public support for European integration: An empirical test of five theories. *Journal of Politics*, 60(2), 333–354.
Garrett, G. (1998). *Partisan Politics in the Global Economy*. New York: Cambridge University Press.
Gasiorowski, M. J. (2000). Democracy and macroeconomic performance in underdeveloped countries: an empirical analysis. *Comparative Political Studies*, 33(3), 319–349.
Gerring, J., Bond, P., Barndt, W. T. and Moreno, C. (2005). Democracy and economic growth: A historical perspective. *World Politics*, 57(3), 323–364.
Gerring, J., Thacker, S. C. and Alfaro, R. (2012). Democracy and human development. *Journal of Politics*, 74(1), 1–17.
Green, D. P., Kim, S. Y. and Yoon, D. H. (2001). Dirty pool. *International Organization*, 55(2), 441–468.
Greene, W. H. (2012). *Econometric Analysis*, 7th edn. New York: Prentice Hall.
Grey, M. and Caul, M. (2000). Declining voter turnout in advanced industrial democracies, 1950 to 1997. The effects of declining group mobilization. *Comparative Political Studies*, 33(9), 1091–1122.
Hamilton, J. D. (1994). *Time Series Analysis*. Princeton, NJ: Princeton University Press.
Hicks, A. M. (1994). Introduction to pooling. In T. Janoski and A. M. Hicks (eds), *The Comparative Political Economy of the Welfare State*. Cambridge: Cambridge University Press.
Hicks, A. M. and Swank, D. H. (1992). Politics, institutions, and welfare spending in industrialized democracies, 1960–1982. *American Political Science Review*, 86(3), 658–674.
Honaker, J. and King, G. (2010). What to do about missing values in time-series cross-section data. *American Journal of Political Science*, 54(2), 561–581.
Hsiao, C. (2003). *Analysis of Panel Data*. Cambridge: Cambridge University Press.
Huber, E. and Stephens, J. D. (2001). *Development and Crisis of the Welfare State: Parties and Policies in Global Markets*. Chicago: University of Chicago Press.
Jackman, R. W. and Miller, R. A. (1995). Voter turnout in the industrial democracies during the 1980s. *Comparative Political Studies*, 25(4), 467–492.
Judge, G. G., Griffiths, W. E., Hill, R. C., Lütkepohl, H. and Lee, T.-C. (1985). *The Theory and Practice of Econometrics*. New York: Wiley.
Judson, R. A. and Owen, A. L. (1999). Estimating dynamic panel data models: A guide for macroeconomists. *Economic Letters*, 69, 9–15.
Keele, L. and Kelly, N. J. (2006). Dynamic models for dynamic theories: The ins and outs of lagged dependent variables. *Political Analysis*, 14(2), 186–205.
Kittel, B. (1999). Sense and sensitivity in pooled analysis of political data. *European Journal of Political Research*, 35(2), 225–259.
Kittel, B. and Winner, H. (2005). How reliable is pooled analysis in political economy? The globalization-welfare state nexus revisited. *European Journal of Political Research*, 44(2), 269–293.
Kmenta, J. (1971). *Elements of Econometrics*. New York: Macmillan.
Kohler, U. and Kreuter, F. (2008). *Data Analysis using Stata*. College Station, TX: Stata Press.
Kristensen, I. P. and Wawro, G. (2003). Lagging the dog? The robustness of panel corrected standard errors in the presence of serial correlation and observation specific effects. Working Paper, Society for Political Methodology.
Leamer, E. E. (1983). Let's take the con out of econometrics. *American Economic Review*, 73(1), 31–43.
Leamer, E. E. (1985). Sensitivity analysis would help. *American Economic Review*, 75(3), 308–313.
Leblang, D. (1997). Political democracy and economic growth: Pooled cross-sectional and time-series evidence. *British Journal of Political Science*, 27(3), 453–472.
Levin, A., Lin, C.-F. and James Chu, C.-S. (2002). Unit root test in panel data: Asymptotic and finite-sample properties. *Journal of Econometrics*, 108, 1–24.
Lipset, S. M. (1959). Some social requisites of democracy: Economic development and political legitimacy. *American Political Science Review*, 53(1), 69–105.
Maddala, G. S. (1977). *Econometrics*. New York: McGraw-Hill.
Maddala, G. S. (1991). To pool or not to pool: That is the question. *Journal of Quantitative Economics*, 7, 255–263.
Maoz, Z. and Russett, B. M. (1993). Normative and structural causes of democratic peace, 1946-1986. *American Political Science Review*, 87(3), 624–638.
Nickel, S. (1981). Biases in dynamic models with fixed effects. *Econometrica*, 49, 1417–1426.

Norris, P. (2009). Democracy timeseries data release 3.0. http://www.pippanorris.com/

Pampel, F. C. and Williamson, J. B. (1988). Welfare spending in advanced industrial democracies, 1950–1980. *American Journal of Sociology*, 93(6), 1424–1456.

Parks, R. (1967). Efficient estimation of a system of regression equations when disturbances are both serially and contemporaneously correlated. *Journal of the American Statistical Association*, 62, 500–509.

Pesaran, M. H. (2007). A simple panel unit root test in the presence of cross-section dependence. *Journal of Applied Econometrics*, 22(2), 265–312.

Plümper, T. and Troeger, V. E. (2007). Efficient estimation of time-invariant and rarely changing variables in finite sample panel analyses with unit fixed effects. *Political Analysis*, 15(2), 124–139.

Plümper, T., Troeger, V. E. and Manow, P. (2005). Panel data analysis in comparative politics: Linking method to theory. *European Journal of Political Research*, 44, 327–354.

Ross, M. L. (2001). Does oil hinder democracy? *World Politics*, 53(3), 325–361.

Sayrs, L. W. (1989). *Pooled Time Series Analysis*. Newbury Park, CA: Sage.

Stimson, J. A. (1985). Regression in space and time: A statistical essay. *American Journal of Political Science*, 29(4), 914–947.

Stone, M. (1974). Cross-validatory choice and assessment of statistical predictions. *Journal of the Royal Statistical Society, Series B*, 36(2), 111–147.

Tavits, M. (2008). Party systems in the making: The emergence and success of new parties in new democracies. *British Journal of Political Science*, 38(1), 113–133.

Wawro, G. (2002). Estimating dynamic panel models in political science. *Political Analysis*, 10(1), 25–48.

Williams, L. K. and Whitten, G. D. (2012). But wait, there's more! Maximizing substantive inferences from tscs models. *Journal of Politics*, 74(3), 685–693.

Wilson, S. E. and Butler, D. M. (2007). A lot more to do: The sensitivity of time-series cross-section analyses to simple alternative specifications. *Political Analysis*, 15(2), 101–123.

Wooldridge, J. M. (2002). *Econometric Analysis of Cross Section and Panel Data*. Cambridge, MA: MIT Press.

Worrall, J. L. and Pratt, T. C. (2004). Estimation issues associated with time-series-cross-section analysis in criminology. *Western Criminology Review*, 5(1), 35–48.

Zorn, C. (2001). Generalized estimating equation models for correlated data: A review with applications. *American Journal of Political Science*, 45(2), 470–490.

Name Index

Subject Index